International
REVIEW OF
Neurobiology
Volume 91

International
REVIEW OF
Neurobiology

Volume 91

SERIES EDITORS

R. ADRON HARRIS
Waggoner Center for Alcohol and Drug Addiction Research
The University of Texas at Austin
Austin, Texas, USA

PETER JENNER
Division of Pharmacology and Therapeutics
GKT School of Biomedical Sciences
King's College, London, UK

EDITORIAL BOARD

Functional Plasticity and Genetic Variation: Insights into the Neurobiology of Alcoholism

EDITED BY

MATTHEW T. REILLY

Program Director-Molecular and Behavioral Genetics
Division of Neuroscience and Behavior
National Institute on Alcohol Abuse & Alcoholism
National Institutes of Health
5635 Fishers Lane
Suite 2065, MSC 9304
Bethesda, MD 20892-1705
USA

and

DAVID M. LOVINGER

Laboratory for Integrative Neuroscience
Division of Intramural Clinical and Biomedical Research
National Institute on Alcohol Abuse and Alcoholism
5625 Fishers Lane, Room TS-11
Rockville, MD 20852

ELSEVIER

AMSTERDAM • BOSTON • HEIDELBERG • LONDON
NEW YORK • OXFORD • PARIS • SAN DIEGO
SAN FRANCISCO • SINGAPORE • SYDNEY • TOKYO
Academic Press is an imprint of Elsevier

AP

Academic Press is an imprint of Elsevier
360 Park Avenue South, New York, NY 10010-1700
525 B Street, Suite 1900, San Diego, California 92101-4495, USA
32 Jamestown Road, London NW1 7BY, UK

This book is printed on acid-free paper.

For information on all Elsevier publications
visit our website at books.elsevier.com

ISBN: 978-0-12-381276-6

Printed and bound by CPI Group (UK) Ltd, Croydon, CR0 4YY
Transferred to Digital Printing, 2013

CONTENTS

Using Expression Genetics to Study the Neurobiology of Ethanol and Alcoholism

Sean P. Farris, Aaron R. Wolen and Michael F. Miles

Genetic Variation and Brain Gene Expression in Rodent Models of Alcoholism: Implications for Medication Development

Karl Björk, Anita C. Hansson and Wolfgang H. Sommer

Identifying Quantitative Trait Loci (QTLs) and Genes (QTGs) for Alcohol-Related Phenotypes in Mice

Lauren C. Milner and Kari J. Buck

Glutamate Plasticity in the Drunken Amygdala: The Making of an Anxious Synapse

Brian A. McCool, Daniel T. Christian, Marvin R. Diaz and Anna K. Läck

Ethanol Action on Dopaminergic Neurons in the Ventral Tegmental Area: Interaction with Intrinsic Ion Channels and Neurotransmitter Inputs

Hitoshi Morikawa and Richard A. Morrisett

Alcohol and the Prefrontal Cortex

Kenneth Abernathy, L. Judson Chandler and John J. Woodward

BK Channel and Alcohol, A Complicated Affair

Gilles Erwan Martin

A Review of Synaptic Plasticity at Purkinje Neurons with a Focus on Ethanol-Induced Cerebellar Dysfunction

C. Fernando Valenzuela, Britta Lindquist and Paula A. Zamudio-Bulcock

CONTRIBUTORS

Numbers in parentheses indicate the pages on which the authors contributions begin.

Kenneth Abernathy (289), Department of Neurosciences and Center for Drug and Alcohol Programs, Medical University of South Carolina, Charleston, SC 29425, USA

Karl Björk (129), Translational Neuropharmacology, Center for Molecular Medicine, Karolinska Institutet, SE-171 77 Stockholm, Sweden

Kari J. Buck (173), Department of Behavioral Neuroscience, VA Medical Center and Oregon Health & Science University, Portland, OR 97239, USA

L. Judson Chandler (289), Department of Neurosciences and Center for Drug and Alcohol Programs, Medical University of South Carolina, Charleston, SC 29425, USA

Daniel T. Christian (205), Department of Physiology and Pharmacology, Wake Forest University School of Medicine, Winston-Salem, NC 27157, USA

Marvin R. Diaz (205), Neuroscience Training Program, Wake Forest University School of Medicine, Winston-Salem, NC 27157, USA

Sean P. Farris (95), Department of Pharmacology and Toxicology, Virginia Commonwealth University, Richmond, VA 23298, USA

Anita C. Hansson (129), Department of Psychopharmacology, Central Institute for Mental Health, D 68159 Mannheim, Germany

Shirley Y. Hill (53), Department of Psychiatry, University of Pittsburgh School of Medicine, Pittsburgh, PA 15213, USA

Anna K. Läck (205), Department of Physiology and Pharmacology, Wake Forest University School of Medicine, Winston-Salem, NC 27157, USA

Britta Lindquist (339), Department of Neurosciences, School of Medicine, University of New Mexico Health Sciences Center, Albuquerque, NM 87131, USA

Gilles Erwan Martin (321), Department of Psychiatry, The Brudnick Neuropsychiatric Research Institute, University of Massachusetts Medical School, Worcester, MA 01605, USA

Brian A. McCool (205), Department of Physiology and Pharmacology, Translational Center for the Neurobehavioral Study of Alcohol, and Neuroscience Training Program, Wake Forest University School of Medicine, Winston-Salem, NC 27157, USA

Michael F. Miles (95), Department of Pharmacology and Toxicology, Virginia Commonwealth University, Richmond, VA 23298, USA

Lauren C. Milner (173), Department of Behavioral Neuroscience, VA Medical Center and Oregon Health & Science University, Portland, OR 97239, USA

Hitoshi Morikawa (235), Waggoner Center for Alcohol and Addiction Research, University of Texas at Austin, Austin, TX 78712, USA

Richard A. Morrisett (235), Waggoner Center for Alcohol and Addiction Research, University of Texas at Austin, Austin, TX 78712, USA

Andrzej Zbigniew Pietrzykowski (1), Department of Animal Sciences, Rutgers University, New Brunswick, NJ 08901-1414, USA

Aylin R. Rodan (25), Division of Nephrology, Department of Psychiatry and Program in Neuroscience, UT Southwestern Medical Center, Dallas, TX, USA

Adrian Rothenfluh (25), Department of Psychiatry and Program in Neuroscience, UT Southwestern Medical Center, Dallas, TX, USA

Wolfgang H. Sommer (129), Department of Psychopharmacology, Central Institute for Mental Health, D 68159 Mannheim, Germany

C. Fernando Valenzuela (339), Department of Neurosciences, School of Medicine, University of New Mexico Health Sciences Center, Albuquerque, NM 87131, USA

Aaron R. Wolen (95), Department of Human and Molecular Genetics, Virginia Commonwealth University, Richmond, VA 23298, USA

John J. Woodward (289), Department of Neurosciences and Center for Drug and Alcohol Programs, Medical University of South Carolina, Charleston, SC 29425, USA

Paula A. Zamudio-Bulcock (339), Department of Neurosciences, School of Medicine, University of New Mexico Health Sciences Center, Albuquerque, NM 87131, USA

PREFACE

The brain, arguably the most complex organ of the body, plays a critical role in the predisposition, expression, and maintenance of alcoholism. This adds an unmatched degree of complexity in the search for the causes of alcoholism as compared to other illnesses with a genetic component such as type 2 diabetes. The brain orchestrates such processes as cognition, emotion, mood, learning, and motivation through circuits in a network of different brain regions, which communicate through synaptic connections. Synaptic communication is spatially and temporally mediated by the cellular expression of genes, regulation of the expression of these genes, and the functioning of the corresponding proteins. However, how individual differences, influenced by genetic variation, affect synaptic communication and functional plasticity of alcohol-sensitive brain circuits is not well understood. This is a burgeoning research area which holds promise for psychiatric research in general (Akil *et al.*, 2010; Koob and Volkow, 2010). Over the last several decades, a great deal of work has examined the consequences of alcohol exposure on the plasticity of the brain. A parallel line of work has begun to uncover the genes and genetic variation that might contribute to or mediate alcohol-induced plasticity. This volume contains critical reviews of the current state of these research domains: Functional plasticity and genetic variation; which we hope will offer novel insights into the neurobiology of alcoholism.

The first chapter discusses the newly identified role of microRNAs in the neuronal mechanisms of alcohol tolerance and the potential involvement in other alcohol phenotypes. The use of Drosophila in understanding the genetics of alcohol-related phenotypes is then reviewed followed by a chapter on neural plasticity, human genetics, and risk for alcohol dependence. In the next two chapters, approaches to understand alcoholism via examination of gene expression are reviewed. The first of these (Farris et al.) is a comprehensive review of the powerful use of expression genetics to study the neurobiology of alcoholism, while the second (Björk et al.) highlights the need to examine gene expression alterations in clinically relevant animal models of alcoholism. The final chapter in this section reviews progress in the identification of genes for alcohol phenotypes in mice. The second half of the volume changes gears with chapters focusing on the cellular and synaptic consequences of alcohol-induced plasticity in specific brain regions. The first of these reviews glutamate plasticity in the amygdala followed by a chapter on ethanol action on dopamine neurons of the ventral tegmental area. The prefrontal cortex and striatum are covered in the next two chapters. Finally, synaptic plasticity at Purkinje neurons in the cerebellum is reviewed.

In summary, this volume has highlighted two important areas in alcohol research which have significantly increased our understanding and given us an appreciation for the complexity of alcoholism as a brain disease. We hope this volume will begin to stimulate some convergence between these areas.

Matthew T. Reilly
David M. Lovinger

References

Akil, H., Brenner, S., Kandel, E., Kendler, K. S., King, M. C., Scolnick, E., Watson, J. D., and Zoghbi, H. Y. (2010). Medicine. The future of psychiatric research: Genomes and neural circuits. *Science* **327**(5973), 1580–1581.

Koob, G. F., and Volkow, N. D. (2010). Neurocircuitry of addiction. *Neuropsychopharmacology* **35**(1), 217–238.

THE ROLE OF microRNAs IN DRUG ADDICTION: A BIG LESSON FROM TINY MOLECULES

Andrzej Zbigniew Pietrzykowski

Department of Animal Sciences, Rutgers University, New Brunswick, NJ 08901-1414, USA

Alcoholism is a multifactorial disease of unclear molecular underpinnings. Currently, we are witnessing a major shift in our understanding of the functional elements of the genome, which could help us to discover novel insights into the nature of alcoholism. In humans, the vast majority of the genome encodes non-protein-coding DNA with unclear function. Recent research has started to unveil this mystery by describing the functional relevance of microRNAs, and examining which genes are regulated by non-protein-coding DNA. Here, I describe alcohol regulation of microRNAs and provide examples of microRNAs that control the expression of alcohol-relevant genes. Emphasis is put on the potential of microRNAs in explaining the polygenic nature of alcoholism and prospects of microRNA research and future directions of this burgeoning field.

INTERNATIONAL REVIEW OF
NEUROBIOLOGY, VOL. 91
DOI: 10.1016/S0074-7742(10)91001-5

I. Introduction

Alcoholism is a complex, human disease, with strong genetic and environmental components (Pinto and Ansseau, 2009). Mounting evidence, accumulated over several decades, leaves practically no doubt that genetic factors contribute significantly to the development and maintenance of alcohol addiction. However, all these data suggest that there is no single gene responsible for alcoholism, but rather that alcoholism is a polygenic disease by nature (Tabakoff *et al.*, 2009). Several loci bear the risk associated with alcoholism, and certain loci are linked to specific phenotypes, such as tolerance, withdrawal, etc. (Tabakoff *et al.*, 2009). It is thought that genes located within these loci express products with altered abundance or products with altered function that puts an organism at higher risk of developing alcoholism upon exposure to alcohol. Human linkage and association studies have identified several candidate genes for susceptibility toward alcoholism (e.g., alcohol dehydrogenase (ADH), aldehyde dehydrogenase (ADLH), gamma-aminobutyric acid (GABA) A receptor subunits, muscarinic receptors, opioid receptors, etc.), which have been replicated in multiple studies. Animal models, including invertebrate models, have identified candidate genes for specific alcohol phenotypes (Davies *et al.*, 2003; Tabakoff *et al.*, 2009). However, much of the genetic variation contributing to alcoholism has yet to be identified. In addition to genetic factors, there is an important contribution of environmental factors to alcoholism. Multiple environmental factors (e.g., stress) substantially contribute to the risk of becoming an alcoholic (Breese *et al.*, 2005; Fox *et al.*, 2007; Madrid *et al.*, 2001; Sinha, 2007). Importantly, alcohol addiction cannot develop without exposure to alcohol and thus susceptible individuals must be in an environment where alcohol is available. Based on epidemiological data, the International Center for Alcohol Policies (ICAP) has created worldwide guidelines for alcohol consumption. In the United States, women should consume no more than one standard drink (14 g EtOH = 12 fl. oz of beer, 5 fl. oz of wine, 1 fl. oz of distilled spirits) per day, while men should consume no more than two standard drinks per day (Ferreira and Willoughby, 2008) to avoid the detrimental effects of alcohol. However, there is surprisingly little data describing molecular underpinnings of alcohol exposure and an associated probability of development of alcoholism. Therefore, it is important to understand how different alcohol exposure patterns can lead to different qualitative and/or quantitative changes in molecular pathways relevant for development of addiction.

In this chapter, we will describe some recent advances in our understanding of alcohol regulation of molecular processes, which should help in answering these questions. We will focus specifically on the role of microRNAs, newly discovered regulators of gene expression, and their potential to explain the effects of alcohol on multiple molecular pathways and the polygenic nature of alcoholism.

II. microRNA

A. Overview

microRNAs are a class of short ~22-nucleotide (nt)-long RNA molecules, which act as regulators of gene expression (Ambros, 2004). In the remarkable short time since their discovery in 1993 (Lee *et al.*, 1993), evidence has accumulated exponentially indicating that microRNAs and other non-coding RNA species play fundamental roles in almost every biological process. Typically, microRNAs bind to a complementary sequence in the 3′-untranslated region (UTR) of mRNA, which they target, causing mRNA silencing or degradation (Bartel, 2009; Filipowicz *et al.*, 2008; Liu *et al.*, 2008; Perron and Provost, 2008). Various target prediction programs indicate that a single microRNA can regulate expression of hundreds of genes. microRNAs are abundantly present in all human cells, where they target at least 60% of all genes encoding proteins. The phylogenetic conservation of microRNA and their presence in all kinds of species from unicellular organisms, e.g., algae (Molnar *et al.*, 2009) to complex, multicellular creatures such as humans (Friedman *et al.*, 2009), indicate that microRNAs have a very ancient origin (Axtell and Bartel, 2005) and that they play a critical role in regulation of gene expression. Indeed, it is now known that microRNAs are essential controllers of cellular proliferation, differentiation, and apoptosis. Rapidly growing evidence indicates a key rôle of microRNAs in many types of physiological functions as well as pathological processes.

Since the discovery of the first microRNA (Lee *et al.*, 1993) in 1993, more than 500 microRNAs (by conservative estimates) have been identified in humans (Saini *et al.*, 2007). Since microRNA genes encode functional RNA, from which proteins are not transcribed, they belong to the non-protein-coding DNA, at one point dubbed "junk DNA" due to its unclear function. Recent examination of the transcriptome revealed that many transcripts encode non-protein-coding RNA, including microRNAs (Birney *et al.*, 2007; Kapranov *et al.*, 2007; Willingham and Gingeras, 2006). The role of this elaborate network of various RNA transcripts is just starting to be comprehended and the biological sciences of the twenty-first century have a tremendous opportunity to solve these mysteries of the genomic "dark matter."

microRNAs can be located in the intergenic regions of the genome and use their own transcription machinery, including their own promoters and regulatory units, or lie within introns (so-called mitrons) of protein and non-protein-coding genes that can exhibit expression concurrent with a host gene using a common promoter (Ason *et al.*, 2006; Berezikov *et al.*, 2007; Li *et al.*, 2007; Lin *et al.*, 2006; Okamura *et al.*, 2007; Saini *et al.*, 2007; Takada *et al.*, 2008). The first microRNA transcript is capped at the 5′-end, polyadenylated at the 3′-end, and spliced similarly to the regular mRNA, producing a so-called pri-miRNA—a double-stranded, hairpin-loop structure, several hundred base pairs (bps) in length (Fig. 1).

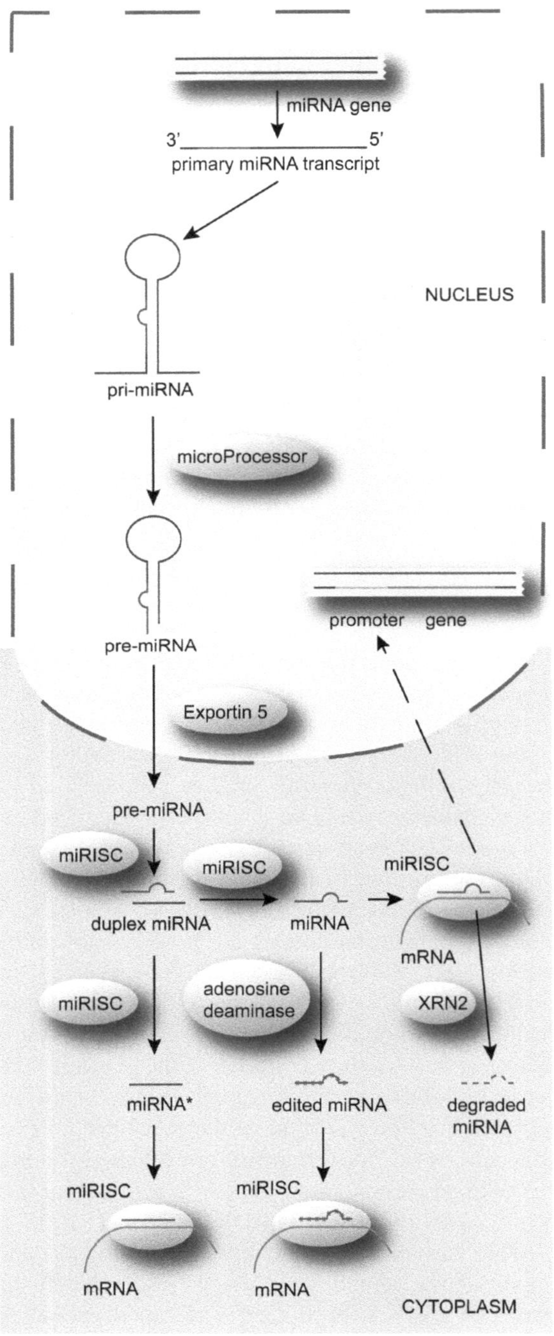

FIG. 1. *(Continued)*

Pri-miRNAs are further cleaved in the nucleus by the microProcessor complex, consisting mainly of Drosha and DGCR8 (DiGeorge syndrome critical region 8) proteins (Han *et al.*, 2004; Yeom *et al.*, 2006). The cleaved product, called pre-miRNA, is ~70 bp in length, consists of a double-stranded structure, contains a phosphorylated 5′-end, and a 2-nt-long 3′-end overhang. The overhang is recognized by the nuclear membrane protein exportin 5, which transports the pre-miRNA into the cytoplasm (Bohnsack *et al.*, 2004; Zeng and Cullen, 2004). In the cytoplasm, pre-miRNA undergoes additional cleavage resulting in a 20–25-bp-long imperfect duplex, containing the mature miRNA strand and its opposite complementary (passenger) strand—miRNA* (Hutvagner *et al.*, 2001). This cleavage is thought to be coupled with unwinding of the duplex to produce single-stranded, active miRNA forms; however, the exact order of events is still under investigation (MacRae *et al.*, 2008). The latter steps take place in an miRISC (the microRNA-induced silencing complex (Williams, 2008), also called microribonucleoprotein (miRNP) complex (Mourelatos *et al.*, 2002)). miRISC is a multiprotein structure, composed of Dicer (Bernstein *et al.*, 2001; Provost *et al.*, 2002), transactivating response RNA binding protein (TRBP; Chendrimada *et al.*, 2005), protein activator of the interferon-induced protein kinase (PACT; Lee *et al.*, 2006), Ago (Baumberger and Baulcombe, 2005), and other proteins, some yet to be discovered. Probably only one strand of mature miRNA guides the miRISC to target mRNA in order to induce its silencing or cleavage. In animals, including humans, miRNA usually targets a site located in the 3′-UTR (Bartel, 2004, 2009; Grimson *et al.*, 2007; Lewis *et al.*, 2003). miRNA:mRNA complementarity is essential for target recognition, with perfect complementarity of the "seed region" (nt 2–7 of the miRNA). Partial complementarity in the rest of miRNA:mRNA complex seems to accelerate deadenylation, which leads to degradation of mRNA (Chen *et al.*, 2009; Eulalio *et al.*, 2009).

FIG. 1. miRNA biogenesis and degradation. Each microRNA gene produces a primary microRNA transcript, which is capped at the 5′-end, polyadenylated at the 3′-end, and can be spliced similarly to the regular mRNA (not shown), and folded into so-called pri-miRNA—a double-stranded, hairpin-loop structure, several hundred base pairs in length. microProcessor protein complex cleaves further pri-miRNAs, producing pre-miRNA—~70 bp in length containing 2-nt overhang at the 3′-end. The overhang is necessary for the nuclear membrane protein, exportin 5, to export pre-miRNA to the cytoplasm. Next, pre-miRNA undergoes additional cleavage resulting in a 20–25-bp-long imperfect duplex miRNA, containing the mature, miRNA strand and its opposite complementary (passenger) strand—miRNA*. Duplex miRNA undergoes unwinding to produce single-stranded, active miRNA forms. The latter steps take place in an miRISC – the microRNA-induced silencing complex. Probably only one strand of mature miRNA guides the miRISC to the 3′-UTR of target mRNA in order to induce its silencing or cleavage. Both strands can be functional, each having its own set of targets. Also, in humans, around 6% of mature microRNAs can undergo editing by adenosine deaminase producing, e.g., microRNAs of different target specificity. Finally, mature miRNA can be degraded by 5′–3′ exoribonuclease (XRN2). Additionally, microRNAs can also act on promoters and regulate transcription of targeted genes.

The passenger strand (miRNA*) is typically presented in lower levels in the steady state. Initially, it was thought that the passenger strand is quickly degraded as a nonfunctional by-product (Matranga *et al.*, 2005). However, recent findings indicate that it can also be functional, acting on its own binding sites and regulating expression of its own set of targets (Ghildiyal *et al.*, 2010; Okamura *et al.*, 2009; Tyler *et al.*, 2008). To further complicate the picture, microRNA can additionally undergo RNA editing (conversion of adenosine to inosine catalyzed by adenosine deaminases that act on RNA (ADAR)), yielding products that differ from genomically encoded DNA (Blow *et al.*, 2006). In humans, it is estimated that around 6% of microRNA species are subject to RNA editing (Blow *et al.*, 2006). Editing of a microRNA can prevent its processing (Yang *et al.*, 2005) or alter its specificity (Kawahara *et al.*, 2007). Another layer of regulation is provided by 5′–3′ exoribonuclease (XRN2; Bail *et al.*, 2010), or the small RNA degrading nuclease (SDN) (Ramachandran and Chen, 2008), both of which can degrade mature miRNA, enabling rapid changes in its expression.

In a very dynamically growing field of microRNAology, every year brings new discoveries. Data published during the last couple of years indicate that microRNA actions are not limited to repression of protein synthesis or the regulation of translation in the cytoplasm as initially thought. As shown for miRNA-369-3, depending on the composition of microRNPs, miRISC can activate translation (Place *et al.*, 2008; Steitz and Vasudevan, 2009; Vasudevan and Steitz, 2007; Vasudevan *et al.*, 2007, 2008). Also, microRNAs can regulate transcription, by causing chromatin remodeling or DNA methylation of promoter sites. These processes result in altered expression of targeted genes (Tan *et al.*, 2009), and can also cause gene activation (Place *et al.*, 2008)—a process called RNAa (small RNA-induced gene activation)—instead of inhibition.

B. Role of microRNA in Brain Function and Neuronal Plasticity

An organism can successfully adapt to ever-changing environmental demands thanks to the plasticity of the nervous system, in which precise and dynamic fine-tuning of gene expression is of major importance. Therefore, a prominent role of miRNAs as master controllers of gene regulation quickly attracted the attention of neurobiologists.

Recent investigations clearly indicate that miRNAs play a crucial role in regulating neurogenesis, neuronal plasticity, and brain homeostasis (Li and Jin, 2010; Schratt, 2009; Smalheiser and Lugli, 2009). miRNAs are abundant in the central nervous system (CNS) with unique expression profiles during development (Krichevsky *et al.*, 2003; Vreugdenhil and Berezikov, 2010) and in the adult brain

(Olsen *et al.*, 2009). They provide a very useful mechanism of dynamical adjustment of protein levels without the necessity of new gene transcription. Moreover, changes in microRNA levels could have a big impact on neuronal and glial function, as each microRNA can regulate hundreds of targets. Expression of several microRNAs is regulated by neuronal activity, and is essential for, e.g., dendritogenesis or morphogenesis of spines on dendrities (Christensen *et al.*, 2010; Khudayberdiev *et al.*, 2009). Recently, it has been indicated that specific microRNAs and other components of the miRISC complex are expressed locally in the dendritic compartment of a neuron, and can control local protein synthesis at synapses, thus contributing to fundamental neuronal functions, such as formation of long-term potentiation/ long-term depression (LTP/LTD) (Park and Tang, 2009), or long-term memory (Ashraf *et al.*, 2006; Batassa *et al.*, 2010; Rajasethupathy *et al.*, 2009). Several working models have been put forward to explain the detailed mechanics of miRNA function at the synapse. In the future, testing of these models should reveal the modus operandi of microRNA in neuronal plasticity.

Increasing evidence provides strong support for an involvement of miRNAs in the development of diseases of the nervous system. Abnormal levels of specific microRNAs have been described in many neurological disorders: Huntington's disease (Packer *et al.*, 2008), Alzheimer's disease (Cogswell *et al.*, 2008; Hebert *et al.*, 2008, 2009; Nunez-Iglesias *et al.*, 2010; Sethi and Lukiw, 2009; Wang *et al.*, 2008, 2009b), autism (Abu-Elneel *et al.*, 2008), depression (Feng *et al.*, 2009), and schizophrenia (Beveridge *et al.*, 2008; Burmistrova *et al.*, 2007; Feng *et al.*, 2009; Guo *et al.*, 2010; Hansen *et al.*, 2007; Mellios *et al.*, 2009; Perkins *et al.*, 2007). The next few years should bring new discoveries delineating the role of miRNAs in the mechanisms of these diseases. It is also likely that miRNAs will help to understand the etiology of alcoholism.

III. microRNA and Alcohol

Recent data from our lab as well as other labs indicate that miRNA may help to understand the etiology of alcoholism. Alcohol exposure changes processes underlying neuronal plasticity in many brain regions, which produce tolerance and dependence. Biomolecules present in the brain, by trying to adapt to chronic and/or the frequent presence of alcohol in their microenvironment, cause permanent changes in neuronal function. Subsequently, this maladaptation drives behavior, such as seeking and obtaining alcohol, regardless of the cost and despite complex, negative consequences to the brain itself, to other organs, or to an individual's function in society.

As described above, microRNAs are major contributors to neuronal plasticity, and they could be a prominent "prong" of alcohol's multipronged attack on the brain. As such, microRNAs could be considered as potentially very important contributors to alcoholism, with a key role in various aspects of alcoholism such as propensity to intoxication, development of tolerance, severity of withdrawal, etc.

A. Regulation of microRNA by Alcohol

One of the first questions that should be addressed is whether alcohol can regulate the expression of microRNAs? Brain regional and developmental expression of microRNAs is well characterized (Krichevsky *et al.*, 2003; Olsen *et al.*, 2009). However, a study examining a genome-wide brain-specific miRNA profile following alcohol exposure has not been conducted. Nevertheless, as demonstrated by Dolganouc *et al.* (2009), alcohol, at least in murine liver, alters expression of ~2% of sampled microRNA species, causing down or upregulation of roughly an equal number of microRNAs. Interestingly, some of these miRNA species are also present in brain (Table I) and are differentially expressed in different regions of the adult CNS (Olsen *et al.*, 2009) with a seemingly critical role in brain development (Lee *et al.*, 2009; Zhang *et al.*, 2010).

We have focused on a specific microRNA species, miR-9. miR-9 is important during the development of the brain, specifically for the formation of the hindbrain–midbrain boundary (Kapsimali *et al.*, 2007; Krichevsky *et al.*, 2003; Leucht *et al.*, 2008; Pietrzykowski *et al.*, 2008; Shibata *et al.*, 2008) and is also expressed in the adult brain (Kapsimali *et al.*, 2007; Krichevsky *et al.*, 2003; Leucht *et al.*, 2008; Pietrzykowski *et al.*, 2008; Shibata *et al.*, 2008). Using organotypic cultures, we observed that alcohol rapidly and specifically upregulated miR-9 in rat adult supraoptic nucleus (SON) neurons, subsequently evoking a molecular tolerance (Pietrzykowski *et al.*, 2008). Within 15 min of exposure to a physiologically relevant alcohol concentration (20 mM), miR-9 levels were almost doubled. According to our single-cell analysis, an individual, "naïve" SON neuron harbors on average 350 miR-9 molecules. We detected over 500 miR-9 molecules per the SON cell after a 15 min exposure to alcohol. We observed even more profound upregulation (from 600 miR-9 molecules before alcohol exposure to 1200 molecules 15 min after the exposure) in murine striatal neurons (Pietrzykowski *et al.*, 2008). Interestingly, although alcohol induced miR-9 expression in adult rodent neurons (Pietrzykowski *et al.*, 2008) and fetal murine brains (Wang *et al.*, 2009a), it suppressed miR-9 expression in murine neurospheres (Sathyan *et al.*, 2007). It is possible that certain factors controlling expression of microRNAs are tightly regulated during development, are differentially expressed among species, or are differentially regulated

Table I
MICRORNA SPECIES REGULATED BY ALCOHOL

microRNA	Role related to alcohol effects	References	# of targets
miR-9	Premature differentation of neutrons	Sathyan *et al.*, (2007)	936
	Developement of molecular tolerance in neurons	Pietrzykowski *et al.*, (2008)	
	Contribution to fetal teratogenesis	Wang *et al.*, (2009)	
miR-10	Contribution to fetal teratogenesis	Wang *et al.*, (2009)	186
miR-21	Antiapoptotic factor in neuronal progenitors	Sathyan *et al.*, (2007)	210
miR-29c	Contribution to fetal teratogenesis	Wang *et al.*, (2009)	851
miR-30	Contribution to fetal teratogenesis	Wang *et al.*, (2009)	1080
miR-145	Contribution to fetal teratogenesis	Wang *et al.*, (2009)	528
miR-152	Contribution to fetal teratogenesis	Wang *et al.*, (2009)	526
miR-153	Function of neuronal progenitors	Sathyan *et al.*, (2007)	550
miR-154	Contribution to fetal teratogenesis	Wang *et al.*, (2009)	84
miR-200a	Contribution to fetal teratogenesis	Wang *et al.*, (2009)	531
miR-296	Contribution to fetal teratogenesis	Wang *et al.*, (2009)	283
miR-335	Proapoptotic factor in neuronal progenitors	Sathyan *et al.*, (2007)	155
miR-339	Contribution to fetal teratogenesis	Wang *et al.*, (2009)	99
miR-362	Contribution to fetal teratogenesis	Wang *et al.*, (2009)	93
miR-496	Contribution to fetal teratogenesis	Wang *et al.*, (2009)	90
miR-199	Affects expression of cytokines in liver	Yeligar *et al.*, (2009)	375
	Contributes to alcoholic fatty liver	Dolganiuc *et al.*, (2009)	
miR-212	Contributes to leakiness of intestinal epithelium	Tang *et al.*, (2008)	283
miR-705	Contributes to alcoholic fatty liver	Dolganiuc *et al.*, (2009)	83
miR-1224	Contributes to alcoholic fatty liver	Dolganiuc *et al.*, (2009)	141

Alcohol Affects the Expression of Several microRNAs in Neuronal Fetal Tissue, Neurospheres, and Adult Neurons. Please Note that Alcohol Effect on miR-199, -212, -705, and -1224 Has Been Tested in Non-Neuronal Tissues So Far; However, As Demonstrated by Lee *Et Al.* (2009) for miR-199, as well as Olsen *Et Al.* (2009) and Zhang *et al.* (2010) for miR-212, these miRNAs Are Expressed in the Brain and Play an Important Role in Brain Development. Expression of miR-705 and -1224 Has Not Yet Been Tested in the Brain. For Each Human miRNA, the Number of Targets Has Been Determined Using Conserved Targets Algorithm of TargetScan.

depending on the concentration of alcohol. However, regardless of whether alcohol causes up or downregulation of microRNAs, it will be of great interest to determine which steps of the complex process of microRNA biosynthesis and degradation (Fig. 1) are sensitive to alcohol.

Differential regulation of miR-9 by alcohol could also shed some light on the molecular mechanisms affected by different drinking patterns. Based on epidemiological data, consumption of one standard drink per day by women and two standard drinks per day by men seems to be "safe" (Ferreira and Willoughby, 2008). This could indicate that a short exposure to small doses of alcohol causes short-living changes in molecular mechanisms without long-term effects, while higher doses or longer exposure to alcohol may create a more permanent molecular footprint, leading subsequently to detrimental consequences.

Interestingly, a short (15 min) exposure to 20 mM alcohol (corresponding to an episode of social drinking) caused miR-9 upregulation (Pietrzykowski *et al.*, 2008), while a longer (at least 24 h) exposure to higher concentration of alcohol (70 mM, corresponding to alcohol levels in alcoholics (Adachi *et al.*, 1991) suppressed miR-9 expression (Sathyan *et al.*, 2007). As drinking pattern matters, it will be of great interest to determine in greater detail how various alcohol exposure paradigms affect expression of miRNA species and their targets, and how these regulations contribute to the development of addiction.

Although understanding the role of microRNAs in alcoholism is in its early stages, it is already apparent that only a specific subset of microRNAs are sensitive to alcohol. Documented microRNA species sensitive to alcohol exposure are depicted in Table I. Interestingly, miR-199 and miR-212 are not only present in the gastrointestinal system, where their expression is profoundly affected by alcohol (Dolganiuc *et al.*, 2009; Tang *et al.*, 2008), but these two microRNA species are also expressed in the brain (Lee *et al.*, 2009; Zhang *et al.*, 2010). It will be of great interest to determine whether alcohol affects expression of these miRNAs in the brain. Alcohol also affects the expression of two other microRNAs in the liver—miR-705 and miR-1224—possibly deregulating the metabolism of lipids in this organ (Dolganiuc *et al.*, 2009; Tang *et al.*, 2008) and contributing to the development of alcoholic fatty liver. As lipid metabolism is pivotal for proper brain function, it will be of great interest to determine whether these two microRNAs are expressed in "naïve" and/or alcoholic brains.

B. microRNA and Alcoholism Susceptibility Loci

It is widely accepted that alcoholism has a strong genetic component (Tyndale, 2003). However, alcoholism is not caused by a mutation in a single gene, but rather it is a polygenic disease, where genetic variation in multiple genes each contributes a small portion of the risk (Johnson *et al.*, 2006). Genetic linkage studies point toward several, specific chromosomal regions that harbor alcohol susceptibility loci. Two of the most frequently appearing regions are located on chromosomes 1 and 15 (Arya *et al.*, 2005; Dick *et al.*, 2002; Ehlers *et al.*, 2004; Guerrini *et al.*, 2005; Hill *et al.*, 2004; Joslyn *et al.*, 2008; Song *et al.*, 2003; Strauch *et al.*, 2005; Windemuth *et al.*, 1999), but other loci are also linked to alcoholism (Ehlers *et al.*, 2004; Hill *et al.*, 2004; Strauch *et al.*, 2005).

For a long time, great emphasis has been put on protein-coding genes, based on the knowledge of their biological functions, located within these loci. Results have showed that with alcoholism there is a strong link of genes encoding ADH and ADLH (Ehlers, 2007), and a more discrete link of several other genes encoding, e.g., dopamine transporter (DAT) (Ueno *et al.*, 1999), GABA A receptor (Dick *et al.*, 2004, 2006; Edenberg *et al.*, 2004; Parsian and Zhang, 1999; Song *et al.*, 2003), dopamine D2 receptor (Sasabe *et al.*, 2007), muscarinic receptor (Dick *et al.*, 2007; Wang *et al.*, 2004), opioid receptor (Ehlers *et al.*, 2008; Franke *et al.*, 1999; Gelernter *et al.*, 1999, 2007; Gscheidel *et al.*, 2000; Loh*et al.*, 2004; Rommelspacher *et al.*, 2001; Xuei *et al.*, 2006; Zhang *et al.*, 2006, 2008), and others. However, all these data do not provide a fully satisfactory explanation of the genetic constituent of alcoholism.

Interestingly, recent sequencing of the human genome indicated that protein-coding genes are only a very small fraction of the human genome (~1.5%), with the vast majority of the genome consisting of non-coding regions (Lander *et al.*, 2001; Venter *et al.*, 2001). Similarly, alcohol susceptibility loci contain large regions lacking protein-coding genes. Recently, there has been an increasing appreciation of the complexity of the genome and the vital role of the non-coding regions of the human genome in the regulation of gene expression. As microRNA genes belong to a group of non-coding genes and are essential regulators of gene expression in the brain, it is of great interest to determine their genetic contribution to susceptibility toward alcoholism.

miR-9 may provide a basis for testing this hypothesis. There are three different miR-9 genes: miR-9-1, miR-9-2, and miR-9-3 located on three different chromosomes, chr1 (1q23.1), chr5 (5q14.3), and chr15 (15q26.1), respectively (Griffiths-Jones, 2004, 2006; Griffiths-Jones *et al.*, 2008). Each gene encodes a different form of miR-9 precursors, but the ultimate, functional mature form of miR-9 produced from each gene has an identical sequence. There are alcoholism susceptibility loci on chromosomes 1 (1q21.3–24.2) and 15 (15q21.2–25.1), respectively (Guerrini *et al.*, 2005; Joslyn *et al.*, 2008; Windemuth *et al.*, 1999). Interestingly, two out of three miR-9 genes (miR-9-1 and miR-9-3) are located near or within these loci. It will be of great interest to determine genetic variations in all miR-9 genes and their frequency in populations of individuals addicted to alcohol. Thus, miR-9 appears to be a very attractive model to test the hypothesis that non-protein-coding regions are linked to susceptibility to alcoholism, which we are currently pursuing in our laboratory.

Obviously, miR-9 genes are not the only microRNA genes located within alcoholism susceptibility loci located on chromosomes 1 and 15. Additionally, there are also miRNAs at other loci located on different chromosomes. This chapter is not intended to provide a comprehensive list of such miRNAs. Rather, we would like to bring to the attention of the scientific and alcohol research communities the possibility that miRNAs may contribute to susceptibility to alcoholism because they are found within alcoholism susceptibility loci, and because of the crucial role of miRNAs in gene function and adaptational processes. Also, performing a screening of all known miRNAs using microarray analysis should help to establish which microRNA species are regulated by alcohol. This subgroup of alcohol-sensitive microRNAs should be considered next in the genetic studies of alcoholism. Hopefully, one of these (or other) approaches will provide evidence for the role of non-coding RNAs in alcoholism. Interestingly, considering that one microRNA typically controls the expression of many targets, this could, at least partially, explain the polygenic and complex nature of alcoholism.

C. Alcohol-Relevant microRNA Targets

Although the number of potential targets varies substantially among microRNAs, target prediction software indicates that each microRNA can anneal to many targets (hundreds of mRNA targets in some cases) (Table I). The ability of silencing expression of such a large number of targets provides microRNAs with a very powerful function as master regulators of gene expression, which can be critical during the course of development of alcoholism (Miranda *et al.*, 2010). This feature is particularly well studied during essential periods of development or in synaptic plasticity in an adult brain. During the progression to alcohol dependence, the entire brain is rewired, triggering changes in the expression of hundreds of genes. Below are several examples of specific targets of alcohol-relevant microRNAs. This list is not, in any way, complete. In upcoming years, exciting new findings are sure to develop.

1. *miR-9 and BK Channel*

We have primarily focused on miR-9 regulation of the main, pore-forming alpha subunit of a potassium channel of large conductance, called BK channel (Butler *et al.*, 1993; Dworetzky *et al.*, 1994). The BK channel (Slopoke, MaxiK) is abundant in the brain (Misonou *et al.*, 2006; Sailer *et al.*, 2006; Sausbier *et al.*, 2006; Wanner *et al.*, 1999) and critical for neuronal function (Benhassine and Berger, 2009; Dopico *et al.*, 1999; Farley and Rudy, 1988; Hu *et al.*, 2001; Loane *et al.*, 2007; Sausbier *et al.*, 2004; Shipston *et al.*, 1999). Due to BK channel sensitivity to both changes in membrane voltage and changes in concentration of intracellular calcium, it can serve as an integrator of cellular processes. The BK channel can be found in all neuronal compartments, where it contributes to many neuronal activities like regulation of neuronal excitability, shaping of action potentials, or control of neurotransmitter release (Dopico *et al.*, 1999; Hu *et al.*, 2001; Martin *et al.*, 2004; Shipston *et al.*, 1999). The BK channel alpha subunit is expressed in many different isoforms, which are translated from several splice variants (Chen *et al.*, 2005; Petrik and Brenner, 2007; Poulsen *et al.*, 2009;

Soom *et al.*, 2008). Interestingly, regulation of BK channel splice variant abundance is controlled by environmental factors (e.g., stress hormones) (Xie and McCobb, 1998). Excessive exposure to stress hormones is known to increase the risk of alcoholism (Campbell *et al.*, 2009; Madrid *et al.*, 2001). The BK channel is an important alcohol target. Worms (*Caenorhabditis elegans*) with a genetic deletion of the BK channel are tolerant to alcohol, while worms constitutively expressing the BK channel show behavioral symptoms of intoxication (Davies *et al.*, 2003). In *Drosophila melanogaster*, deletion of the BK channel leads to a loss of behavioral rapid tolerance (Cowmeadow *et al.*, 2005, 2006; Wang *et al.*, 2009c). In mammals, we have previously observed that alcohol evokes tolerance of BK channels (Crowley *et al.*, 2005; Knott *et al.*, 2002; Martin *et al.*, 2004, 2008; Pietrzykowski *et al.*, 2004). Our recent results indicate that posttranscriptional regulation of BK channels by one of the alcohol-sensitive microRNAs, miR-9, can provide at least a partial explanation of the molecular mechanism of tolerance to alcohol.

We observed that neurons contain a specific subset of BK splice variants, each encoding a different isoform of the channel. Various assemblies of splice variants into tetramers can further create several forms of the BK channel, with each form expressing different characteristics such as gating or calcium sensitivity (Chen *et al.*, 2005; Erxleben *et al.*, 2002; Shipston, 2001; Tian *et al.*, 2001a, 2001b, 2004). This abundance of splice variants possibly enables the BK channel to act as a molecular integrator of cellular signaling, and to be "plastic" and respond quickly to the rapid changes in external or internal environment. Interestingly, we observed that BK channel isoforms differed in their sensitivity to alcohol. The open probability of some isoforms was substantially increased by alcohol, while the open probability of others was only moderately enhanced by the drug. Some isoforms were not potentiated by alcohol at all, exhibiting innate tolerance to alcohol. Temporal quantification of individual variants at various times of alcohol exposure allowed us to determine dynamic changes of expression of isoforms evoked by alcohol. We observed a radical reduction of splice variant diversity. Computation modeling revealed that an ultimate outcome of the relative change in channel splice variants is tolerance to alcohol.

But what was the mechanism of the observed change in the ratio of alcohol-sensitive to alcohol-tolerant isoforms? miR-9 seemed to play a critical role in this regulation. Upregulation of miR-9 by alcohol led subsequently to the binding of miR-9 to its binding site located on some, but not all, 3′-UTRs of the BK channel. Interestingly, 3′-UTRs containing an miR-9 binding site have been primarily associated with the mRNA transcripts encoding BK channel isoforms highly sensitive to alcohol potentiation. In contrast, 3′-UTRs lacking miR-9 binding sites were connected to transcripts encoding alcohol-tolerant channels. Hence, upon miR-9 binding to specific transcripts, these transcripts underwent rapid degradation, skewing the ratio toward the transcripts encoding tolerant channels.

2. *Other miR-9 Targets*

miR-9 can target several other molecules (Fig. 2A), including vital elements of neuronal function, synaptic plasticity, or neurotransmitter release (Pietrzykowski *et al.*, 2008). Moreover, many of these targets have been indicated to play a role in the development of alcoholism. One of the miR-9 targets is dopamine receptor D2 (DRD2). Several studies indicate that mutations in DRD2 can predispose to alcoholism (Luo *et al.*, 2005; Noble *et al.*, 2000; Pinto *et al.*, 2009; Sasabe *et al.*, 2007). Reduction of DRD2 expression has been associated with higher risk of alcohol abuse (Volkow *et al.*, 2006). Interestingly, alcohol also alters the alternative splicing of DRD2 (Oomizu *et al.*, 2003). Another alcoholism-relevant miR-9 target is the beta-2 subunit of the GABA A receptor (GABA(A)). Human genetic association studies indicate that the beta-2 subunit of GABA(A) receptor contributes to the development of alcohol dependence, at least in some ethnic groups (Loh *et al.*, 1999; Sander *et al.*, 1999). Another miR-9 target important in alcohol actions on the brain is synaptojanin 1 (SYNJ1). SYNJ1 is predominantly

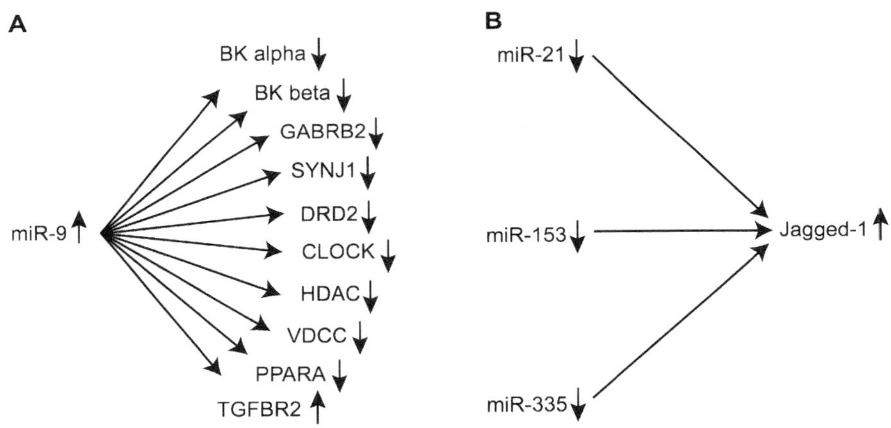

FIG. 2. Divergent and convergent regulation of gene expression by miRNA. (A) An example of a divergent regulation of gene expression by miRNA. miR-9 can target several molecules vital for neuronal function, synaptic plasticity, or neurotransmitter release. Alcohol-induced increased expression of miR-9 causes mainly downregulation of targets, but can also result in their upregulation. BK alpha- and beta-2 – alpha- and beta-2 subunits of calcium-sensitive potassium channel of large conductance (BK), respectively; GABRB2 – beta-2 subunit of gamma-aminobutyric acid receptor; SYNJ1 – synaptojanin 1; DRD2 – dopamine receptor D2; HDAC – histone deacetylase; VDCC – voltage-dependent calcium channel; PPARA – peroxisome proliferator-activated receptor alpha; TGFBR2 – beta subunit of transforming growth factor receptor 2. (B) An example of a convergent regulation of gene expression by miRNA. According to the findings of researchers Sathyan *et al.* (2007), only concomitant suppression of miR-21, miR-153, and miR-335 results in the increased expression of Jagged-1.

expressed in the CNS and concentrated in neuronal terminals (Nemoto *et al.*, 2001). It regulates the recycling of synaptic vesicles and is important in synaptic plasticity. Interestingly, a new microarray analysis points to synaptojanin as an important alcohol target (Uddin and Singh, 2007). Other miR-9 targets in the brain include CLOCK, histone deacetylase (HDAC), voltage-dependent calcium channel (VDCC), peroxisome proliferator-activated receptor alpha (PPARA), beta-2 receptor for the transforming growth factor (TGFBR2), and the beta 2 subunit of the BK channel (Fig. 2A). These are just a handful of examples of miR-9 targets with an already established role in alcohol actions. A thorough examination of their regulation by miR-9 including functional consequences of this regulation, as well as discovery of new miR-9 targets regulated by alcohol, should help to further strengthen the role of miR-9 in mediating alcohol-induced neuronal plasticity and alcoholism.

3. *miR-21, miR-153, miR-335, and Their Common Targets*

Recently, Sathyan *et al.* (2007) took an opposite, "convergent" approach. Because they discovered that in neurospheres at least four microRNAs are regulated by alcohol, they looked for targets common to these miRNAs. They predicted and confirmed experimentally that the ligand for the Notch receptor, Jagged-1, and the neuron-specific RNA-binding protein, ELAVL2/HuB, were targeted by three of these miRNAs: miR-21, miR-153, and miR-335 (Fig. 2B). Both targeted genes are associated with neuronal identity and differentiation, suggesting that alcohol could trigger early maturation of neurospheres by releasing the block of multiple miRNAs imposed on factors controlling neuronal identity and differentiation. This is a very interesting example of alcohol effect on a specific target via simultaneous inhibition of several miRNAs.

4. *Targets of Other miRNAs*

It is likely that the effects of alcohol on miRNAs and the networks of targets they control are much more intricate. Each of the microRNAs regulated by alcohol could target hundreds of genes (Table I), and, as shown above, some of these targets can be common to several miRNAs. Hopefully, the described examples are a prelude to a much more elaborate research, aiming to understanding the role of non-coding RNA on brain function and the development of alcoholism.

Modern techniques that use systems biology approaches and combine genetics, molecular biology, and informatics can allow a comprehensive, integrated analysis of all of the alcohol-relevant microRNA species and their targets, and to determine their change in expression upon alcohol exposure and withdrawal.

IV. Future Directions

The discovery of non-coding RNAs, including miRNA, made us realize that gene expression is a much more intricate phenomenon then initially thought. Alcohol exposure alters gene expression, which significantly contributes to the development of alcoholism. Newly emerging research indicates that miRNA-related mechanisms could constitute core pathways of alcohol regulation of neuronal functions. There are many new, exciting questions arising, and some of them have been asked in this chapter: Can different alcohol exposure patterns affect different molecular elements of miRNA biogenesis and degradation pathways? Could microRNA play a role in transition from transient to permanent alcohol effects? Can the effects of alcohol on miRNA explain, at least partially, the polygenic nature of alcoholism? What is the full list of miRNAs regulated by alcohol in the brain? Are miRNA genes potential candidate genes associated with susceptibility to alcoholism? Is there an association between genetic variation in miRNA genes and different alcohol-related phenotypes (tolerance, craving, withdrawal, etc.)? Finally, can environmental factors, such as stress, increase susceptibility to alcoholism via miRNAs and their targets? There are many questions, but, hopefully, studying miRNAs will help us to deepen our understanding of the molecular mechanisms of alcoholism, and create novel possibilities for successful therapies.

V. Conclusions

The sequencing of the genomes of several species including humans and the discovery of the complex regulatory role of non-coding RNAs, including miRNAs, have unearthed novel mechanisms of gene regulation. It became apparent pretty quickly not only that this additional level of gene expression regulation is an integral part of almost any biological process but also that its complexity is astonishing. As described in this chapter, microRNA can cause both gene activation and gene suppression; its biogenesis and degradation are highly controlled and complex processes, involving many proteins, some probably still awaiting their discovery. We have just started to address the role of miRNA in alcohol effects on neuronal function and development as well as susceptibility to alcoholism.

The discovery of miRNA and novel regulation of gene expression should challenge our current thinking of the molecular underpinnings of alcohol actions, help in creating new concepts regarding the development of addiction to alcohol, and stimulate further research to test these ideas. Hopefully, we will be able to further our understanding of the pathophysiological mechanisms of alcoholism and to develop novel therapeutic strategies for this devastating disease.

References

Abu-Elneel, K., *et al.* (2008). Heterogeneous dysregulation of microRNAs across the autism spectrum. *Neurogenetics* **9**, 153–161.

Adachi, J., *et al.* (1991). Degrees of alcohol intoxication in 117 hospitalized cases. *J. Stud. Alcohol.* **52**, 448–453.

Ambros, V. (2004). The functions of animal microRNAs. *Nature* **431**, 350–355.

Arya, R., *et al.* (2005). Effect of genotype x alcoholism interaction on linkage analysis of an alcoholism-related quantitative phenotype. *BMC Genet.* **6**(Suppl. 1), S120.

Ashraf, S. I., *et al.* (2006). Synaptic protein synthesis associated with memory is regulated by the RISC pathway in *Drosophila*. *Cell* **124**, 191–205.

Ason, B., *et al.* (2006). Differences in vertebrate microRNA expression. *Proc. Natl. Acad. Sci. USA* **103**, 14385–14389.

Axtell, M. J., and Bartel, D. P. (2005). Antiquity of microRNAs and their targets in land plants. *Plant Cell* **17**, 1658–1673.

Bail, S., *et al.* (2010). Differential regulation of microRNA stability. *RNA* **16**, 1032–1039.

Bartel, D. P. (2004). MicroRNAs: Genomics, biogenesis, mechanism, and function. *Cell* **116**, 281–297.

Bartel, D. P. (2009). MicroRNAs: Target recognition and regulatory functions. *Cell* **136**, 215–233.

Batassa, E. M. *et al.*, (2010). RISC activity in hippocampus is essential for contextual memory. *Neurosci. Lett.* **471**, 185–188.

Baumberger, N., and Baulcombe, D. C. (2005). *Arabidopsis* ARGONAUTE1 is an RNA slicer that selectively recruits microRNAs and short interfering RNAs. *Proc. Natl. Acad. Sci. USA* **102**, 11928–11933.

Benhassine, N., and Berger, T. (2009). Large-conductance calcium-dependent potassium channels prevent dendritic excitability in neocortical pyramidal neurons. *Pflugers Arch.* **457**, 1133–1145.

Berezikov, E., *et al.* (2007). Mammalian mirtron genes. *Mol. Cell* **28**, 328–336.

Bernstein, E., *et al.* (2001). Role for a bidentate ribonuclease in the initiation step of RNA interference. *Nature* **409**, 363–366.

Beveridge, N. J., *et al.* (2008). Dysregulation of miRNA 181b in the temporal cortex in schizophrenia. *Hum. Mol. Genet.* **17**, 1156–1168.

Birney, E., *et al.* (2007). Identification and analysis of functional elements in 1% of the human genome by the ENCODE pilot project. *Nature* **447**, 799–816.

Blow, M. J., *et al.* (2006). RNA editing of human microRNAs. *Genome Biol.* **7**, R27.

Bohnsack, M. T., *et al.* (2004). Exportin 5 is a RanGTP-dependent dsRNA-binding protein that mediates nuclear export of pre-miRNAs. *RNA* **10**, 185–191.

Breese, G. R., *et al.* (2005). Stress enhancement of craving during sobriety: A risk for relapse. *Alcohol. Clin. Exp. Res.* **29**, 185–195.

Burmistrova, O. A., *et al.* (2007). MicroRNA in schizophrenia: Genetic and expression analysis of miR-130b (22q11). *Biochemistry (Mosc)* **72**, 578–582.

Butler, A., *et al.* (1993). mSlo, a complex mouse gene encoding "maxi" calcium-activated potassium channels. *Science* **261**, 221–224.

Campbell, J. C., *et al.* (2009). Contribution of early environmental stress to alcoholism vulnerability. *Alcohol* **43**, 547–554.

Chen, C. Y., *et al.* (2009). Ago-TNRC6 triggers microRNA-mediated decay by promoting two deadenylation steps. *Nat. Struct. Mol. Biol.* **16**, 1160–1166.

Chen, L., *et al.* (2005). Functionally diverse complement of large conductance calcium- and voltage-activated potassium channel (BK) alpha-subunits generated from a single site of splicing. *J. Biol. Chem.* **280**, 33599–33609.

Chendrimada, T. P., *et al.* (2005). TRBP recruits the Dicer complex to Ago2 for microRNA processing and gene silencing. *Nature* **436**, 740–744.
Christensen, M., *et al.* (2010). Recombinant adeno-associated virus-mediated microRNA delivery into the postnatal mouse brain reveals a role for miR-134 in dendritogenesis *in vivo*. *Front Neural Circuits* **3**, 16.
Cogswell, J. P., *et al.* (2008). Identification of miRNA changes in Alzheimer's disease brain and CSF yields putative biomarkers and insights into disease pathways. *J. Alzheimers Dis.* **14**, 27–41.
Cowmeadow, R. B., *et al.* (2005). The slowpoke gene is necessary for rapid ethanol tolerance in *Drosophila*. *Alcohol. Clin. Exp. Res.* **29**, 1777–1786.
Cowmeadow, R. B., *et al.* (2006). Ethanol tolerance caused by slowpoke induction in *Drosophila*. *Alcohol. Clin. Exp. Res.* **30**, 745–753.
Crowley, J. J., *et al.* (2005). Distinct structural features of phospholipids differentially determine ethanol sensitivity and basal function of BK channels. *Mol. Pharmacol.* **68**, 4–10.
Davies, A. G., *et al.* (2003). A central role of the BK potassium channel in behavioral responses to ethanol in *C. elegans*. *Cell* **115**, 655–666.
Dick, D. M., *et al.* (2007). Alcohol dependence with comorbid drug dependence: Genetic and phenotypic associations suggest a more severe form of the disorder with stronger genetic contribution to risk. *Addiction* **102**, 1131–1139.
Dick, D. M., *et al.* (2006). The role of GABRA2 in risk for conduct disorder and alcohol and drug dependence across developmental stages. *Behav. Genet.* **36**, 577–590.
Dick, D. M., *et al.* (2004). Association of GABRG3 with alcohol dependence. *Alcohol. Clin. Exp. Res.* **28**, 4–9.
Dick, D. M., *et al.* (2002). Suggestive linkage on chromosome 1 for a quantitative alcohol-related phenotype. *Alcohol. Clin. Exp. Res.* **26**, 1453–1460.
Dolganiuc, A., *et al.* (2009). MicroRNA expression profile in Lieber–DeCarli diet-induced alcoholic and methionine choline deficient diet-induced nonalcoholic steatohepatitis models in mice. *Alcohol. Clin. Exp. Res.* **33**, 1704–1710.
Dopico, A. M., *et al.* (1999). Rat supraoptic magnocellular neurones show distinct large conductance, Ca^{2+}-activated K^+ channel subtypes in cell bodies versus nerve endings. *J. Physiol.* **519**(Pt 1), 101–114.
Dworetzky, S. I., *et al.* (1994). Cloning and expression of a human large-conductance calcium-activated potassium channel. *Brain Res. Mol. Brain Res.* **27**, 189–193.
Edenberg, H. J., *et al.* (2004). Variations in GABRA2, encoding the alpha 2 subunit of the GABA(A) receptor, are associated with alcohol dependence and with brain oscillations. *Am. J. Hum. Genet.* **74**, 705–714.
Ehlers, C. L. (2007). Variations in ADH and ALDH in southwest California Indians. *Alcohol Res. Health* **30**, 14–17.
Ehlers, C. L., *et al.* (2004). Genomic screen for loci associated with alcohol dependence in Mission Indians. *Am. J. Med. Genet. B Neuropsychiatr. Genet.* **129**, 110–115.
Ehlers, C. L., *et al.* (2008). Association between single nucleotide polymorphisms in the mu opioid receptor gene (OPRM1) and self-reported responses to alcohol in American Indians. *BMC Med. Genet.* **9**, 35.
Erxleben, C., *et al.* (2002). Interacting effects of N-terminal variation and strex exon splicing on slo potassium channel regulation by calcium, phosphorylation, and oxidation. *J. Biol. Chem.* **277**, 27045–27052.
Eulalio, A., *et al.* (2009). Deadenylation is a widespread effect of miRNA regulation. *RNA* **15**, 21–32.
Farley, J., and Rudy, B. (1988). Multiple types of voltage-dependent Ca^{2+}-activated K^+ channels of large conductance in rat brain synaptosomal membranes. *Biophys. J.* **53**, 919–934.
Feng, J., *et al.* (2009). Evidence for X-chromosomal schizophrenia associated with microRNA alterations. *PLoS One* **4**, e6121.

Ferreira, M. P., and Willoughby, D. (2008). Alcohol consumption: The good, the bad, and the indifferent. *Appl. Physiol. Nutr. Metab.* **33**, 12–20.

Filipowicz, W., *et al.* (2008). Mechanisms of post-transcriptional regulation by microRNAs: Are the answers in sight? *Nat. Rev. Genet.* **9**, 102–114.

Fox, H. C., *et al.* (2007). Stress-induced and alcohol cue-induced craving in recently abstinent alcohol-dependent individuals. *Alcohol. Clin. Exp. Res.* **31**, 395–403.

Franke, P., *et al.* (1999). Human delta-opioid receptor gene and susceptibility to heroin and alcohol dependence. *Am. J. Med. Genet.* **88**, 462–464.

Friedman, R. C., *et al.* (2009). Most mammalian mRNAs are conserved targets of microRNAs. *Genome Res.* **19**, 92–105.

Gelernter, J., *et al.* (1999). Genetics of two mu opioid receptor gene (OPRM1) exon I polymorphisms: Population studies, and allele frequencies in alcohol- and drug-dependent subjects. *Mol. Psychiatry* **4**, 476–483.

Gelernter, J., *et al.* (2007). Opioid receptor gene (OPRM1, OPRK1, and OPRD1) variants and response to naltrexone treatment for alcohol dependence: Results from the VA Cooperative Study. *Alcohol Clin. Exp. Res.* **31**, 555–563.

Ghildiyal, M., *et al.* (2010). Sorting of *Drosophila* small silencing RNAs partitions microRNA* strands into the RNA interference pathway. *RNA* **16**, 43–56.

Griffiths-Jones, S. (2004). The microRNA registry. *Nucleic Acids Res.* **32**, D109–111.

Griffiths-Jones, S. (2006). miRBase: The microRNA sequence database. *Methods Mol. Biol.* **342**, 129–138.

Griffiths-Jones, S., *et al.* (2008). miRBase: Tools for microRNA genomics. *Nucleic Acids Res.* **36**, D154–158.

Grimson, A., *et al.* (2007). MicroRNA targeting specificity in mammals: Determinants beyond seed pairing. *Mol. Cell* **27**, 91–105.

Gscheidel, N., *et al.* (2000). Five exon 1 variants of mu opioid receptor and vulnerability to alcohol dependence. *Pol. J. Pharmacol.* **52**, 27–31.

Guerrini, I., *et al.* (2005). Genetic linkage analysis supports the presence of two susceptibility loci for alcoholism and heavy drinking on chromosome 1p22.1–11.2 and 1q21.3–24.2. *BMC Genet.* **6**, 11.

Guo, A. Y., *et al.* (2010). A novel microRNA and transcription factor mediated regulatory network in schizophrenia. *BMC Syst. Biol.* **4**, 10.

Han, J., *et al.* (2004). The Drosha–DGCR8 complex in primary microRNA processing. *Genes Dev.* **18**, 3016–3027.

Hansen, T., *et al.* (2007). Brain expressed microRNAs implicated in schizophrenia etiology. *PLoS One* **2**, e873.

Hebert, S. S., *et al.* (2008). Loss of microRNA cluster miR-29a/b-1 in sporadic Alzheimer's disease correlates with increased BACE1/beta-secretase expression. *Proc. Natl. Acad. Sci. USA* **105**, 6415–6420.

Hebert, S. S., *et al.* (2009). MicroRNA regulation of Alzheimer's amyloid precursor protein expression. *Neurobiol. Dis.* **33**, 422–428.

Hill, S. Y., *et al.* (2004). A genome wide search for alcoholism susceptibility genes. *Am. J. Med. Genet. B Neuropsychiatr. Genet.* **128**, 102–113.

Hu, H., *et al.* (2001). Presynaptic Ca^{2+}-activated K^{+} channels in glutamatergic hippocampal terminals and their role in spike repolarization and regulation of transmitter release. *J. Neurosci.* **21**, 9585–9597.

Hutvagner, G., *et al.* (2001). A cellular function for the RNA-interference enzyme Dicer in the maturation of the let-7 small temporal RNA. *Science* **293**, 834–838.

Johnson, C., *et al.* (2006). Pooled association genome scanning for alcohol dependence using 104,268 SNPs: Validation and use to identify alcoholism vulnerability loci in unrelated individuals from the collaborative study on the genetics of alcoholism. *Am. J. Med. Genet. B Neuropsychiatr. Genet.* **141**, 844–853.

Joslyn, G., *et al.* (2008). Chromosome 15q25.1 genetic markers associated with level of response to alcohol in humans. *Proc. Natl. Acad. Sci. USA* **105**, 20368–20373.

Kapranov, P., *et al.* (2007). Genome-wide transcription and the implications for genomic organization. *Nat. Rev. Genet.* **8**, 413–423.

Kapsimali, M., *et al.* (2007). MicroRNAs show a wide diversity of expression profiles in the developing and mature central nervous system. *Genome Biol.* **8**, R173.

Kawahara, Y., *et al.* (2007). RNA editing of the microRNA-151 precursor blocks cleavage by the Dicer–TRBP complex. *EMBO Rep.* **8**, 763–769.

Khudayberdiev, S., *et al.* (2009). MicroRNA as modulators of neuronal responses. *Commun. Integr. Biol.* **2**, 411–413.

Knott, T. K., *et al.* (2002). Integrated channel plasticity contributes to alcohol tolerance in neurohypophysial terminals. *Mol. Pharmacol.* **62**, 135–142.

Krichevsky, A. M., *et al.* (2003). A microRNA array reveals extensive regulation of microRNAs during brain development. *RNA* **9**, 1274–1281.

Lander, E. S., *et al.* (2001). Initial sequencing and analysis of the human genome. *Nature* **409**, 860–921.

Lee, R. C., *et al.* (1993). The *C. elegans* heterochronic gene lin-4 encodes small RNAs with antisense complementarity to lin-14. *Cell* **75**, 843–854.

Lee, Y., *et al.* (2006). The role of PACT in the RNA silencing pathway. *EMBO J.* **25**, 522–532.

Lee, Y. B., *et al.* (2009). Twist-1 regulates the miR-199a/214 cluster during development. *Nucleic Acids Res.* **37**, 123–128.

Leucht, C., *et al.* (2008). MicroRNA-9 directs late organizer activity of the midbrain–hindbrain boundary. *Nat. Neurosci.* **11**, 641–648.

Lewis, B. P., *et al.* (2003). Prediction of mammalian microRNA targets. *Cell* **115**, 787–798.

Li, S. C., *et al.* (2007). Intronic microRNA: Discovery and biological implications. *DNA Cell. Biol.* **26**, 195–207.

Li, X., Jin, P. (2010). Roles of small regulatory RNAs in determining neuronal identity. *Nat. Rev. Neurosci.* **11**, 329–338.

Lin, S. L., *et al.* (2006). Intronic microRNA (miRNA). *J. Biomed. Biotechnol.* **2006**, 26818.

Liu, X., *et al.* (2008). MicroRNAs: Biogenesis and molecular functions. *Brain Pathol.* **18**, 113–121.

Loane, D. J., *et al.* (2007). Co-assembly of N-type Ca^{2+} and BK channels underlies functional coupling in rat brain. *J. Cell Sci.* **120**, 985–995.

Loh, E. W., *et al.* (1999). Association between variants at the GABAAbeta2, GABAAalpha6 and GABAAgamma2 gene cluster and alcohol dependence in a Scottish population. *Mol. Psychiatry.* **4**, 539–544.

Loh, el W., *et al.*. (2004). Endogenous opioid receptor genes and alcohol dependence among Taiwanese Han. *Alcohol. Clin. Exp. Res.* **28**, 15–19.

Luo, H. R., *et al.* (2005). Evolution of the DRD2 gene haplotype and its association with alcoholism in Mexican Americans. *Alcohol* **36**, 117–125.

MacRae, I. J., *et al.* (2008). *In vitro* reconstitution of the human RISC-loading complex. *Proc. Natl. Acad. Sci. USA* **105**, 512–517.

Madrid, G. A., *et al.* (2001). Stress as a mediating factor in the association between the DRD2 TaqI polymorphism and alcoholism. *Alcohol* **23**, 117–122.

Martin, G., *et al.* (2004). Somatic localization of a specific large-conductance calcium-activated potassium channel subtype controls compartmentalized ethanol sensitivity in the nucleus accumbens. *J. Neurosci.* **24**, 6563–6572.

Martin, G. E., *et al.* (2008). Identification of a BK channel auxiliary protein controlling molecular and behavioral tolerance to alcohol. *Proc. Natl. Acad. Sci. USA* **105**, 17543–17548.

Matranga, C., *et al.* (2005). Passenger-strand cleavage facilitates assembly of siRNA into Ago2-containing RNAi enzyme complexes. *Cell* **123**, 607–620.
Mellios, N., *et al.* (2009). Molecular determinants of dysregulated GABAergic gene expression in the prefrontal cortex of subjects with schizophrenia. *Biol. Psychiatry* **65**, 1006–1014.
Miranda, R. C., *et al.* (2010). MicroRNAs: Master regulators of ethanol abuse and toxicity? *Alcohol. Clin. Exp. Res.* **34**, 575–587.
Misonou, H., *et al.* (2006). Immunolocalization of the Ca^{2+}-activated K^{+} channel Slo1 in axons and nerve terminals of mammalian brain and cultured neurons. *J. Comp. Neurol.* **496**, 289–302.
Molnar, A., *et al.* (2009). Highly specific gene silencing by artificial microRNAs in the unicellular alga *Chlamydomonas reinhardtii. Plant J.* **58**, 165–174.
Mourelatos, Z., *et al.* (2002). miRNPs: A novel class of ribonucleoproteins containing numerous microRNAs. *Genes Dev.* **16**, 720–728.
Nemoto, Y., *et al.* (2001). Identification and characterization of a synaptojanin 2 splice isoform predominantly expressed in nerve terminals. *J. Biol. Chem.* **276**, 41133–41142.
Noble, E. P., *et al.* (2000). Haplotypes at the DRD2 locus and severe alcoholism. *Am. J. Med. Genet.* **96**, 622–631.
Nunez-Iglesias, J., *et al.* (2010). Joint genome-wide profiling of miRNA and mRNA expression in Alzheimer's disease cortex reveals altered miRNA regulation. *PLoS One* **5**, e8898.
Okamura, K., *et al.* (2007). The mirtron pathway generates microRNA-class regulatory RNAs in *Drosophila. Cell* **130**, 89–100.
Okamura, K., *et al.* (2009). Distinct mechanisms for microRNA strand selection by *Drosophila* Argonautes. *Mol. Cell* **36**, 431–444.
Olsen, L., *et al.* (2009). MicroRNAs show mutually exclusive expression patterns in the brain of adult male rats. *PLoS One* **4**, e7225.
Oomizu, S., *et al.* (2003). Ethanol and estradiol modulate alternative splicing of dopamine D2 receptor messenger RNA and abolish the inhibitory action of bromocriptine on prolactin release from the pituitary gland. *Alcohol. Clin. Exp. Res.* **27**, 975–980.
Packer, A. N., *et al.* (2008). The bifunctional microRNA miR-9/miR-9* regulates REST and CoREST and is downregulated in Huntington's disease. *J. Neurosci.* **28**, 14341–14346.
Park, C. S., and Tang, S. J. (2009). Regulation of microRNA expression by induction of bidirectional synaptic plasticity. *J. Mol. Neurosci.* **38**, 50–56.
Parsian, A., and Zhang, Z. H. (1999). Human chromosomes 11p15 and 4p12 and alcohol dependence: Possible association with the GABRB1 gene. *Am. J. Med. Genet.* **88**, 533–538.
Perkins, D. O., *et al.* (2007). microRNA expression in the prefrontal cortex of individuals with schizophrenia and schizoaffective disorder. *Genome Biol.* **8**, R27.
Perron, M. P., and Provost, P. (2008). Protein interactions and complexes in human microRNA biogenesis and function. *Front. Biosci.* **13**, 2537–2547.
Petrik, D., and Brenner, R. (2007). Regulation of STREX exon large conductance, calcium-activated potassium channels by the beta4 accessory subunit. *Neuroscience* **149**, 789–803.
Pietrzykowski, A. Z., *et al.* (2008). Posttranscriptional regulation of BK channel splice variant stability by miR-9 underlies neuroadaptation to alcohol. *Neuron* **59**, 274–287.
Pietrzykowski, A. Z., *et al.* (2004). Alcohol tolerance in large-conductance, calcium-activated potassium channels of CNS terminals is intrinsic and includes two components: Decreased ethanol potentiation and decreased channel density. *J. Neurosci.* **24**, 8322–8332.
Pinto, E., and Ansseau, M. (2009). [Genetic factors of alcohol-dependence]. *Encephale* **35**, 461–469.
Pinto, E., *et al.* (2009). The TaqI A DRD2 polymorphism in type II alcohol dependence: A marker of age at onset or of a familial disease? *Alcohol* **43**, 271–275.
Place, R. F., *et al.* (2008). MicroRNA-373 induces expression of genes with complementary promoter sequences. *Proc. Natl. Acad. Sci. USA* **105**, 1608–1613.

Poulsen, A. N., *et al.* (2009). Differential expression of BK channel isoforms and beta-subunits in rat neuro-vascular tissues. *Biochim. Biophys. Acta* **1788**, 380–389.

Provost, P., *et al.* (2002). Ribonuclease activity and RNA binding of recombinant human Dicer. *EMBO J.* **21**, 5864–5874.

Rajasethupathy, P., *et al.* (2009). Characterization of small RNAs in aplysia reveals a role for miR-124 in constraining synaptic plasticity through CREB. *Neuron* **63**, 803–817.

Ramachandran, V., and Chen, X. (2008). Degradation of microRNAs by a family of exoribonucleases in *Arabidopsis*. *Science* **321**, 1490–1492.

Rommelspacher, H., *et al.* (2001). Genetic analysis of the mu-opioid receptor in alcohol-dependent individuals. *Alcohol* **24**, 129–135.

Sailer, C. A., *et al.* (2006). Immunolocalization of BK channels in hippocampal pyramidal neurons. *Eur. J. Neurosci.* **24**, 442–454.

Saini, H. K., *et al.* (2007). Genomic analysis of human microRNA transcripts. *Proc. Natl. Acad. Sci. USA* **104**, 17719–17724.

Sander, T., *et al.* (1999). Association analysis of sequence variants of GABA(A) alpha6, beta2, and gamma2 gene cluster and alcohol dependence. *Alcohol. Clin. Exp. Res.* **23**, 427–431.

Sasabe, T., *et al.* (2007). Association analysis of the dopamine receptor D2 (DRD2) SNP rs1076560 in alcoholic patients. *Neurosci. Lett.* **412**, 139–142.

Sathyan, P., *et al.* (2007). Competing interactions between micro-RNAs determine neural progenitor survival and proliferation after ethanol exposure: Evidence from an *ex vivo* model of the fetal cerebral cortical neuroepithelium. *J. Neurosci.* **27**, 8546–8557.

Sausbier, M., *et al.* (2004). Cerebellar ataxia and Purkinje cell dysfunction caused by Ca^{2+}-activated K^+ channel deficiency. *Proc. Natl. Acad. Sci. USA* **101**, 9474–9478.

Sausbier, U., *et al.* (2006). Ca^{2+} -activated K^+ channels of the BK-type in the mouse brain. *Histochem. Cell Biol.* **125**, 725–741.

Schratt, G. (2009). Fine-tuning neural gene expression with microRNAs. *Curr. Opin. Neurobiol.* **19**, 213–219.

Sethi, P., and Lukiw, W. J. (2009). Micro-RNA abundance and stability in human brain: Specific alterations in Alzheimer's disease temporal lobe neocortex. *Neurosci. Lett.* **459**, 100–104.

Shibata, M., *et al.* (2008). MicroRNA-9 modulates Cajal–Retzius cell differentiation by suppressing Foxg1 expression in mouse medial pallium. *J. Neurosci.* **28**, 10415–10421.

Shipston, M. J. (2001). Alternative splicing of potassium channels: A dynamic switch of cellular excitability. *Trends Cell Biol.* **11**, 353–358.

Shipston, M. J., *et al.* (1999). Molecular components of large conductance calcium-activated potassium (BK) channels in mouse pituitary corticotropes. *Mol. Endocrinol.* **13**, 1728–1737.

Sinha, R. (2007). The role of stress in addiction relapse. *Curr. Psychiatry Rep.* **9**, 388–395.

Smalheiser, N. R., and Lugli, G. (2009). microRNA regulation of synaptic plasticity. *Neuromol. Med.* **11**, 133–140.

Song, J., *et al.* (2003). Association of GABA(A) receptors and alcohol dependence and the effects of genetic imprinting. *Am. J. Med. Genet. B Neuropsychiatr. Genet.* **117**, 39–45.

Soom, M., *et al.* (2008). A mutually exclusive alternative exon of slo1 codes for a neuronal BK channel with altered function. *Channels (Austin.)* **2**, 278–282.

Steitz, J. A., and Vasudevan, S. (2009). miRNPs: Versatile regulators of gene expression in vertebrate cells. *Biochem. Soc. Trans.* **37**, 931–935.

Strauch, K., *et al.* (2005). Linkage analysis of alcohol dependence using MOD scores. *BMC Genet.* **6** (Suppl 1), S162.

Tabakoff, B., *et al.* (2009). Genetical genomic determinants of alcohol consumption in rats and humans. *BMC Biol.* **7**, 70.

Takada, S., *et al.* (2008). MicroRNA expression profiles of human leukemias. *Leukemia* **22**, 1274–1278.

Tan, Y., *et al.* (2009). Transcriptional inhibiton of Hoxd4 expression by miRNA-10a in human breast cancer cells. *BMC Mol. Biol.* **10**, 12.

Tang, Y., *et al.* (2008). Effect of alcohol on miR-212 expression in intestinal epithelial cells and its potential role in alcoholic liver disease. *Alcohol. Clin. Exp. Res.* **32**, 355–364.

Tian, L., *et al.* (2001). Alternative splicing switches potassium channel sensitivity to protein phosphorylation. *J. Biol. Chem.* **276**, 7717–7720.

Tian, L., *et al.* (2001). Alternative splicing determines sensitivity of murine calcium-activated potassium channels to glucocorticoids. *J. Physiol.* **537**, 57–68.

Tian, L., *et al.* (2004). Distinct stoichiometry of BKCa channel tetramer phosphorylation specifies channel activation and inhibition by cAMP-dependent protein kinase. *Proc. Natl. Acad. Sci. USA* **101**, 11897–11902.

Tyler, D. M., *et al.* (2008). Functionally distinct regulatory RNAs generated by bidirectional transcription and processing of microRNA loci. *Genes Dev.* **22**, 26–36.

Tyndale, R. F. (2003). Genetics of alcohol and tobacco use in humans. *Ann. Med.* **35**, 94–121.

Uddin, R. K., and Singh, S. M. (2007). Ethanol-responsive genes: Identification of transcription factors and their role in metabolomics. *Pharmacogenomics J.* **7**, 38–47.

Ueno, S., *et al.* (1999). Identification of a novel polymorphism of the human dopamine transporter (DAT1) gene and the significant association with alcoholism. *Mol. Psychiatry.* **4**, 552–557.

Vasudevan, S., and Steitz, J. A. (2007). AU-rich-element-mediated upregulation of translation by FXR1 and Argonaute 2. *Cell* **128**, 1105–1118.

Vasudevan, S., *et al.* (2007). Switching from repression to activation: MicroRNAs can up-regulate translation. *Science* **318**, 1931–1934.

Vasudevan, S., *et al.* (2008). Cell-cycle control of microRNA-mediated translation regulation. *Cell Cycle* **7**, 1545–1549.

Venter, J. C., *et al.* (2001). The sequence of the human genome. *Science* **291**, 1304–1351.

Volkow, N. D., *et al.* (2006). High levels of dopamine D2 receptors in unaffected members of alcoholic families: Possible protective factors. *Arch. Gen. Psychiatry* **63**, 999–1008.

Vreugdenhil, E., and Berezikov, E. (2010). Fine-tuning the brain: MicroRNAs. *Front. Neuroendocrinol.* **31**, 128–133.

Wang, J. C., *et al.* (2004). Evidence of common and specific genetic effects: Association of the muscarinic acetylcholine receptor M2 (CHRM2) gene with alcohol dependence and major depressive syndrome. *Hum. Mol. Genet.* **13**, 1903–1911.

Wang, L. L., *et al.* (2009). Ethanol exposure induces differential microRNA and target gene expression and teratogenic effects which can be suppressed by folic acid supplementation. *Hum. Reprod.* **24**, 562–579.

Wang, W. X., *et al.* (2008). The expression of microRNA miR-107 decreases early in Alzheimer's disease and may accelerate disease progression through regulation of beta-site amyloid precursor protein-cleaving enzyme 1. *J. Neurosci.* **28**, 1213–1223.

Wang, X., *et al.* (2009). miR-34a, a microRNA up-regulated in a double transgenic mouse model of Alzheimer's disease, inhibits bcl2 translation. *Brain Res. Bull.* **80**, 268–273.

Wang, Y., *et al.* (2009). CREB regulation of BK channel gene expression underlies rapid drug tolerance. *Genes Brain Behav.* **8**, 369–376.

Wanner, S. G., *et al.* (1999). High-conductance calcium-activated potassium channels in rat brain: Pharmacology, distribution, and subunit composition. *Biochemistry* **38**, 5392–5400.

Williams, A. E. (2008). Functional aspects of animal microRNAs. *Cell Mol. Life Sci.* **65**, 545–562.

Willingham, A. T., and Gingeras, T. R. (2006). TUF love for "junk" DNA. *Cell* **125**, 1215–1220.

Windemuth, C., *et al.* (1999). Linkage analysis in alcohol dependence. *Genet. Epidemiol.* **17**(Suppl. 1), S403–407.

Xie, J., and McCobb, D. P. (1998). Control of alternative splicing of potassium channels by stress hormones. *Science* **280**, 443–446.

Xuei, X., *et al.* (2006). Association of the kappa-opioid system with alcohol dependence. *Mol. Psychiatry* **11**, 1016–1024.
Yang, W., *et al.* (2005). ADAR1 RNA deaminase limits short interfering RNA efficacy in mammalian cells. *J. Biol. Chem.* **280**, 3946–3953.
Yeom, K. H., *et al.* (2006). Characterization of DGCR8/Pasha, the essential cofactor for Drosha in primary miRNA processing. *Nucleic Acids Res.* **34**, 4622–4629.
Zeng, Y., and Cullen, B. R. (2004). Structural requirements for pre-microRNA binding and nuclear export by Exportin 5. *Nucleic Acids Res.* **32**, 4776–4785.
Zhang, H., *et al.* (2006). Association between two mu-opioid receptor gene (OPRM1) haplotype blocks and drug or alcohol dependence. *Hum. Mol. Genet.* **15**, 807–819.
Zhang, H., *et al.* (2008). The OPRD1 and OPRK1 loci in alcohol or drug dependence: OPRD1 variation modulates substance dependence risk. *Mol. Psychiatry.* **13**, 531–543.
Zhang, Z., *et al.* (2010). MicroRNAs: Potential regulators involved in human anencephaly. *Int. J. Biochem. Cell Biol.* **42**, 367–374.

THE GENETICS OF BEHAVIORAL ALCOHOL RESPONSES IN *DROSOPHILA*

Aylin R. Rodan[1] and Adrian Rothenfluh[2]

[1]Division of Nephrology, Department of Psychiatry and Program in Neuroscience, UT Southwestern Medical Center, Dallas, TX, USA

[2]Department of Psychiatry and Program in Neuroscience, UT Southwestern Medical Center, Dallas, TX, USA

Drosophila melanogaster is commonly found near rotting or fermenting fruit, reflected in its name pomace, or vinegar fly. In such environments, flies often encounter significant levels of ethanol. Three observations have made Drosophila a very promising model organism to understand the genetic contributions to the behavioral responses to alcohol. First, similar to higher vertebrates, flies show hyperactivation upon exposure to a low to medium dose of alcohol, while high doses can lead to sedation. In addition, when given a choice, flies will actually prefer alcohol-containing food over regular food. Second, the genes and biochemical pathways implicated in controlling these behavioral responses in flies are also participating in determining alcohol responses, and drinking behavior in mammals. Third, the fact that flies have been studied genetically for over one hundred years means that an exceptional repertoire of genetic tools are at our disposal. Here, we will review some of these tools and experimental approaches, survey the methods for, and measures after Drosophila ethanol exposure, and

DOI: 10.1016/S0074-7742(10)91002-7

0074-7742/10 $35.00

discuss the different molecular components and functional pathways involved in these behavioral responses to alcohol.

I. Introduction

A. Alcohol and Alcoholism

Alcohol is one of the most widely abused drugs in the world, and alcohol use disorders have grave consequences for both the affected individual and society at large. Our understanding of the changes in the brain that cause a person to go from casual consumption to uncontrolled use is still lacking. One of the reasons for this is the fact that unlike most other drugs of abuse, ethanol does not have a primary high-affinity target in the brain, but rather, alcohol binds to many proteins at relatively low affinities and thus many neurochemical systems may be affected by ethanol and contribute to alcoholism.

Many studies have indicated that alcoholism has a large genetic component (reviewed by Hill, this issue). This makes genetic model organisms such as the mouse (Buck, this issue) or *Drosophila* very relevant as tools to isolate candidate genes altering behavioral alcohol responses and drinking patterns, and to understand how these genes work and how they might act to alter brain function in addicted individuals. This chapter focuses on the advances in our understanding of the genes and biochemical pathways that mediate the behavioral responses to alcohol in flies.

B. *Drosophila* as a Model Organism

Drosophila melanogaster vinegar flies have been used for over a 100 years as a model organism to study the laws and mechanisms of heredity. In this chapter, the terms *Drosophila* and flies will refer to this species, unless specifically noted otherwise. The main reasons why Thomas Hunt Morgan's research group decided to focus on *Drosophila* a 100 years ago were threefold: first, flies are easily and inexpensively cultured in glass bottles with banana pulp. Second, their life cycle is fast, requiring only about 12 days at room temperature to go from freshly laid egg to reproducing adult. And third, a single female can have over a hundred offspring. In addition to that, *Drosophila* larvae have giant salivary gland chromosomes, allowing the visualization of subsegments of the four chromosomes. This was the basis for the accumulation and characterization of many fly strains with cytologically defined chromosomal aberrations, such as deficiencies and duplications. These were, and still are, invaluable tools for genetic mapping and stock maintenance techniques. All these reasons have made *Drosophila* an excellent model organism for the study of the basics of genetic inheritance.

Many fundamental discoveries were made in flies, including genetic recombination, or the fact that X-rays are mutagenic (Sturtevant, 1967). The genomic sequence of *D. melanogaster* was completed in 2000 (Adams *et al.*, 2000), and, currently, 192 wild-derived recombinant inbred lines are being fully sequenced, some of which have already been used to associate genetic variation with behavioral alcohol responses (Morozova *et al.*, 2009) (see Section III.B). Current estimates are that ~75% of genes associated with a human disease have an obvious ortholog in flies (Chien *et al.*, 2002). This high degree of conservation became obvious in the 1980s when many developmental genes found in *Drosophila* were shown to have similar function in humans (Gehring *et al.*, 2009). However, even in such complex behaviors as circadian rhythm, there is a very high degree of conservation between flies and mammals, both in gene structure and in molecular function (Collins and Blau, 2007).

While forward genetics, that is, going from phenotype to gene, has traditionally been the approach taken in *Drosophila*, genetic transformation was established in 1982 (Rubin and Spradling, 1982). This allows both the reintroduction of a gene to confirm that its mutation causes a phenotype, and also allows for introduction of transgenes mis- or overexpressing specific proteins or their altered derivatives. One of the most widely used tools in that regard is the binary Gal4/UAS system (Brand and Perrimon, 1993). One transgene carries the yeast transcriptional activator Gal4 under the control of a specific promoter, allowing for spatially controlled expression. The second transgene carries a cDNA of interest under the control of the Gal4-responsive upstream activating sequence (UAS). This allows testing of hypotheses such as: Is central nervous system (CNS)-specific expression of a cDNA sufficient to rescue the phenotype caused by a given mutation? In addition, certain proteins can be expressed in neurons that induce temperature-sensitive neuronal silencing (Kitamoto, 2001), or action potentials (Pulver *et al.*, 2009), allowing questions such as: Is a given set of neurons necessary or even sufficient for a certain behavioral response? Given that a substantial effort is being made to produce 5000 unique Gal4 drivers expressing in a small subset of CNS neurons (Pfeiffer *et al.*, 2008), we might well learn much more about the neurons and circuits mediating these behaviors in the near future.

In addition to the binary Gal4/UAS system, other reverse genetic tools have been developed in flies that allow going from gene to phenotype. One of them is the systematic generation of UAS lines expressing interfering RNA constructs for knock-down of every *Drosophila* gene (Dietzl *et al.*, 2007). And in the last few years, it has also become easier, and more common, to use homologous recombination to generate targeted gene knock-outs or replacement knock-ins of any given fly gene (Maggert *et al.*, 2008). Finally, large-scale efforts have been undertaken to generate a library of fly strains that contain mutations in every single gene (Matthews *et al.*, 2005).

II. Ethanol Exposure in Flies

A. Ethanol Toxicity Studies

Many early studies, reviewed in Guarnieri and Heberlein (2003), have focused on the effects of ethanol on larval and adult fly survival, and increases in survival were often labeled tolerance. In this chapter, the term tolerance is exclusively applied to a decreased ethanol response caused by prior ethanol exposure. Many of these studies focused on the frequency of the ethanol catabolizing enzyme alcohol dehydrogenase, ADH, and two naturally occurring alleles: Adh^F, a more active, and Adh^S, a less active variant. While ADH-deficient larvae are very sensitive to ethanol toxicity (David *et al.*, 1976), very little difference was observed between these flies and wild-type controls in ethanol-induced sedation in the inebriometer (Singh and Heberlein, 2000). Furthermore, the more active allele Adh^F did not increase in frequency during a 12-generation selection for sedation resistance in the inebriometer (Cohan and Graf, 1985), nor was *Adh* differentially expressed in a different set of selectively bred resistant versus sensitive strains (Morozova *et al.*, 2007). Thus *Adh* plays a significant role in ethanol toxicity in flies, but has little effect on ethanol-induced sedation.

B. Exposing Flies to Alcohol

Flies can be exposed to ethanol in a number of different ways. Mixing alcohol in food has been the traditional way of exposure to study ethanol resistance and toxicity. Up until recently, feeding had the distinct disadvantage that the amounts of alcohol consumed by the flies was an unknown quantity. With the introduction of the CAFE assay, this has changed (Ja *et al.*, 2007). CAFE stands for capillary feeder, and in this paradigm the only available food to the flies is contained within calibrated microcapillaries that allow the experimenter to precisely measure the amounts of ethanol-containing food that has been consumed by the flies. This assay has been used to great success to show that *Drosophila* like drinking alcohol (Devineni and Heberlein, 2009) (see Section II.F).

Subcutaneous injection of ethanol is routine in rodents, yet even mice are still enormous when compared with *Drosophila*, a small species even for flies in general. Given that an adult *Drosophila* has a volume of about 2 μl, it is not surprising that very few studies have used intraabdominal injection as a means of delivery. Nevertheless, direct injection of ethanol into adult flies has been reported and used to show the involvement of gamma-aminobutyric acid (GABA) receptors in acute responses to alcohol (Dzitoyeva *et al.*, 2003) (see Section III.C).

The most common way to expose flies to alcohol is to vaporize it and place the flies into a stream of defined ethanol saturation. This is most commonly done by bubbling an air stream through alcohol in a gas washing bottle, and combining this stream at predetermined flow rates with a second air stream that has been "washed" in pure water (Singh and Heberlein, 2000). In this way, flies will passively breathe-in alcohol via their tracheal system in this continuous exposure paradigm. A variant of this exposure paradigm is to simply add a large drop of ethanol onto the cotton plug that keeps flies within their vial (Bhandari *et al.*, 2009).

C. Measuring Alcohol-Induced Behaviors

To learn about the acute behavioral effect that alcohol has on flies, a number of assays have been developed that measure their behavioral response immediately following the (start) of the alcohol exposure. These assays measure the effect that alcohol has on four parameters: flies' position, posture, locomotion, and their consummatory behavior.

1. *Positional Choice*

One acute behavior that can be measured is whether flies are attracted to the smell of alcohol. One simple way to measure this is with an olfactory trap assay, where alcohol is placed in a vial sealed with a funnel (Reed, 1938). This allows flies to easily enter the vial, but few will escape. In this way the percentage of flies that is attracted by a given odor can be measured, or if two funneled traps are presented at the same time, a preference index (PI) can be calculated. An olfactory trap assay is commonly run over many hours, because the trap presents a small volume within a much larger receptacle. A more acute response to, and choice for or against, alcohol can be measured in a Y- or T-maze. In this assay, the flies are motivated to run toward a choice point by their natural inclination for either negative geotaxis, or phototaxis. At the choice point they decide between one of two directions, and enter a vapor-containing tube. This brief assay, generally run for around 1 min, allows for the calculation of an attraction index, or when two odors are presented, for a PI.

Positional choice is also measured in the context of oviposition preference and is measured in two ways (Joseph *et al.*, 2009). First, adult females' likelihood to spend time on ethanol-containing food can simply be filmed and compared to the time spent on a patch of control food. Second, their propensity to lay their eggs onto ethanol-containing food is measured by counting the number of eggs laid on this, versus control food.

2. *Postural Control*

When flies are exposed to a substantial dose of ethanol, they stop moving, start to lose postural control, and then fall over onto their backs unable to right themselves. This loss of postural control is the basis for the inebriometer. This apparatus was originally designed to measure the effects of volatile anesthetics on fly posture, but was subsequently applied to measure the response to ethanol vapor as well (Cohan and Graf, 1985). The inebriometer consists of a vertical column containing angled mesh baffles, into which typically 100 flies are introduced. Due to their natural inclination they tend to remain in the upper end of the cylinder, but as they are exposed to ethanol vapor, they start losing postural control, and drop down from one baffle onto the next, ultimately eluting from the column as a function of their sensitivity to ethanol. In this way, a mean elution time can be calculated for a given fly strain, representing a simple measure of that strains' sensitivity to ethanol. In addition, the inebriometer has also been used to selectively breed sensitive and resistant flies from a heterogeneous population (Cohan and Graf, 1985).

Because the inebriometer is a large and bulky apparatus, other ways to measure flies' loss of postural control have also been developed. Exposed flies can be challenged with a mechanical stimulus to determine how many have lost their righting reflex (Rothenfluh *et al.*, 2006), or inspected visually to determine whether they are still standing or lying sedated on their backs (Cowmeadow *et al.*, 2005). Sometimes loss of locomotion activity has been used as a proxy for sedation (Park *et al.*, 2000; Wen *et al.*, 2005). However, it is worth noting that at an intermediate dose of alcohol the loss of locomotion activity precedes the loss of postural control by several minutes (A. Rothenfluh, unpublished observation), and certain fly strains seem to dissociate locomotion inactivity from loss of postural control (Rodan *et al.*, 2002). Thus alcohol-induced inactivity should not be regarded as equivalent to sedation and loss of postural control.

If after a sedating dose of ethanol flies are left to recover in regular air, they will do so within 5–30 min. As the time it takes to recover is longer if flies sedate early in a given exposure regimen, the recovery time has also been used as an indirect measure for fly strains' sensitivity to alcohol (Berger *et al.*, 2004), similar to ethanol-induced sleep time measured in rodents. Because this recovery time is measured after ethanol exposure is finished, it may be that this measure will be more impacted by changes in ethanol pharmacokinetics, for example mutations in *Adh* that catabolize ethanol at a different rate. It is worth noting that *Adh* mutant flies showed only a minor effect on the time to ethanol-induced loss of posture as measured from the inebriometer as well as the internal alcohol concentration accrued during that time. Conversely, large differences in internal ethanol concentration were measured during recovery from the exposure (Singh and Heberlein, 2000).

3. *Measures of Locomotion Activity*

The natural tendency of flies for negative geotaxis is especially prominent after a mechanical startle; therefore, the extent to which ethanol interferes with this reflexive response has been used as a measure of ethanol-induced akinesia and/or loss of postural control (Bhandari *et al.*, 2009). In this case, it is the absence of negative geotaxis locomotion that reflects the drug effect, and therefore this assay does not measure any activating effects that ethanol might have on spontaneous locomotion, as has been seen in many organisms. Early assays tackling ethanol-induced locomotion activation included manual counting of flies' line crossing (Singh and Heberlein, 2000) and the inebriactometer (Parr *et al.*, 2001), a device that can measure up to 128 single flies, and how often they break an infrared beam as ethanol vapor is passed around them. Both of these assays indicated that flies show ethanol-induced hyperlocomotion, and the internal ethanol concentration is about 20 mM at times of hyperactivity (corresponding to a blood alcohol concentration (BAC) of 0.09%) and 45 mM at the time of sedation (corresponding to 0.21%) (Singh and Heberlein, 2000). While both of these assays give some information regarding the activity of the flies, the temporal and spatial resolution of the locomotion analysis was not very detailed, and for that reason a video-based locomotion tracking system was designed (Wolf *et al.*, 2002). In it, groups of 20 flies are exposed and their locomotion is tracked with an analysis software, which reveals four phases of the behavioral ethanol response: seconds after the start of the ethanol exposure, flies react by increasing their locomotion response. This is an olfactory-mediated startle response, since no measurable ethanol has accumulated inside flies yet, and since removal of the antennae, the primary olfactory sensory organ, abrogates this response (but not later hyperactivity) (Wolf *et al.*, 2002). Within about 1 min, the flies calm down again, most likely due to habituation to the startling ethanol odor (Cho *et al.*, 2004). After a few more minutes, the flies begin to become hyperactive, and this is followed later by lack of movement, akinesia, and then sedation. Note that loss of postural control and akinesia cannot be distinguished with most video tracking setups, but flies can be visually inspected while they are being filmed, and the time to loss of righting can be determined simultaneously (Rothenfluh *et al.*, 2006).

A careful examination of ethanol-induced locomotion response revealed that there is an ethanol concentration-dependent hyperactivity, with faster and earlier movement at higher alcohol concentrations. Locomotion occurs in bouts in flies, and as internal alcohol concentrations increased, individual flies showed faster locomotion during a bout, and longer bouts, until sedation started setting in, when those values both declined. Bout frequency, on the other hand, did not change as a function of exposure time (Wolf *et al.*, 2002).

4. *Consummatory Behavior*

As mentioned above, the CAFE assay allows for the precise measurement of ethanol-containing food consumed by flies over days. This can be combined with video tracking to allow for the measurement of both meal/drink size and their frequency.

D. Rapid Ethanol Tolerance

Repeat exposures to ethanol cause tolerance, that is, a reduction in the effect of the same drug dose. Human alcoholics can have such high acquired alcohol tolerance that their blood alcohol levels would cause severe impairment, or even lethality, in a naive, unexposed person (Urso *et al.*, 1981). Since tolerance is one of the seven criteria for substance addiction (APA, 1994), it is thought to contribute to the development of alcoholism and is therefore of considerable interest to researchers. Different types of tolerance can be distinguished: metabolic tolerance refers to changes in ethanol pharmacokinetics due to changes in catabolism or drug distribution, while functional or pharmacodynamic tolerance reflects changes in the CNS that lead to a decreased drug response upon re-exposure. Different ways of inducing tolerance are also differentiated: acute tolerance occurs during alcohol experience, while rapid tolerance is induced after a single ethanol exposure that is followed by clearance of the drug from the organism's system. Chronic tolerance is achieved by exposing the subject for multiple or prolonged times.

Adult *Drosophila* flies are a particularly useful organism to study functional alcohol tolerance. Unlike *Drosophila* larvae, they show no changes in enzymatic ADH levels when fed ethanol-containing food (Geer *et al.*, 1988), and after the induction of rapid behavioral tolerance no changes in ethanol absorption or metabolism are observed (Berger *et al.*, 2004; Cowmeadow *et al.*, 2005; Scholz *et al.*, 2000). Therefore, changes in ethanol-induced behavior are caused by changes in the nervous system (i.e., pharmacodynamic tolerance).

A single sedating exposure of flies in the inebriometer increased their mean elution time from 20 to 26 min when re-exposed 4 h later (Scholz *et al.*, 2000). This was also seen with a video tracking system, where pre-exposed flies showed a prolonged phase of hyperactivity and delayed akinesia. The longer the first exposure, the more marked was the increase in mean elution time for the second exposure, up to a certain point, when tolerance reached a plateau. This plateau was also seen with repeat sedating exposures given every 2 h.

Maximum tolerance is seen 2 h after the first exposure, after which time the amount of tolerance drops sharply, but remains at a steady lower level from 6 to 24 h, still showing significant levels of tolerance a day later (Scholz *et al.*, 2000). Inhibition of protein synthesis with cycloheximide did not change the extent of rapid tolerance 4 h after the initial exposures, but it was not ascertained whether the lower level tolerance 24 h after exposures was also unaffected (Berger *et al.*, 2004).

The relationship between sensitivity or resistance to an acute alcohol exposure and tolerance is unclear. Berger *et al.* (2008) found that some acutely sensitive, normal, and resistant fly strains could all have lowered tolerance, while others could all have no effect on tolerance. In other words, the acute sensitivity of a fly strain does not predict its tolerance phenotype. In this context a word of caution is appropriate. If a given strain has an acute alcohol exposure phenotype, it is unclear what kind of exposure these flies should get to induce tolerance. Is it appropriate to expose sensitive and normal flies for the same duration, or do they need to be exposed until they each reach the same behavioral endpoint, that is, just to sedation as in the inebriometer, but no longer? As mentioned above, the amount of tolerance induced is a variable depending on a number of parameters, including the dose of the first exposure, and unfortunately few studies investigating tolerance measure a tolerance dose–response curve to ascertain that any tolerance defect found is consistently observed irrespective of initial exposure dose.

E. Chronic Tolerance

As discussed above, a single sedating exposure to alcohol induces rapid tolerance. In addition to such a regimen, a chronic exposure to low levels of alcohol (inducing neither hyperactivity nor sedation) for 2 days also causes flies to be more resistant to a high-concentration ethanol challenge when compared with flies chronically exposed to air (Berger *et al.*, 2004). The amount of tolerance induced by chronic exposure was slightly lower compared with a rapid induction

(~30% sedation time increase compared with 50%), but it was longer lasting, and as opposed to rapid tolerance, significant chronic tolerance could still be observed 48 h after the end of the ethanol pre-exposure. Given this long-lasting effect of chronic pre-exposure, it is not surprising that the establishment of chronic tolerance required protein synthesis, while rapid tolerance did not. Furthermore, unlike rapid tolerance, chronic tolerance was not dependent on the biogenic amine octopamine, indicating that these two types of tolerance depend on different mechanisms.

F. Alcohol Preference

The common name of *D. melanogaster* is vinegar or pomace fly, and this implies that they would be found around rotting fruit with high likelihood. Ethanol levels present in fruit have been measured to be from 0.6% in ripe hanging fruit to 4.5% in fallen rotten ones (Dudley, 2004). In addition, ripe fruit contain many other compounds and odors such as acetic acid (Chapman, 2003).

1. *Female Oviposition Preference*

Numerous studies have therefore investigated whether adult female vinegar flies would prefer to lay their eggs on food containing ethanol, given a choice. Richmond and Gerking showed that *D. melanogaster* strongly preferred to lay eggs on 9%-ethanol-containing food, compared to regular food. This is in contrast to *Drosophila willistoni*, which avoided laying on alcohol-containing food (Richmond and Gerking, 1979). When given a choice to lay on 15%-ethanol-containing food however, normal *D. melanogaster* flies preferred to lay on normal food. This was in contrast to flies kept on 12–18%-ethanol-containing food for 80 generations. These females preferred to lay on 15%-ethanol-containing food (van Delden and Kamping, 1990). When flies were given a three-way choice between regular, 3% acetic acid, and 9% ethanol food, over 80% of the eggs were laid on the acetic acid food (Eisses, 1997). Thus, overall it appears that ethanol is mildly attractive for females to lay their eggs on, but not nearly as much as other fruit-produced compounds such as acetic acid.

2. *Adult Olfactory Responses to Ethanol*

It has been appreciated for a long time that the smell of 5–15% ethanol solution is attractive to vinegar flies, while higher than 25% solutions are repellent. This is true for flies assayed over hours with a single odor-baited trap (Reed, 1938), or within 30 min in an odor versus solvent choice paradigm (Fuyama, 1976). Given the presence of ethanol in rotting fruit, it might be a significant attractant in the wild. However, the concentrations of ethanol need to be orders of magnitude higher to be attractive compared with other odorants. Acetic acid for example is attractive at 0.1–0.5% (Fuyama, 1976), while other odorants are attractive at even lower concentrations.

3. *Adult Preference and Alcohol Drinking*

To ask whether adult vinegar flies like ethanol, Cadieu *et al.* (1999) measured proboscis extension on 3%-alcohol-containing versus control food. They found that flies had an innate PI of about 0.2 (where −1 is total avoidance, 0 is equal chance, and +1 is full preference), and this was increased to about 0.45 when they were kept on ethanol food for 24 h prior to the test. These results suggest that flies like the taste of ethanol, but the correlation between proboscis extension time and amount consumed was not established; thus, it remained unclear whether flies prefer to actually consume alcohol. This question was tackled by Devineni and Heberlein (2009), and they found that flies' PI toward 15%-ethanol-containing sucrose solution (over sucrose alone) would increase from 0.15 on the first to 0.4 on the fifth day. When after establishing their drinking preference the flies underwent 1 or 3 days of forced abstinence, they immediately returned, or relapsed, to a high preference after having alcohol as a choice again. The authors also showed that flies would learn to overcome adversity to consume alcohol. When quinine was added to sucrose, flies avoided it (PI = −0.1 to −0.2), and this avoidance remained constant over the 5 days tested. Adding ethanol to this sucrose/quinine solution did not alter aversion on the first day, but by day 5 the flies preferred this mix over sucrose alone (PI = 0.1–0.2). Thus, overall these authors showed that flies would overcome adversity, or at least adverse taste, to consume alcohol, and that they would immediately relapse to pre-established high levels of drinking after a period of forced abstinence.

III. Genes and Biochemical Pathways Involved in Alcohol Responses

A. Forward Genetic Approaches

One of the first approaches to genetic differences in ethanol-induced behavior was to look at natural populations of *Drosophila* from the West Coast of the United States. Using the inebriometer, a significant trend was found that populations from more northern clines were more resistant to the sedating effects of ethanol (Cohan and Graf, 1985). Five subpopulations were then selectively bred for increased resistance in the inebriometer, and after 12 generations the mean sedation time increased from 14 to 28 min, indicating strong genetic contribution to acute ethanol sensitivity.

Singh and Heberlein (2000) used three rounds of selection of a chemically randomly mutagenized population to set up individual resistant or sensitive fly strains with unique isolated X chromosomes. In this way they were able to isolate true-breeding strains with mean inebriometer elution times ranging from 15 to 30 min (with wild-type eluting in 20 min). Because of the semi-dominance of many of these mutations, and due to large effects by the genetic background of mapping strains, the mutant genes causing these phenotypes have never been isolated (Guarnieri and Heberlein, 2003). For that reason mutations with altered ethanol resistance were then induced with transposable P-elements, which allow for a quick determination of the genes affected, as the element's full sequence is known and sequence flanking the insertion site can easily be obtained. This approach has led to the isolation of mutations with altered acute ethanol sensitivity which affect cyclic AMP (cAMP) signaling (Moore *et al.*, 1998), dynamic control of the actin cytoskeleton (Rothenfluh *et al.*, 2006), or the mitogen-activated protein (MAP) kinase/extracellular signal-related kinase (ERK) pathway (Corl *et al.*, 2005), as well as mutations affecting tolerance (Scholz *et al.*, 2005) (see Section III.C, Table I).

B. Genomic Approaches

Numerous studies have taken advantage of the genome-wide approaches that can be performed with microarray technology. Most of those studies compared gene expression levels before and after alcohol exposure (Kong *et al.*, 2010a), sometimes including repeat exposures resulting in the development of tolerance (Morozova *et al.*, 2006; Urizar *et al.*, 2007). It is worth noting that in these three studies, only 14% of 1669 significantly changed transcripts were found in at least two of the three studies (Kong *et al.*, 2010a). This may be due to differences in the exact design of ethanol exposure, the data capture and analysis, or the respective wild-type flies used for exposure. Nevertheless, numerous functional groups of genes were repeatedly overrepresented, and many candidate genes were found that had mutations with altered alcohol-induced behaviors.

In two studies, Morozova, Mackay, and colleagues combined genetic variation with genomic approaches to find genes and pathways involved in alcohol responses. The first study used wild-type fly strains selected over 35 generations

TABLE I

Behavioral Ethanol Phenotypes of Selected *Drosophila* Mutations

Gene	Encoded protein	Pathway signaling	Phenotype	References
Signaling				
GABA(B)	Metabotropic receptor	GABA	lof: AR	Dzitoyeva *et al.* (2003)
TH	Tyrosin hydroxylase	Dopamine	lof: AYD	Bainton *et al.* (2000)
DopR	Dopamine D1 receptor	Dopamine	lof: AYD	Kong *et al.* (2010b)
Tbh	Tyramine β-hydroxylase	Octopamine	lof: A+, RTD	Scholz *et al.* (2000)
InR	Insulin receptor	Insulin	lof: AS	Corl *et al.* (2004)
Chico	Insulin receptor substrate	Insulin	lof: AS	Corl *et al.* (2004)
FOXO	Transcription factor	Insulin	gof: AS	Corl *et al.* (2004)
NPF	Neuropeptide F (NPY)	NPY	gof: AS	Wen *et al.* (2005)
Cheapdate	Amnesiac/PACAP	Cyclic AMP	lof: AS	Moore *et al.* (1998)
Rutabaga	Adenylate cyclase	Cyclic AMP	lof: AS	Moore *et al.* (1998)
DCO	Protein kinase A	Cyclic AMP	lof: AS	Moore *et al.* (1998), Rodan *et al.* (2002)
DCO	Protein kinase A	Cyclic AMP	lof: AR[a]	Rodan *et al.* (2002)
Pka-RII	PKA regulation	Cyclic AMP	lof: AS	Park *et al.* (2000)
happy hour	Ste20 kinase	EGF	lof: AR, gof: AS	Corl *et al.* (2009)
Egfr	EGF receptor	EGF	lof: AS, gof: AR	Corl *et al.* (2009)
Spitz	EGF ligand	EGF	gof: AR	Corl *et al.* (2009)
Rhomboid	Ligand-activating peptidase	EGF	lof: AS	Corl *et al.* (2009)
Rolled	ERK kinase	EGF	gof: AR	Corl *et al.* (2009)
Elm	P22 calcineurin B homolog	Ca-binding	lof: AS	LaFerriere *et al.* (2008)
Synapse function				
Slowpoke	BK channel	Membrane potential	lof: RTD, gof: AR	Cowmeadow *et al.* (2005, 2006)
Synapsin	Presynaptic scaffolding	Synaptic plasticity	lof: A+, RTI	Godenschwege *et al.* (2004)
Homer	postsynaptic scaffolding	Synaptic plasticity	lof: AS, RTD	Urizar *et al.* (2007)
Cell structure				
fasII	fasciclinII/NCAM2	Cell adhesion	lof: AS	Cheng *et al.* (2001)
Myospheroid	β-integrin	Cell adhesion	lof: AS, RTI	Bhandari *et al.* (2009)
Scabrous	a-integrin	Cell adhesion	lof: AS, RTI	Bhandari *et al.* (2009)
white rabbit-RA	RhoGAP18B-PA[b]	Actin dynamics	lof: AYD, gof: AYI	Rothenfluh *et al.* (2006)
white rabbit-RC	RhoGAP18B-PC[b]	Actin dynamics	lof: AR	Rothenfluh *et al.* (2006)

(*continued*)

TABLE I (*continued*)

Gene	Encoded protein	Pathway signaling	Phenotype	References
Rho1	RhoGTPase	Actin dynamics	gof: AR	Rothenfluh *et al.* (2006)
Rac1	RhoGTPase	Actin dynamics	lof: AS, gof: AR	Rothenfluh *et al.* (2006)
Cdc42	RhoGTPase	Actin dynamics	lof: AR, gof: AS	Rothenfluh *et al.* (2006)
Jwa	ARL6IP5	Microtubule assoc.	lof: RTD, gof: RTI	Li *et al.* (2008)
DNA regulation				
Sir2	Histone deacetylase	Chromatin regulation	lof: AR, RTD	Kong *et al.* (2010)
pipsqueak	Helix-turn-helix protein	Transcription factor	gof: AS	Morozova *et al.* (2009)
hangover	Zinc-finger protein	Transcription factor	lof: A+, RTD	Scholz *et al.* (2005)
Metabolism				
Men	Malic enzyme	Intermediate metab.	lof: AR	Morozova *et al.* (2009)
AcCoAS	Acetyl CoA synthase	Intermediate metab.	lof: A+, AYD	Kong *et al.* (2010a)

The table includes mutant alleles that have been studied in some detail, and where there is some understanding on the molecular effects of the mutant. Abbreviations: loss-of-function (lof), gain-of-function (gof), acute sedation sensitivity (AS), acute sedation resistance (AR), no change in acute sedation (A+), decreased acute hyperactivation (AYD), increased acute hyperactivation (AYI), decreased rapid tolerance (RTD), increased rapid tolerance (RTI). [a]Ubiquitous inhibition of PKA causes acute sensitivity, while inhibition in a subset of neurons can also lead to acute resistance. [b]Two different isoforms from the same gene, sharing the GTPase-activating domain, but with different N-termini.

for sensitivity or resistance to ethanol to assay transcriptional profiles (Morozova *et al.*, 2007). Currently, there is a greater than 90% chance that for any given gene a mutant strain exists that can be easily obtained from public stock centers. For that reason, many genes that show altered expression levels in the above transcription profiling approaches can quickly be tested for any behavioral ethanol phenotypes. Morozova and colleagues, for example, tested 37 candidate gene mutations of genes differentially expressed in sensitive versus resistant strains and found 32 of those had significantly different sensitivity to an acute ethanol exposure. Thus, in flies, these approaches can quickly lead to the isolation of new genes and mutations that are behaviorally relevant.

In an exciting new approach, Morozova and colleagues assayed 40 isogenic lines derived from wild populations for their alcohol sensitivity (Morozova *et al.*, 2009). The interesting thing about those 40 strains is that they are all fully sequenced, and transcriptional profiles are available for all 40. This allowed for the correlation between ethanol sensitivity, genomic polymorphisms, and mRNA abundance, including the identification of potential *cis*-regulatory elements shared between co-regulated genes. In addition, this approach can highlight the *trans*-regulators which act upon a given set of co-regulated genes.

C. Functional Pathways in Alcohol-Induced Behavior

1. *Overrepresentation in Transcriptional Profiling*

Three studies have investigated changes in transcript levels as a result of acute alcohol exposures. While a surprisingly small number of the ethanol-regulated transcripts was found in all three studies (2%), or even two of the three (14%), some gene ontology (GO) annotations were significantly overrepresented in the ethanol-regulated transcripts in all three studies. These categories were regulators of signal transduction, transcription factors, and metabolic enzymes. Indeed, of the 25 genes found in all three studies as similarly ethanol regulated, 12 are annotated with the GO terms "metabolic and biosynthesis processes" and 3 are involved in serine biosynthesis (Kong *et al.*, 2010a). Yet, no mutations of these genes have been tested for their behavioral ethanol phenotype.

Based on transcriptional profiling, a number of mutations in intermediary metabolism and fatty acid biosynthesis enzymes were found to alter acute ethanol sensitivity, including pyruvate dehydrogenase kinase, pyruvate carboxylase, and malic enzyme (Morozova *et al.*, 2006). Genes associated with pyruvate metabolism were also found differentially expressed in alcohol preferring and non-preferring mice (Mulligan *et al.*, 2006). Malic enzyme serves as a link between the glycolytic pathway, tri-carboxylic acid cycle, and fatty acid synthesis, allowing for conversion of excess ethanol-derived energy into lipid biosynthesis. Indeed, alcohol-induced fatty acid synthesis is well documented in alcoholics (Lieber, 2004). Interestingly, a polymorphism in human malic enzyme associated with cocktail drinking was then found in the Framingham Offspring cohort, albeit with a small effect (Morozova *et al.*, 2009).

It is noteworthy that many genes that affect the response to alcohol (see next sections) showed no change in these transcriptome analyses (Morozova *et al.*, 2006), indicating that this approach only captures a subset of the genes involved in, and relevant for, mediating behavioral alcohol responses.

2. *Stress and Ethanol Tolerance*

Many genes involved in stress responses such as heat shock proteins, cytochrome P450 proteins, and glutathione transferases are upregulated after an acute ethanol exposure. Indeed, Kong *et al.* (2010a) found that 5 of the 20 genes with the highest ethanol-induced induction and 9 of the 21 *Hsp* genes overall had significantly increased expression levels after ethanol exposure.

The *hangover* mutation was found in a forward genetic screen to greatly reduce rapid tolerance. When flies were stressed by a 30 min exposure to a 37°C heat shock, instead of a sedating ethanol pre-exposure, they developed cross-tolerance to ethanol exposure, similar in magnitude as if they were pre-exposed to ethanol. *Hangover* mutants showed a reduction in this stress-induced cross-tolerance, while the rapid tolerance mutants *Tbh* (see next section) were completely normal. This suggests that the development of ethanol tolerance in flies is composed of a *Tbh*- and octopamine-dependent, but stress-independent, component, as well as a *hangover*-dependent stress component. Indeed, when *hangover* mutants were exposed to oxidative stress, they were more susceptible than wild-type flies, arguing that *hangover* is a gene required for normal stress responses (Scholz *et al.*, 2005). Interestingly, there is another connection between stress resistance and ethanol resistance. Flies that were selectively bred for alcohol resistance also showed increased resistance to such stressors as heat, desiccation, and other chemicals (Cohan and Hoffmann, 1986), and flies selected to tolerate high levels of desiccation also co-selected resistance to ethanol (Hoffmann and Parsons, 1989).

It is unclear what the exact molecular function of the *hangover* gene is, but it encodes a zinc-finger transcription factor, and is involved in the regulation of synaptic growth in the larval neuromuscular junction (Schwenkert *et al.*, 2008). It is currently not known whether central synapses are also affected in *hangover* mutants. The human *hangover* ortholog is called *ZNF699*, and polymorphisms in this gene were found associated with alcoholic cases in a study of Irish sibling pairs. In addition, one risk haplotype was associated with decreased gene expression in the dorsolateral prefrontal cortex in postmortem cases (Riley *et al.*, 2006).

3. *Octopamine, Dopamine, and Cyclic AMP Signaling*

One of the first mutations with altered ethanol tolerance was found in the *Tbh* gene, encoding tyramine β-hydroxylase. This enzyme converts tyramine to octopamine, a biogenic amine thought to have similar functions in invertebrates as norepinephrine has in mammals (Monastirioti, 1999). The latter has been implicated in the development of ethanol tolerance in rats (Ritzmann and Tabakoff, 1976). Flies with a mutation in *Tbh* showed a reduction of 50–60% in the development of tolerance 4 h after a single sedating alcohol exposure.

Dopamine is an important neurotransmitter in the mammalian reward system, and is strongly implicated in the development of addictions. Bainton *et al.* (2000) showed that inhibiting tyrosine hydroxylase (TH), the rate-limiting synthesis enzyme for dopamine, resulted in decreased locomotor hyperactivation by ethanol, and this effect could be rescued by feeding flies L-DOPA, the product of TH activity and immediate precursor of dopamine. The different dopamine receptors couple to adenyl cyclase and increase or decrease cAMP levels and protein kinase A (PKA) activity, depending on the receptor subtype. The first evidence that cAMP signaling was involved in *Drosophila* ethanol responses came from a forward genetic transposon screen that yielded the sensitive mutant *cheapdate* (Moore *et al.*, 1998). This mutation turned out to be an allele of *amnesiac,* a gene which is important in learning and memory, and encodes the fly ortholog of pituitary adenylate cyclase-activating peptide (PACAP). This study also found that mutations in genes resulting in reduced cAMP signaling, such as mutations in genes encoding adenylate cyclase or PKA, caused increased ethanol sensitivity. Activation of the cAMP pathway by forskolin rescued the sensitivity caused by *cheapdate.* In addition, a mutation in the PKA regulatory subunit, thought to increase cAMP signaling, caused resistance to the sedating effects of ethanol (Park *et al.*, 2000). Olfactory learning and memory critically depends on proper cAMP signaling and the genes encoding adenylate cyclase (*rutabaga*), or cAMP phosphodiesterase (*dunce,* hence the gene names). Since both behavioral alcohol responses and associative learning depend on cAMP, it is therefore not surprising that in a substantial collection of learning mutants, many had ethanol sedation or tolerance phenotypes as well (Berger *et al.*, 2008).

Consistent with the above results, Rodan and colleagues (2002) found that inhibition of PKA by globally expressing a UAS–PKA inhibitor construct caused sensitivity to ethanol-induced sedation. However, when the inhibitor was expressed only in certain subsets of the brain, it caused resistance (Rodan *et al.*, 2002), while in others it caused sensitivity (Corl *et al.*, 2005). This indicates that different neurons and circuits in the fly brain utilize cAMP signaling to mediate either resistance or sensitivity to ethanol.

4. *Neuropeptide Signaling*

The neurons that mediated ethanol sensitivity after PKA inhibition contained peptidergic neurosecretory cells, among other peptides. Demonstrating that insulin-like peptide is involved in behavioral ethanol responses was the finding that insulin receptor mutants also caused sensitivity, as did inhibition of phosphoinositide 3-kinase (PI3K) or overexpression of constitutively active transcription factor forkhead box O (FOXO), two members of the downstream insulin-signaling cascade.

Another neuropeptide involved in *Drosophila* ethanol responses is neuropeptide F (NPF), the invertebrate ortholog of neuropeptide Y (NPY). Overexpression of NPF caused sensitivity to ethanol-induced sedation, while ablating NPF-expressing neurons with diphtheria toxin caused resistance to ethanol (Wen *et al.*, 2005). Mice lacking NPY displayed increased resistance to ethanol sedation and increased ethanol consumption, while overexpression of NPY caused the opposite phenotypes (Thiele *et al.*, 1998).

5. *Synaptic Transmission and Ion Channels*

A candidate gene that came from an ethanol responsive transcript analysis was the postsynaptic scaffolding and signaling molecule homer. Transcript levels of homer decreased after ethanol exposure, and mutations in the gene caused sensitivity (Urizar *et al.*, 2007). Interestingly, homer knock-out mice also showed sensitivity to ethanol, and decreased voluntary drinking and ethanol-induced place preference (Szumlinski *et al.*, 2005).

In vertebrates, the $GABA_B$ receptor is involved in ethanol-mediated behavior (Maccioni and Colombo, 2009). This metabotropic receptor is conserved in flies, and knock-down of the receptor with intraabdominal small interfering RNA caused flies to show reduced sedation upon ethanol injection. Similarly, co-injection of ethanol and the $GABA_B$ antagonist CGP 54626 also decreased sedation. Pretreatment with the agonist 3-aminopropyl-(methyl)phosphinic acid (APMPA), on the other hand, abolished the development of rapid ethanol tolerance (Dzitoyeva *et al.*, 2003).

The presynaptic vesicle scaffolding protein synapsin has also been implicated in ethanol responses. Flies lacking their only synapsin gene showed an increase in rapid ethanol tolerance (Godenschwege *et al.*, 2004). Interestingly, mammalian synapsins I and II showed increased phosphorylation upon an exposure of ethanol, which was adenylate cyclase, and thus, presumably, PKA dependent (Conti *et al.*, 2009).

The pore-forming subunits of the Ca^{2+}-activated BK potassium channel are encoded by the *slowpoke* gene in flies. This channel is highly conserved between worms, flies, and vertebrates and its activity is potentiated by direct ethanol binding (Treistman and Martin, 2009). Worms with mutations in the BK channel showed resistance to ethanol (Davies *et al.*, 2003). In flies, the *slo* gene has five different tissue-specific promoters (Bohm *et al.*, 2000), and a nervous system-specific mutant abolished ethanol-induced rapid tolerance (Cowmeadow *et al.*, 2005). Interestingly, overexpressing the CNS-specific isoform of *slo* induced tolerance (Cowmeadow *et al.*, 2006), and *slo* was differentially expressed in selectively bred-resistant versus sensitive strains (Morozova *et al.*, 2007).

6. *Transcription Factors and Epigenetics*

The anesthetic benzyl alcohol also induces rapid tolerance in flies (Ghezzi *et al.*, 2004), and benzyl alcohol-mediated sedation induced *slo* expression via the transcription factor CREB2 (Wang *et al.*, 2007). Flies without CREB2 were unable to acquire tolerance to benzyl alcohol and did not upregulate *slo* expression after alcohol exposure (Wang *et al.*, 2009). In addition, subsets of the *slo* promoter showed altered histone H4 acetylation patterns after benzyl alcohol-induced sedation, as demonstrated with chromatin immunoprecipitation, and the histone deacetylase inhibitor sodium butyrate induced *slo* transcripts and resistance to benzyl alcohol sedation (Wang *et al.*, 2007). Considerable evidence indicates that epigenetic regulation is an important contributor to mammalian addiction as well (reviewed in Renthal and Nestler, 2008), including responses to ethanol (Pandey *et al.*, 2008).

7. *Cytoskeleton*

In a meta-analysis of transcript changes between mouse strains with different alcohol consumption patterns, the two most overrepresented GO categories were "regulation of the actin cytoskeleton" and the MAP kinase signaling pathway (Mulligan *et al.*, 2006).

It has been known for a while that an acute ethanol exposure of cultured cells results in rearrangement of the actin cytoskeleton (Allansson *et al.*, 2001), and in cultured rat hippocampal cells, chronic ethanol caused an increase in dendritic spine number and levels of filamentous actin (Carpenter-Hyland and Chandler, 2006). Exposing cultured cells to alcohol also led to a reorganization of their actin cytoskeleton. In astrocytes this reorganization is mediated by an ethanol-induced decrease of GTP-loaded RhoA, which can be reversed by exogenous RhoA activation via lysophosphatic acid (Martinez *et al.*, 2007). Does this effect of ethanol on the actin cytoskeleton have any significance in alcohol-related behaviors, or is it merely a reflection of its cytotoxicity? A mutation in the fly *RhoGAP18B* gene caused resistance to the sedating effects of ethanol (Rothenfluh *et al.*, 2006). This gene encodes a GTPase-activating protein that likely exerts its effect on ethanol-induced sedation via the Rho or Rac GTPases, key regulators of the actin cytoskeleton. Interestingly, *RhoGAP18B* also encodes a smaller isoform, which specifically affected ethanol-induced hyperactivation and not sedation. Thus, proper actin regulation is important for both ethanol-induced hyperactivation and sedation. Coincident with the discovery of RhoGAP18B in ethanol responses, Offenhäuser and colleagues found that a genetic knock-out of an actin-capping protein, Eps8, led to mice that are resistant to ethanol's sedative effects and that also drank more in a two-bottle choice paradigm (Offenhauser *et al.*, 2006). Primary cell cultures of normal cerebellar granule cells lost filamentous F-actin and decreased *N*-methyl-d-aspartate receptor current when exposed to ethanol, and these two ethanol responses were lost in Eps8 knock-out mice, indicating a strong correlation between the dynamics of the actin cytoskeleton and drinking behavior.

Some other mutations in fly genes involved in actin dynamics have alcohol behavior phenotypes. Berger *et al.* (2008) found that mutations in the actin nucleator *formin3* caused increased acute ethanol sensitivity, and reduced rapid tolerance, while a mutation in the actin-binding filamin protein cheerio also caused increased acute sensitivity. On the other hand, mutations in the RAB11-FIP3/4-encoding gene *nuf*, which is involved in vesicle trafficking and actin organization, caused resistance to acute ethanol sedation (Morozova *et al.*, 2007). Finally, the mouse *addicsin* gene encodes ARL6IP5, which is involved in the development of morphine tolerance and dependence (Ikemoto *et al.*, 2002). RNAi-mediated knock-down of the fly ortholog caused a decrease while over-expression caused an increase in rapid ethanol tolerance (Li *et al.*, 2008a). The molecular function of ARL6IP5 is not exactly known, but it is microtubule associated, and involved in F-actin rearrangement and MAP kinase activation (Chen and Olsen, 2007). It has also been shown to regulate the excitatory amino acid carrier-1 (Watabe *et al.*, 2007) and inhibit the trafficking molecule Rab1 (Maier *et al.*, 2009).

8. *Growth Factors*

As mentioned above, the MAP kinase signaling pathway was overrepresented in a transcriptome meta-analysis between drinking and non-drinking mice (Mulligan *et al.*, 2006). In addition, a meta-analysis of over 1000 publications about various drug addictions also highlighted this pathway as significantly overrepresented (Li *et al.*, 2008b). In flies, an unbiased genetic screen yielded mutations in the Ste20-family kinase gene *happy hour*, which caused resistance to ethanol-induced sedation (Corl *et al.*, 2009). This kinase is a negative regulator of the epidermal growth factor (EGF) receptor pathway, and loss-of-function mutations in pathway members, such as the receptor ligand Spitz or the Spitz-activating protease rhomboid, caused sensitivity, while activating the pathway by over-expressing the receptor or constitutive active ERK kinase caused resistance (Corl *et al.*, 2009). This study also showed that the EGF receptor inhibitor erlotinib caused ethanol sensitivity in flies and mice and reduced voluntary ethanol intake in rats for 24 h after inhibitor administration.

9. *Cell Adhesion*

Numerous cell adhesion molecules have been found associated with addiction disorders, including alcoholism (Li *et al.*, 2009). In flies, the best studied group of cell adhesion molecules and their effects on ethanol-induced behaviors is the integrin family of cell adhesion and signaling molecules. Mutations in the β-integrin encoding genes *mys* (Bhandari *et al.*, 2009) and *βint-v* (Morozova *et al.*, 2007) as well as the a-integrin encoding gene *scab* (Bhandari *et al.*, 2009; Kong *et al.*, 2010a) all caused acute ethanol sensitivity, and both *mys* and *scab* mutants showed increased rapid ethanol tolerance (Bhandari *et al.*, 2009). The neural-cell adhesion molecule (NCAM)-encoding gene *fasII* was also required for normal acute resistance to alcohol and for normal olfactory learning (Cheng *et al.*, 2001). Mutations in the Notch-regulated cell adhesion gene *klingon* caused defects both in long-term memory (Matsuno *et al.*, 2009) and acute ethanol sensitivity, and in decreased chronic and rapid tolerance (Berger *et al.*, 2008), again highlighting the molecular overlap between learning and memory, and behavioral ethanol responses.

IV. Neural Circuits Involved in Ethanol Responses

In the mammalian brain, a number of brain regions and circuits are well known to play key roles in addiction, including alcohol drinking. These include the mesolimbic dopamine pathway, which includes dopaminergic neurons in the ventral tegmental area that mediate a reinforcement signal, and their targets in the forebrain, notably the nucleus accumbens. Are there functionally equivalent circuits in the fly brain? And which other regions in the fly brain are involved in behavioral responses to alcohol?

Numerous studies have shown that an alcohol-induced behavioral phenotype could be rescued by expressing the mutated gene's wild-type cDNA specifically in the CNS. This was true for Ste20-family kinase encoded by *happy hour* (Corl *et al.*, 2009), the tolerance gene *hangover* (Scholz *et al.*, 2005), or the postsynaptic protein homer (Urizar *et al.*, 2007). The *homer* gene is more specifically required in the ellipsoid body of the brain, a region that is also involved in rapid tolerance (Scholz *et al.*, 2000). This brain region is known to be important in the control of locomotion behavior and is also involved in GABA-mediated spatial working-type memory (Neuser *et al.*, 2008). Other neurochemical systems involved in behavioral ethanol responses mentioned above include dopaminergic and octopaminergic neurons, as well as NPF- or insulin-like peptide-producing neurosecretary cells.

Neurons expressing the amnesiac peptide project to the mushroom bodies, which are crucial for amnesiac-dependent learning and memory (Waddell *et al.*, 2000). While the *amnesiac* allele *cheapdate* leads to ethanol sensitivity (Moore *et al.*, 1998), the mushroom bodies are not required for normal ethanol-induced sedation (Rodan *et al.*, 2002). Thus the PACAP ortholog amnesiac mediates normal ethanol sedation via different neurons. Rodan *et al.* (2002) assayed 64 Gal4-expressing lines to test which neurons required PKA signaling to mediate normal ethanol sedation. The three Gal4 lines that showed significant ethanol resistance when expressing a PKA inhibitor unfortunately did not show significant overlap in the expressing neurons. These studies did, however, indicate five things: first, the Gal4-binary system is a viable approach to learn about the anatomical requirements of ethanol-induced behaviors. Second, multiple types of neurons, and possibly circuits, are involved in mediating ethanol-induced sedation. This finding is highlighted by the fact that PKA inhibition in insulin-like peptide-expressing neurosecretory cells caused sensitivity to ethanol (Corl *et al.*, 2005), unlike the three Gal4 lines mentioned above, which caused resistance. Third, the neurons involved in ethanol-induced sedation are not necessarily the same neurons mediating ethanol-induced hyperactivity. Indeed, none of these three Gal4 lines causing ethanol resistance altered the amplitude of ethanol-induced hyperactivation (Rodan *et al.*, 2002), which was reduced in flies that had reduced dopamine levels (Bainton *et al.*, 2000). Fourth, while there is a substantial overlap in the molecules required for behavioral alcohol responses and for associative memory, the neural substrates of these behaviors seem distinct. And fifth, even though hundreds of differentially expressing Gal4 lines are available in flies, there is a need for both more spatially restricted and better characterized lines. As mentioned in the introduction, this very problem is currently being tackled on a large scale (Pfeiffer *et al.*, 2008), and it will be fascinating to see whether the functional organization of the fly brain bears any resemblance to the mammalian brain.

V. Conclusions

A variety of approaches and assays have been used to isolate many genes that are involved in the behavioral responses to alcohol. Many of these genes are also involved in mammalian ethanol behaviors, including allelic associations in human alcoholic cohorts. Therefore, the fly has proved to be an exceptionally useful model organism to find the molecular components and therefore the potential risk factors and therapeutic targets, which mediate these behaviors. The genes mentioned in this review, affecting behavioral responses to alcohol in the fly, fall into many diverse functional groups. Since addiction hijacks endogenous reward and reinforcement learning pathways, it is not surprising that many of the mentioned genes would be involved in synaptic plasticity, learning, and memory. The accumulated evidence so far indicates that there are no single gene mutations that by themselves are sufficient to cause addiction. Rather, many genetic variants are likely involved that each contributes a small risk to the development of alcohol abuse. For this reason the fly is an excellent model organism to identify many of those genes, and because of its experimental accessibility, it is possible to mechanistically dissect the function of those genes and their associated molecular partners. With the recent development of a reliable alcohol drinking and preference assay (Devineni and Heberlein, 2009), together with the development of new tools allowing the precise spatial (Pfeiffer *et al.*, 2008) and dynamic (Pulver *et al.*, 2009) control of small groups of neurons, *Drosophila* is poised to generate yet more insights into the basic biology and the molecular and functional mechanisms that underlie the transition from alcohol exposure to excessive consumption of this widely abused drug.

References

Adams, M. D., Celniker, S. E., Holt, R. A., Evans, C. A., Gocayne, J. D., Amanatides, P. G., Scherer, S. E., Li, P. W., Hoskins, R. A., Galle, R. F., *et al.* (2000). The genome sequence of *Drosophila melanogaster. Science (New York, NY)* **287**, 2185–2195.

Allansson, L., Khatibi, S., Olsson, T., and Hansson, E. (2001). Acute ethanol exposure induces [Ca2+]i transients, cell swelling and transformation of actin cytoskeleton in astroglial primary cultures. *J. Neurochem.* **76**, 472–479.

APA (1994). Diagnostic and Statistical Manual of Mental Disorders IV.American Psychiatric Association, Washington, DC.

Bainton, R., Tsai, L. T., Singh, C. M., Moore, M. S., Neckameyer, W. S., and Heberlein, U. (2000). Dopamine modulates acute responses to cocaine, nicotine and ethanol in *Drosophila. Curr. Biol.* **10**, 187–194.

Berger, K. H., Heberlein, U., and Moore, M. S. (2004). Rapid and chronic: Two distinct forms of ethanol tolerance in *Drosophila. Alcohol. Clin. Exp. Res.* **28**, 1469–1480.

Berger, K. H., Kong, E. C., Dubnau, J., Tully, T., Moore, M. S., and Heberlein, U. (2008). Ethanol sensitivity and tolerance in long-term memory mutants of *Drosophila melanogaster. Alcohol. Clin. Exp. Res.* **32**, 895–908.

Bhandari, P., Kendler, K. S., Bettinger, J. C., Davies, A. G., and Grotewiel, M. S. (2009). An assay for evoked locomotor behavior in *Drosophila* reveals a role for integrins in ethanol sensitivity and rapid ethanol tolerance. *Alcohol. Clin. Exp. Res.* **10**, 1–12.

Bohm, R. A., Wang, B., Brenner, R., and Atkinson, N. S. (2000). Transcriptional control of Ca(2+)-activated K(+) channel expression: Identification of a second, evolutionarily conserved, neuronal promoter. *J. Exp. Biol.* **203**, 693–704.

Brand, A. H., and Perrimon, N. (1993). Targeted gene expression as a means of altering cell fates and generating dominant phenotypes. *Development (Cambridge, England)* **118**, 401–415.

Cadieu, N., Cadieu, J., El Ghadraoui, L., Grimal, A., and Lamboeuf, Y. (1999). Conditioning to ethanol in the fruit fly – a study using an inhibitor of ADH. *J. Insect Physiol.* **45**, 579–586.

Carpenter-Hyland, E. P., and Chandler, L. J. (2006). Homeostatic plasticity during alcohol exposure promotes enlargement of dendritic spines. *Eur. J. Neurosci.* **24**, 3496–3506.

Chapman, R. F. (2003). Contact chemoreception in feeding by phytophagous insects. *Annu. Rev. Entomol.* **48**, 455–484.

Chen, Z. W., and Olsen, R. W. (2007). GABAA receptor associated proteins: A key factor regulating GABAA receptor function. *J. Neurochem.* **100**, 279–294.

Cheng, Y., Endo, K., Wu, K., Rodan, A. R., Heberlein, U., and Davis, R. L. (2001). *Drosophila* fasciclinII is required for the formation of odor memories and for normal sensitivity to alcohol. *Cell* **105**, 757–768.

Chien, S., Reiter, L. T., Bier, E., and Gribskov, M. (2002). Homophila: Human disease gene cognates in *Drosophila*. *Nucleic Acids Res.* **30**, 149–151.

Cho, W., Heberlein, U., and Wolf, F. W. (2004). Habituation of an odorant-induced startle response in *Drosophila*. *Genes Brain Behav.* **3**, 127–137.

Cohan, F. M., and Graf, J. (1985). Latitudinal cline in *Drosophila melanogaster* for knockdown resistance to ethanol fumes and for rates of response to selection for further resistance. *Evolution* **39**, 278–293.

Cohan, F. M., and Hoffmann, A. A. (1986). Genetic divergence under uniform selection: II. Different responses to selection for knockdown resistance to ethanol among *Drosophila melanogaster* populations and their replicate lines. *Genetics* **114**, 145–164.

Collins, B., and Blau, J. (2007). Even a stopped clock tells the right time twice a day: Circadian timekeeping in *Drosophila*. Pflugers Archiv : *Eur. J. Appl. Physiol.* **454**, 857–867.

Conti, A. C., Maas, J. W., Moulder, K. L., Jiang, X., Dave, B. A., Mennerick, S., and Muglia, L. J. (2009). Adenylyl cyclases 1 and 8 initiate a presynaptic homeostatic response to ethanol treatment. *PLoS ONE* **4**, e5697.

Corl, A., Rodan, A. R., and Heberlein, U. (2005). Insulin signaling in the nervous system regulates ethanol intoxication in *Drosophila melanogaster*. *Nat. Neurosci.* **8**, 18–19.

Corl, A. B., Berger, K. H., Ophir-Shohat, G., Gesch, J., Simms, J. A., Bartlett, S. E., and Heberlein, U. (2009). Happyhour, a Ste20 family kinase, implicates EGFR signaling in ethanol-induced behaviors. *Cell.* **137**, 949–960.

Cowmeadow, R. B., Krishnan, H. R., and Atkinson, N. S. (2005). The *slowpoke* gene is necessary for rapid ethanol tolerance in *Drosophila*. *Alcohol. Clin. Exp. Res.* **29**, 1777–1786.

Cowmeadow, R. B., Krishnan, H. R., Ghezzi, A., Al'Hasan, Y. M., Wang, Y. Z., and Atkinson, N. S. (2006). Ethanol tolerance caused by slowpoke induction in *Drosophila*. *Alcohol. Clin. Exp. Res.* **30**, 745–753.

David, J. R., Bocquet, C., Arens, M. F., and Fouillet, P. (1976). Biological role of alcohol dehydrogenase in the tolerance of *Drosophila melanogaster* to aliphatic alochols: Utilization of an ADH-null mutant. *Biochem. Genet.* **14**, 989–997.

Davies, A. G., Pierce-Shimomura, J. T., Kim, H., VanHoven, M. K., Thiele, T. R., Bonci, A., Bargmann, C. I., and McIntire, S. L. (2003). A central role of the BK potassium channel in behavioral responses to ethanol in *C. elegans*. *Cell* **115**, 655–666.

Devineni, A. V., and Heberlein, U. (2009). Preferential ethanol consumption in *Drosophila* models features of addiction. *Curr. Biol.* **19**, 2126–2132.

Dietzl, G., Chen, D., Schnorrer, F., Su, K. C., Barinova, Y., Fellner, M., Gasser, B., Kinsey, K., Oppel, S., Scheiblauer, S., *et al.* (2007). A genome-wide transgenic RNAi library for conditional gene inactivation in *Drosophila*. *Nature* **448**, 151–156.

Dudley, R. (2004). Ethanol, fruit ripening, and the historical origins of human alcoholism in primate frugivory. *Integr. Comp. Biol.* **44**, 315–323.

Dzitoyeva, S., Dimitrijevic, N., and Manev, H. (2003). Gamma-aminobutyric acid B receptor 1 mediates behavior-impairing actions of alcohol in *Drosophila*: Adult RNA interference and pharmacological evidence. *Proc. Natl. Acad. Sci. USA* **100**, 5485–5490.

Eisses, K. T. (1997). The influence of 2-propanol and acetone on oviposition rate and oviposition site preference for acetic acid and ethanol of *Drosophila melanogaster*. *Behav. Genet.* **27**, 171–180.

Fuyama, Y. (1976). Behavior genetics of olfactory responses in *Drosophila*: I. Olfactometry and strain differences in *Drosophila melanogaster*. *Behav. Genet.* **6**, 407–420.

Geer, B. W., McKechnie, S. W., Bentley, M. M., Oakeshott, J. G., Quinn, E. M., and Langevin, M. L. (1988). Induction of alcohol dehydrogenase by ethanol in *Drosophila melanogaster*. *J. Nutr.* **118**, 398–407.

Gehring, W. J., Kloter, U., and Suga, H. (2009). Evolution of the Hox gene complex from an evolutionary ground state. *Curr. Top. Dev. Biol.* **88**, 35–61.

Ghezzi, A., Al-Hasan, Y. M., Larios, L. E., Bohm, R. A., and Atkinson, N. S. (2004). slo K(+) channel gene regulation mediates rapid drug tolerance. *Proc. Natl. Acad. Sci. USA* **101**, 17276–17281.

Godenschwege, T. A., Reisch, D., Diegelmann, S., Eberle, K., Funk, N., Heisenberg, M., Hoppe, V., Hoppe, J., Klagges, B. R., Martin, J. R., *et al.* (2004). Flies lacking all synapsins are unexpectedly healthy but are impaired in complex behaviour. *Eur. J. Neurosci.* **20**, 611–622.

Guarnieri, D. J., and Heberlein, U. (2003). *Drosophila melanogaster*, a genetic model system for alcohol research. *Int. Rev. Neurobiol.* **54**, 199–228.

Hoffmann, A. A., and Parsons, P. A. (1989). Selection for increased desiccation resistance in *Drosophila melanogaster*: Additive genetic control and correlated responses for other stresses. *Genetics* **122**, 837–845.

Ikemoto, M. J., Inoue, K., Akiduki, S., Osugi, T., Imamura, T., Ishida, N., and Ohtomi, M. (2002). Identification of addicsin/GTRAP3-18 as a chronic morphine-augmented gene in amygdala. *Neuroreport* **13**, 2079–2084.

Ja, W., Carvalho, G., Mak, E., De La Rosa, N., Fang, A., Liong, J., Brummel, T., and Benzer, S. (2007). Prandiology of Drosophila and the CAFE assay. *Proc. Natl. Acad. Sci. USA* **104**, 8253–8256.

Joseph, R. M., Devineni, A. V., King, I. F., and Heberlein, U. (2009). Oviposition preference for and positional avoidance of acetic acid provide a model for competing behavioral drives in *Drosophila*. *Proc. Natl. Acad. Sci. USA* **106**, 11352–11357.

Kitamoto, T. (2001). Conditional modification of behavior in *Drosophila* by targeted expression of a temperature-sensitive *shibire* allele in defined neurons. *J. Neurobiol.* **47**, 81–92.

Kong, E. C., Allouche, L., Chapot, P. A., Vranizan, K., Moore, M. S., Heberlein, U., and Wolf, F. W. (2010). Ethanol-regulated genes that contribute to ethanol sensitivity and rapid tolerance in *Drosophila*. *Alcohol. Clin. Exp. Res.* **34**, 1–14.

Kong, E. C., Woo, K., Li, H., Lebetsky, T., Mayer, N., Sniffen, M. R., Heberlein, U., Bainton, R. J., Hirsh, J., and Wolf, F. W. (2010). A pair of dopamine neurons target the D1-like dopamine receptor DopR in the central complex to promote ethanol-stimulated locomotion in *Drosophila*. *PLoS ONE* **5**, e9954.

LaFerriere, H., Guarnieri, D. J., Sitaraman, D., Diegelman, S., Heberlein, U., and Zars, T. (2008). Genetic dissociation of ethanol sensitivity and memory formation in *Drosophila melanogaster*. *Genetics* **178**, 1895–1902.

Li, C., Zhao, X., Cao, X., Chu, D., Chen, J., and Zhou, J. (2008). The *Drosophila* homolog of *jwa* is required for ethanol tolerance. *Alcohol Alcohol.* **43**, 529–536.

Li, C. Y., Liu, Q. R., Zhang, P. W., Li, X. M., Wei, L., and Uhl, G. R. (2009). OKCAM: An ontology-based, human-centered knowledgebase for cell adhesion molecules. *Nucleic Acids Res.* **37**, D251–260.

Li, C. Y., Mao, X., and Wei, L. (2008). Genes and (common) pathways underlying drug addiction. *PLoS Comput. Biol.* **4**, e2.

Lieber, C. S. (2004). Alcoholic fatty liver: Its pathogenesis and mechanism of progression to inflammation and fibrosis. *Alcohol* **34**, 9–19.

Maccioni, P., and Colombo, G. (2009). Role of the GABA(B) receptor in alcohol-seeking and drinking behavior. *Alcohol* **43**, 555–558.

Maggert, K. A., Gong, W. J., and Golic, K. G. (2008). Methods for homologous recombination in *Drosophila*. *Methods Mol. Biol. (Clifton, NJ)* **420**, 155–174.

Maier, S., Reiterer, V., Ruggiero, A. M., Rothstein, J. D., Thomas, S., Dahm, R., Sitte, H. H., and Farhan, H. (2009). GTRAP3-18 serves as a negative regulator of Rab1 in protein transport and neuronal differentiation. *J Cell.Mol. Med.* **13**, 114–124.

Martinez, S., Lazaro-Dieguez, F., Selva, J., Calvo, F., Piqueras, J., Crespo, P., Claro, E., and Egea, G. (2007). Lysophosphatidic acid rescues RhoA activation and phosphoinositides levels in astrocytes exposed to ethanol. *J. Neurochem.* **102**, 1044–1052.

Matsuno, M., Horiuchi, J., Tully, T., and Saitoe, M. (2009). The *Drosophila* cell adhesion molecule *klingon* is required for long-term memory formation and is regulated by Notch. *Proc. Natl. Acad. Sci. USA* **106**, 310–315.

Matthews, K. A., Kaufman, T., and Gelbart, W. M. (2005). Research resources for *Drosophila*: The expanding universe. *Nat. Rev. Genet.* **6**, 179–193.

Monastirioti, M. (1999). Biogenic amine systems in the fruit fly *Drosophila melanogaster*. *Microsc. Res. Tech.* **45**, 106–121.

Moore, M. S., DeZazzo, J., Luk, A. Y., Tully, T., Singh, C. M., and Heberlein, U. (1998). Ethanol intoxication in *Drosophila*: Genetic and pharmacological evidence for regulation by the cAMP signaling pathway. *Cell* **93**, 997–1007.

Morozova, T. V., Anholt, R. R., and Mackay, T. F.C. (2006). Transcriptional response to alcohol exposure in *Drosophila melanogaster*. *Genome Biol.* **7**, R95.

Morozova, T. V., Anholt, R. R., and Mackay, T. F.C. (2007). Phenotypic and transcriptional response to selection for alcohol sensitivity in *Drosophila melanogaster*. *Genome Biol.* **8**, R231.

Morozova, T. V., Ayroles, J. F., Jordan, K. W., Duncan, L. H., Carbone, M. A., Lyman, R. F., Stone, E. A., Govindaraju, D. R., Ellison, C. R., Mackay, T. F.C., *et al.* (2009). Alcohol sensitivity in *Drosophila*: Translational potential of systems genetics. *Genetics*. **183**, 733–745.

Mulligan, M. K., Ponomarev, I., Hitzemann, R. J., Belknap, J. K., Tabakoff, B., Harris, R. A., Crabbe, J. C., Blednov, Y. A., Grahame, N. J., Phillips, T. J., *et al.* (2006). Toward understanding the genetics of alcohol drinking through transcriptome meta-analysis. *Proc. Natl. Acad. Sci. USA* **103**, 6368–6373.

Neuser, K., Triphan, T., Mronz, M., Poeck, B., and Strauss, R. (2008). Analysis of a spatial orientation memory in *Drosophila*. *Nature* **453**, 1244–1247.

Offenhauser, N., Castelletti, D., Mapelli, L., Soppo, B. E., Regondi, M. C., Rossi, P., D'Angelo, E., Frassoni, C., Amadeo, A., Tocchetti, A., *et al.* (2006). Increased ethanol resistance and consumption in Eps8 knockout mice correlates with altered actin dynamics. *Cell* **127**, 213–226.

Pandey, S. C., Ugale, R., Zhang, H., Tang, L., and Prakash, A. (2008). Brain chromatin remodeling: A novel mechanism of alcoholism. *J. Neurosci.* **28**, 3729–3737.

Park, S. K., Sedore, S. A., Cronmiller, C., and Hirsh, J. (2000). Type II cAMP-dependent protein kinase-deficient *Drosophila* are viable but show developmental, circadian, and drug response phenotypes. *J. Biol. Chem.* **275**, 20588–20596.

Parr, J., Large, A., Wang, X., Fowler, S. C., Ratzlaff, K. L., and Ruden, D. M. (2001). The inebriactometer: A device for measuring the locomotor activity of *Drosophila* exposed to ethanol vapor. *J. Neurosci. Methods*. **107**, 93–99.

Pfeiffer, B. D., Jenett, A., Hammonds, A. S., Ngo, T. T., Misra, S., Murphy, C., Scully, A., Carlson, J. W., Wan, K. H., Laverty, T. R., *et al.* (2008). Tools for neuroanatomy and neurogenetics in *Drosophila*. *Proc. Natl. Acad. Sci. USA* **105**, 9715–9720.

Pulver, S. R., Pashkovski, S. L., Hornstein, N. J., Garrity, P. A., and Griffith, L. C. (2009). Temporal dynamics of neuronal activation by Channelrhodopsin-2 and TRPA1 determine behavioral output in *Drosophila* larvae. *J. Neurophysiol.* **101**, 3075–3088.

Reed, M. R. (1938). The olfactory reactions of *Drosophila melanogaster* Meigen to the products of fermenting banana. *Physiol. Zool.* **11**, 317–325.

Renthal, W., and Nestler, E. J. (2008). Epigenetic mechanisms in drug addiction. *Trends Mol. Med.* **14**, 341–350.

Richmond, R. C., and Gerking, J. L. (1979). Oviposition site preference in *Drosophila*. *Behav. Genet.* **9**, 233–241.

Riley, B. P., Kalsi, G., Kuo, P. H., Vladimirov, V., Thiselton, D. L., Vittum, J., Wormley, B., Grotewiel, M. S., Patterson, D. G., Sullivan, P. F., *et al.* (2006). Alcohol dependence is associated with the ZNF699 gene, a human locus related to *Drosophila* hangover, in the Irish Affected Sib Pair Study of Alcohol Dependence (IASPSAD) sample. *Mol. Psychiatry* **11**, 1025–1031.

Ritzmann, R. F., and Tabakoff, B. (1976). Dissociation of alcohol tolerance and dependence. *Nature* **263**, 418–420.

Rodan, A. R., Kiger, J. A., and Heberlein, U. (2002). Functional dissection of neuroanatomical loci regulating ethanol sensitivity in *Drosophila*. *J. Neurosci.* **22**, 9490–9501.

Rothenfluh, A., Threlkeld, R. J., Bainton, R., Tsai, L., Lasek, A., and Heberlein, U. (2006). Distinct behavioral responses to ethanol are regulated by alternate RhoGAP18B isoforms. *Cell* **127**, 199–211.

Rubin, G. M., and Spradling, A. C. (1982). Genetic transformation of *Drosophila* with transposable element vectors. *Science (New York, NY)* **218**, 348–353.

Scholz, H., Franz, M., and Heberlein, U. (2005). The *hangover* gene defines a stress pathway required for ethanol tolerance development. *Nature* **436**, 845–847.

Scholz, H., Ramond, J., Singh, C. M., and Heberlein, U. (2000). Functional ethanol tolerance in *Drosophila*. *Neuron* **28**, 261–271.

Schwenkert, I., Eltrop, R., Funk, N., Steinert, J. R., Schuster, C. M., and Scholz, H. (2008). The *hangover* gene negatively regulates bouton addition at the *Drosophila* neuromuscular junction. *Mech. Dev.* **125**, 700–711.

Singh, C. M., and Heberlein, U. (2000). Genetic control of acute ethanol-induced behaviors in *Drosophila*. *Alcohol. Clin. Exp. Res.* **24**, 1127–1136.

Sturtevant, A. (1967). A History of Genetics. Cold Spring Harbor Laboratory Press, New York, NY.

Szumlinski, K. K., Lominac, K. D., Oleson, E. B., Walker, J. K., Mason, A., Dehoff, M. H., Klugmann, M., Klugman, M., Cagle, S., Welt, K., *et al.* (2005). Homer2 is necessary for EtOH-induced neuroplasticity. *J. Neurosci.* **25**, 7054–7061.

Thiele, T. E., Marsh, D. J., Ste Marie, L., Bernstein, I. L., and Palmiter, R. D. (1998). Ethanol consumption and resistance are inversely related to neuropeptide Y levels. *Nature* **396**, 366–369.

Treistman, S. N., and Martin, G. E. (2009). BK channels: Mediators and models for alcohol tolerance. *Trends Neurosci.* **32**, 629–637.

Urizar, N. L., Yang, Z., Edenberg, H. J., and Davis, R. L. (2007). *Drosophila* homer is required in a small set of neurons including the ellipsoid body for normal ethanol sensitivity and tolerance. *J. Neurosci.* **27**, 4541–4551.

Urso, T., Gavaler, J. S., and Van Thiel, D. H. (1981). Blood ethanol levels in sober alcohol users seen in an emergency room. *Life Sci.* **28**, 1053–1056.

van Delden, W., and Kamping, A. (1990). Genetic variation for oviposition behavior in *Drosophila melanogaster*. II. Oviposition preferences and differential survival. *Behav. Genet.* **20**, 661–673.

Waddell, S., Armstrong, J. D., Kitamoto, T., Kaiser, K., and Quinn, W. G. (2000). The amnesiac gene product is expressed in two neurons in the *Drosophila* brain that are critical for memory. *Cell* **103**, 805–813.

Wang, Y., Ghezzi, A., Yin, J. C., and Atkinson, N. S. (2009). CREB regulation of BK channel gene expression underlies rapid drug tolerance. *Genes Brain Behav.* **8**, 369–376.

Wang, Y., Krishnan, H. R., Ghezzi, A., Yin, J. C., and Atkinson, N. S. (2007). Drug-induced epigenetic changes produce drug tolerance. *PLoS Biol.* **5**, e265.

Watabe, M., Aoyama, K., and Nakaki, T. (2007). Regulation of glutathione synthesis via interaction between glutamate transport-associated protein 3-18 (GTRAP3-18) and excitatory amino acid carrier-1 (EAAC1) at plasma membrane. *Mol. Pharmacol.* **72**, 1103–1110.

Wen, T., Parrish, C. A., Xu, D., Wu, Q., and Shen, P. (2005). *Drosophila* neuropeptide F and its receptor, NPFR1, define a signaling pathway that acutely modulates alcohol sensitivity. *Proc. Natl. Acad. Sci. USA* **102**, 2141–2146.

Wolf, F. W., Rodan, A. R., Tsai, L. T., and Heberlein, U. (2002). High-resolution analysis of ethanol-induced locomotor stimulation in *Drosophila*. *J. Neurosci.* **22**, 11035–11044.

NEURAL PLASTICITY, HUMAN GENETICS, AND RISK FOR ALCOHOL DEPENDENCE

Shirley Y. Hill

Department of Psychiatry, University of Pittsburgh School of Medicine, Pittsburgh, PA 15213, USA

Opportunities for advances in the neurobiology of alcohol dependence have been facilitated by the development of sophisticated neurophysiological and neuroimaging techniques that allow us to have a window on developmental changes in brain structure and function. The search for genes that may increase

INTERNATIONAL REVIEW OF
NEUROBIOLOGY, VOL. 91
DOI: 10.1016/S0074-7742(10)91003-9

susceptibility to alcohol dependence has been greatly facilitated by the recognition that intermediate phenotypes, sometimes referred to as endophenotypes, may be closer to the genetic variation than is the more complex alcohol dependence phenotype. This chapter will review the evidence that the brain is highly plastic, exhibiting major postnatal changes, especially during adolescence, in neural circuits that appear to influence addiction susceptibility. This chapter will suggest that heritable aspects of brain structure and function that are seen developmentally may be an important endophenotypic characteristic associated with familial risk for developing alcohol dependence. Finally, a review of studies showing associations between brain structural and functional characteristics and specific genes will be offered.

I. Introduction

Alcohol dependence (AD) is a major health problem in many parts of the world including the United States (Murray and Lopez, 1997). Data from the National Comorbidity Survey, a survey of respondents aged 15–54, find that AD is a common condition in the United States, with 20.1% of men and 8.2% of women affected (Kessler *et al.*, 1997).

The scientific study of AD began to take shape when Jellinek (1960) first described the variants of alcoholism subtypes he had observed. As with any disorder for which we seek to find improvements in prevention, treatment, and intervention, defining the characteristics that typify the "alcoholic" person and delineating the major subtypes is a critical part of identifying the genetic and environmental determinants of the condition. As an example, identification of variants of diabetes (type I and type II) was an important step in better understanding the genetic underpinnings of both, and providing guidelines for treatment that led to identification of genes conferring susceptibility to type I diabetes (see Permutt *et al.*, 2005, for review).

The recognition that genes have pleiotropic effects so that behavioral manifestations of the disorder at one age or stage in a person's life may be quite different than they are at another stage has also enhanced the scientific study of AD. The relationship between particular child and adolescent disorders and later development of AD and related phenotypes makes it possible to identify risk factors useful in targeted intervention and prevention efforts. As one example, conduct disorder of adolescence has been identified as a precursor of adult substance use disorders (SUDs) in a number of studies (Robins, 1966; Hill and Muka, 1996; Hill *et al.*, 1999a, 2008; Ohannessian *et al.*, 2004; Kuperman *et al.*, 2005).

A. Family Studies and the Search for Genes

One historical antecedent of the scientific study of AD occurred when family studies began to emerge. Among the earliest family studies that showed increased likelihood of developing AD if one's relatives were alcohol dependent was conducted in St. Louis, Missouri, USA (Winokur and Clayton, 1968). This study along with many other more recent family studies (Merikangas *et al.*, 1998; Hill *et al.*, 2008) established that transmission of susceptibility to developing AD occurs within families. Based on the evidence that AD runs in families, along with more recent sophisticated statistical analyses of family data including segregation analysis (Yuan *et al.*, 1996) and biometrical modeling of covariance among relatives (Kendler *et al.*, 1992), most researchers in the addiction field now view AD as a complex, heritable disorder. Heritability in males has been reported to range between 0.49 (Caldwell and Gottesman, 1991) and 0.64 (Heath *et al.*, 1997). For women, heritability has been reported to be in the same general range (0.56–0.59) (Prescott *et al.*, 1999; Kendler *et al.*, 1992).

B. Clinical Heterogeneity and Alcohol Dependence

An intensive genome-wide search for genes that alter susceptibility for AD and related phenotypes has been ongoing in the past 20 years facilitated by marked improvements in genetic technologies (Reich *et al.*, 1998; Hill *et al.*, 2004). However, determining the genetic underpinnings of this complex disorder is a daunting task because there are multiple clinical subtypes and variable expression across a lifespan. Existence of subtypes occurs because of the presence of comorbid conditions (mood disorders, drug dependence, anxiety disorders) that introduce clinical heterogeneity, which can lead to reduced statistical power to uncover a significant association between a genetic variation and the AD phenotype. Therefore, increased attention has been focused on finding biological variation associated with the AD phenotype that can be used to identify genes conferring increased susceptibility to the disorder. Recent advances in neuroscience have made it possible to study endophenotypes, sometimes referred to as intermediate phenotypes, that are potentially closely linked to AD and presumably closer to the genetic variation that is sought. The initiation of studies focusing on endophenotypes was an important breakthrough because most brain disorders show developmental trajectories. This plasticity of the human brain probably explains why some endophenotypes of childhood are such salient predictors of adult addictive disorders.

C. Genetic Heterogeneity: Implications for Gene Finding

AD is considered to be a complex disorder because its development is thought to be controlled by multiple genes. In contrast, there are many human diseases such as cystic fibrosis that are thought to be the result of mutations in a single gene, though even the expression of this single-gene disorder is influenced by additional modifier genes (Merlo and Boyle, 2003). Complex diseases often display allelic and locus heterogeneity (Botstein and Risch, 2003; Goldstein *et al.*, 2003). In the case of allelic heterogeneity, multiple mutations within the same gene lead to the illness, while in the case of locus heterogeneity one sees multiple genes with mutations that increase risk for the disease. AD is undoubtedly the result of both processes and other more elusive mechanisms as well. These include germ line mutations, which are the result of the parent of origin and influenced by parental age (Crow, 2000). Additionally, epigenetic effects or heritable changes in DNA expression that are not the result of changes in the DNA are of increasing interest (Bird, 2007).

As noted by McClellan *et al.* (2007), current research in the genetics of schizophrenia has been guided primarily by the "common disease–common alleles" model advanced by Chakravarti (1999). This view as originally stated by Gottesman and Shields (1982) argues that illness results from the cumulative impact of multiple common genetic variants each of small effect interacting with environmental exposure that exceeds some disease threshold. McClellan *et al.* (2007) argue that a "common disease–rare allele" model may explain many cases of schizophrenia. This would appear to be equally true for AD.

Although AD is a common disorder in the population, there are families with multiple family members with the disorder (multiplex families) where it appears likely that a few large-effect alleles may be segregating. As recently noted by McClellan and King (2010), it is now recognized that large-effect mutations are causal in many common medical conditions. Evidence is offered that rare severe alleles have been implicated in all forms of inherited susceptibility to cancer, are present in inherited hearing loss, extreme levels of lipids, and severe mental illness such as schizophrenia. They note that this is a major theoretical shift in human genetics because the last decade has focused on the common disease/common variant model promulgated by the Risch and Merikangas (1996) publication which presumed that additive or multiplicative effects of combinations of common risk variants of small effect contributed to the emergence of a disease. No one variant was thought to cause the disorder.

McClellan *et al.* (2007) also persuasively argue that if many cases of schizophrenia arise from individually rare large-effect alleles, then current approaches will fail to identify critical genes. This is the result of the fact that most current psychiatric genetic research is designed to identify common alleles or haplotypes associated with an increased risk of disease and shared by large numbers of unrelated patients (cases) in comparison to unrelated controls (Merikangas and Risch, 2003). This approach may yield limited information based on the results of a large number of genome-wide association (GWAS) studies of nonpsychiatric complex disorders, where the majority of variants identified have no established biological relevance to the disease of interest or its outcome (see McClellan and King, 2010 for review). In contrast, studies of families that harbor informative genomic events may reveal multiple mutations, some of a more modest effect as well as those having more severe consequences for disease likelihood and progression. In short, rare severe mutations may occur in genes that also harbor more common genetic variants with modest effects on disease risk. The foregoing review will focus on data collected for endophenotypes of brain plasticity as they are related to AD susceptibility with special emphasis on studies that have collected such information in families where rare and important mutations may be occurring.

In this chapter, I will (1) review the evidence that the brain is a highly plastic organ that exhibits major postnatal changes in structure and function, especially during adolescence, in specific circuits that appear to influence addiction susceptibility; (2) present findings suggesting that salient endophenotypes for AD have been identified; (3) provide a review of the evidence that brain morphology is heritable and that developmental changes in morphology during adolescence and young adulthood is an important endophenotypic characteristic associated with familial risk for developing AD; (4) summarize the evidence showing that brain functional differences have been identified in those with familial risk for AD; and (5) provide a review of studies showing associations between brain structural and/or functional characteristics and specific genes.

II. Brain Plasticity

A. Developmental Issues

During adolescence and young adulthood, marked changes in brain structure occur, which are accompanied by refinement of brain organization, which in turn leads to changes in cognitive, social, and emotional behavior (Casey *et al.*, 2005; Yurgelun-Todd, 2007). Cortical development including that seen during adolescence and young adulthood generally follows a pattern that supports the changing needs of the organism with primary motor, sensory, and visual areas maturing earlier than those supporting more complex cognitive functions such as the association areas (Gogtay *et al.*, 2004). Accompanying the changes in regional volumes are changes in white and gray matter volumes. White matter (WM) increases well into adulthood while gray matter volume tends to increase in childhood and adolescence followed by a decrease (Giedd *et al.*, 1996a,b; Jernigan *et al.*, 1991; Pfefferbaum *et al.*, 1994; Sowell *et al.*, 2004), with females reaching their peak 1–2 years earlier than males (Lenroot and Geidd, 2006). These morphological changes during adolescence and young adulthood appear to vary by familial risk group status. Results for WM volume of the orbitofrontal cortex (OFC) in the right hemisphere differ by familial risk group (Hill *et al.*, 2009b) (Fig. 1). Similarly, gray matter volume appears to vary by risk group for the amygdala (Fig. 2) (Hill *et al.*, unpublished). These morphological differences, which vary by whether or not one has multiple relatives with AD (multiplex families) or no first-degree relatives with AD (control families), suggests that brain regions showing differing developmental trajectories by familial risk are the result of genetic variation.

Morphological changes in brain development during childhood and adolescence are accompanied by changes in cognitive abilities including development of mature decision-making strategies that have been associated with electrophysiological (Crone and van der molen, 2007) and neurophysiological concomitants (Hill *et al.*, 1999a). Changes in cognitive abilities are often assessed using neuropsychological tests though cognitive tasks are frequently administered while

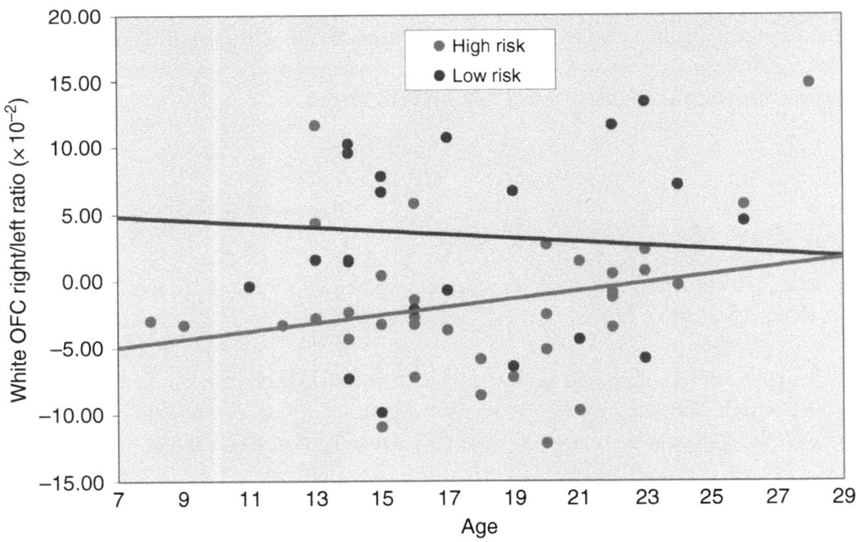

Fig. 1. White matter volume by age is shown for high-risk males. White matter volumes show a statistically significant correlation with Multidimensional Personality Questionnaire (MPQ) Control scale scores (reduced white matter being associated with greater impulsivity).

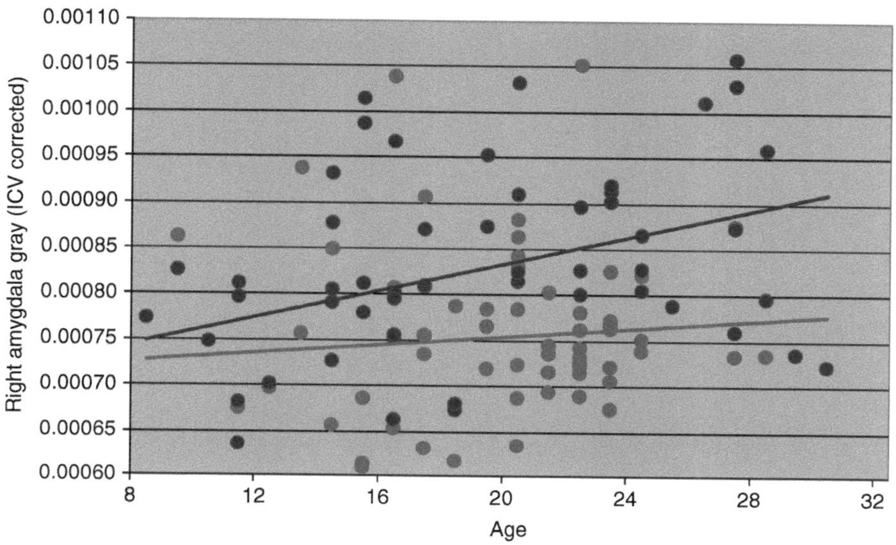

FIG. 2. Gray matter volume of the amygdala in the right hemisphere is shown by age and risk status.

participants are monitored for changes in electroencephalographic (EEG) or event-related potential (ERP) activity. Studies using these methods provide dimensions of inquiry that are not possible using neuropsychological testing alone. These include the ability to determine if the participant is attending to the task and often whether this may have influenced their performance of the task.

One component of the ERP that has received increased interest in the alcoholism field is the P300 component. P300 is a positive wave that occurs approximately 300 ms after the onset of a stimulus and is maximal over the parietal electrode in paradigms designed to measure the P3b variant of this wave. Although paradigms used to elicit the P300 component vary across studies, many utilize an oddball paradigm in which the participant is asked to attend to a "target" stimulus and ignore a "nontarget." Both auditory and visual modalities are used. Many early cross-sectional studies had shown that P300 amplitude is reduced in children from alcoholic families (Begleiter *et al.*, 1984; Berman *et al.*, 1993; Hill *et al.*, 1990a; Hill and Steinhauer 1993; Steinhauer and Hill, 1993; Hill *et al.*, 1995). However, the author's laboratory has the only large-scale follow-up using repeated measures of P300 across childhood, adolescence, and young adulthood. Using data acquired for offspring seen through childhood and adolescence, we performed growth curve modeling for 635 P300 assessments in which two-thirds of the sample were tested five or more times (Fig. 3). This study

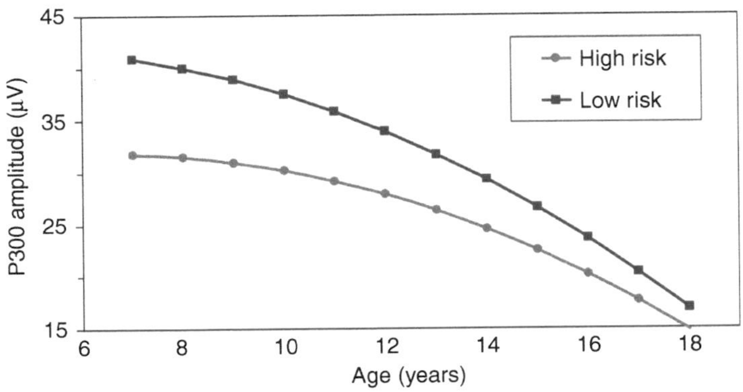

FIG. 3. Quadratic growth curves of visual P300 amplitude obtained for high- and low-risk males. High-risk males had lower P300 amplitude than low-risk males and displayed a slower rate of change with age than did the low-risk males.

established that P300 was not a static quantity but was related to development. These developmental changes in P300 are most likely due to brain morphological changes associated with child/adolescent development, though the specific region is currently unknown. These data also showed that high-risk children exhibit altered patterns of age-appropriate P300 amplitude (Hill *et al.*, 1999b).

Clearly, the human brain is a highly plastic organ that undergoes considerable postnatal change extending into young adulthood. With such notable changes being the rule, the search for genes that may control these processes requires considerable effort. Significant progress has been made in identifying genetic variation that is associated with regional differences in brain volume. A future challenge is identifying alterations not only in specific structures but also in the entire circuit that is involved in addiction (see Tessner and Hill, 2010, for review). Coupled with this challenge is a need to identify genetic variation associated with altered brain circuitry. This is a challenge not only for the addiction field but also for psychiatric disorders in general (Akil *et al.*, 2010).

B. Plasticity of Neural Circuits Potentially Involved in Dependence

It has been hypothesized that drug addiction involves dysfunction of brain circuitry that leads to alterations in both impulsive and compulsive behavior (Koob and Kreek, 2007). It has further been hypothesized that the process of addiction initially involves behavioral characteristics that are more impulsive and in time the more compulsive aspects of addiction appear to take over. In this view, a transition from more positively reinforced behavior becomes behavior that is driven by negative reinforcement or avoidance of an aversive state such as drug withdrawal.

In the Koob and Kreek model, two distinct circuits are proposed as the underlying substrate for negative and positive reinforcement. One circuit, characterized as the "reward neurocircuit," includes connections between the extended amygdala and lateral hypothalamus with input to the ventral medial ventral pallidum and output to the medial dorsal thalamus with feedback to the dorsal prelimbic cortex and basolateral amygdala. The other circuit, termed the "OCD neurocircuit," by Koob and Kreek (2007) includes the dorsal prefrontal cortex (DPFC), cingulate and OFC, which provide feedback to the striatum, and, in turn, to the pallidum and thalamus with returning loop to the OFC, and dorsolateral prefrontal cortex (DLPFC) and cingulate. A fronto-striato-cerebellar network has also been proposed to account for developmental differences in response inhibition (Rubia *et al.*, 2007) that overlaps the OCD circuit described by Koob and Kreek.

III. The Case for Intermediate Phenotypes or Endophenotypes

Because psychiatric disorders including AD and its related phenotypes such as substance use disorders (SUDs) may represent more distal and variable effects of genes, it has been suggested that molecular genetic studies not be performed on psychiatric diagnoses alone but rather include quantitative biological markers that reflect more proximal effects of genes that may be involved in such complex disorders (Tsuang and Faraone, 2000). If one is interested in relating genetic variation directly to the AD phenotype, candidate endophenotypes can also be used as covariates to reduce heterogeneity and better define the trait (Gottesman and Gould, 2003). This approach was utilized in our laboratory to perform a genome-wide linkage analysis in which P300 and personality variants were used as covariates in models in which genetic polymorphisms were related to the AD phenotype (Hill *et al.*, 2004). A number of endophenotypes have now been recognized by the alcoholism field as valid intermediate phenotypes. Some of these show developmental trajectories that make them especially interesting in relation to brain plasticity.

A. Age of Onset to Begin Regular Drinking as an Intermediate Phenotype

The age of onset to begin regular drinking during adolescence is an important predictor of age of first alcohol problem and subsequent AD (Hawkins *et al.*, 1997;

Grant and Dawson, 1997), and greater severity and persistence of problems with illicit drugs (Kandel *et al.*, 1992). Using a large-scale population sample, Grant and Dawson (1997) showed that age of onset of regular drinking predicted the likelihood of adult AD. For those individuals younger than 14 years the rate was 40%; for those aged 20 and older, it was only 10%. Using data from offspring studied longitudinally from a pedigree series identified through familial loading for male alcoholism, we have modeled variables across multiple domains that could be potentially influential in this process. We find that temperament (greater extraversion), familial density of alcoholism (number of alcohol-dependent members in the family), and markers of altered neurodevelopment (P300 amplitude trajectories and acquisition of age-appropriate postural sway) are important mediators of the relationship between age of onset and familial loading for AD (Hill and Yuan, 1999; Hill *et al.*, 2000). These studies illustrate how a clinical phenotype that is related to familial risk for AD such as the age of onset to begin regular drinking can be related to biological endophenotypes and, in turn, to genetic variation.

B. Disinhibition as an Intermediate Phenotype for AD and Other SUDs

Although age at first drink (AFD) has been repeatedly associated with significant elevations in AD, the mechanism underlying this relationship continues to remain unclear. Importantly, McGue and colleagues (2001a,b) found that disinhibited behavioral problems at age 11, including oppositionality, hyperactivity/impulsivity, and inattention, predict earlier onset of drinking and numerous other indicators of externalizing problems (early onset for cigarette use, sexual activity, and drug use). Kuperman *et al.* (2005) also report greater externalizing problems (conduct symptoms) indicative of a behavioral orientation for disinhibition related to AFD. Although the connection between disinhibition and SUD is now well established, the neurobiological underpinnings of this trait are just beginning to emerge.

C. P300 Amplitude and Disinhibition

A consensus seems to have emerged that the amplitude of the P300 component of ERP is a measure of the disinhibited temperament that is related to the risk for SUD (Iacono and McGue, 2006; Porjesz and Rangaswamy, 2007).

The P300 component of ERP reflects active stimulus processing that is affected by attention and memory and is genetically mediated (Van Beijersterveldt and Boosma, 1994). A lower amplitude of P300 appears to be associated with generalized disinhibition, which is associated with the early onset of a number of deviant behaviors (McGue *et al.*, 2001a,b).

D. Developmental Trajectories of P300 Amplitude as an Intermediate Phenotype

Because developmental trajectories of the P300 component of ERP show marked change during childhood and adolescence (Hill *et al.*, 1999b; Hill and Shen, 2002), and differ by familial risk for AD, this suggests that these trajectories are indicators of brain plasticity, which may explain why some individuals have a greater risk for developing AD and related phenotypes. Trajectory class membership is of interest in attempting to find risk and protective factors among those with familial risk to determine if P300 trajectories can predict individual cases with psychopathology including SUDs. When growth curve analysis is applied to P300 data obtained at regular intervals for youngsters followed longitudinally, one sees that there is a significant amount of variation among individuals. Nevertheless, application of a statistical technique known as mixture model analysis reveals that three distinct patterns occur (Hill and Shen, 2002) (Fig. 4). Each

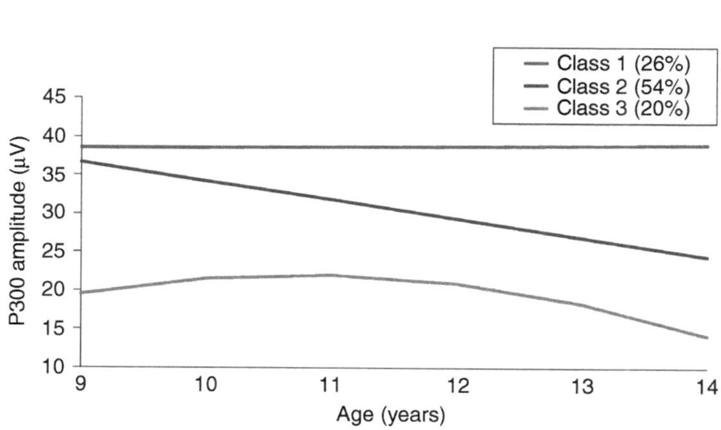

FIG. 4. Estimated growth curves of visual P300 are illustrated for the three-class solution that provided the best fit to the data obtained for children and adolescents who had each provided five annual assessments. Class 2 pattern is the most commonly occurring pattern, with Classes 1 and 3 less common.

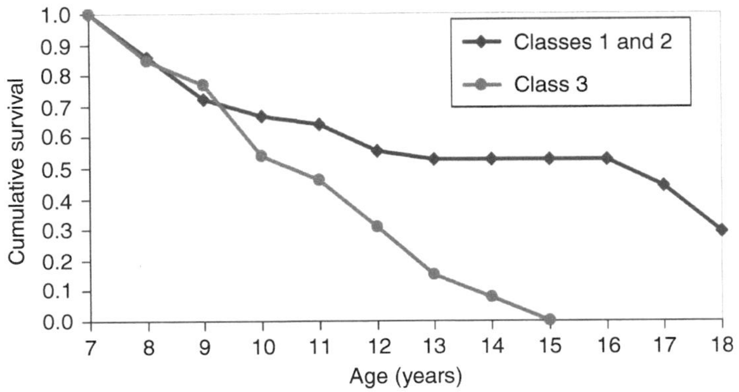

FIG. 5. Survival curves for age of onset at first childhood psychiatric diagnosis among high-risk offspring can be seen. Membership in the Class 3 developmental trajectory group increases the likelihood that a child with a familial/genetic risk for AD will exhibit a childhood diagnosis by a given age.

individual's trajectory class placement is based on a probability estimate so that any given individual may have a lower or higher probability of being in one class than another. However, most individuals' data can be placed in a particular class with reasonable certainty. These three trajectory patterns of P300 are linked to risk for childhood psychiatric disorders (Hill and Shen, 2002) (Fig. 5). Specifically, those with a pattern characterized by low P300 amplitude across childhood and adolescence have a greater risk for externalizing disorders.

With the established link between risk for externalizing disorders and SUD outcome already well known, we followed youngsters to young adulthood to determine if P300 amplitude could predict SUD outcome at annual intervals for an average of 11 years, finding significant associations with outcome (Hill *et al.*, 2009a). A similar finding has been reported with 7-year follow-up (Carlson *et al.*, 2004). ERP tracings by age group for participants seen in our longitudinal follow-up may be seen in Fig. 6. What the figure shows is that those who later developed an SUD by young adulthood could be differentiated by P300 amplitude at age 9. When the child's P300 amplitude was combined with assessment of body sway as measured by a computerized movement platform, we found an 8-fold increase in risk for developing SUD by young adulthood. Those in the lowest tercile for P300 amplitude and the highest tercile for body sway (children who were less capable of standing steady on the platform) had the highest risk for SUD outcome.

In summary, intermediate phenotypes have been identified that aid in recognizing those who may be at the highest risk for developing AD by adulthood.

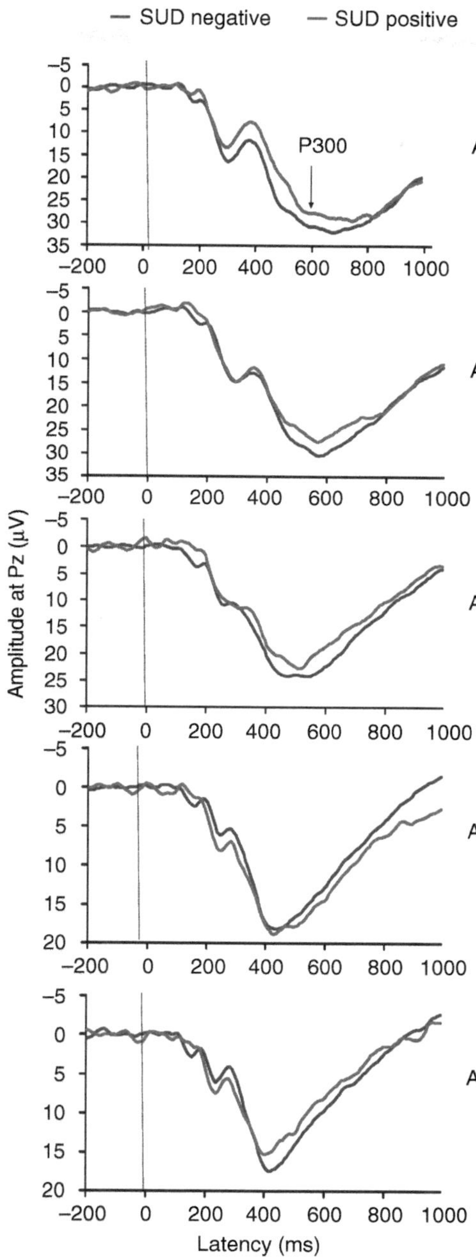

FIG. 6. Grand averages for visual P300 amplitude by SUD outcome groups are illustrated by age at P300 recording. Only the amplitude difference at age 9 differentiates those who will later develop SUD.

These include having an early onset to begin drinking and indicators of disinhibition across several behaviors including initiation of smoking, sexual activity, and drug use. These disinhibited behaviors appear to be related to the amplitude of the P300 component of ERP. Because P300 amplitude changes during child and adolescent maturation, trajectories of amplitude change have been identified that appear to confer greater or lesser risk for developing AD and related phenotypes.

IV. Brain Morphology as an Endophenotype for Alcohol Dependence Risk

A. The Heritability of Brain Volumes

There is substantial evidence that the volume of specific brain structures is heritable (Pfefferbaum *et al.*, 2004; Schmitt *et al.*, 2007). Based on twin and family data, the heritability of large brain regions appears to be consistently high and suggests that genetic factors account for at least half of the phenotypic variance. This suggests that offspring from families where AD has a strong recurrence risk over multiple generations would be likely to have altered brain morphology that might predispose them to greater risk for developing SUD. Although numerous studies point to a relationship between atypical behavior and neuroanatomic phenotypes, few have attempted to understand how genetic and environmental effects on typical cognitive and behavioral measures are mediated through brain morphology.

B. Neurological Underpinnings of the Familial Diathesis for AD

The study of brain circuitry that may be specifically involved in addiction is still relatively uncommon (Volkow *et al.*, 2004) and especially so in samples of adolescent/young adults at risk by familial/genetic background or personal alcohol/drug exposure. Also, it has been suggested that dysfunction of a neurologically based system designed for decision-making processes is the underlying factor in the profound drug/alcohol craving that leads to continued use in spite of adverse consequences (Goldstein and Volkow, 2002). Because brain morphology appears to be heritable, it would appear important to control for familial background in studies that address effects of alcohol and drug exposure and their effects on brain morphology.

C. The Amygdala and Hippocampus

In mammals, the amygdala has connections to the hypothalamus and brainstem that modulate visceral responses to emotional stimuli and connections with a forebrain circuit that appears to be directly involved in emotional processing and mood (Price, 2003). A ventral system that includes the ventral striatum, ventral anterior cingulate gyrus, and the prefrontal (PFC), particularly the orbital and medial PFC, has received extensive attention for its role in the etiology and consequences of major depressive disorder (Drevets, 2000). The possible role of this circuit in addiction is beginning to be recognized (Hill *et al.*, 2001; Glahn *et al.*, 2007).

The hippocampus is of interest because of its well-known role in working memory. Based on data from animal studies, the hippocampus appears to be especially vulnerable to effects of alcohol (Lundqvist *et al.*, 1995). Adults with AD have been reported to show reduction in hippocampal volume (Agartz *et al.*, 1999). DeBellis and colleagues (2000) were among the first to perform MRI scans of adolescent alcohol-dependent participants, finding a reduced volume of the hippocampus in these individuals. However, the DeBellis study could not disentangle the possible effects of genetic predisposition to altered brain morphology and subsequent exposure to alcohol. Utilizing the offspring from our multiplex set of pedigrees, we investigated whether hippocampal volume might differ in those not exposed to alcohol and explored possible differences in the amygdala. We found a reduction in right amygdala volume in high-risk (HR) subjects relative to controls with no differences seen in the hippocampus (Hill *et al.*, 2001). Recent results (Hill *et al.*, unpublished) from a larger sample ($N = 115$) support the right amygdala finding in showing reduced volume among those with multiplex for AD familial background.

A more recent study of hippocampal volume performed in adolescents (ages 15–18) who were either users of alcohol, alcohol and marijuana, or controls without use of either reported that alcohol users had reduced volume of the left hippocampus compared to the other two groups (Medina *et al.*, 2007). Also, the authors concluded that heavy alcohol use contributes to aberrations in hippocampal asymmetry, with left hippocampal volume being the most affected. However, this study did not address familial risk for AD, so it is unclear if the effects might have been present before heavy drinking was initiated.

D. Volume of the Orbitofrontal Cortex as an Intermediate Phenotype

The OFC has been a candidate region for addiction studies because it is thought to be involved in inhibitory decision-making processes (Bechara and

Damasio, 2002). Methamphetamine abusers show reduced functional magnetic resonance imaging (fMRI) blood oxygen level–dependent (BOLD) response in the right OFC region (Paulus *et al.*, 2002, 2003). Without task presentation, substance users in the period beyond acute withdrawal exhibit hypofunctionality of the OFC (Dom *et al.*, 2005). Our work with offspring from multiplex families shows reduced volume in the right hemisphere (Hill *et al.*, 2009b).

E. The Possible Importance of Laterality of the Orbitofrontal Cortex

Because of the functional asymmetry of the cerebral cortex, Bechara and colleagues (2002) set out to determine if the decision-making deficit associated with damage to the ventromedial (VM) PFC is caused mostly by unilateral right or left lesions. Utilizing rare patients with either right or left VM lesions, the left VM patients performed well on the Iowa Gambling Task (IGT) and were not impaired in real-life decision making as all were gainfully employed. In contrast, right VM patients were severely impaired in real-life decision making and performed as poorly on the IGT as bilaterally damaged patients. These results suggest the importance of considering laterality in hypotheses concerning a possible role for the OFC in addiction. Our reduced right/left OFC ratios in HR males may provide a neurological basis for poor impulse control and poorer decision-making capacity (Hill *et al.*, 2009b).

F. Cerebellum

Although the cerebellum has not been formally included in models of reinforcement such as that suggested by Koob and Kreek (2007), there is reason to believe that cerebellar involvement in compulsive behaviors such as those seen in OCD is quite likely. Neuroimaging of a consecutive series of adult OCD patients has revealed relative increases in gray matter volume in the anterior cerebellum (Pujol *et al.*, 2004). Functional imaging studies have been consistent in showing increased resting cerebral blood flow in the cerebellum of OCD patients (Busatto *et al.*, 2000; Nakao *et al.*, 2005). We have reported increased gray matter for cerebellum in offspring from multiplex AD families (Hill *et al.*, 2007b). Further comparison to the Pujol *et al.* (2004) study in which increased gray matter volume in anterior cerebellum were found shows that the OCD patients also exhibited smaller volumes of the amygdala in the right hemisphere. With similar findings in high-risk offspring for AD (Hill *et al.*, 2001, 2007b), it is possible that a neural diathesis for compulsivity may be present before individuals develop an addiction. If so, it may suggest that individuals at highest risk for developing an addiction may be more prone to compulsive behaviors that set the stage for disinhibited behavior, which fuels the drive to ignore negative reinforcers and continue the addicted behavior.

G. White Matter Volume, Fractional Anistropy, and Limbic Connectivity

Robust increases in WM are seen in adolescence through early young adulthood (Pfefferbaum *et al.*, 1994). This is also a period when age-related improvements in executive functioning are thought to occur. This is probably due to functional reorganization of the frontal cortex during this period along with improved functional WM connectivity within and between brain regions (Pfefferbaum *et al.*, 1994). Increased axonal diameter, thicker myelin sheaths, or better organization of tracts can lead to better signal conduction. Improved connectivity may alter behavioral functioning as well, decreasing impulsivity with age. Changes in behavior have been documented in association with WM microstructure. WM microstructure of the corpus callosum has been shown to be related to impulsivity in adolescence (Silveri *et al.*, 2006). Similarly, we have reported that WM volume of the OFC is related to impulsivity as measured by Multidimensional Personality Questionnaire (MPQ) scale scores (Hill *et al.*, 2009b).

Although WM volume has been investigated across adolescence and young adulthood (Giedd *et al.*, 1999; Paus *et al.*, 1999; Jernigan *et al.*, 1991; Pfefferbaum *et al.*, 1994; DeBellis *et al.*, 2001; Sowell *et al.*, 2002, 2003; Lenroot and Geidd, 2006; Gogtay *et al.*, 2004; Nagel *et al.*, 2006), extension of these developmental studies of WM volume to those varying by familial risk group status has not been undertaken very often. As previously shown in Fig. 1, analysis of WM volume of the OFC by hemisphere does show differing patterns by age and family risk for AD (Hill *et al.*, 2009b).

Magnetic resonance diffusion tensor imaging (DTI) is a method for quantifying brain WM microstructure using the sensitivity of DTI to the linear structure of WM (Sullivan and Pfefferbaum, 2006). Diffusion tensor image of a normal brain can be seen in Fig. 7. The scan shown illustrates the good resolution that is possible for collecting fractional anisotropy (FA) maps.

The diffusion of water molecules in brain tissue is anisotropic because motion is restricted by fibers. By quantifying the magnitude and orientation of diffusion, DTI can provide information about a structure. Anistropy measures can estimate the extent to which water is diffusing unidirectionally, as it diffuses more along

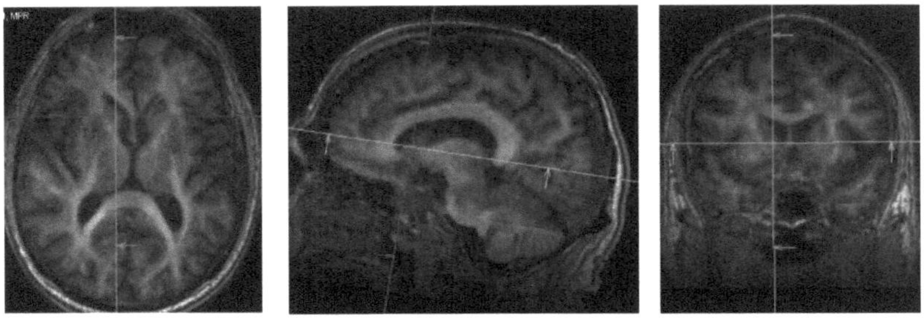

FIG. 7. Diffusion tensor image of a normal adult brain depicting the white matter tracts. Thirty diffusion directions were used to obtain the data for the measurement of the components of the diffusion tensor. DTI can be used to assess differing patterns of connectivity in relation to familial/ genetic predisposition to AD and related phenotypes.

the length of axonal fibers than perpendicular to them. FA is an intra-voxel measure that yields values between 0 (equal diffusion in all directions) and 1 (all in one direction, e.g., along an axon). Using DTI techniques, greater WM anistropy is seen with increasing age (Klingberg *et al.*, 1999; Schmithorst *et al.*, 2002; Barnea-Goraly *et al.*, 2005). Klingberg and colleagues reported greater WM anisotropy in adults than in children. Schmithorst *et al.* (2002) reported decreased trace (a measure of overall diffusivity) throughout WM and greater FA with increasing age (ages 5–18) in 33 subjects. Barnea-Goraly *et al.* (2005) found a linear increase in FA on a voxel-by-voxel basis in 6–19-year-old subjects ($N = 30$) in several regions including the prefrontal region, though many did not show concurrent increases in WM density. One significant cluster extended into the orbitofrontal gyrus. This suggests that the significant changes seen may indicate increased axonal diameter or axonal density, or increased myelination, all of which may indicate increased connectivity with other brain regions with development.

Although a number of developmental differences have been shown for WM volume and FA, few have considered familial risk group differences in their analyses. Studies of this type are essential to better understanding of altered functional connectivity between structures within specific circuits thought to influence addiction liability.

V. Brain Functional Differences—Familial Risk and Developmental Stage

Brain function has been addressed using neuropsychological tests, fMRI BOLD response, and electrophysiological response in those with exposure to a variety of substances. In understanding the underlying predisposition to AD susceptibility, the following review will focus on studies that have specifically contrasted those with and without a family history. Also, brain regions that have been implicated in addiction will be highlighted. Studies involving neuroimaging or neurophysiology will be the central focus of this section. Neuropsychological studies will be included in this review where specifically related to familial risk for AD.

A. fMRI Studies Related to Alcohol Dependence Susceptibility

1. *Amygdala*

Among the structures postulated to be part of the Reward and OCD circuits (Koob and Kreek, 2007), the amygdala, as part of the Reward circuitry, stands out as a structure with abundant fMRI paradigms that have variously been used to promote a BOLD response of the amygdala. These include presentation of stimuli designed to elicit an emotional response such as presentation of happy, sad, or angry faces (Hariri *et al.*, 2002) or stimuli as simple as presentation of single-letter stimulus with informative positive feedback (presentation of the word CORRECT) (Bischoff-Grethe, 2009). However, only one study has specifically addressed amygdala activation specifically in relation to familial risk for addiction (Glahn *et al.*, 2007). In that study, subjects viewed trios of faces (Ekman and Friesen, 1976) and geometric shapes as control stimuli. The family history negative (FHN) group showed greater percent signal change in response to faces than to geometric shapes, while the family history positive (FHP) individuals did not. Conclusions that can be drawn from that study must be tempered by the small number of subjects ($N=8$ per group). Nevertheless, the findings suggest that amygdala hyporesponsiveness seen in those with a family history of addiction (AD) may have implications for signaling from the amygdala to the hypothalamus and PFC, potentially preventing the formation of aversive associations to the negative consequences of alcohol use.

2. *Orbitofrontal Cortex*

The OFC proposed by Koob and Kreek (2007) as an integral part of the OCD neurocircuit that maintains the negatively reinforced aspects of long-term effects of addiction has been studied using a variety of functional imaging techniques. These studies have addressed functional effects in various addicted subject groups (e.g., alcohol, cocaine, methamphetamine). Of the 18 studies using this methodology, only 3 were found to not show involvement of this region (see Dom, 2005, for review).

The OFC has been a candidate region for addiction studies because it is thought to be involved in inhibitory decision-making processes (Bechara and

Damasio, 2002). However, only a few imaging studies have investigated BOLD activation during a decision-making task. These include studies of methamphetamine abusers showing reduced fMRI BOLD response in the right OFC region (Paulus *et al.*, 2002, 2003). The Paulus studies used a two-choice prediction task in which participants did not know at the onset of the experiment which action would lead to the best outcome, and a neutral control task (two-choice response task). In comparison to controls, the methamphetamine abusers showed less task-related activation in the OFC (Brodmann areas (BA) 10 and 11), the DLPFC (BA 9), and the anterior cingulate (BA 32) during the prediction task relative to the response task. These results suggest that drug-dependent individuals may have OFC dysfunction, which is the most prominent in tasks where choices are associated with uncertain outcomes.

Without task presentation, substance users in the period beyond acute withdrawal exhibit hypofunctionality of the OFC (see Dom *et al.*, 2005, for review). A few structural studies provide evidence of changes in OFC in cocaine-dependent subjects (Franklin *et al.*, 2002; Lim *et al.*, 2002; Matochik *et al.*, 2003). Our work with offspring from multiplex alcohol-dependent families shows reduced volume of the OFC in the right hemisphere in offspring that is independent of alcohol or drug exposure.

3. *Cerebellum*

Studies of cerebellar activation in groups with drug reinforcement histories are relatively uncommon. Chang *et al.* (2006) utilized a set of nonverbal visual attention tasks with variable levels of difficulty to show that relative to active marijuana users, abstinent marijuana users had normalized activation in the frontal and medial portions of the cerebellum. Schneider *et al.* (2001) investigated cue-induced craving in alcoholic patients before undergoing treatment reported finding activation in the cerebellum and in the right amygdala/hippocampus. Interestingly, administration of alcohol suppresses activity in frontal and parietal regions along with cerebellar nuclei, all of which are believed to be involved in internal cognitive models of motor representation and action (Van Horn *et al.*, 2006). Using a Go/No-Go response inhibition task, Hester and Garavan (2004) have demonstrated that cocaine-addicted individuals find it difficult to inhibit their behavior when the working memory load becomes greater and when cue-induced craving for the drug increases. Cerebellar activation identified during response inhibition showed increases in the cocaine users while controls showed decreases as the working memory load increased. Cerebellar vermis BOLD activation has been reported to be greater in crack cocaine users than comparison subjects (Anderson *et al.*, 2006). This is consistent with earlier studies showing cerebellar activation in users exposed to morphine cues (London *et al.*, 1990).

B. fMRI Studies Focusing on Traits Related to Alcohol Dependence

1. *Theory of Mind Task*

The clinical characteristics of alcohol-dependent persons suggest that many may have difficulties in emotional learning. It has been known for some time now that alcoholics show poorer social skills than nonalcoholics (Van Hasselt *et al.*, 1978). Consistent with this observation, alcohol-dependent individuals show impaired performance on emotional facial expression tasks which may be seen as long as 3 months after achieving abstinence (Foisy *et al.*, 2007).

Based on these observations, we hypothesized (Hill *et al.*, 2007a) that judgment of facial expression, being a neurobiologically based trait, might be altered in offspring from multiplex AD families. We expected that deficits observed in alcohol-dependent individuals might have preceded the onset of AD. Our hypothesis was based in part on the observation that adult alcoholics and their high-risk nonalcoholic adult relatives report higher levels of Alienation than do controls (Hill *et al.*, 1990b; McGue *et al.*, 1997). Persons scoring higher on the Alienation scale from the Multidimensional Personality Questionnaire (Tellegen, 1985; Tellegen *et al.*, 1988) report believing that they are mistreated, that others wish them harm, and often feel betrayed and used by "friends." Because there had been no fMRI studies of offspring from alcohol-dependent families to test for emotion recognition, we viewed this as an important goal. We chose the Reading the Mind in the Eyes Task modeled after the theory of mind (ToM) test to measure fMRI BOLD response to stimuli varying in emotional content. It was expected that differences in BOLD activation in regions previously identified by Baron-Cohen *et al.* (1999a) would show enhanced activation in response to processing of the Eyes Task. Regions that are typically activated by the Eyes Task include primarily the medial prefrontal cortex (mPFC), the temporal lobe, particularly the middle and superior regions, the left inferior frontal cortex, and amygdala (Baron-Cohen *et al.*, 1999).

Participants included eight individuals with multiplex family history of AD (four female and four male) and eight normal controls without a personal or family history (at least first-degree relatives) of AD or other psychiatric illness (four female and four male). Following the method described by Baron-Cohen *et al.* (1999), two tasks were presented that required the participants to extract visual information considered to be socially relevant. In one task the participant simply judged the gender of faces presented in the scanner. In the other, the subject was asked to judge the inferred mental state of persons shown in photographs (24 faces of men and women). The experiment alternated between three conditions. In Task A (Gender) the subject indicated whether the photograph viewed was male or female, and in Task B (Emotion) the subject indicated the inferred mental state of the person in the photograph, followed by a rest condition. For Task A, the subject was instructed to decide for each stimulus which of two simultaneously

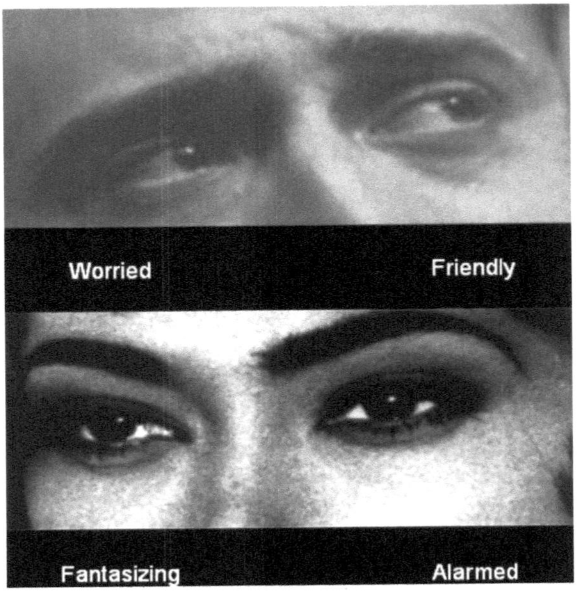

FIG. 8. An example of the eyes stimuli used in Task B. Participants were presented with a choice of mental state words that best described the mental state of the person whose eyes are depicted. The correct response is "worried" in the upper figure and "fantasizing" in the lower figure. Brain regions typically activated in the ToM task include middle temporal gyrus, superior frontal gyrus, and inferior frontal gyrus (see Baron-Cohen *et al.*, 1999a).

presented words, "male" or "female," best described the eyes presented. Response for the emotion condition (Task B) was a two-alternative forced choice (modal response vs. a randomly picked alternative) based on the original Baron-Cohen *et al.* (1999b) norms. For this task, the subject was required to make a forced-choice decision between two simultaneously presented words (Fig. 8).

BOLD fMRI analysis of the high risk ($n = 8$) versus low-risk comparison subjects ($n = 8$) was conducted at a preset threshold of uncorrected $p = .005$ with no extent threshold. An activation difference between the BOLD response to Task B (emotion task) and the rest epoch defined the individual subject contrast. These contrast images were then used to perform a between-group analysis. Using SPM5 running on MATLAB version 7.0.4, small volume correction analysis was applied to the regions showing overall uncorrected values of 0.001 or less and for which an *a priori* reason could be justified such as reports of significant BOLD activation in controls relative to clinical populations. Masks were created using the Wake Forest Pick Atlas (Maldjian *et al.*, 2003). The results may be seen in Fig. 9.

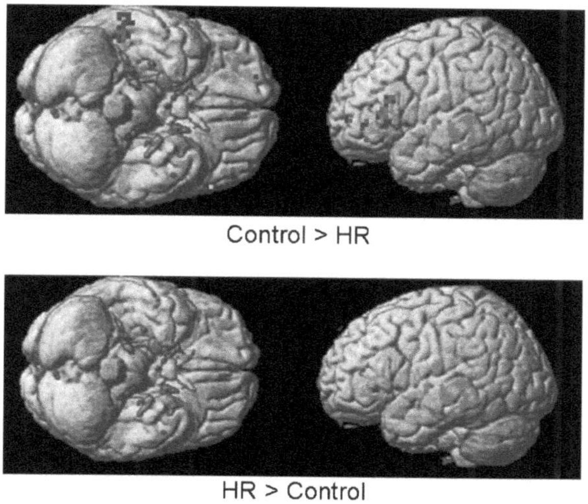

FIG. 9. Blood oxygen level–dependent (BOLD) fMRI analysis of high-risk-for-AD participants from multiplex families ($N = 8$) conducted at a preset threshold of uncorrrected $p = .005$ and no extent threshold. Note reduced activations in high-risk subjects in the right middle temporal gyrus, left middle frontal gyrus, and left inferior frontal gyrus.

As may be seen, diminished BOLD response was seen in the high-risk offspring relative to control participants in the right middle temporal gyrus (RMTG), right superior frontal gyrus (RSFG), and the left inferior frontal gyrus (LIFG) (uncorrected significance). Small volume correction analysis was performed for the three regions showing diminished BOLD response in the uncorrected analysis and for which there was *a priori* evidence to suggest ToM-associated activation. This analysis revealed significant reduction in BOLD response in the RMTG that survived our small volume correction and corrections for multiple comparisons, along with a trend for reduction in the RSFG and the LIFG. Also depicted is the Control minus High-Risk comparison. Reduced activations may be seen in HR subjects in left inferior frontal and right temporal regions.

This pilot fMRI study suggests that adolescent/young adult individuals at high genetic risk for developing AD (members of a multiplex for AD families) may inherit deficits in social intelligence that are reflected in reduced BOLD response in brain regions previously associated with performance of theory of mind tasks. These results showing reduced activation in areas previously associated with performance of ToM tasks suggest that High-Risk offspring may experience deficiencies in emotion recognition that may make it more difficult for them to interpret the actions of others in terms of mental states, such as their thoughts or beliefs, to empathize with them, and to predict how others will react or feel. These deficits may have important implications for life satisfaction as it relates to successful social skill and absence of an alienated personality style.

Importantly, the reduction in activation seen in regions associated with performance of the ToM task occurred in individuals who showed equivalent IQ and educational attainment to the controls. Also, the High- and Low-Risk subjects performed equally well on the task administered. This further suggests that the high-risk individuals were processing emotional material differently than controls though clearly unimpaired in their cognitive performance. This dissociation between cognitive performance and social intelligence has been addressed by Baron-Cohen *et al.* (1999a, 2001), noting that diminished social intelligence, including capacity for ToM tasks, can often be seen among individuals with higher levels of general intelligence such as mathematicians and scientists.

2. *Response Inhibition*

Disinhibition may be a characteristic of those with a family history of AD and/or may be the result of long-term use of alcohol. Schweinsburg *et al.* (2004) studied 12 FHP and 14 FHN youth between the ages of 12 and 14 years and contrasted their response to a Go/No-Go task during fMRI acquisition. The FHP youth appeared to show less BOLD response in frontal regions than did the FHN youth. These results appear to differ from those of Hester and Garavan (2004) for cocaine-addicted individuals. When cocaine-addicted individuals begin to find it difficult to inhibit their behavior as the working memory load increases and when cue-induced craving for the drug increases, cerebellar activation increases. In contrast, the controls showed decreases as the working memory load increased. However, the studies are not directly comparable because Schweinsburg *et al.* (2004) studied frontal regions and Hester and Garavan (2004) the cerebellum. Nevertheless, these studies are important in emphasizing the value of using fMRI paradigms to investigate traits or conditions thought to be related to risk for developing alcohol abuse and dependence and their related phenotypes.

3. *Alcohol Cue Reactivity*

Reactivity to alcohol cues in adult alcohol-dependent individuals has shown that those with a history of alcohol abuse respond differently than those without such a history. Two fMRI studies in which alcohol-related cues have been administered have reported that adolescents and young adults with a history of alcohol use disorders (AUDs) show greater BOLD response to alcohol-related words (Tapert *et al.*, 2004) and to pictures of alcoholic and nonalcoholic beverages (Tapert *et al.*, 2003). In the task involving presentation of alcohol-related words, alcohol-dependent women ($N = 8$) and female light social drinkers ($n = 9$) aged 18–24 were administered the cue reactivity paradigm during an fMRI assessment. Comparison of the response to alcohol-related words versus neutral words for the two groups revealed significantly more BOLD response in several regions (subcallosal, anterior cingulate, left prefrontal, and bilateral insular regions) for the alcohol-dependent women. In contrast, controls showed greater response to alcohol words in some right cortical regions. Similarly, presentation of alcohol-related pictures to adolescents with an AUD show greater BOLD activation in left anterior, limbic, and visual system areas in comparison to adolescents without a history of AUD. The greatest response was found in youth who reported greater craving for alcohol and a greater number of drinks consumed per month.

These studies illustrate the plasticity of brain functioning by demonstrating that experience using alcohol alters the response of specific brain regions. This phenomenon appears to be related to gender and age and degree of exposure and/or craving for alcohol. An important addition to studies of this type would be to vary the familial/genetic loading for AD. As was the case for individuals with a family history of AD who showed diminished BOLD response in the RMTG in the ToM task involving emotionally cued faces (Hill *et al.*, 2007a), possibly an even greater BOLD response to alcohol cues might be seen in those with a family history even before they begin drinking regularly.

VI. Brain Structural and Functional Effects and Genetic Variation

A. Structural Characteristics and Genes

A number of genes have been associated with the AD phenotype (see Hill, 2000, for review), but only a few genes have been investigated with respect to brain structures that appear to be good intermediate phenotypes for AD or related phenotypes. The serotonin transporter (5-HTT) regulates the reuptake of serotonin to the presynaptic neuron for recycling or degradation after serotonin has been released. One polymorphism that has been studied frequently is the serotonin transporter–linked promoter region (5-HTTLPR). Two primary variants have been identified, a 44-base-pair insertion (L allele) or deletion (S allele). Variation in 5-HTTLPR alleles is associated with variation in 5-HTT expression. Lymphoblast cell lines homozygous for the L allele have been reported to have a 2-fold increase in serotonin reuptake (Lesch *et al.*, 1996). Individuals who are homozygous for the L allele have been reported to have reduced hippocampal

(Frodl *et al.*, 2004, 2008) and amygdala volume (Scherk *et al.*, 2009). Also, alterations in the microstructure of frontal limbic WM tracts have been reported to be associated with 5-HTTLPR polymorphic variation (Pacheco *et al.*, 2009). Although associations between variation in 5-HTTLPR and structural aspects of the brain have been reported, studies involving alcohol-dependent persons or their offspring are rare.

Another gene that has been studied frequently with respect to brain morphometry is the brain-derived neurotrophic factor (BDNF) gene. BDNF is a member of the neurotrophin superfamily and plays a role in differentiation during development (Engelhardt *et al.*, 2007) and survival of neurons in the adult brain (Morse *et al.*, 1993). It has also been implicated in regulation of synaptic function (Lessmann *et al.*, 2003). One frequently occurring single nucleotide polymorphism (SNP) (rs6265) at nucleotide 196 (G/A) has been identified in the human BDNF gene. This SNP produces an amino acid substitution of valine to methionine at codon 66 (val66met) in the 5′-region. This polymorphism is responsible for decreased distribution of BDNF val66met to neuronal dendrites.

Variation in this polymorphism has been found to be related to volumetric differences in a number of brain structures. Reduced volume of the PFC (Pezawas *et al.*, 2004), hippocampus (Pezawas *et al.*, 2004; Bueller *et al.*, 2006; Frodl *et al.*, 2007), caudate and cerebellar vermis (Agartz *et al.*, 2006), OFC in the right hemisphere (Hill *et al.*, 2009b), and total gray matter of the temporal and occipital lobes (Ho *et al.*, 2006) has been reported. This gene is especially interesting because variation in this gene influences the behavioral effects of alcohol in laboratory animals (Janak *et al.*, 2006). As noted previously, a study in the author's laboratory demonstrated that offspring from families with multiple cases of AD differed from controls by having lesser volume of the OFC in the right hemisphere. The outline of the structure and its right and left components can be seen in Fig. 10. The volume of this structure was determined by using manual tracings guided by a software program, BRAINS2, which allows the operator to consistently draw regions of interest for multiple brain slices to arrive at brain volumes. The lesser amount of OFC volume in the right hemisphere relative to the left for high-risk individuals may be seen in Fig. 11. The difference is especially pronounced in high-risk males. The morphological data obtained were then studied with respect to variation in the 5-HTT and BDNF genes. Results of the analysis may be seen in Fig. 12. Here, it can be seen that there was interaction between the two polymorphisms. Genotyping variation in the 5-HTT and BDNF were then evaluated with respect to these OFC ratios. As may be seen, there was a significant association between the presence of the S allele of the 5-HTT gene, the Met allele of the VAL/MET variation of the BDNF gene, and volume of the OFC in the right hemisphere (Right − Left)/(Right + Left), corrected for intracranial volume (ICV), for the 87 participants. For total volume, the BDNF by 5-HTT by risk interaction was significant ($F = 5.55$, $df = 1, 77$, $p = .021$). This

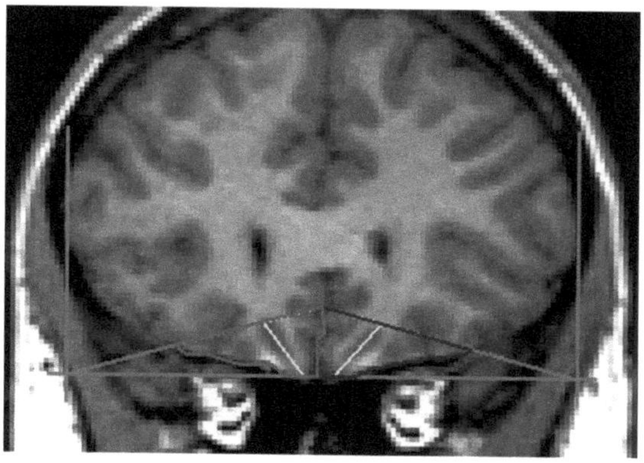

FIG. 10. The boundaries and landmarks for the OFC: right, left, and total. Outlines for the left OFC are seen in blue, with the right OFC in red. The yellow line depicts the lateral and medial portions of the OFC in each hemisphere (See color plate 1).

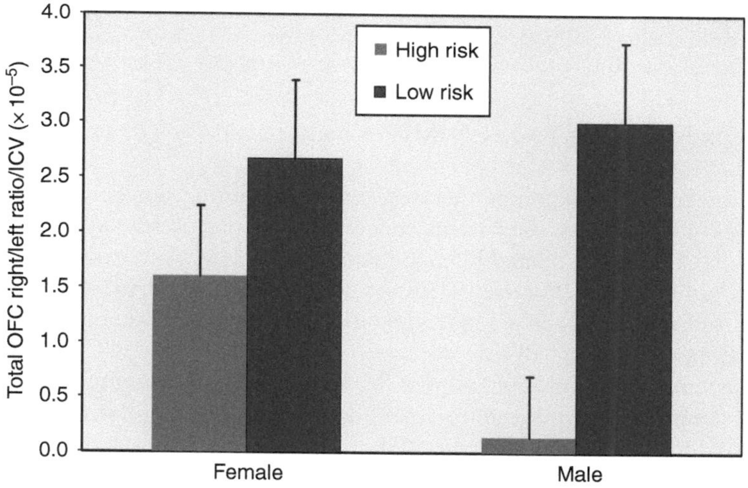

FIG. 11. OFC ratios were determined for each participant using the formula (Right − Left)/(Right + Left). Volumes were adjusted for intra-cranial volume before statistical analyses were performed. Depicted here are the adjusted means (adjustment for age) and standard deviations for the male and female High- and Low-Risk participants.

interaction was also significant for white ($F = 4.92$, $df = 1, 77$, $p = .03$) and for gray ($F = 4.16$, $df = 1, 77$, $p = .045$) volumes. An analysis restricted to the 57 High-Risk participants was also significant ($F = 5.31$, $df = 1, 51$, $p = .025$), without ICV correction, and with ICV correction ($F = 6.32$, $df = 1, 51$, $p = .015$).

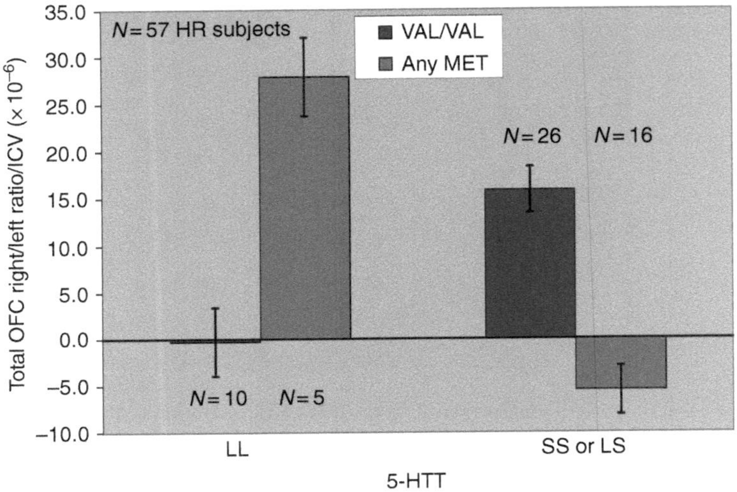

FIG. 12. A statistically significant association between the presence of the S allele of the 5-HTT gene, the Met allele of the VAL/Met variation of the BDNF gene, and volume of the OFC in the right hemisphere (Right − Left /Right + Left)(ICV corrected) was seen for the 87 participants genotyped.

B. Brain Function Assessed by fMRI Neuroimaging and Genes

Although there are numerous studies now suggesting functional differences among individuals who use psychoactive substances extensively or have a family history of AD and its related phenotypes, inclusion of genetic variation in these studies has not been addressed. This would appear to be a fruitful approach in future studies based on an emerging literature suggesting that genotypic variation is associated with fMRI BOLD response.

Functional neuroimaging studies have placed a major emphasis on two neurotransmitter systems (serotonin and dopamine). The emphasis on serotonin comes from animal and human studies which have implicated variation in serotonin (5-HT) neurotransmission as a major determinant in individual differences in trait characteristics, especially in the continuum of negative affect (Lucki, 1998). Pharmacological manipulation of the serotonergic system indicates a potential for modifying stress reactivity peripherally and for alteration of negative affect (Roth, 2006). Dopaminergic transmission is critical to motivational and reward-related functions of the brain including reinforcement learning (Tobler *et al.*, 2003) and decision making (Montague *et al.*, 2004). Because there is considerable variation among individuals with respect to these characteristics (Gonzalez and Wu, 1999), it has been suggested that genetic variation may be responsible (Trepel *et al.*, 2005). Individual variation in dopaminergic functioning has been suggested as a major contributing factor to personality traits (Benjamin *et al.*, 2000) and addiction (Volkow *et al.*, 1997).

There is evidence that genetic variation may predict alterations in BOLD response in the ventral striatum and PFC (Yacubian *et al.*, 2007), middle and inferior temporal gyrus, superior frontal gyrus and cerebellum (Boettiger *et al.*, 2007), and amygdala (Fisher *et al.*, 2009; Munafò *et al.*, 2008). Using a guessing task that is sensitive to reward-related activation in the PFC and ventral striatum, Yacubian *et al.* (2007) found that the dopamine transporter (DAT) and catechol *O*-methyltransferase (COMT) influences reward processing, as reflected in changes in BOLD activation. Because of the established relationship between reduced reward sensitivity and addiction, these findings suggest that genetic variation in these two genes is related to fMRI BOLD activation in the PFC and ventral striatum.

Additionally, a relationship between threat-related amygdala reactivity and variation in the serotonin 2A receptor gene (5-HT 2A) has been shown (Fisher *et al.*, 2009). This relationship was found using BOLD fMRI response and focusing on the density of 5-HT2A receptors in the mPFC. A remarkable 25–37% of variability in amygdala variability was explained by mPFC 2A density. Because there is feedback inhibition of the amygdala via the mPFC, finding genetic variation that can influence complex emotional behaviors that are influenced by this circuit is of some significance.

There are a number of studies that have related variation in the serotonin transporter (5-HTTLPR) genotype and amygdala activation (see Munafò *et al.*, 2008, for review). Munafò and colleagues conducted a meta-analysis of 17 studies in which the 5-HTTLPR polymorphism was genotyped and either fMRI or positron emission tomography (PET) was used. They concluded that the 5-HTTLPR polymorphism may contribute up to 10% in phenotypic variance in the BOLD response. However, they also concluded that most studies reviewed were not of sufficient size to provide adequate statistical power.

In summary, there are clear indications that genetic variation contributes to functional effects as measured by fMRI and PET. The challenge now is to bring these observations together with studies designed to assess variations in familial risk for AD to begin to uncover how altered brain functioning may influence susceptibility to AD and related phenotypes.

C. Brain Function Assessed with Electrophysiological Variants—Genetic Variation

Genetic association has been found between the amplitude of visual P300 in high-risk children and a marker locus in the D2 dopamine receptor on chromosome 11 (Taq1 A restriction fragment length polymorphism (RFLP) near the DRD2 receptor locus) (Hill *et al.*, 1998). Although this finding has not been replicated in normal young women (Lin *et al.*, 2001) or in patients with depression (Chen *et al.*, 2002), important differences across samples may have contributed to different outcomes. As noted previously in the discussion of P300 trajectories of childhood and adolescence, there are substantial differences in P300 amplitude by age and gender that can be expected to influence results. Search for genes associated or linked to P300 need to control for these important sources of variation.

Linkage analysis of P300 data from the Collaborative Study on the Genetics of Alcoholism (COGA) has revealed significant linkage on chromosomes 2, 5, 6, 13, and 17 using P300 recorded during a visual oddball task (Begleiter *et al.*, 1998). In more recent studies, the COGA group has reported on specific candidate genes in their linkage and linkage disequilibrium analyses. These have included study of a cholingeric receptor gene (CHRM2), a gene that encodes the muscarinic acetycholine receptor M2 (Jones *et al.*, 2006), a glutamate receptor gene (GRM8) (Chen *et al.*, 2008), and GABRA2 (Porjesz and Rangaswamy, 2007).

Although earlier studies from the COGA group had looked for linkage to the P300 component of the ERP wave, more recent research from this group has focused on event-related oscillations (EROs). The basis for this shift has been in the emphasis on the superposition of multiple EROs that comprise the ERP rather than emphasis on the ERP as a unitary transient phenomenon (Basar, 1980; Gruber *et al.*, 2005). This view continues to be open to debate (Yeung *et al.*, 2004; Makinen *et al.*, 2005). The P300 component is thought to be primarily composed of delta (1–3 Hz) and theta (4–7 Hz) band energy. The delta component is concentrated in more posterior regions while theta is more fronto-central (Basar-Eroglu *et al.*, 1992). The underlying multiple source generators of P300 can be studied using time–frequency distribution analysis as an alternative measure of P300 amplitude. The important point is that studies of the P300 component of ERP and its associated amplitude and study of EROs are not fundamentally different; rather, they are alternative forms of the same brain process that can be utilized to measure aspects of brain functioning.

In summary, ERPs or EROs provide an important source for understanding functional differences in individuals with familial/genetic risk for AD and to elucidate mechanisms involved in alcohol exposure. These electrophysiological variants provide a complementary method to fMRI techniques and are especially useful in addressing response to stimuli within 1 s after presentation. The P300 response occurs approximately 300 ms after presentation of a stimulus, though this can be somewhat later in children. Nevertheless, the typical ERP occurs within 1 s. Although fMRI BOLD can be measured within a few seconds after a stimulus is presented, it requires sufficient time for blood flow to redistribute following presentation. The typical fMRI response signal changes above baseline at approximately 2 s, growing to maximum value at about 5 s for a short stimulus (Huettel *et al.*, 2004).

VII. Conclusions

Alcoholism is a complex disease that is influenced by genetic and environmental factors and their interaction. The search for genes that may increase susceptibility to AD has been greatly facilitated by the recognition that intermediate phenotypes, sometimes referred to as endophenotypes, may be closer to the genetic variation than is the more complex AD phenotype.

Although the genetic makeup of the individual is fixed at birth, the response of the genome to environmental triggers (both external and internal) means that expression of genetic variation can be expected to vary across developmental stages. Consequently, major changes can be expected to occur by developmental stage, which in turn affect the brain structure and function. These structural and functional effects can be expected to have a major influence on the individual's risk for developing a variety of psychopathological conditions including AD. Development of sophisticated neurophysiological and neuroimaging techniques over the past several years has allowed us to identify neural circuits involved in addictive behaviors and provided a window on developmental changes in brain structure and function. Identification of neural plasticity within specific components of these addiction circuits that are related to genetic variation will offer numerous opportunities for treatment and intervention.

We can expect to see a growth in the number of studies addressing genetic influences on brain function in the future. Because it is clear that environmental variation influences genetic mechanisms, the challenge will be in identifying the relevant environmental triggers and specifying these by developmental stage. It is quite likely that critical periods will be identified where environmental influences will be more salient than at others. In addition, defining the relevant genetic variation will be a challenge in future studies. Because there are over 30,000 genes in the human genome, and many of these show complex temporal and spatial expression in the brain, selection of the most appropriate candidate genes for study is a daunting task. Moreover, the number of gene–gene interactions that might be responsible for a given mechanism is beyond the number that could be realistically assessed. It has been suggested that an alternative approach to this problem might be to sequence the entire genome of individuals from families where there is evidence that extraordinary genetic vulnerability is segregating (McClellan *et al.*, 2007). The idea behind this view comes from linkage studies where tracking genetic variants within families shows that some families show evidence that a given variant is highly linked to the psychiatric disorder of interest while others do not. In conventional linkage analysis, the linkage value for a study is the sum of the LOD scores obtained for each family. Experience with linkage studies shows that this variation is considerable even when endophenotypes are utilized rather than the more complex phenotype.

In the very near future, it is expected that the cost of sequencing and analyzing single personal genomes will be in the range of US$5000–10,000 (Akil *et al.*, 2010). Nevertheless, it would appear prohibitively expensive to do a large-scale epidemiological investigation using case-control methodologies. Recent reports from collaborative efforts to identify genetic loci associated with type 2 diabetes required genotyping over 200,000 separate individuals to identify five loci (Dupuis *et al.*, 2010; Saxena *et al.*, 2010). Even if the extraordinary cost of sequencing that many individuals is not considered, there would undoubtedly be an even greater cost associated with acquiring endophenotypic data such as MRI scans on such a scale. Finally, the issue of whether or not this approach would lead to useful scientific data is in question if as McClellan *et al.* (2007) suggest the common disease–rare allele model is the correct model for studying psychiatric disorders. They note that "a current mantra in schizophrenia genetic research is the need for ever-larger sample sizes in order to detect common small-effect variants" (Devon *et al.*, 2001). Quite possibly, this mantra has widened to include many psychiatric disorders including AD.

A highly appealing alternative is to target families with known ultrahigh loading for the disease of interest, such as those seen in multiplex families, a strategy that could provide an economical method for complete sequencing of targeted families, and to combine this data with structural and functional imaging and neurophysiological assessment.

References

Agartz, I., Momenan, R., Rawlings, R. R., Kerich, M. J., and Hommer, D. W. (1999). Hippocampal volume in patients with alcohol dependence. *Arch. Gen. Psychiatry* **56**, 356–363.

Agartz, I., Sedvall, G. C., Terenius, L., Kulle, B., Frigessi, A., Hall, H., and Jönsson, E. G. (2006). BDNF gene variants and brain morphology in schizophrenia. *Am. J. Med. Gen. Part B* **141**, 513–523.

Akil, H., Brenner, S., Kandel, E., Kendler, K. S., King, M.-C., Scolnick, E., Watson, J. D., and Zoghbi, H. Y. (2010). The future of psychiatric research: Genomes and neural circuits. *Science* **327**, 1580–1581.

Anderson, C. M., Maas, L. C., Frederick, B. D., Bendor, J. T., Spencer, T. J., Livni, E., Lukas, S. E., Fischman, A. J., Madras, B. K., Renshaw, P. F., and Kaufman, M. J. (2006). Cerebellar vermis involvement in cocaine-related behaviors. *Neuropsychopharmacology* **31**, 1318–1326.

Barnea-Goraly, N., Menon, V., Eckert, M., Tamm, L., Bammer, R., Karchemskiy, A., Dant, C. C., and Reiss, A. L. (2005). White matter development during childhood and adolescence: A cross-sectional diffusion tensor imaging study. *Cereb. Cortex* **15**, 1848–1854.

Baron-Cohen, S., Ring, H. A., Wheelwright, S., Bullmore, E. T., Brammer, M. J., Simmons, A., and Williams, S. C. R. (1999a). Social intelligence in the normal and autistic brain: An fMRI study. *Eur. J. Neurosci.* **11**, 1891–1898.

Baron-Cohen, S., Wheelwright, S., Stone, V., and Rutherford, M. (1999b). A mathematician, a physicist, and a computer scientist with Asperger syndrome: Performance on folk psychology and folk physics test. *Neurocase* **5**, 475–483.

Baron-Cohen, S., Wheelwright, S., Skinner, R., Martin, J., and Clubley, E. (2001). The autism spectrum quotient (AQ): Evidence from Asperger syndrome high functioning autism, males and females, scientists and mathematicians. *J. Autism Dev. Disord.* **31**, 5–17.

Basar, E. (1980). EEG–Brain Dynamics: Relation between EEG and Brain Evoked Potentials. Elsevier, New York, NY.

Basar-Eroglu, C., Basar, E., Demiralp, T., and Schurmann, M. (1992). P300-response: Possible psychophysiological correlates in delta and theta frequency channels. A review. *Int. J. Psychophysiol.* **13**, 161–179.

Bechara, A., and Damasio, H. (2002). Decision-making and addiction (part I): Impaired activation of the somatic states in substance dependent individuals when pondering decisions with negative future consequences. *Neuropsychologia* **40**, 1675–1689.

Begleiter, H., Porjesz, B., Bihari, B., and Kissin, B. (1984). Event-related brain potentials in boys at risk for alcoholism. *Science (Washington, DC, USA)* **225**, 1493–1496.

Begleiter, H., Porjesz, B., Reich, T., Edenberg, H. J., Goate, A., Blangero, J., Almasy, L., Foroud, T., van Eerdewegh, P., Polich, J., Rohrbaugh, J., Kuperman, S., Bauer, L. O., O'Connor, S. J., Chorlian, D. B., Li, T-K, Conneally, P. M., Hesselbrock, V., Rice, J., Schuckit, M., Cloninger, R., Nurnberger, J., Crowe, R., and Bloom, F. E. (1998). Quantitative trait loci analysis of human event-related brain potentials: P3 voltage. *Electroencephalogr. Clin. Neurophysiol.* **108**, 244–250.

Benjamin, J., Osher, Y., Kotler, M., Gritsenko, I., Nemanov, L., Belmaker, R. H., and Ebstein, R. P. (2000). Association between the tridimensional personality questionnaire (TPQ) traits and three functional polymorphisms: Dopamine receptor D4 (DRD4), sertonin transporter promoter region (5HTTLPR) and catechol *O*-methyltransferase (COMT). *Mol. Psychiatry* **5**, 96–100.

Berman, S. M., Whipple, S. C., Fitch, R. J., and Noble, E. P. (1993). P3 in young boys as a predictor of adolescent substance abuse. *Alcohol* **10**, 69–76.

Bueller, J. A., Aftab, M., Sen, S., Gomez-Hassan, D., Burmeister, M., and Zubieta, J. K. (2006). BDNF Val66Met allele is associated with reduced hippocampal volume in healthy subjects. *Biol. Psychiatry* **59**, 812–815.

Bird, A. (2007). Perceptions of epigenetics. *Nature* **447**, 396–398.

Bischoff-Grethe, A., Hazeltine, E., Bergren, L., Ivry, R. B., and Grafton, S. T. (2009). The influence of feedback valence in associative learning. *Neuroimage* **44**, 243–251.

Boettiger, C. A., Mitchell, J. M., Tavares, V. C., Robertson, M., Joslyn, G., D'Esposito, M., and Fields, H. L. (2007). Immediate reward bias in humans: Fronto-parietal networks and a role for the catechol-*O*-methyltransferase $158^{Val/Val}$ genotype. *J. Neurosci.* **27**, 14383–14391.

Botstein, D., and Risch, N. (2003). Discovering genotypes underlying human phenotypes: Past successes for Mendelian disease. *Nat. Genet.* **33**, 229–237.

Busatto, G. F., Zamignani, D. R., Buchpiguel, C. A., Garrido, G. E., Glabus, M. F., Rocha, E. T., Maia, A. F., Rosario-Campos, M. C., Campi Castro, C., Furuie, S. S., Gutierrez, M. A., McGuire, P. K., and Miguel, E. C. (2000). A voxel-based investigation of regional cerebral blood flow abnormalities in obsessive-compulsive disorder using single photon emission computed tomography (SPECT). *Psychiatry Res.* **99**, 15–27.

Caldwell, C. B., and Gottesman, I. I. (1991). Sex differences in the risk for alcoholism: A twin study. *Behav. Genet.* **21**, 563 .

Carlson, S. R., Iacono, W. G., and McGue, M. (2004). P300 amplitude in nonalcoholic adolescent twin pairs who become discordant for alcoholism as adults. *Psychophysiology* **41**, 841–844.

Casey, B. J., Tottenham, N., Liston, C., and Durston, S. (2005). Imaging the developing brain: What have we learned about cognitive development? *Trends Cogn. Sci.* **9**(3), 104–110.

Chakravarti, A. (1999). Population genetics – making sense out of sequence. *Nat. Genet.* **21**(Suppl.), 56–60.

Chang, L., Yakupov, R., Cloak, C., and Ernst, T. (2006). Marijuana use is associated with a reorganized visual-attention network and cerebellar hypoactivation. *Brain* **129**, 1096–1112.

Chen, A. C.H., Tang, Y., Rangaswamy, M., Wang, J. C., Almasy, L., Foroud, T., Edenberg, H. J., Hesselbrock, V., Nurnberger, J., Kuperman, S., O'Connor, S. J., Schuckit, M. A., Bauer, L. O., Tischfield, J., Rice, J. P., Bierut, L., Goate, A., and Porjesz, B. (2008). Association of single nucleotide polymorphisms in a glutamate receptor gene (GRM8) with theta power of event-related oscillations and alcohol dependence. *Am. J. Med. Genet., Part B* **150**, 359–368.

Chen, T. J., Yu, Y. W., Chen, J. Y., Wang, Y. C., Chen, M. C., Hong, C. J., and Tsai, S. J. (2002). Association analysis of two dopamine D2 receptor gene polymorphisms and p300 event-related potential in depressive patients. *Neuropsychobiology* **46**, 141–144.

Crone, E. A., and van der Molen, M. W. (2007). Development of decision making in school aged children and adolescents: Evidence from heart rate and skin conductance analysis. *Child Dev.* **78**, 1288–1301.

Crow, J. F. (2000). The origins, patterns and implications of human spontaneous mutation. *Nat. Rev. Genet.* **1**, 40–47.

DeBellis, M. D., Clark, D. B., Beers, S. R., Soloff, P. H., Boring, A. M., Hall, J., Kersh, A., and Keshavan, M. S. (2000). Hippocampal volume in adolescent-onset alcohol use disorders. *Am. J. Psychiatry* **57**, 737–744.

DeBellis, M. D., Keshavan, M. S., Beers, S. R., Hall, J., Frustaci, K., Masalehdan, A., Noll, J., and Boring, A. M. (2001). Sex differences in brain maturation during childhood and adolescence. *Cereb. Cortex* **11**, 552–557.

Devon, R. S., Anderson, S., Teague, P. W., Burgess, P., Kipari, T. M., Semple, C. A., Millar, J. K., Muir, W. J., Murray, V., Pelosi, A. J., Blackwood, D. H., and Porteous, D. J. (2001). Identification of polymorphisms within disrupted in schizophrenia 1 and disrupted in schizophrenia 2, and an investigation of their association with schizophrenai and bipolar affective disorder. *Psychiatr. Genet.* **11**, 71–78.

Dom, G., Sabbe, B., Hulstijn, W., and Van Den Brink, W. (2005). Substance use disorders and the orbitofrontal cortex: Systematic review of behavioural decision-making and neuroimaging studies. *Br. J. Psychiatry* **187**, 209–220.

Dupuis, J., Langenberg, C., Prokopenko, I., Saxena, R., Soranzo, N., Jackson, A. U., Wheeler, E., Glazer, N. L., Bouatia-Naji, N., Gloyn, A. L., Lindgren, C. M., Magi, R., Morris, A. P., Randall, J., Johnson, T., Elliott, P., Rybin, D., Thorleifsson, G., Steinthorsdottir, V., Henneman, P., Grallert, H., Dehghan, A., Hottenga, J. J., Franklin, C. S., Navarro, P., Song, K., Goel, A., Perry, J. R., Egan, J. M., Lajunen, T., Grarup, N., Sparso, T., Doney, A., Voight, B. F., Stringham, H. M., Li, M., Kanoni, S., Shrader, P., Cavalcanti-Proenca, C., Kumari, M., Qi, L., Timpson, N. J., Gieger, C., Zabena, C., Rocheleau, G., Ingelsson, E., An, P., O'Connell, J., Luan, J., Elliott, A., McCarroll, S. A., Payne, F., Roccasecca, R. M., Pattou, F., Sethupathy, P., Ardlie, K., Ariyurek, Y., Balkau, B., Barter, P., Beilby, J. P., Ben-Shlomo, Y., Benediktsson, R., Bennett, A. J., Bergmann, S., Bochud, M., Boerwinkle, E., Bonnefond, A., Bonnycastle, L. L., Borch-Johnsen, K., Bottcher, Y., Brunner, E., Bumpstead, S. J., Charpentier, G., Chen, Y. D., Chines, P., Clarke, R., Coin, L. J., Cooper, M. N., Cornelis, M., Crawford, G., Crisponi, L., Day, I. N., de Geus, E. J., Delplanque, J., Dina, C., Erdos, M. R., Fedson, A. C., Fischer-Rosinsky, A., Forouhi, N. G., Fox, C. S., Frants, R., Franzosi, M. G., Galan, P., Goodarzi, M. O., Graessler, J., Groves, C. J., Grundy, S., Gwilliam, R., Gyllensten, U., Hadjadj, S., Hallmans, G., Hammond, N., Han, X., Hartikainen, A. L., Hassanali, N., Hayward, C., Heath, S. C., Hercberg, S., Herder, C., Hicks, A. A., Hillman, D. R., Hingorani, A. D., Hofman, A., Hui, J., Hung, J., Isomaa, B., Johnson, P. R., Jorgensen, T., Jula, A., Kaakinen, M., Kaprio, J., Kesaniemi, Y. A., Kivimaki, M., Knight, B., Koskinen, S., Kovacs, P., Kyvik, K. O., Lathrop, G. M., Lawlor, D. A., Le Bacquer, O., Lecoeur, C., Li, Y., Lyssenko, V., Mahley, R., Mangino, M., Manning, A. K., Martinez-Larrad, M. T., McAteer, J. B., McCulloch, L. J., McPherson, R., Meisinger, C., Melzer, D., Meyre, D., Mitchell, B. D., Morken, M. A., Mukherjee, S., Naitza, S., Narisu, N., Neville, M. J., Oostra, B. A., Orru, M., Pakyz, R., Palmer, C. N., Paolisso, G., Pattaro, C.,

Pearson, D., Peden, J. F., Pedersen, N. L., Perola, M., Pfeiffer, A. F., Pichler, I., Polasek, O., Posthuma, D., Potter, S. C., Pouta, A., Province, M. A., Psaty, B. M., Rathmann, W., Rayner, N. W., Rice, K., Ripatti, S., Rivadeneira, F., Roden, M., Rolandsson, O., Sandbaek, A., Sandhu, M., Sanna, S., Sayer, A. A., Scheet, P., Scott, L. J., Seedorf, U., Sharp, S. J., Shields, B., Sigurethsson, G., Sijbrands, E. J., Silveira, A., Simpson, L., Singleton, A., Smith, N. L., Sovio, U., Swift, A., Syddall, H., Syvanen, A. C., Tanaka, T., Thorand, B., Tichet, J., Tonjes, A., Tuomi, T., Uitterlinden, A. G., van Dijk, K. W., van Hoek, M., Varma, D., Visvikis-Siest, S., Vitart, V., Vogelzangs, N., Waeber, G., Wagner, P. J., Walley, A., Walters, G. B., Ward, K. L., Watkins, H., Weedon, M. N., Wild, S. H., Willemsen, G., Witteman, J. C., Yarnell, J. W., Zeggini, E., Zelenika, D., Zethelius, B., Zhai, G., Zhao, J. H., Zillikens, M. C., DIAGRAM Consortium, GIANT Consortium, Global BPgen Consortium, Borecki, I. B., Loos, R. J., Meneton, P., Magnusson, P. K., Nathan, D. M., Williams, G. H., Hattersley, A. T., Silander, K., Salomaa, V., Smith, G. D., Bornstein, S. R., Schwarz, P., Spranger, J., Karpe, F., Shuldiner, A. R., Cooper, C., Dedoussis, G. V., Serrano-Rios, M., Morris, A. D., Lind, L., Palmer, L. J., Hu, F. B., Franks, P. W., Ebrahim, S., Marmot, M., Kao, W. H., Pankow, J. S., Sampson, M. J., Kuusisto, J., Laakso, M., Hansen, T., Pedersen, O., Pramstaller, P. P., Wichmann, H. E., Illig, T., Rudan, I., Wright, A. F., Stumvoll, M., Campbell, H., Wilson, J. F., Anders Hamsten on behalf of Procardis Consortium, MAGIC investigators, Bergman, R. N., Buchanan, T. A., Collins, F. S., Mohlke, K. L., Tuomilehto, J., Valle, T. T., Altshuler, D., Rotter, J. I., Siscovick, D. S., Penninx, B. W., Boomsma, D. I., Deloukas, P., Spector, T. D., Frayling, T. M., Ferrucci, L., Kong, A., Thorsteinsdottir, U., Stefansson, K., van Duijn, C. M., Aulchenko, Y. S., Cao, A., Scuteri, A., Schlessinger, D., Uda, M., Ruokonen, A., Jarvelin, M. R., Waterworth, D. M., Vollenweider, P., Peltonen, L., Mooser, V., Abecasis, G. R., Wareham, N. J., Sladek, R., Froguel, P., Watanabe, R. M., Meigs, J. B., Groop, L., Boehnke, M., McCarthy, M. I., Florez, J. C., and Barroso, I. (2010). New genetic loci implicated in fasting glucose homeostasis and their impact on type 2 diabetes risk. *Nat. Genet.* **42**, 105–116.

Drevets, W. C. (2000). Functional anatomical abnormalities in limbic and prefrontal cortical structures in major depression. *Prog. Brain Res.* **126**, 413–431.

Ekman, P., and Friesen, W. V. (1976) Pictures of Facial Affect. Consulting Psychologists Press, Palo Alto, CA.

Engelhardt, M., Di Cristo, G., Berardi, N., and Wahle, P. (2007). Differental effects of NT-4, NGF and BDNF on development of neurochemical architecture and cell size regulation in rat visual cortex during the critical period. *Eur. J. Neurosci.* **25**, 529–540.

Fisher, P. M., Meltzer, C. C., Price, J. C., Coleman, R. L., Ziolko, S. K., Becker, C., Moses-Kolko, E. L., Berga, S. L., and Hariri, A. R. (2009). Medial prefrontal cortex 5-HT_{2A} density is correlated with amygdala reactivity, response habituation, and functional coupling. *Cereb. Cortex* **19**, 2499–2507.

Foisy, M.-L., Kornreich, C., Fobe, A., D'Hondt, L., Pelc, I., Hanak, C., Verbanck, P., and Phillippot, P. (2007). Impaired emotional facial expression recognition in alcohol dependence: Do these deficits persist in abstinence. *Alcohol. Clin. Exp. Res.* **31**, 404–410.

Franklin, T. R., Acton, P. D., Maldjian, J. A., Gray, J. D., Croft, J. R., Dackis, C. A., O'Brien, C. P., and Childress, A. R. (2002). Decreased gray matter concentration in the insular, orbitofrontal, cingulate, and temporal cortices of cocaine patients. *Biol. Psychiatry* **51**, 134–142.

Frodl, T., Meisenzahl, E. M., Zill, P., Baghai, T., Rujescu, D., Leinsinger, G., Bottlender, R., Schüle, C., Zwanzger, P., Engel, R. R., Rupprecht, R., Bondy, B., Reiser, M., and Möller, H.-J. (2004). Reduced hippocampal volumes associated with the long variant of the serotonin transporter polymorphism in major depression. *Arch. Gen. Psychiatry* **61**, 177–183.

Frodl, T., Schule, C., Schmitt, G., Born, C., Baghai, T., Zill, P., Bottlender, R., Rupprecht, R., Bondy, B., Reiser, M., Möller, H.-J., and Meisenzahl, E. M. (2007). Association of the brain-derived neurotrophic factor Val66Met polymorphism with reduced hippocampal volumes in major depression. *Arch. Gen. Psychiatry* **64**, 410–416.

Frodl, T., Zill, P., Baghai, T., Schüle, C., Rupprecht, R., Zetzsche, T., Bondy, B., Reiser, M., Möller, H.-J., Meisenzahl, E. M. (2008). Reduced hippocampal volumes associated with the long variant of the tri- and diallelic serotonin transporter polymorphism in major depression. *Am. J. Med. Gen. Part B* **147**, 1003–1007.

Giedd, J. N., Blumenthal, J., Jeffries, N. O., Castellanos, F. X., Liu, H., Zijdenbos, A., Paus, T., Evans, A. C., and Rapoport, J. L. (1999). Brain development during childhood and adolescence: A longitudinal MRI study. *Nat. Neurosci.* **2**, 861–863.

Giedd, J. N., Snell, J. W., Lange, N., Rajapakse, J. C., Casey, B. J., Kozuch, P. L., Vaituzis, A. C., Vauss, Y. C., Hamburger, S. D., Kaysen, D., and Rapoport, J. L. (1996a). Quantitative magnetic resonance imaging of human brain development: Ages 4–18. *Cereb. Cortex* **6**, 551–560.

Giedd, J. N., Vaituzis, A. C., Hamburger, S. D., Lange, N., Rajapakse, J. C., Kaysen, D., Vauss, Y. C., and Rapoport, J. L. (1996b). Quantitative MRI of the temporal lobe, amygdala, and hippocampus in normal human development: Ages 4–18. *J. Comp. Neurol.* **366**, 223–230.

Glahn, D. C., Lovallo, W. R., and Fox, P. T. (2007). Reduced amygdala activation in young adults at high risk of alcoholism: Studies from the Oklahoma Family Health Patterns Project. *Biol. Psychiatry* **61**, 1306–1309.

Gogtay, N., Giedd, J. N., Lusk, L., Hayashi, K. M., Greenstein, D., Vaituzis, A. C., Nugent, T. F., 3rd., Herman, D. H., Clasen, L. S., Toga, A. W., Rapoport, J. L., and Thompson, P. M. (2004). Dynamic mapping of human cortical development during childhood through early adulthood. *Proc. Natl. Acad. Sci. USA* **101**, 8174–8179.

Goldstein, D. B., Cavalleri, G. L., and Ahmadi, K. R. (2003). The genetics of common diseases: 10 million times as hard. *Cold Spring Harb Symp Quant Biol* **68**, 395–402.

Goldstein, R. Z., and Volkow, N. D. (2002). Drug addiction and its underlying neurobiological basis: Neuroimaging evidence for the involvement of the frontal cortex. *Am. J. Psychiatry* **159**, 1642–1652.

Gonzalez, R., and Wu, G. (1999). On the shape of the probability weighting function. *Cogn. Psychol.* **38**, 129–166.

Gottesman, I. I., and Gould, T. D. (2003). The endophenotype concept in psychiatry: Etymology and strategic intentions. *Am. J. Psychiatry* **160**, 636–645.

Gottesman, I. I., and Shields, J. (1982). Schizophrenia: The Epigenetic Puzzle. Cambridge University Press, Cambridge, UK.

Grant, B. F., and Dawson, D. A. (1997). Age at onset of alcohol use and its association with DSM-IV alcohol abuse and dependence: Results from the national longitudinal alcohol epidemiologic survey. *J. Subst. Abuse* **9**, 103–110.

Gruber, W. R., Klimesch, W., Sauseng, P., and Doppelmayr, M. (2005). Alpha phase synchronization predicts P1 and N1 latency and amplitude size. *Cereb. Cortex* **15**, 371–377.

Hariri, A. R., Tessitore, A., Mattay, V. S., Fera, F., and Weinberger, D. R. (2002). The amygdala response to emotional stimuli: A comparison of faces and scenes. *Neuroimage* **17**, 317–323.

Hawkins, J. D., Graham, J. W., Maguin, E., Abbott, R., Hill, K. G., and Catalano, R. F. (1997). Exploring the effects of age of alcohol use initiation and psychosocial risk factors on subsequent alcohol misuse. *J. Stud. Alcohol* **58**, 280–290.

Heath, A. C., Bucholz, K. K., Madden, P. A.F., Dinwiddie, S. H., Slutske, W. S., Bierut, L. J., Statham, D. J., Dunne, M. P., Whitfield, J. B., and Martin, N. G. (1997). Genetic and environmental contributions to alcohol dependence risk in a national twin sample: Consistency of findings in women and men. *Psychol. Med.* **27**, 1381–1396.

Hester, R., and Garavan, H. (2004). Executive dysfunction in cocaine addiction: Evidence for discordant frontal, cingulate, and cerebellar activity. *J. Neurosci.* **24**, 11017–11022.

Hill, S. Y. (2000). Biologic phenotypes associated with individuals at high risk for developing alcohol-related disorders. *Addict. Biol.* **5**, 5–22.

Hill, S. Y., DeBellis, M. D., Keshavan, M. S., Lowers, L., Shen, S., Hall, J., and Pitts, T. (2001). Right amygdala volume in adolescent/young adult offspring from families at high risk for developing alcoholism. *Biol. Psychiatry* **49**, 894–905.

Hill, S. Y., Kostelnik, B., Holmes, B., Goradia, D., McDermott, M. D., Diwadkar, V., and Keshavan, M. S. (2007a). fMRI BOLD response to the eyes task in offspring from multiplex alcohol dependence families. *Alcohol. Clin. Exp. Res.* **31**, 2028–2035.

Hill, S. Y., Locke, J., Lowers, L., and Connolly, J. (1999a). Psychopathology and achievement in children at high risk for developing alcoholism. *J. Am. Acad. Child. Adolesc. Psychiatry* **38**, 883–891.

Hill, S. Y., Locke, J., Zezza, N., Kaplan, B., Neiswanger, K., Steinhauer, S. R., Wipprecht, G., and Xu, J. (1998). Genetic association between reduced P300 amplitude and the DRD2 dopamine receptor A1 allele in children at high risk for alcoholism. *Biol. Psychiatry* **43**, 40–51.

Hill, S. Y., Muddasani, S., Prasad, K., Steinhauer, S., Scanlon, J., McDermott, M., and Keshavan, M. (2007b). Cerebellar volume in offspring from multiplex alcohol dependence families. *Biol. Psychiatry* **61**, 41–47.

Hill, S. Y., Muka, D. H. (1996). Childhood psychopathology in children from families of alcoholic female probands. *J. Am. Acad. Child Adolesc. Psychiatry* **35**, 725–733.

Hill, S. Y., Muka, D. H., Steinhauer, S. R., and Locke, J. (1995). P300 amplitude decrements in children from families of alcoholic female probands. *Biol. Psychiatry* **38**, 622–632.

Hill, S. Y., and Shen, S. (2002). Patterns of visual P3b in association with familial risk and childhood diagnosis. *Biol. Psychiatry* **51**, 621–631.

Hill, S. Y., Shen, S., Locke-Wellman J., and Lowers, L. (2008). Psychopathology in childhood and adolescence: A prospective study of offspring from multiplex alcoholism families. *Psychiatry Res.* **160**, 155–166.

Hill, S. Y., Shen, S., Locke, J., Steinhauer, S. R., Konicky, C., Lowers, L., and Connolly, J. (1999b). Developmental delay in P300 production in children at high risk for developing alcohol-related disorders. *Biol. Psychiatry* **46**, 970–981.

Hill, S. Y., Shen, S., Lowers, L., and Locke, J. (2000). Factors predicting the onset of adolescent drinking in families at high-risk for developing alcoholism. *Biol. Psychiatry* **48**, 265–275.

Hill, S. Y., Shen, S., Zezza, N., Hoffman, E. K., Perlin, M., and Allan, W. (2004). A genome-wide search for alcoholism susceptibility genes. *Am. J. Med. Genet. Part B* **128**, 102–113.

Hill, S. Y., and Steinhauer, S. R. (1993). Assessment of prepubertal and postpubertal boys and girls at risk for developing alcoholism with P300 from a visual discrimination task. *J. Stud. Alcohol* **54**, 350–358.

Hill, S. Y., Steinhauer, S. R., Locke-Wellman, J., and Ulrich, R. (2009a). Childhood risk factors for young adult substance dependence outcome in offspring from multiplex alcohol dependence families. *Biol. Psychiatry* **66**, 750–757.

Hill, S. Y., Steinhauer, S. R., Park, J., and Zubin, J. (1990a). Event-related potential characteristics in children of alcoholics from high density families. *Alcohol. Clin. Exp. Res.* **14**, 6–16. (Reprinted in *Ann. Rev. of Addict. Res. Treatment* 177–192, 1992.)

Hill, S. Y., Wang, S., Kostelnik, B., Carter, H., Holmes, B., McDermott, M., Zezza, N., Stiffler, S., and Keshavan, M. S. (2009b). Disruption of orbitofrontal cortex laterality in offspring from multiplex alcohol dependence families. *Biol. Psychiatry* **65**, 129–136.

Hill, S. Y., and Yuan, H. (1999). Familial density of alcoholism and onset of adolescent drinking. *J. Stud. Alcohol* **60**, 7–17.

Hill, S. Y., Zubin, J., and Steinhauer, S. R. (1990b). Personality resemblance in relatives of male alcoholics: A comparison with families of male control cases. *Biol. Psychiatry* **27**, 1305–1322.

Ho, B. C., Millev, P., O'Leary, D. S., Librant, A., Andreasen, N. C., and Wassink, T. H. (2006). Cognitive and magnetic resonance imaging brain morphometric correlates of brain neurotrophic factor val66met gene polymorphism in patients with schizophrenia and healthy volunteers. *Arch. Gen. Psychiatry* **63**, 731–740.

Huettel, S. A., Song, A. W., and McCarthy, G. (2004). Functional Magnetic Resonance Imaging. Sinauer Associates, Sunderland, MA.

Iacono, W. G., and McGue, M. (2006). Association between P3 event-related brain potential amplitude and adolescent problem behavior. *Psychophysiology* **43**, 465–469.

Janak, P. H., Wolf, F. W., Herberlein, U., Pandey, S. C., Logrip, M. L., and Ron, D. (2006). BIG news in alcohol addiction: New findings on growth factor pathways BDNF, insulin, and GDNF. *Alcoholism: Clin. and Exper. Res.* **30**, 214–221.

Jellinek, E. M. (1960). The Disease Concept of Alcoholism. Hillhouse, New Haven, CT.

Jernigan, T. L., Trauner, D. A., Hesselink, J. R., and Tallal, P. A. (1991). Maturation of human cerebrum observed *in vivo* during adolescence. *Brain* **114**, 2037–2049.

Jones, K. A., Porjesz, B., Almasy, L., Bierut, L., Dick, D., Goate, A., Hinrichs, A., Rice, J. P., Wang, J. C., Bauer, L. O., Crowe, R., Foroud, T., Hesselbrock, V., Kuperman, S., Nurnberger, J. Jr., O'Connor, S. J., Rohrbaugh, J., Schuckit, M. A., Tischfield, J., Edenberg, H. J., and Begleiter, H. (2006). A cholinergic receptor gene (CHRM2) affects event-related oscillations. *Behav. Genet.* **36**, 627–639.

Kandel, D. B., Yamaguchi, K., and Chen, K. (1992). Stages of progression in drug involvement from adolescence to adulthood: Further evidence for the gateway theory. *J. Stud. Alcohol* **53**, 447–457.

Kendler, K. S., Heath, A. C., Neale, M. C., Kessler, R. C., and Eaves, L. J. (1992). A population-based twin study of alcoholism in women. *JAMA* **268**, 1877–1882.

Kessler, R. C., Crum, R. M., Warner, L. A., Nelson, C. B., Schulenberg, J., and Anthony, J. (1997). Lifetime co-occurrence of DSM-III-R alcohol abuse and dependence with other psychiatric disorder in the national comorbidity survey. *Arch. Gen. Psychiatry* **54**, 3123–3321.

Klingberg, T., Vaidya, C. J., Gabrieli, J. D.E., Moseley, M. E., and Hedehus, M. (1999). Myelination and organization of the frontal white matter in children: A diffusion tensor MRI study. *Neuroreport* **10**, 2817–2821.

Koob, G., and Kreek, M. J. (2007). Stress, dysregulation of drug reward pathways, and the transition to drug dependence. *Am. J. Psychiatry* **164**, 1149–1159.

Kuperman, S., Chan, G., Kramer, J. R., Bierut, L., Bucholz, K. K., Fox, L., Hesselbrock, V., Nurnberger, J. I. Jr., Reich, T., Reich, W., and Schuckit, M. A. (2005). Relationship of age of first drink to child behavioral problems and family psychopathology. *Alcohol. Clin. Exp. Res.* **29**, 1869–1876.

Lenroot, R. K., and Giedd, J. N. (2006). Brain development in children and adolescents: Insights from anatomical magnetic resonance imaging. *Neuroscience and Biobehavioral Reviews* **30**, 718–729.

Lesch, K. P., Benegal, D., Heils, A., Sabol, S. Z., Greenberg, B. D., Petri, S., Benjamin, J., Müller, C. R., Hammer, D. H., and Murphy, D. L. (1996). Association of anxiety-related traits with a polymorphism in the serotonin transporter gene regulatory region. *Science (Washington, DC, USA)* **29**, 1527–1531.

Lessmann, V., Gottmann, K., and Malcangio, M. (2003). Neurotrophin secretion: Current facts and future prospects. *Prog. Neurobiol. (Amsterdam, Neth.)* **69**, 341–374.

Lim, K. O., Choi, S. J., Pomara, N., Wolkin, A., and Rotrosen, J. P. (2002). Reduced frontal white matter integrity in cocaine dependence: A controlled diffusion tensor imaging study. *Biol. Psychiatry* **51**, 890–895.

Lin, C. H., Yu, Y. W., Chen, T. J., Tsa, S. J., and Hong, C. J. (2001). Association analysis for dopamine D2 receptor Taq1 polymorphism with P300 event-related potential for normal young females. *Psychiatr. Genet.* **11**, 165–168.

London, E. D., Broussolle, E. P., Links, J. M., Wong, D. F., Cascella, N. G., Dannals, R. F., Sano, M., Herning, R., Snyder, F. R., Rippetoe, L. R., Toung, T. J.K., Jaffe, J. H., Wagner, H. N. Jr. (1990). Morphine-induced metabolic changes in human brain. Studies with positron emission tomography and [fluorine 18] fluorodeoxyglucose. *Arch. Gen. Psychiatry* **47**, 73–81.

Lucki, I. (1998). The spectrum of behaviors influenced by serotonin. *Biol. Psychiatry* **44**, 151–162.

Lundqvist, C., Ailing, C., Knoth, R., and Volk, B. (1995). Intermittent ethanol exposure of adult rats: Hippocampal cell loss after one month of treatment. *Alcohol* **30**, 737–748.

Makinen, V., Tiitinen, H., and May, P. (2005). Auditory event-related responses are generated independently of ongoing brain activity. *Neuroimage* **24**, 961–968.

Maldjian, J. A., Laurienti, P. J., Kraft, R. A., and Burdette, J. H. (2003). An automated method for neuroanatomic and cytoarchitectonic atlas-based interrogation of fMRI data sets. *Neuroimage* **19**, 1233–1239.

Matochik, J. A., London, E. D., Eldreth, D. A., Cadet, J. L., and Bolla, K. I. (2003). Frontal cortical tissue composition in abstinent cocaine abusers: A magnetic resonance imaging study. *Neuroimage* **19**, 1095–1102.

McClellan, J. M., and King, M.-C. (2010). Genetic heterogeneity in human disease. *Cell* **141**, 210–217.

McClellan, J. M., Susser, E., and King, M.-C. (2007). Schizophrenia: A common disease caused by multiple rare alleles. *Br. J. Psychiatry* **190**, 194–199.

McGue, M., Iacono, W. G., Legrand, L. N., and Elkins, I. (2001a). Origins and consequences of age at first drink: II. Familial risk and heritability. *Alcohol. Clin. Exp. Res.* **25**, 1166–1173.

McGue, M., Iacono, W. G., Legrand, L. N., Malone, S., and Elkins, I. (2001b). Origins and consequences of age at first drink: I. Associations with substance-use disorders, disinhibitory behavior and psychopathology, and P3 amplitude. *Alcohol. Clin. Exp. Res.* **25**, 1156–1165.

McGue, M., Slutske, W., Taylor, J., and Iacono, W. G. (1997). Personality and substance use disorders: I. Effects of gender and alcoholism subtype. *Alcohol. Clin. Exp. Res.* **21**, 513–520.

Medina, K. L., Schweinsburg, A. D., Cohen-Zion, M., Nagel, B. J., Tapert, S. F., (2007). Effects of alcohol and combined marijuana and alcohol use during adolescence on hippocampal volume and asymmetry. *Neurotoxicol. Teratol.* **29**, 141–152.

Merikangas, K. R., Stevens, D. E., Fenton, B., Stolar, M., O' Malley, S., Woods, S. W., and Risch, N. (1998). Comorbidity and familial aggregation of alcoholism and anxiety disorders. *Psychol. Med.* **28**, 773–788.

Merikangas, K. R., and Risch, N. (2003). Genomic priorities and public health. *Science* **302**, 599–601.

Merlo, C. A., and Boyle, M. P. (2003). Modifier genes in cystic fibrosis lung disease. *J. Lab. Clin. Med.* **141**, 237–241.

Montague, P. R., Hyman, S. E., and Cohen, J. D. (2004). Computational roles for dopamine in behavioural control. *Nature* **431**, 760–767.

Morse, J. K., Wiegand, S. J., Anderson, K., You, Y., Cai, N., Carnahan, J., Miller, J., DiStefano, P. S., Alter, C. A., and Lindsay, R. M. (1993). Brain-derived neurotrophic factor (BDNF) prevents the degeneration of medial septal cholinergic neurons following fimbria transaction. *J. Neurosci.* **13**, 4146–4156.

Munafò, M. R., Brown, S. M., and Hariri, A. R. (2008). Serotonin transporter (5-HTTLPR) genotype and amygdala activation: A meta-analysis. *Biol. Psychiatry* **63**, 852–857.

Murray, C. J., and Lopez, A. D. (1997). Global mortality, disability and the contribution of risk factors. *Lancet* **349**, 1436–1442.

Nagel, B. J., Medina, K. L., Yoshii, J., Schweinsburg, A. D., Moadab, I., and Tapert, S. F., (2006). Age-related changes in prefrontal white matter volume across adolescence. *Neuroreport* **17**, 1427–1431.

Nakao, T., Nakagawa, A., Yoshiura, T., Nakatani, E., Nabeyama, M., Yoshizato, C., Kudoh, A., Tada, K., Yoshioka, K., Kawamoto, M., Togao, O., and Kanba, S. (2005). Brain activation of patients with obsessive-compulsive disorder during neuropsychological and symptom provocation tasks before and after symptom improvement: A functional magnetic resonance imaging study. *Biol. Psychiatry* **57**, 901–910.

Ohannessian, C. M., Hesselbrock, V. M., Kramer, J., Kuperman, S., Bucholz, K. K., Schuckit, M. A., and Nurnberger, J. I. (2004). The relationship between parental alcoholism and adolescent psychopathology: A systematic examination of parental comorbid psychopathology. *J. Abnorm. Child Psychol.* **32**, 519–533.

Pacheco, J., Beevers, C. G., Benavides, C., McGeary, J., Stice, E., and Schnyer, D. M. (2009). Frontal-limbic white matter pathway associations with the serotonin transporter gene promoter region (5-HTTLPR) polymorphism. *J. Neurosci.* **29**, 6229–6233.

Paulus, M. P., Hozack, N., Frank, L., Brown, G. G., and Schuckit, M. A. (2003). Decision making by methamphetamine-dependent subjects is associated with error-related-independent decrease in prefrontal and parietal activation. *Biol. Psychiatry* **53**, 65–74.

Paulus, M. P., Hozack, N. E., Zauscher, B. E., Frank, L., Brown, G. G., Braff, D. L., and Schuckit, M. A. (2002). Behavioral and functional neuroimaging evidence for prefrontal dysfunction in methamphetamine dependent subjects. *Neuropsychopharmacology* **26**, 53–63.

Paus, T., Zijdenbos, A., Worsley, K., Collins, D. L., Blumenthal, J., Giedd, J. N., Rapoport, J. L., and Evans, A. C. (1999). Structural maturation of neural pathways in children and adolescents: *In vivo* study. *Science (Washington, DC, USA)* **283**, 1908–1911.

Permutt, M. A., Wasson, J., and Cox, N. (2005). Genetic epidemiology of diabetes. *J. Clin. Invest.* **115**, 1431–1439.

Pezawas, L., Verchinski, B. A., Mattay, V. S., Callicott, J. H., Kolachana, B. S., Straub, R. E., Egan, M. F., Meyer-Lindenberg, A., and Weinberger, D. R. (2004). The brain-derived neurotrophic factor val66met polymorphism and variation in human cortical morphology. *J. Neurosci.* **24**, 10099–10102.

Pfefferbaum, A., Mathalon, D. H., Sullivan, E. V., Rawles, J. M., Zipursky, R. B., and Lim, K. O. (1994). A quantitative magnetic resonance imaging study of changes in brain morphology from infancy to late adulthood. *Arch. Neurol.* **51**, 874–887.

Pfefferbaum, A., Sullivan, E. V., and Carmelli, D. (2004). Morphological changes in aging brain structures are differentially affected by time-linked environmental influences despite strong genetic stability. *Neurobiol Aging* **25**, 175–183.

Porjesz, B., and Rangaswamy, M. (2007). Neurophysiological endophenotypes, CNS disinhibition, and risk for alcohol dependence and related disorders. *Sci. World J.* **7**, 131–141.

Prescott, C. A., Aggen, S. H., and Kendler, K. S. (1999). Sex differences in the sources of genetic liability to alcohol abuse and dependence in a population based sample of US twins. *Alcohol. Clin. Exp. Res.* **23**, 1136–1144.

Price, J. L. (2003). Comparative aspects of amygdala connectivity. *Ann. NY Acad. Sci.* **985**, 50–58.

Pujol, J., Soriano-Mas, C., Alonso, P., Cardoner, N., Menchon, J. M., Deus, J., and Vallejo, J. (2004). Mapping structural brain alterations in obsessive-compulsive disorder. *Arch. Gen. Psychiatry.* **61**, 720–730.

Reich, T., Edenberg, H. J., Goate, A., Williams, J. T., Rice, J. P., Van Eerdewegh, P., Foroud, T., Hesselbrock, V., Schuckit, M. A., Bucholz, K., Porjesz, B., Li, T-K, Conneally, P. M., Nurnberger, J. I., Tischfield, J. A., Crowe, R. R., Cloninger, C. R., Wu, W., Shears, S., Carr, K., Crose, C., Willig, C., and Begleiter, H. (1998). Genomewide search for genes affecting the risk for alcohol dependence. *Am. J. Med. Genet. Part B* **81**, 207–215.

Risch, N., and Merikangas, K. (1996). The future of genetic studies of complex human diseases. *Science* **273**, 1516–1517.

Robins, L. N. (1966). Deviant Children Grown Up – A Sociological and Psychiatric Study of Sociopatic Personality. Williams [[amp]] Wilkins, Baltimore, MD.

Roth, B. L. (2006). The Serotonin Receptor: From Molecular Pharmacology to Human Therapeutics. Humana Press, Totowa, NJ.

Rubia, K., Smith, A. B., Taylor, E., and Brammer, M. (2007). Linear age-correlated functional development of right inferior fronto-striato-cerebellar networks during response inhibition and anterior cingulate during error-related processes. *Human Brain Mapping* **28**, 1163–1177.

Saxena, R., Hivert, M. F., Langenberg, C., Tanaka, T., Pankow, J. S., Vollenweider, P., Lyssenko, V., Bouatia-Naji, N., Dupuis, J., Jackson, A. U., Kao, W. H., Li, M., Glazer, N. L., Manning, A. K., Luan, J., Stringham, H. M., Prokopenko, I., Johnson, T., Grarup, N., Boesgaard, T. W., Lecoeur, C., Shrader, P., O'Connell, J., Ingelsson, E., Couper, D. J., Rice, K., Song, K., Andreasen, C. H., Dina, C., Kottgen, A., Le Bacquer, O., Pattou, F., Taneera, J., Steinthorsdottir, V., Rybin, D., Ardlie, K., Sampson, M., Qi, L., van Hoek, M., Weedon, M. N., Aulchenko, Y. S., Voight, B. F., Grallert, H., Balkau, B., Bergman, R. N., Bielinski, S. J., Bonnefond, A., Bonnycastle, L. L., Borch-Johnsen, K., Bottcher, Y., Brunner, E., Buchanan, T. A., Bumpstead, S. J., Cavalcanti-Proenca, C., Charpentier, G., Chen, Y. D., Chines, P. S., Collins, F. S., Cornelis, M., Crawford, G., Delplanque, J., Doney, A., Egan, J. M., Erdos, M. R., Firmann, M., Forouhi, N. G., Fox, C. S., Goodarzi, M. O., Graessler, J., Hingorani, A., Isomaa, B., Jorgensen, T., Kivimaki, M., Kovacs, P., Krohn, K., Kumari, M., Lauritzen, T., Levy-Marchal, C., Mayor, V., McAteer, J. B., Meyre, D., Mitchell, B. D., Mohlke, K. L., Morken, M. A., Narisu, N., Palmer, C. N., Pakyz, R., Pascoe, L., Payne, F., Pearson, D., Rathmann, W., Sandbaek, A., Sayer, A. A., Scott, L. J., Sharp, S. J., Sijbrands, E., Singleton, A., Siscovick, D. S., Smith, N. L., Sparso, T., Swift, A. J., Syddall, H., Thorleifsson, G., Tonjes, A., Tuomi, T., Tuomilehto, J., Valle, T. T., Waeber, G., Walley, A., Waterworth, D. M., Zeggini, E., Zhao, J. H., GIANT consortium, MAGIC investigators, Illig, T., Wichmann, H. E., Wilson, J. F., van Duijn, C., Hu, F. B., Morris, A. D., Frayling, T. M., Hattersley, A. T., Thorsteinsdottir, U., Stefansson, K., Nilsson, P., Syvanen, A. C., Shuldiner, A. R., Walker, M., Bornstein, S. R., Schwarz, P., Williams, G. H., Nathan, D. M., Kuusisto, J., Laakso, M., Cooper, C., Marmot, M., Ferrucci, L., Mooser, V., Stumvoll, M., Loos, R. J., Altshuler, D., Psaty, B. M., Rotter, J. I., Boerwinkle, E., Hansen, T., Pedersen, O., Florez, J. C., McCarthy, M. I., Boehnke, M., Barroso, I., Sladek, R., Froguel, P., Meigs, J. B., Groop, L., Wareham, N. J., and Watanabe, R. M. (2010). Genetic variation in GIPR influences the glucose and insulin responses to an oral glucose challenge. *Nat. Genet.* **42**, 142–148.

Scherk, H., Gruber, O., Menzel, P., Schneider-Axmann, T., Kemmer, C., Usher, J., Reith, W., Meyer, J., and Falkai, P. (2009). 5-HTTLPR genotype influences amygdala volume. *Eur. Arch. Psychiatry Clin. Neurosc.* **259**, 212–217.

Schmithorst, V. J., Wilke, M., Dardzinski, B. J., and Holland, S. K. (2002). Correlation of white matter diffusivity and anisotropy with age during childhood and adolescence: A cross-sectional diffusion-tensor MR imaging study. *Radiology* **222**, 212–218.

Schmitt, E. J., Eyler, L. T., Giedd, J. N., Kremen, W. S., Kendler, K. S., and Neale, M. C. (2007). Review of twin and family studies on neuroanatomic phenotypes and typical neurodevelopment. *Twin Res. Hum. Genet.* **10**, 683–694.

Schneider, F., Habel, U., Wagner, M., Franke, P., Salloum, J. B., Shah, N. J., Toni, I., Sulzbach, C., Hönig, K., Maier, W., Gaebel, W., and Zilles, K. (2001). Subcortical correlates of craving in recently abstinent alcoholic patients. *Am. J. Psychiatry* **158**, 1075–1083.

Schweinsburg, A. D., Paulus, M. P., Barlett, V. C., Killeen, L. A., Caldwell, L. C., Brown, S. A., and Tapert, S. F. (2004). An fMRI study of response inhibition in youths with a family history of alcoholism. *Ann. NY Acad. Sci.* **1021**, 391–394.

Silveri, M. M., Rohan, M. L., Pimentel, P. J., Gruber, S. A., Rosso, I. M., and Yurgelin-Todd, D. A. (2006). Sex differences in the relationship between white matter microstructure and impulsivity in adolescents. *Magn. Reson. Imaging* **24**, 833–841.

Sowell, E. R., Peterson, B. S., Thompson, P. M., Welcome, S. E., Henkenius, A. L., and Toga, A. W. (2003). Mapping cortical change across the human life span. *Nat. Neurosci.* **6**, 309–315.

Sowell, E. R., Trauner, D. A., Gamst, A., and Jernigan, T. L. (2002). Development of cortical and subcortical brain structures in childhood and adolescence: A structural MRI study. *Dev. Med. Child Neurol.* **44**, 4–16.

Sowell, E. R., Thompson, P. M., Leonard, C. M., Welcome, S. E., Kan, E., and Toga, A. W. (2004). Longitudinal mapping of cortical thickness and brain growth in normal children. *J. Neurosci.* **38**, 8223–8231.

Steinhauer, S. R., and Hill, S. Y. (1993). Auditory event-related potentials in children at high risk for alcoholism. *J. Stud. Alcohol* **54**, 408–421.

Sullivan, E. V., and Pfefferbaum, A. (2006). Diffusion tensor imaging and aging. *Neurosci. Biobehav. Rev.* **30**, 749–761.

Tapert, S. F., Brown, G. G., Baratta, M. V., and Brown, S. A. (2004). fMRI BOLD response to alcohol stimuli in alcohol dependent young women. *Addict. Behav.* **29**, 33–50.

Tapert, S. F., Cheung, E. H., Brown, G. G., Frank, L. R., Paulus, M. P., Schweinsburg, A. D., Meloy, M. J., and Brown, S. A. (2003). Neural response to alcohol stimuli in adolescents with alcohol use disorder. *Arch. Gen. Psychiatry* **60**, 727–735.

Tellegen, A. (1985). Multidimensional Personality Questionnaire. unpublished manual.

Tellegen, A., Lykken, D. T., Bouchyard, T. J., Jr., Wilcox, K. J., Segal, N. L., and Rich, S. (1988). Personality similarity in twins reared apart and together. *J. Pers. Soc. Psychol.* **54**, 1031–1039.

Tessner, K., and Hill, S. Y. (2010). Neural circuitry associated with risk for alcohol dependence. *Neuropsychol. Rev.* **20**, 1–20.

Tobler, P. N., Fiorillo, C. D., Tobler, P. N., and Schultz, W. (2003). Adaptive coding of reward value by dopamine neurons. *Science (Washington, DC, USA)* **299**, 1898–1902.

Trepel, C., Fox, C. R., and Poldrack, R. A. (2005). Prospect theory on the brain? Toward a cognitive neuroscience of decision under risk. *Brain Res. Cogn. Brain Res.* **23**, 34–50.

Tsuang, M. T., and Faraone, S. V. (2000). The future of psychiatric genetics. *Curr. Psychiatry Rep.* **2**, 133–136.

Van Beijersterveldt, C. E.M., and Boosma, D. I. (1994). Genetics of the human electroencephalogram (EEG) and event-related potentials (ERPs): A review. *Hum. Genet.* **94**, 319–330.

Van Hasselt, V. B., Hersen, M., and Milliones, J. (1978). Social skills training for alcoholics and drug addicts: A review. *Addict. Behav.* **3**, 221–233.

Van Horn, J. D., Yanos, M., Schmitt, P. J., and Grafton, S. T. (2006). Alcohol-Induced suppression of BOLD activity during goal-directed visuomoto performance. *Neuroimage* **31**, 1209–1221.

Volkow, N. D., Fowler, J. S., and Wang, G. J. (2004). The addicted brain viewed in the light of imaging studies: Brain circuits and treatment strategies. *Neuropharmacology* **47**(Suppl. 1), 3–13.

Volkow, N. D., Wang, G. J., Fowler, J. S., Logan, J., Gatley, S. J., Hitzemann, R., Chen, A. D., Dewey, S. L., and Pappas, N. (1997). Decreased striatal dopaminergic responsiveness in detoxified cocaine-dependent subjects. *Nature* **386**, 830–833.

Winokur, G., and Clayton, P. J. (1968). Family history studies: IV. Comparison of male and female alcoholics. *Q. J. Stud. Alcohol* **29**, 885–891.

Yacubian, J., Sommer, T., Schroeder, K., Gläscher, J., Kalisch, R., Leuenberger, B., Braus, D. F., and Büchel, C. (2007). Gene–gene interaction associated with neural reward sensitivity. *Proc. Natl. Acad. Sci. USA* **104**, 8125–8130.

Yeung, N., Bogacz, R., Holroyd, C. B., and Cohen, J. D. (2004). Detection of synchronized oscillations in the electroencephalogram: An evaluation of methods. *Psychophysiology* **41**, 822–832.

Yuan, H., Marazita, S. Y., and Hill, S. Y. (1996). Segregation analysis of alcoholism in high density families: A replication. *Am. J. Med. Genet.: Neuropsychiatric Genet* **67**, 71–76.

Yurgelun-Todd, D. (2007). Emotional and cognitive changes during adolescence. *Curr. Opin. Neuorbiol.* **17**, 251–257.

USING EXPRESSION GENETICS TO STUDY THE NEUROBIOLOGY OF ETHANOL AND ALCOHOLISM

Sean P. Farris,* Aaron R. Wolen,† and Michael F. Miles*

*Department of Pharmacology and Toxicology, Virginia Commonwealth University, Richmond, VA 23298, USA

†Department of Human and Molecular Genetics, Virginia Commonwealth University, Richmond, VA 23298, USA

Recent simultaneous progress in human and animal model genetics and the advent of microarray whole genome expression profiling have produced prodigious data sets on genetic loci, potential candidate genes, and differential gene expression related to alcoholism and ethanol behaviors. Validated target genes or gene networks functioning in alcoholism are still of meager proportions. Genetical genomics, which combines genetic analysis of both traditional phenotypes

INTERNATIONAL REVIEW OF
NEUROBIOLOGY, VOL. 91
DOI: 10.1016/S0074-7742(10)91004-0

0074-7742/10 $35.00

and whole genome expression data, offers a potential methodology for characterizing brain gene networks functioning in alcoholism. This chapter will describe concepts, approaches, and recent findings in the field of genetical genomics as it applies to alcohol research.

I. Introduction

Rapid progress in human and animal model genetics over the last 10–15 years has identified multiple genetic regions or candidate genes as associated with risk for alcoholism or ethanol-related behavioral responses. Simultaneously, the advent of high-throughput whole genome expression profiling with microarrays has produced prodigious data sets on differential gene expression related to alcoholism and ethanol behaviors. These expression profiling results, combined with modern bioinformatics tools, often produce insight into functional gene networks relevant to ethanol rather than simple lists of genes. Despite these parallel marked advances in our molecular databases relevant to alcoholism, there remains considerable difficulty in providing an "endgame" regarding identifying genes or gene networks that actually improve our ability to treat alcoholism or identify risk factors for becoming an alcoholic. Both genetics and genomics appear to have their individual weaknesses. Modern genetic approaches such as genome-wide association studies (GWAS) provide huge lists of questionably significant genes associated with alcoholism traits. Adequate statistical power to detect a large portion of the genetic variance underlying complex traits such as alcoholism seems difficult or impossible to achieve. Conversely, microarray studies provide dizzyingly complex lists of genes or gene network diagrams but most often without a causal link to ethanol behaviors or other complex phenotypes.

This chapter will describe a novel approach to solving the issues raised above. Genetical genomics, a term coined by Jansen and Nap (2001), combines the use of complex genetic studies on both behavior and gene expression. This represents a form of molecular triangulation whereby genetic variance in both gene expression and ethanol behaviors is used to define causal links between the two fields. The underlying premise for this approach is that gene expression *per se* can be treated as a quantitative trait just as with behavioral phenotypes, and that genetically driven alterations in gene expression might underlie much of the genetic contribution to complex traits. By using genetic linkage to anchor both gene expression networks and behavioral sub-phenotypes, the actual goal of defining an "alcoholism gene" or quantitative trait gene (QTG) becomes, at least in part, irrelevant. In short, it is "the network, stupid," that becomes the focus of attention with this approach. We define such a focus with the operative term "quantitative trait gene network" (QTGN).

Numerous reviews and original papers concerning genetical genomics have been published over the last few years and many of these concern applications to ethanol behavioral genetics. This chapter will strive to describe the key elements and resources in each aspect of genetical genomics and summarize recent work relevant to ethanol responses and alcoholism. In particular, we will try to provide an overview of how genetical genomics in general can provide enormous power in ranking candidate genes underlying ethanol behavioral quantitative trait loci (QTLs). We will also introduce the concept of studying the genetics of ethanol-responsive gene expression networks (as opposed to basal gene expression) in mouse brain regions. This recent work from our own laboratory and others serves to provide novel insight for both the mechanisms and functional consequences associated with brain ethanol-responsive gene expression networks. It is our hope that the reader will be left with both a clear understanding of the goals and promise for ethanol genetical genomics studies, as well as the basic principles and shortcomings for this complex but exciting area of research.

II. Genetic and Genomic Approaches in Alcohol Research

A. Behavioral QTL Mapping and Identifying QTGs by Traditional Mapping Strategies

As the field of genetics moved beyond the challenge of identifying the genetic basis of binary phenotypes and began to shift its focus to complex traits, new tools were required. Like most human diseases, alcoholism is a complex trait, influenced by a combination of multiple genetic and environmental factors, and interactions thereof. The weapon of choice for identifying genes that influence complex diseases has become quantitative trait locus (QTL) mapping. A more complete description of QTL mapping will be found in the chapter by Buck and colleagues within this volume. Here we include a brief description of the concepts for reference to expression QTL mapping discussed below.

The strategy for QTL mapping a phenotype of interest is similar to that of classic linkage analysis, with a common goal of identifying alleles that co-segregate with a disease trait more frequently than would be expected by chance. Working with complex traits that typically vary in their manifestation across a continuous distribution, in contrast to the binary nature of monogenic traits, QTLs are discovered by simply identifying loci with alleles that consistently covary with a phenotype across a population. Genomic regions that show a sufficiently strong association with a phenotype are considered QTLs. The simplest, or most hopeful, interpretation of a mapped QTL is that the implicated region harbors a single gene affecting manifestation of the associated phenotype. However, it is quite possible that a QTL is driven by multiple genes, non-coding RNA species, epigenetic mechanisms (including environmental interactions), or a combination thereof.

While QTL mapping studies may be carried out in human populations, the inflated non-genetic variance contributed to a phenotype by each subject's unique environment and life experience introduces a tremendous amount of noise, making results difficult to interpret (Broman, 2009). These issues, plus the potential population stratification issues and tremendous costs associated with assembling sufficiently large human samples, explain why most QTL mapping studies are conducted with rodents, and primarily mice.

The mouse is a very attractive subject for dissecting complex traits by QTL analysis for a variety of reasons. Apart from its small size, relatively low maintenance cost, and short gestation period (Peters *et al.*, 2007), the mouse provides an incredibly deep and ever-expanding arsenal of genetic tools. Laboratory mouse providers such as Jackson Laboratories and Charles Rivers Laboratories make available hundreds of inbred mice that vary greatly in genotype and phenotype. Perhaps the greatest advantage to using mice for QTL mapping regards the fact that the entire genome from multiple mouse strains has been fully sequenced with 17 strains currently being completed by the Wellcome Trust Sanger Institute (http://www.sanger.ac.uk/resources/mouse/genomes/).

Because these inbred animals are completely isogenic, meaning they carry two effectively identical haploid genomes, each generation of progeny is a genetic clone of its forebears. As such, these genetic lines are essentially immortal, an incredible boon for scientific reproducibility because an experiment conducted in laboratory "A" can be directly compared against the results published by laboratory "B" when the same inbred strain is used, even if laboratory B's results were published years earlier. However, some confounding interactions with the laboratory environment have been reported for mouse genetic studies (Crabbe *et al.*, 1999). Nevertheless, the popularity of inbred mice has greatly encouraged interlab collaborations and more open data sharing.

Of course QTLs can only be mapped in the presence of genetic variation; therefore, QTL studies are usually conducted using derivatives of inbred strains. A typical experimental design might involve characterizing a panel of F2 progeny for a phenotype where the inbred parental strains differ significantly. QTL mapping could then commence after genotyping each of the F2 progeny. A special derivative of inbred strains, recombinant inbred (RI) strains, are produced much the same as an F2 panel but includes an additional phase of multiple generations of sibling inbreeding (Bailey, 1971). The result is a panel of new inbred strains, each one carrying a unique combination of the progenitor genomes.

Established RI panels carry several significant advantages over other gene mapping populations. Being inbred, each strain needs to be genotyped only once, provided that breeding strategies are designed to minimize genetic drift. In practice, this has meant that larger laboratories specializing in genotyping, such as the Wellcome Trust Sanger Institute and the Jackson Laboratory (see http://cgd.jax.org/cgdsnpdb/), have genotyped many inbred strains or RI panels and generously made the results publicly available. This exemplifies the true power of utilizing RI panels in QTL mapping studies; all acquired data is cumulative and directly relatable, regardless of where it originates. A bold new extension of the use of RI panels will soon be available with the generation of a new panel, termed the Collaborative Cross, containing up to 1000 inbred lines derived from eight progenitor inbred lines (Churchill *et al.*, 2004). This enormous resource should allow the fine mapping of complex traits with detection of loci having even very low contributions to genetic variance of a trait. Furthermore, the use of multiple strains in developing the collaborative cross will enrich the genetic variation of the panel compared to traditional RI batteries that are derived from two progenitor strains.

At present, the most widely used RI panels for alcohol research have been the BXD and LXS batteries of RI lines. The BXD lines, derived from progenitor C57BL/6J (B6) and DBA2/J (D2), currently have over 80 inbred strains (Peirce *et al.*, 2004). The LXS strains were derived from the ILS and ISS inbred strains that were originally derived by selective breeding for sensitivity to ethanol sedation (Badcock *et al.*, 2004). A large collection of behavioral, anatomical and neurochemical phenotypes derived from the BXD and LXS RI lines is maintained at the GeneNetwork web resource (www.GeneNetwork.org), with many of the >2000 phenotypes being related to ethanol or drugs of abuse.

The molecular and genetic resources outlined above serve to greatly increase the power and resolution of QTL mapping for various behaviors or other traits of interest. The major barriers to mapping are the availability of a trait amenable to high-throughput study with low technical or environmental variance and showing significant genetic variance across the reference population used for QTL mapping. However, despite tremendous advances in genetic resources and identification of a large number of behavioral, anatomical, or neurochemical QTLs, validation of corresponding QTGs has greatly lagged. This difficulty stems largely from lack of sufficient recombination events in existing mouse panels to reduce haplotype block size for fine mapping. Additionally, small effect sizes and epistatic interactions complicate detection of a QTL as fine mapping efforts proceed. The effect size issue may be due in part to the existence of multiple QTGs underlying QTL detected by initial screens. Strategies such as derivation of congenic lines have been successful for fine-mapping a number of ethanol traits and for identifying *Mpdz* as one of the first QTGs mapped for a mammalian behavioral phenotype (Buck *et al.*, 1999; Fehr *et al.*, 2002; Shirley *et al.*, 2004).

However, such approaches take large investments in time, animals, and research expenditures. In many cases, even with the derivation of congenic lines, the investigator is left with a genetic interval comprising millions of base pairs and possibly hundreds of candidate genes. As described below, the use of whole genome expression profiling has provided a powerful approach for obviating some of the difficulties presented by traditional genetic QTL mapping approaches.

B. Genome-Wide Analysis of mRNA

Through an unbiased, parallel examination of mRNA expression from virtually the entire genome, DNA microarrays identify genes or gene networks associated with a given phenotype (drug response, disease, developmental stage, etc.). Gene networks provided by such studies can provide novel information regarding the mechanisms underlying a disease state. Thus, microarrays provide an alternative or complementary approach to genetic mapping for the non-biased identification of disease-related genes. As described below, a large number of studies have indeed applied microrarray analysis to the study of alcoholism or responses to ethanol in animal models. In this section we provide a brief overview of mRNA expression microarrays as this is important to understanding the limitations of this technology for QTG identification. Many reviews have been written regarding the use of microarrays and provide a rich source for more detailed information (Brown and Botstein, 1999; Kerns and Miles, 2008; Lyons, 2003; Miles, 2001). In addition to mRNA expression analysis, a large number of microarray platforms now exist for conducting assays of microRNA abundance, mRNA splicing (exon arrays), transcription factor binding (ChIP on chip assays), and DNA methylation. Additionally, recent technological advances in DNA sequencing are enabling a whole new approach to transcriptome profiling. Such approaches, termed NextGen RNASeq, rely on ultrafast sequencing to provide a transcript count across the entire genome. This approach, not discussed further in this chapter, offers the advantage of profiling all splice variants of transcripts as well as not being susceptible to hybridization artifacts on microarrays due to single nucleotide polymorphisms (SNP effects; see below).

Multiple platforms now exist for whole genome–wide mRNA expression profiling. However, the existing literature mainly utilizes oligonucleotide microarrays produced by photolithography (Chee *et al.*, 1996) or spotted cDNA microarrays (Schena *et al.*, 1995). Recent work has turned to the use of oligonucleotides (50–70 nucleotides (nt)) either spotted on glass slides or immobilized on beads as preferable to the initial "Brown arrays" that used cDNA spotting. Thus, most investigators currently use either short (25 nt), high-density oligonucleotide arrays (www.affymetrix.com), bead-based oligonucleotide arrays (www.illumina.com), or "home-grown" spotted oligonucleotide arrays.

All microarray platforms depend on similar underlying molecular fundamentals. Fluorescent labeled probes are generated from RNA isolated from a desired tissue or cell type. Such "target" molecules may be either RNA or DNA based, depending on the particular microarray platform. A "target" gene of interest, through the basic principle of complementary base pairing of nucleotides, adheres to small "probes" attached to a solid-phase support such as a glass slide, bead, or nylon membrane. The probes are generally DNA oligonucleotides 25–70 nt in length as mentioned above. Following hybridization and washing, fluorescent signals are detected and these correlate directly to specific mRNA abundance.

The experimental design for microarray experiments is relatively straightforward, but at each individual stage of sample processing, considerable attention to detail is needed to minimize batch effects. Batch effects are non-biological changes in gene expression occurring from a systematic grouping of samples throughout the protocol. Supervised randomization techniques are needed at every step in the protocol to ensure that changes in gene expression are due to the independent variables of the experiment and not from misleading factors such as different lots of reagents. A number of methods exist to "correct" batch effects (Alter *et al.*, 2000; Benito *et al.*, 2004; Johnson *et al.*, 2007); however, in our opinion, it is best to avoid these problems as much as possible in the initial experimental design. As discussed below, such batch effect issues are very important considerations when performing very large expression profiling experiments such as those used across RI panels.

The analysis of microarray data is an extremely detailed topic beyond the scope of this chapter. Investigators should be aware that there are multiple algorithms available for analyzing virtually all microarray platforms. Major factors to be considered include the particular approach used for background subtraction, normalization, and probe summarization. Following the initial calculation of expression values for individual probes or genes, a variety of statistical approaches are used for detection of expression differences across independent variables. Given the huge multiple testing confound of microarray analyses, statistical approaches utilizing a false-discovery rate calculation rather than a family-wise error rate analysis are generally most appropriate (Tusher *et al.*, 2001). Perhaps the most challenging and potentially rewarding aspect of analyzing microarray data comes from the bioinformatic approaches that are needed to identify and interpret gene network information contained within microarray results (Kerns and Miles, 2008). This important topic is discussed in Section IV and represents one of the most powerful aspects of applying microarray technology to identification of ethanol-responsive gene expression or identification of QTGs for ethanol traits.

Another critical factor affecting microarray expression results concerns the genetic background of both the array design itself and the mRNA under study. This is an important factor since most microarrays in use today depend on hybridization of probes to relatively short oligonucleotide targets. The stability of such hybrids is very sensitive to secondary structure as can be introduced by probe/target mismatching caused by sequence polymorphism differences between the probe and target. For example, SNPs have been shown to produce false-positive strain-specific (B6 mice vs. D2 mice) differences in expression for up to 10–20% of the probes on Affymetrix oligonucleotide arrays (Walter *et al.*, 2007). Any microarray experiment detailing expression differences across strains or individuals with differing genetic backgrounds has to take such SNP effects into account when interpreting microarray results. Such SNP effects can generate devastating false-positive results in genetical genomics studies as discussed in Section V.A.

C. Genomic Analysis of Alcoholism and Ethanol Behavioral Responses

Alcoholism is a complex disease influenced by both genetic and environmental factors. The genetic predisposition for the risk of developing alcoholism has been well documented through a host of publications on the familial origin of substance abuse (Hill *et al.*, 2008; Merikangas *et al.*, 1998; Prescott *et al.*, 1999). Human and animal model genetic studies, however, have to date only identified a few well-documented genes contributing to the genetic risk for alcoholism or genetic variation in behavioral responses to ethanol. Although the genes involved in the development or progression of this complex trait are largely unknown, genomic studies are currently elucidating gene expression patterns relevant to the neurobiology of this disease. Understanding the molecular genetic basis of an alcoholic phenotype requires defining the genes underlying the predisposition for abuse, as well as the genes altered through acute and chronic ethanol exposure. These sets of genes likely overlap but are not identical. Microarray studies in human or animal models of ethanol responses offer the promise of "ranking/identifying" candidate genes for genetic variance in risk for alcoholism or ethanol behavioral responses. These genomic studies also provide a mechanism for identifying the larger network of genes that contribute to the overall behavioral and toxic responses to ethanol in both humans and animal models.

Microarray studies on ethanol can be subdivided into human versus animal model, acute versus chronic, basal versus treated, and whole brain versus brain regional. In general, for the animal model studies, two different experimental approaches have been used. The first entails profiling of "basal gene expression" across two or more lines of animals that are known to have important differences in ethanol drinking or other behavioral responses to ethanol. The second approach looks at responses to acute or chronic ethanol exposure or ethanol withdrawal. The two approaches have been merged in a limited number of studies where ethanol responses across different lines of animals are also studied. As a general conclusion, all of these approaches have produced significant findings, sometimes with very large lists of genes being identified.

Microarray studies have been done on naïve inbred strains of mice differing in one or more ethanol behavioral phenotypes to define gene networks contributing to such behavioral differences. For example, whole-brain expression profiling of ILS and ISS mice, bred for their initial sensitivity to a sedative-hypnotic dose of ethanol, shows differences in their expression of approximately 81 cDNA clones (Xu *et al.*, 2001). The candidate genes from this study suggested the involvement of several gene ontologies including glycolysis through differential expression of pyruvate dehydrogenase E1-a subunit, a key enzyme in the conversion of pyruvate to acetyl coenzyme A (acetyl-CoA). Follow-up reverse transcriptase polymerase chain reaction (RT-PCR) of ILS and ISS samples suggested potential splice variants for unknown cDNA clones that may contribute to ILS and ISS phenotypic differences.

Mulligan *et al.* (2006) recently defined a set of genes involved in the predisposition for abuse through a meta-analysis of whole-brain samples of 13 ethanol-naïve inbred strains of mice that are known to differ in voluntary ethanol drinking behavior. Their results pointed to a diverse array of molecular pathways with those genes of largest effect size related to cellular homeostasis and neuronal function. However, expression differences suggested the potential involvement of thousands of genes for a wide array of functional categories across distinct genotypes of alcohol "preferring" and "non-preferring" genotypes. The subtle complexities of functional neuroanatomical differences in the brain may further complicate these genotypic disparities between alcohol phenotypes because of the regional variation in gene expression across heterogeneous populations of inbred mice. Complex behavioral traits are likely the result of numerous genomic differences encompassing integrative processes across multiple brain regions.

A number of laboratories including our own are interested in brain region–specific differences contributing to ethanol behavioral phenotypes. Genotype-specific differences have been demonstrated within individual brain sections using microarray analysis across five distinct brain regions for eight different strains of inbred mice from priority group A of the Mouse Phenome Project (Bogue and Grubb, 2004; Letwin *et al.*, 2006). Correlation of these expression profiling results in treatment-naïve mice identified a subset of genes solely within the ventral striatum related to the *N*-methyl-d-aspartate (NMDA)/glutamate signaling pathway for distance traveled following 1.5 g/kg ethanol, suggesting a brain region–specific expression network associated with this behavioral feature. This correlation of basal gene expression with an ethanol behavioral phenotype across multiple inbred mouse lines suggested the potential power of combining genomic profiling with genetic analysis of ethanol behaviors.

Human and animal studies continue to demonstrate that the acute behavioral responses to alcohol may serve as a functional barometer indicative of the risk for long-term drinking behavior (Metten *et al.*, 1998b; Schuckit, 1994). Progression toward the full onset of alcoholism, exemplified by abuse and dependence, may thus be affected by acute ethanol sensitivity and concurrent acute ethanol-provoked signaling events. Such acute signaling events are reflected in downstream changes in gene expression. Expression patterns responding to acute ethanol might thus serve as surrogate measures of signal transduction events underlying experience-dependent plasticity that leads to long-term behavioral responses to ethanol and other drugs of abuse (Miles, 1995; Nestler and Aghajanian, 1997). Thus, several microarray studies have profiled expression responses to acute ethanol so as to define the full extent of the ethanol transcriptome.

B6 and D2 are two inbred strains of mice commercially available from the Jackson Laboratory (http://www.jax.org/) that differ in a number of ethanol behavioral phenotypes (McClearn, 1959). For example, D2 mice demonstrate greater acute locomotor responses, withdrawal-induced seizures, and reduced ethanol preference/consumption compared to B6 mice (Belknap *et al.*, 1993; Goldstein, 1973; Metten *et al.*, 1998a; Phillips *et al.*, 1994). An initial global whole-brain expression profiling experiment of B6 and D2 mice acutely treated with 6 g/kg ethanol revealed a small number of genes responding to ethanol and related to cell signaling, gene regulation, and homeostasis/stress response (Treadwell and Singh, 2004). Several of their ethanol-responsive genes resided within known ethanol behavioral QTLs; however, only 16 genes were identified with a differential ethanol response between B6 and D2 mice. The authors attributed this low extent of differential expression between the two strains to heterogeneity of the brain tissue in conducting genomic comparisons on whole brain.

Our laboratory has previously characterized basal and acute ethanol-responsive neurogenomic differences within the ventral tegmental area (VTA), nucleus accumbens (NAC), and prefrontal cortex (PFC) between D2 and B6 mice (Kerns *et al.*, 2005). Extensive multivariate analysis of VTA, NAC, and PFC uncovered basal and ethanol-evoked region-specific differences in gene expression patterns. Acute ethanol induced coordinated changes in gene expression patterns that suggested an overall functional involvement with neuroplasticity. Region-specific functional relationships were determined for retinoic acid signaling (VTA), *Bdnf* signaling and neuropeptide expression (NAC), and glucocorticoid signaling and myelination (PFC). A subset of differentially expressed genes mapped to QTLs that have been previously implicated in acute ethanol behaviors. This overlap of ethanol-responsive genes and genes contained within support intervals for behavioral QTLs provided a glimpse of the power that microarrays might have for prioritizing candidate QTGs. Overall, this study strongly suggested that the activity of divergent brain region–specific gene expression networks, and their cognate signaling mechanisms, might play an important role in the different ethanol behaviors seen with D2 and B6 mice.

The studies from Kerns *et al.* (2005) also defined myelin gene expression results in PFC that were particularly interesting. A set of genes, all relating to myelin function, showed divergent basal and ethanol-responsive effects between the two strains. Ethanol regulation of myelin gene expression is further supported by genomic analysis of postmortem human brain tissue, which demonstrates a coordinate regulation of myelin gene expression in the PFC (Iwamoto *et al.*, 2004; Lewohl *et al.*, 2000; Mayfield *et al.*, 2002). Similar results, demonstrating white matter abnormalities have been obtained at the protein level (Lewohl *et al.*, 2005) as well as with neuropathology and neuroimaging studies (Kril and Harper, 1989; Rosenbloom *et al.*, 2003). Microarray evidence of myelin gene dysregulation has also been identified in studies on schizophrenia and cocaine addiction, suggesting that PFC myelin gene expression may be sensitive to dopaminergic signaling (Sokolov, 2007).

A number of other functional categories, such as glutamate signaling, protein trafficking, and cyclic adenosine monophosphate (cAMP) signaling, have also been revealed through genomic studies of postmortem alcoholic brain tissue (Mayfield *et al.*, 2002; Sokolov, 2007). These changes reflect a large cohort of cell signaling mechanisms cooperating in a highly orchestrated series of central nervous system (CNS) plastic events occurring with prolonged ethanol exposure. The cAMP signaling pathway, for example, has been implicated in a large number of experimental models relevant to alcohol abuse (for review, please see Diamond and Gordon, 1997). A potential weakness of these human autopsy microarray studies on alcoholism, however, is that they only represent the final end point of the disorder, and not necessarily the neuroadaptations occurring over time which lead to the development of abuse and dependence. The possibility of other confounding environmental or population stratification factors affecting the microarray results also cannot be discounted.

Chronic ethanol exposure in animal models perhaps represents the most valid and yet challenging experimental design for understanding the neurobiology of alcoholism. Animal models, although lacking many of the functional criteria for clinical diagnoses of alcoholism, are able to mimic different phenotypic components of the disorder. Brain region–specific differences in gene expression have been detected in the NAC shell and central nucleus of the amygdala of alcohol-preferring rats (P rats) following an 8-week model of alcohol binge–like behavior (McBride *et al.*, 2010). Interestingly, there was little overlap in gene expression results between this study and a previous analysis from the same research group of the NAC within the same strain following ethanol operant self-administration, suggesting that differences in behavioral procedures and the neural circuits they activate may contribute to deviations in expression-profiling results (McBride *et al.*, 2010; Rodd *et al.*, 2008).

A particularly attractive model for progressive levels of high ethanol intake involves chronic intermittent exposure to ethanol vapor. This model produces chronic intermittent episodes of ethanol withdrawal and has been used by multiple investigators to produce high levels of ethanol intake in mice or rats. Rimondini *et al.* performed expression profiling on cingulate cortex and amygdala in rats that had been exposed to the chronic intermittent ethanol exposure (Rimondini *et al.*, 2002). These investigators found striking alterations in expression for genes involved with glutamate neurotransmission, synaptic plasticity, and mitogen-activated protein kinase (MAPK) signaling in the animals exposed to intermittent ethanol. Most intriguingly, these expression changes were detected 3 weeks after withdrawal from the last ethanol vapor exposure. These genomic responses could thus represent important molecular adaptations underlying the dramatic increases in ethanol intake seen with this experimental model. However, these studies used only a single time point and only two microarrays per treatment/brain region group, and thus may be plagued by type I and type II errors.

Ethanol withdrawal–induced changes in gene expression have been demonstrated in a model of acute and chronic withdrawal using expression profiling of hippocampi for both B6 and D2 mice (Daniels and Buck, 2002). These two strains exhibit significant differences in ethanol withdrawal behavioral responses. Consistent with other reports of ethanol's effects on gene expression, this study showed strain- and treatment-specific changes, some of which were consistent with the behavioral phenotypes under study. In a similar study on protein kinase C gamma (PKCγ) null mutants that do not develop tolerance to the sedative-hypnotic effects of ethanol following chronic ethanol exposure (Bowers *et al.*, 1999), Bowers *et al.* showed by microarray analysis of striatal and cerebellar tissue that a variety of functional categories could be associated with tolerance; however, both studies used a liberal statistical approach to their analysis and confirmed only two candidate genes (Bowers *et al.*, 2006; Smith *et al.*, 2006).

Overall, the results to date on expression profiling in alcholism or animal models of ethanol behavioral responses provide unique insight into the genomic response to ethanol. The data may include gene networks contributing to the predisposition for alcoholism, as well as the acute and chronic CNS plastic events associated with the development of the disease. However, validation of these multifaceted gene networks and their direct relationship to behavioral phenotypes remains a considerable challenge. Most of these expression results could be likened to defining a correlation coefficient for a line drawn through two points—treated versus untreated, sensitive strain versus insensitive, or alcoholic versus control. Defining a causal connection between gene expression patterns and behavior requires a much more powerful experimental design.

III. Genetic Analysis of Gene Expression

A. Genetical Genomics: Using Gene Expression as a Quantitative Trait

QTL mapping is, of course, not limited to dissecting classical phenotypes. Any quantitative trait influenced by genetic factors is amenable to QTL analysis, including transcript, protein, or metabolite abundance. Analyzing the genetic regulation of such molecular phenotypes offers a much closer look at the cellular biochemical processes driving the variation in standard quantitative traits (Schadt *et al.*, 2003). By extending these analyses to include high-throughput molecular phenotypes, such as genome-wide microarrays, it becomes possible to map out entire molecular networks or signaling pathways that underlie complex traits.

Several reports of microarray expression data discussed in Section II were done across two or more lines of mice, allowing low-resolution genetic correlation analysis of gene expression traits and cross-correlation with behavioral (or other) phenotypes. Applying robust genetic approaches to a multitude of molecular phenotypes was perhaps first conducted in a study of a subset of the maize proteome (Damerval *et al.*, 1994). Two-dimensional polyacrylamide gels were used to separate 72 proteins and measure their relative abundance levels across a population of 60 F2 individuals. Looking for associations between these measurements and a panel of 100 genetic markers, the authors identified QTLs significantly influencing the abundance of over half of the analyzed proteins. Furthermore, this study effectively demonstrated the potential of this approach to provide unprecedented insight into structural complexities of quantitative trait regulation, by not only determining the number of QTLs influencing a given trait but also characterizing the dominance effects and epistatic interactions between QTLs and uncovering genetic regulators driving the co-expression between proteins with highly similar expression patterns.

Years later, this strategy of performing genetic linkage analysis on genome-wide molecular profiles was formalized and deemed "genetical genomics," a term coined by Jansen and Nap (2001). The proposed methodology outlined by that paper primarily focused on gene expression microarrays, and predicted that mapping expression QTL (or eQTL) would make it possible to construct gene networks and elucidate the roles of genes in metabolic and regulatory pathways. The authors also asserted that eQTL mapping could greatly benefit the search for causative candidate genes underlying classical QTL for disease traits.

Less than a year after Jansen's paper, the first study to carry out QTL analysis across genome-wide gene expression microarrays was published using an experimental cross between two strains of *Saccharomyces cerevisiae* (Brem *et al.*, 2002). Results from this landmark paper shed a great deal of light on the genetic basis of gene expression in a complex organism. Of the 6215 genes measured, 1528 were differentially expressed between the progenitor strains and 570 showed significant linkage to at least one locus. Importantly, the authors noted that the power to detect eQTL for a gene is a direct function of the number of loci regulating that gene and the relative contribution made by each locus. By comparing results derived from empirical computer simulations to their observed data, they determined the majority of differentially expressed genes were likely being regulated by at least five eQTLs.

B. Identifying Gene Networks and Regulatory Mechanisms by Genetical Genomics

Strictly speaking, because microarrays measure steady-state mRNA abundance, eQTLs concern a gene's mRNA level and not the rate of transcription *per se*. There are also two distinct types of eQTLs, classified based on their physical location relative to that of the linked gene. An eQTL regulating a gene that is genetically unlinked, for example by being located on an entirely different chromosome, is considered to be a *trans*-eQTL. Interestingly, Brem *et al.* (2002) identified several loci regulating the expression of many genes simultaneously: over 40% of identified eQTLs mapped to just 8 loci. They noted that the collection of genes mapping to a shared eQTL, sometimes referred to as a *trans*-band, appeared to be functionally related. For example, over half of the genes that comprise the largest *trans*-band are mitochondrial ribosomal proteins. This suggests that functionally related genes may share common genetic regulators. While there are many possible explanations for such an observation, it is easy to imagine that a transcription factor with a polymorphic DNA binding region could exert such drastic downstream effects.

The complement to a *trans*-eQTL is a *cis*-eQTL, which indicates a gene is regulated by a local sequence polymorphism. An obvious example of this would be an SNP located within a gene's promoter sequence, affecting its ability to initiate transcription. However, an apparent *cis*-eQTL may actually represent a gene being regulated in *trans* by a neighboring gene carrying a functional polymorphism. It is not possible to distinguish between such mechanisms without performing follow-up molecular assays (Doss *et al.*, 2005). For a more complete discussion of such nuances, an excellent review has been written by Rockman and Kruglyak (2006). Williams and colleagues have also contributed multiple primary publications and reviews on the general area of expression genetics (Chesler *et al.*, 2003, 2004), as well as providing the GeneNetwork resource (www.genenetwork.org) for pursuing such studies (see discussion below).

C. Combining Phenotypic and Expression Genetics to Identify Genes or Gene Networks Linked to Complex Traits

The first mammalian transcriptome genetical genomics study of a given complex trait used 111 F2 mice placed on high-fat diets to induce an obesity-like phenotype (Schadt *et al.*, 2003). Microarray analysis of liver tissue from these animals identified 280 genes that changed significantly with the high-fat diet. Similar to the yeast transcriptome, eQTL anlalysis of the mouse liver identified several large *trans*-bands that exert enormous control over downstream genes. In fact, 25% of the high-fat diet–responsive genes were regulated primarily by a single eQTL, a so-called "eQTL hotspot." However, this paper is notable for recognizing the potential in intersecting genetical genomics with classical phenotypic QTL, as a way to greatly assist in identifying causative candidate genes underlying disease traits. The authors discovered two high-priority candidates for their obesity phenotype by screening for genes with *cis*-eQTL that overlapped with the QTL for fat pad mass, the phenotype used as a marker for obesity. Their hypothesis is consistent with a number of recent studies suggesting that genetic variation in the expression of QTGs rather than genetic variation in the function of proteins coded for by the QTGs might be important to the biology of complex traits.

The increased genetic resolution afforded by the rapid evolution of genotyping technology has made it possible to obtain continuously denser genetic maps and associate disease traits with correspondingly smaller regions. Additional information provided by eQTL mapping greatly benefits the process of prioritizing candidate genes localized to the disease-associated region by screening for genes whose expression patterns are regulated by local polymorphisms and strongly correlate with the disease trait (see Fig. 1). This will typically yield a small handful of interesting candidate genes. However, the final steps necessary to definitively identify the susceptibility gene in a linked region are still tremendously difficult and remain largely elusive. The experiments necessary to validate a gene's role in disease are expensive and time consuming enough that even a relatively small list of genes may be too large to evaluate each individually. As such, more information is required to further evaluate and prioritize genes that are, empirically, equally strong candidates.

Constructing molecular networks from high-throughput molecular data associated with a disease trait could very well provide the additional context necessary to distinguish the causal gene from a list of positional candidates by providing a more comprehensive view of the involved biological pathways and how they are affected by the genetic polymorphism driving the linkage with an implicated QTL (Chen and Charness, 2008). Studies of gene co-expression networks have already revealed much about the structural architecture of gene regulation by demonstrating that gene networks conform to a scale-free toplogy (van Noort

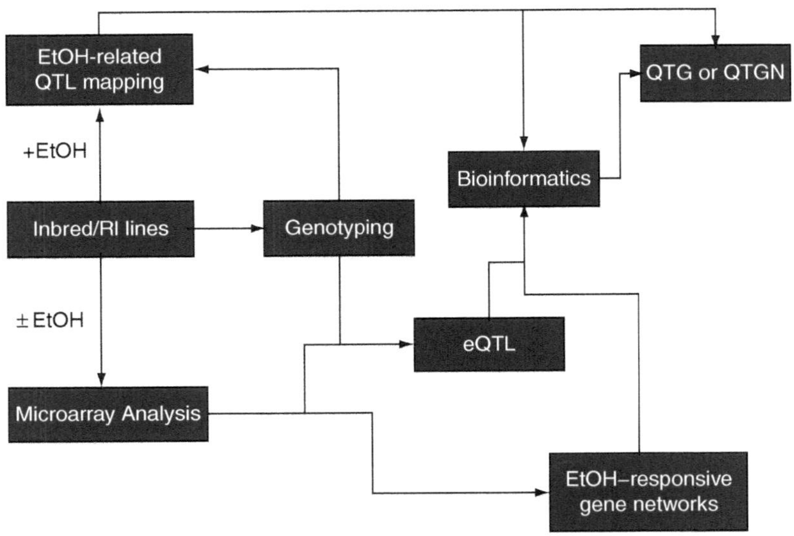

FIG. 1. Overall schematic for identifying a QTG or QTGN for ethanol and alcoholism. Inbred and RI lines (or other animal models) undergo a series of congruent steps for identifying a gene or gene network underlying a specific phenotype relevant to ethanol or risk for the development of alcoholism. Microarray analysis with and without ethanol is used to identify differentially expressed gene expression patterns and ethanol-responsive gene networks. QTL mapping is conducted to narrow genomic regions contributing to an ethanol response. Combining gene expression data and nucleotide information from microarray analysis and genotyping assists in identifying pertinent *cis*- or *trans*-eQTLs. Multilayered bioinformatic analysis of all the data, as well as that of public data and literature resources, is used to identify relevant QTG(s) or QTGN(s).

et al., 2004). Scale-free networks are characterized by the power distribution they are named for, with many sparsely connected nodes and a few that are highly interconnected, and also provide accurate models for protein and metabolite interaction networks, the neural networks of *Caenorhabditis elegans*, and even the World Wide Web (Junker and Schreiber, 2008). Armed with this knowledge, we can begin to ask interesting questions, such as: does the topology of gene networks change in response to repeated drug use or can better drug targets be discovered by identifying genes that represent major hubs of communication within a network?

Network analysis of microarray expression data is a rapidly progressing field of study, fueled by a seemingly exponential increase in the rate of microarray data being accumulated and a surge in the number of computational biologists cultivating it. Novel methods for constructing and analyzing gene expression networks are being released continuously, using increasingly sophisticated methods (Baldwin *et al.*, 2005; Horvath and Dong, 2008). However, gene co-expression networks can be constructed by calculating a simple Pearson product moment for all pairwise gene expression data across a segregating population (Fig. 3C). Identifying clusters of densely intercorrelated genes has proven to be an effective method for identifying gene product interactions that are evolutionarily conserved across multiple species and may participate in common biological pathways or form a protein complex (Stuart *et al.*, 2003). Identifying coregulated genes from microarray expression across segregating populations has been used to identify novel protein interactions (Scott *et al.*, 2005).

Robert Williams and colleagues have greatly facilitated the process of identifying eQTLs, correlating the eQTL with phenotypic QTL and generating co-expression networks through the development of the comprehensive GeneNetwork web resource (www.genenetwork.org). This site includes databases of phenotypes and many different microarray expression data sets from a variety of species, tissues, and brain regions. This allows an exhaustive investigation of links between genetic variation, gene expression, and phenotypes. The data displayed in Fig. 3A–C are examples of output from GeneNetwork analyses.

D. Genetical Genomics Analysis for Ethanol QTG Mapping

As discussed in Section II, identifying QTGs underlying ethanol behavioral QTL has been frustratingly slow despite the mapping and confirmation of many QTLs over the last 15 years. Applying expression profiling to ethanol behaviors or mouse/rat lines selected for ethanol phenotypes, followed by superimposing these results on behavioral QTL data, has been one approach used to prioritize candidate QTGs (Fig. 1). This essentially is an extension of the analysis of ethanol-related microarray data discussed in Section II.C. As an example, Mulligan *et al.* reported a meta-analysis of basal microarray expression data derived from several inbred lines divergent in ethanol drinking behaviors. This resulted in a large set of ~3800 genes theorized to be involved in ethanol preference. To narrow this gene set and also provide candidate QTGs for previously mapped QTLs on ethanol preference drinking, Mulligan *et al.* superimposed the meta-analysis data set on genes contained within the support interval of a Chr 9 ethanol preference QTL (Belknap and Atkins, 2001; Mulligan *et al.*, 2006). Additionally, by using eQTL data from GeneNetwork, a subset of high-priority candidates was identified among genes within the support interval having strong correlations with the ethanol drinking phenotype across BXD lines (and having a significant *cis*-eQTL) and also being positive in the meta-analysis for ethanol preference–related genes. This combination of approaches resulted in a testable set of fewer than six candidate genes rather than the hundreds of genes that resulted from the behavioral genetics study alone.

A similar approach combining known QTLs, gene expression data, and selective breeding strategies in mice was used to identify potential QTGs and signaling pathways involved in acute functional tolerance to ethanol (Tabakoff *et al.*, 2003). A congenic specific approach has also been implemented for a QTL on rat Chr 4 associated with alcohol preference using selectively inbred rats denoted alcohol-preferring (iP) and alcohol-non-preferring (iNP) (Carr *et al.*, 2007). Using congenic approaches largely decreases differential expression between strains because they are essentially isogenic at all loci except for the QTL region of interest, thus allowing for a method of QTG identification restricted to a chromosomal region mapped for an ethanol behavioral feature (e.g., preference). The study on iP and iNP rats identified ~34 genes with expression differences correlating with the QTL region, and residing within the Chr 4 QTL influencing ethanol preference, with suggested involvement in PKC signaling pathways. All of the studies described above used ethanol-naïve samples for either whole brain or a single measure from an average of results across multiple brain regions. Therefore, the results may be limited to genes involved in the predisposition for the ethanol phenotype in question rather than direct differences in ethanol-related signaling events and may fail to delineate brain region–specific QTGs of these complex traits.

Kerns *et al.* (also discussed in Section II.C) performed microarray studies across multiple brain regions in ethanol-treated versus control mice and thus identified expression patterns regulated by ethanol as well as strain-specific expression differences (Kerns *et al.*, 2005). These results suggested that expression analysis across genetically different strains tends to produce similar expression patterns across multiple brain regions in contrast to the more regionally selective expression changes caused by ethanol treatment. Thus, whole-brain expression profiling might be adequate for detection of at least some important QTGs showing eQTL across multiple brain regions. These authors also superimposed these brain regional ethanol-responsive expression results with behavioral and eQTL data from GeneNetwork. The net result of this layered analysis was the identification of several candidate QTGs or QTGNs for previously identified behavioral QTLs. For example, brain-derived neurotrophic factor (*Bdnf*) was identified as a possible candidate gene for a Chr 2 ethanol locomotor activation QTL (*Actre3*) on the basis of ethanol-responsive expression changes in the NAC. Importantly, multiple genes related to Bdnf signaling were also shown to be ethanol responsive within the same brain region. Bdnf signaling could thus be considered a QTGN for the *Actre3* QTL. This illustrates an additional power of merging expression studies and behavioral genetics because it suggests that identifying a QTG may become irrelevant if a QTGN of defined function can be mapped to the QTL. Thus, superimposing *trans*-eQTL bands with behavioral

QTL might pose a more powerful approach than just looking for *cis*-eQTL within a QTL support interval.

Direct genetic analysis of ethanol-responsive gene expression has not been reported on a scale larger than several inbred lines at a time. However, our laboratory has recently completed analysis of ethanol-responsive gene expression across the PFC, NAC, and VTA for over 30 BXD strains (see Virginia Commonwealth University (VCU) databases in GeneNetwork). Analysis of this enormous data set is ongoing (Wolen and Miles, in preparation) but it is already evident that this eQTL analysis of ethanol-responsive genes is uncovering previously unappreciated aspects of brain gene regulation by ethanol. As has been seen for some eQTL mapping of basal gene expression, we have identified a very limited set of genetic loci influencing the ethanol responsiveness for hundreds of genes (Fig. 2). This suggests that a limited number of signaling mechanisms may underlie the bulk of genomic responses to acute ethanol. This has important implications for strategies to prevent or reverse ethanol-related neuroplasticity related to alcoholism.

Our ethanol-responsive expression genetics analysis has already identified potentially important individual gene networks responding to ethanol in a coordinated manner. Figure 3A illustrates ethanol-responsive eQTL for the *Gabrb2* gene coding for a subunit of GABA-A receptors, an important molecular target for ethanol action. The interval map in Fig. 3A shows several *trans*-eQTLs for ethanol regulation of *Gabrb2* but no significant *cis*-eQTLs at the site of the *Gabrb2* gene (Chr 11). The Chr 7 trans eQTL is also the site of a suggestive QTL for several basal anxiety phenotypes in the GeneNetwork database. Figure 3B shows the correlation between one of these basal anxiety phenotypes and ethanol-regulated *Gabrb2* expression. Importantly, Gabrb2 showed strong correlation with ethanol regulation of an entire network of genes (Fig. 3C). As expected, these genes showed Chr 7 ethanol-responsive eQTL responses similar to *Gabrb2* (not shown), and thus this group of genes represents a potential ethanol-responsive QTGN with relevance to anxiety phenotypes.

Putative ethanol behavioral QTLs and QTGs identified in animal models may not be directly paralleled in humans; however, associations do exist between the two approaches. The multiple PDZ domain protein (*Mpdz)* was identified as a QTG for ethanol-induced withdrawal activity through a succession of studies involving F_2 intercrosses, RI line, and interval-specific congenic strains (Buck *et al.*, 1997, 2002; Fehr *et al.*, 2002; Shirley *et al.*, 2004). *Mpdz* was the only gene among five to demonstrate genotype-dependent differences in coding sequence and gene expression which co-segregated with withdrawal severity in mice. *Mpdz* also showed regional specific regulation by ethanol and genotype-specific expression differences in microarray studies by Kerns *et al.* (2005). In humans *MPDZ* does not demonstrate an association with alcohol-induced withdrawal seizures, but haplotype and single-SNP association analyses do suggest a possible

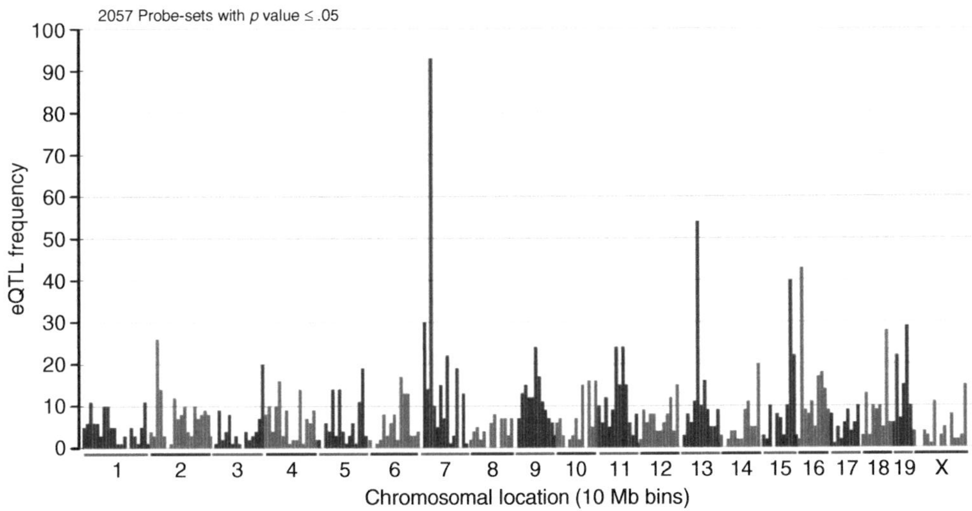

Fig. 2. Ethanol-responsive eQTL from the PFC of BXD RI lines. Mice were treated with saline or ethanol (1.8 g/kg × 4 h) and then brain region RNA isolated for microarray analysis. *S*-score analysis was used to identify significant ethanol-regulated gene expression (Kerns *et al.*, 2003). There are 2057 probe sets from the BXD PFC saline vs. ethanol *S*-score expression data with a significant eQTL, using a genome-wide corrected *p*-value ≤ .05 as the significance threshold. Each bar indicates the number eQTL located within a 10-megabase region of the genome. There are large ethanol responsive *trans*-bands on chromosomes 7, 13, 15, and 16.

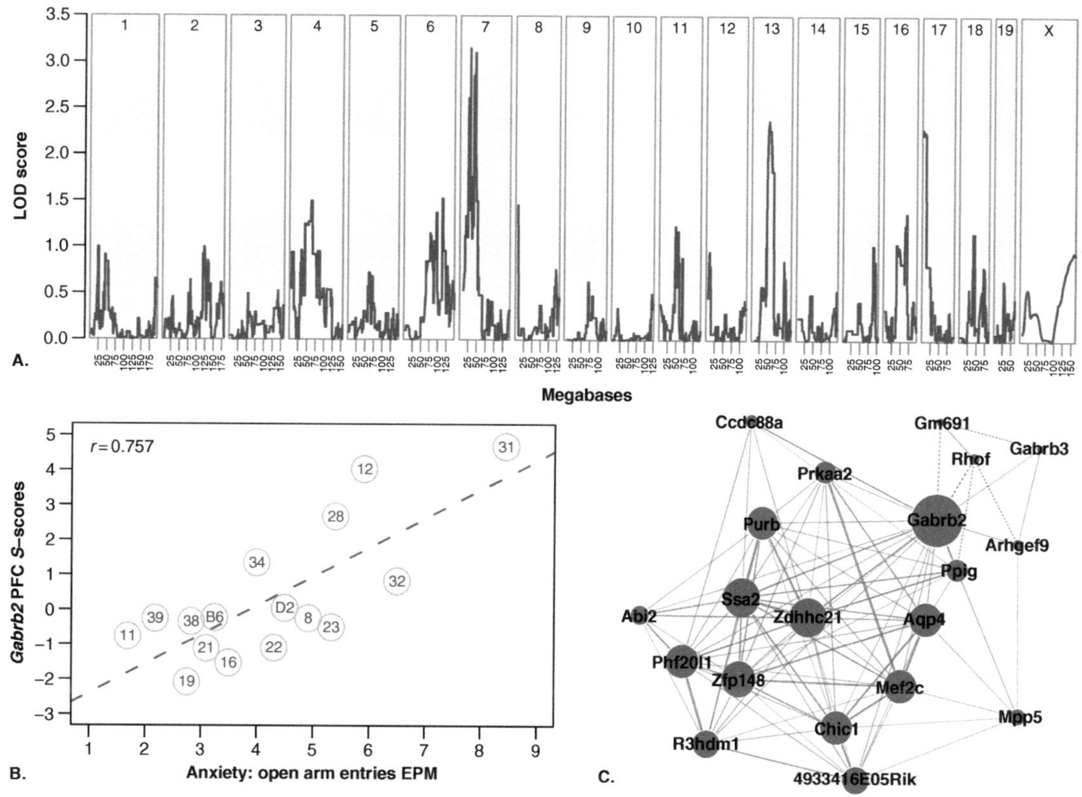

FIG. 3. Ethanol-responsive eQTL for Gabrb2. PFC *S*-score microarray data described in Fig. 2 was searched for Gabrb2 responses within GeneNetwork. (A) QTL tracing for GABA-A receptor subunit beta 2 (*Gabrb2*). The transcript abundance of this gene is being regulated in *trans* by loci on chromosomes 7, 13, and proximal 17. (B) The change in Gabrb2's expression induced by acute ethanol (1.8 g/kg) is significantly correlated with anxiety as measured by elevated plus maze (Yang *et al.*, 2008). (C) Gabrb2 gene coexpression network. The ethanol response of all plotted genes correlate significantly with that of Gabrb2 (p-value $\leq 5 \times 10^{-9}$).

association with alcohol dependence (Karpyak *et al.*, 2009). Disconnects existing between human and animal research may merely represent our lack of understanding for cross-species comparisons or demonstrate the need for more relevant animal models and phenotypes (Crabbe, 2008). Nevertheless, the identification of ethanol-related QTGs or QTGNs through genetical genomics continues to emerge as a powerful approach for the discovery of fundamental processes central to the risk, progression, and full onset of alcoholism and may help resolve some discrepancies between human and animal studies because of the focus on gene networks rather than an individual gene.

IV. Recognizing Biological Function in Genetical Genomics Studies

Interpreting results with such complexity as those from DNA microarrays is a daunting task that has spawned entirely new bioinformatic approaches for assigning biological function to coordinated changes in gene expression. The process of using genetical genomics does serve to organize and focus the often bewildering mass of data derived from microarray studies. Bioinformatic efforts, however, can further reduce the data to manageable sizes and provide a phenotypic or functional "trace" to a particular gene network or *trans*-eQTL band. Associating a presumed biological function to a gene network for a given behavioral phenotype might lead to new hypothesis generation regarding underlying molecular mechanisms and, eventually, avenues for intervention. Such an approach is obviously the hope of using genetical genomics and rigorous bioinformatics analysis with studies on ethanol phenotypes.

A. Over-Representation Analysis

The first step in organizing microarray data is often to determine whether a given list of genes (for example, a *trans*-band) has some coherent function. This is derived by essentially interrogating the biological function of each gene in the list and asking whether any particular "functional group" is over-represented beyond what would be expected by chance alone. Such analysis requires databases of genes categorized in some fashion. This can be done with a number of different categorization schemes: overall biological function (gene ontology), biochemical pathways, protein–protein binding partners, genes containing common transcription factors or miRNA binding sites, and a host of other such methods. The gene ontology consortium (http://www.geneontology.org/) is a collection of investigators working in a collaborative effort to assign function to gene products based on three broad categories—cellular component, biological process, and molecular function—which are further segmented into numerous functional subcategories. A number of tools and browsers are available to determine if a particular data set is enriched for a functional category. For a list of available tools, please see http://www.geneontology.org/GO.tools.shtml. The Database for Annotation, Visualization and Integrated Discovery (DAVID) (Dennis *et al.*, 2003; Huang *et al.*, 2009) is likely the most popular method for assessing gene ontologies with greater than 1600 citations currently among the literature. Hosted by the Laboratory of Immunopathogenesis and Bioinformatics (LIB), SAIC-Frederic, Inc., DAVID is a free online resource for conducting functional over-representation analysis.

The Web-based tool ToppGene (Chen *et al.*, 2009) is a similar bioinformatics program for conducting over-representation analyses but contains several additional databases for querying than does DAVID. ToppGene, for example, will determine whether there are particular miRNA or transcription factor binding sites that are enriched within a given gene set. A very important additional database utilized by ToppGene concerns sets of experimentally derived gene lists available through the Broad Institute Gene Set Enrichment Analysis (GSEA) database (http://www.broadinstitute.org/gsea/index.jsp). ToppGene also has additional tools for identifying and prioritizing candidate genes for further study when compared to a training set of genes known to be associated with a given phenotype of interest.

B. Literature Association Analysis

A published literature association analysis is an additional tool often used for querying functional information in genomics data. These approaches often use expert user–curated databases derived from natural language processing interrogation of the biomedical literature or other sources of gene/gene interaction data. The Ingenuity Pathway Analysis (www.ingenuity.com) is a popular commercial bioinformatics program for interpreting the network structure of data sets based on curated literature associations, biological function, and cell signaling mechanisms. The network structure of gene lists can reveal innovative associations between genes outside of predetermined functional categories. Other commercial programs such as Bibliosphere (Genomatrix Software, Munich, Germany) and GeneGo (www.genego.com) and free programs such as Chilibot (Chen and Sharp, 2004) are available for "literature mining" and drawing network diagrams among gene sets. Unfortunately, all of these programs will draw some sort of network given enough input genes. Therefore, care must be taken in over-interpreting the results of such network associations and verifying the implied literature co-citations. Deriving the same or similar network structure from different bioinformatics resources or through combinations of literature association, functional group over-representation, and expression correlation is a reassuring indication of some meaningful functional correlation to the gene set being interrogated.

C. Gene Set Correlation and Ontology Analysis

Finally, comparing a given genetical genomics–derived gene set or network with other microarray or functional gene sets can often provide valuable information about the biological function of the new gene list. As mentioned previously, GeneNetwork (www.genenetwork.org) is a collaborative Web-based resource equipped with tools and features for studying gene/gene and exploring genetic correlates to neurobehavioral phenotypes (Chesler *et al.*, 2003, 2004). The Web site is home to a growing collection of gene expression and phenotypic data from a variety of species and brain regions, with a host of links to external resources for tracing the interrelationships of a gene among multiple Web-based resources. GeneNetwork also offers a number of correlation and mapping strategies for assessing associations among multiple genes and QTLs. GeneNetwork aims to make the study of complex traits through the use of systems genetics widely available to the scientific community. A powerful tool that can be integrated with GeneNetwork or used on gene sets from a wide variety of sources is the Ontological Discovery Environment (ODE; http://ontologicaldiscovery.org). This public analysis platform provides an ever-growing list of gene sets derived from gene ontology categories, mouse genome informatics definitions, or experimentally derived genomic data sets (Baker *et al.*, 2009). Users can input their own data sets and look for intersection with other gene sets having previously defined ontologies. In this regard, ODE has features of both ToppGene and GSEA but allows users great flexibility in organizing their own gene sets among defined user groups.

As genome-wide research continues to produce ever enlarging data sets from genetic analysis of gene expression and phenotypic traits, there is considerable need for interactive bioinformatics resources that can quickly aid in characterizing biological function underlying the whole genome analyses. Here, we highlighted just a fraction of the bioinformatics tools available to researchers for extracting meaningful information from these complex data sets. Such investigations provide a hypothesis-generating framework for leveraging genetical genomics data on the neurobiology of alcoholism and other complex diseases.

V. Current Problems and Future Solutions

A. SNP Effects Causing False Positive *cis-e*QTL

While the potential of genetical genomics to provide insight into complex diseases continues to grow as high-throughput technologies improve and become more commonplace, the field is not without problems. We have discussed the value added by genetical genomics in the pursuit of causal disease genes by identifying *cis*-eQTL coincident with QTL for a disease trait. However, an important caveat is that putative *cis*-eQTL can suffer from a high false-positive rate. A thorough examination of this issue was performed using microarray expression data from the liver of B6xD2 F2 mice, and revealed a significant bias in the additive effects of *cis*-eQTL in the direction of the B6 allele (Doss *et al.*, 2005). That is, the B6 allele was associated with higher transcript abundance more frequently than would be expected by chance, suggesting that undetected polymorphisms within microarray probe target binding regions were affecting hybridization. This hypothesis is well supported by facts; as the microarray probe sequences were designed against the B6 genome, any undetected probe polymorphisms would bias measurement in favor of the B6 allele. This issue was discussed briefly in Section II.C because such "SNP effects" also affect traditional expression profiling studies comparing expression across a few strains.

In order to obtain an estimate of the proportion of true versus spurious *cis*-eQTL in their data, Doss *et al.* (2005) performed a *cis–trans*-test in a B6xD2 F1 population for a semi-randomly selected subset of probe sets with putative *cis*-eQTL, and found 35% (10/28) to be potentially false positives. This emphasizes the importance of validating a *cis*-eQTL before resources are spent attempting to confirm a gene's role in a complex disease. However, even before performing a *cis–trans*-test, spurious *cis*-eQTL can often be flagged by applying a statistical screen that eliminates deviating probes and repeats the QTL analysis with the remaining probes (Alberts *et al.*, 2007). If the significant association between a probe set and a locus is lost after excluding the deviating probes, the corresponding gene's *cis*-linkage should be considered highly suspect. Conversely, probe target SNPs may also diminish the significance of a potentially genuine *cis*-eQTL by reversing the expression/allele relationship for a subset of probes (see Fig. 4).

Even after applying the above methods and excluding genes with dubious *cis*-linkage, the QTL region for a classical phenotype may still harbor a daunting number of candidate genes with high-confidence *cis*-eQTL. When the QTL region is characterized by strong linkage disequilibrium, many of the local genes' expression patterns will be tightly correlated with each other as well as the linked phenotype, making it very difficult to discern which gene (or genes) is actually

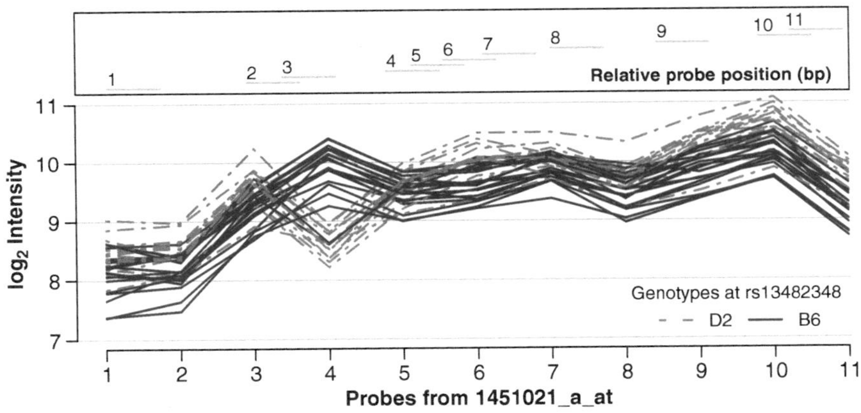

FIG. 4. False eQTL analysis. Probe-level $\log_2$ intensities for Kruppel-like factor 5 (Klf5) measured in NAC tissue from 36 BXD RI strains. Line color indicates strain genotype at the locus underlying Klf5's significant *cis*-eQTL. There are consistently higher levels of Klf5 transcript in strains carrying the D2 allele, with the exception of probe 4, where the trend is completely reversed. This suggests that an SNP effect may be hindering probe 4's ability to properly hybridize; thus it is contaminating the probe set's expression summary value and should be excluded. The plot was generated using code adapted from the affyGG package for R (Alberts *et al.*, 2008).

contributing to the phenotypic variation. If the region also contains a number of *trans*-eQTLs, it may be possible to tease apart the effects of colocalized genes using a partial correlation analysis to determine which local gene most strongly correlates with the *trans*-band after removing the genetic variance contributed by the underlying locus (Liu *et al.*, 2008; Mozhui *et al.*, 2008). Others have proposed a structural equation modeling approach to distinguish between a string of *cis*-eQTLs by identifying the genetic model that best fits the data (Schadt *et al.*, 2005).

B. Batch Effects Generating Spurious Linkage or Correlation

Another important issue to consider is systematic batch effects that can introduce considerable bias into microarray data. In studies seeking to differentiate between two or more classes of samples by identifying unique expression patterns, these hidden factors may cause expression variation that dilutes differences driven by the variable of interest, or masks them completely (Johnson *et al.*, 2007; Leek and Storey, 2007). In microarray studies across genetic mapping panels, batch effects can produce a biologically meaningless correlation between samples. If alleles for a particular SNP happen to segregate among the samples in a manner consistent with the spurious correlation, this may produce what appears to be major *trans*-band, potentially driven by an important genetic regulator (Kang *et al.*, 2008). Such an exciting result is difficult to ignore, and substantial resources could be spent chasing a technical artifact masquerading as an important discovery, causing potentially irrevocable heartbreak to the involved investigators. While there are available novel statistical methods for mapping expression QTLs designed to correct for inter-sample correlations and avoid generating potentially spurious *trans*-bands, these methods are currently unable to distinguish between artifactual and biologically genuine *trans*-bands. This makes the use of such statistical filter prone to high false-negative rates insofar as *trans*-eQTL detection is concerned (Kang *et al.*, 2008).

Of course batch effects are best dealt with by avoiding them as much as possible, using carefully executed randomization schemes during the stages of producing microrarray data. In our own laboratory we conduct careful experimental design studies before initiating genetical genomics or behavioral genetics studies so as to balance and randomize grouping of strains as much as possible. An optimal design to prevent batch effects regarding strain grouping during phenotypic (e.g., behavioral) testing might be to test as many strains as possible at one time, even if only a single or few animals from each strain are tested in a given batch. The process is then repeated until the desired number of animals is tested per strain but with randomization of strains tested at one time if all strains cannot be tested together. We perform supervised re-randomization of samples at each stage of the microarray analysis: RNA extraction, probe preparation, and chip hybridization. While it probably is not possible to resolve all potentially confounding factors (Churchill, 2002), their influence can be mitigated in a properly designed microarray experiment.

C. Defining Causality for Genes or Gene Networks in Behavioral Responses to Ethanol

Establishing causality among genes or gene networks in ethanol behavioral phenotypes is an important feature in discerning their prospective role in the neurobiology of alcoholism. The process is extremely difficult and potentially nearly impossible for genes having small effects and multiple epistatic interactions in regard to their final influence on a given behavior. Other chapters in this volume also cover this topic, so we will only discuss these issues briefly so as to frame the "endgame" for genetical genomics studies. It should be emphasized that such validation studies are crucial because combining expression and phenotypic genetics, although capable of detecting functionally relevant correlations between gene expression and phenotype, still only represents a correlation study and does not represent causality *per se*.

Gene targeting techniques have advanced our understanding of how specific genes may influence behavioral responses to ethanol and modulate risk for alcoholism. Multiple reports exist where genetically engineered transgenic, null knockout, and conditional knockout mice have determined the role(s) that individual genes play in the context of different ethanol behavioral models. For example, transgenic overexpression of the serotonin receptor 5-HT_3 and the neuropeptide Y (*Npy*) decreases ethanol preference and consumption (Engel *et al.*, 1998; Thiele *et al.*, 1998). Interestingly, *Npy* knockout mice show increased ethanol preference and low sensitivity to high doses of ethanol, clearly suggesting that expression of *Npy* may have an important role on ethanol behaviors (Thiele *et al.*, 1998). However, such transgenic or knockout approaches suffer several potential weaknesses related mainly to underlying physiological factors and experimental techniques. Transgenes can often integrate at a random chromosomal region not associated with the gene of interest, and creating a null mutation (knockout) can cause developmental compensation that may contribute to the observed phenotype. Yet these animals are still powerful models for investigating ethanol traits as evidenced by the number of genetically modified mice that have been established and evaluated for the role of the manipulated gene in acute and chronic ethanol behavioral phenotypes (for review, see Crabbe *et al.*, 2006). Cre/loxP or TetR-based transactivators avoid some of the weaknesses of traditional knockout approaches and confirm that a given phenotype is due to the specific gene mutation and not caused by developmental compensation in an adult animal model (Lewandoski, 2001). For example, *PKCε* knockout mice exhibit decreased ethanol preference and acute ethanol sensitivity (Hodge *et al.*, 1999). This is not a developmental effect because conditional expression under control of the *tet* operator sequences coupled to a minimal cytomegalovirus (CMV) promoter subsequently rescued acute ethanol sensitivity and ethanol preference in *PKCε* knockout mice (Choi *et al.*, 2002).

Viral vector–mediated transfection techniques are a complementary approach to traditional genetic mouse models for investigating the behavioral responses of alcohol and other drugs of abuse. Viruses (e.g., adenovirus and adeno-associated virus) and virions (i.e., herpes simplex virus and lentivirus) infect mammalian cells by binding to specific cell surface receptor elements. Packaging a gene of interest inside a viral vector allows for site-specific injection into a targeted brain region with spatiotemporal controlled expression. There are now numerous examples whereby viral vector gene delivery has been used to validate a causal role of given candidate genes for ethanol or other drug abuse–related behaviors. For example, viral-mediated gene transfer of RGS9-2, a regulator of

G-protein signaling, into the NAC of RGS9 knockout mice reversed the increased sensitivity to morphine seen in knockout mice (Zachariou *et al.*, 2003). Additionally, viral-mediated downregulation of *BDNF* in the dorsal lateral striatum, and not the dorsal medial striatum, increases ethanol self-administration in rats (Jeanblanc *et al.*, 2009).

D. Sharing of Resources

High-throughput methods of analyses, such as microarrays and sequencing technologies, have created the need for public repositories to keep pace with the sharing of data sets generated from individual experiments. In recent years, the sharing of such high-throughput data has become a condition for publication and funding. The National Center for Biotechnology Information (NCBI) and the European Bioinformatics Institute (EBI) have respectively created Gene Expression Omnibus (GEO) (Barrett *et al.*, 2009; Edgar *et al.*, 2002) and ArrayExpress (Parkinson *et al.*, 2009) to assist in the sharing of data sets from numerous gene expression studies, as well as some sequencing and epigenetic information. The repositories seek to remain flexible to the constantly evolving nature of genomics, while conforming to the standards of the Minimum Information About a Microarray Experiment (MIAME) (Brazma *et al.*, 2001), which provides colleagues enough information to replicate experiments. Collectively, the two repositories have greater than 20,000 experiments logged to date with tools available for searching and downloading data sets for subsequent analyses. The logistics of quality control and intellectual property rights are obviously of concern and a topic of considerable debate, but the precedent of data sharing in this post-genomic era continues to drive the field forward.

In the case of alcoholism research, genetic/genomic investigations are identifying a large number of genes that may contribute to a variety of ethanol phenotypic traits. Research scientists have created independent gene expression information warehouses equipped with simple tools for understanding the genetical genomics of complex traits. As already discussed, GeneNetwork has a large collection of behavioral, genetic marker, and expression data relevant to alcohol research and available for online analysis. PhenoGen Informatics (Bhave *et al.*, 2007) is a Web-based tool for investigating complex traits originally developed for sharing information among the Integrative Neuroscience Initiative on Alcoholsim (INIA). PhenoGen awards investigators a place to store data, conduct common methods of analyses, research candidate genes of interest, and investigate their relationships to QTLs. Ontological Discovery Environment (ODE) (Baker *et al.*, 2009) is another example of a gene list repository and Web-based tool created through the INIA consortium, which also allows investigators to draw associations between gene networks and phenotypic traits. Undoubtedly the creation of repositories such as PhenoGen and ODE, which are combined with methods for integrating and analyzing differential transcript expression across copious data sets, will make genomic investigations more accessible to the scientific community. Additionally, the Ethanol-Related Gene Resource (ERGR, http://bioinfo.vipbg.vcu.edu/ERGR/) is a public database integrating published ethanol-related genetic and genomic information across multiple species and platforms (Guo *et al.*, 2009). The existence of publicly and privately shared data sets envisions disease-specific management libraries that allow investigators to quickly identify how a given gene or network of genes respond under different experimental parameters and across species.

Acknowledgments

The authors thank Nate Bruce and Paul Vorster for their contributions in generating some of the microarray data discussed in this chapter, and Robert Williams at University of Tennessee Health Sciences Center for his assistance with data entry and analysis within GeneNetwork. This work was supported in part by grants from the National Institute on Drug Abuse (5T32DA007027, SPF), National Institute of Mental Health (MH-20030, ARW), and the National Institute on Alcohol Abuse and Alcoholism (F31AA018615 to SPF; U01AA016662, U01AA016667, P20AA017828, and R01AA014717 to MFM).

References

Alberts, R., *et al.* (2007). Sequence polymorphisms cause many false *cis* eQTLs. *PLoS ONE* **2**, e622.

Alberts, R., *et al.* (2008). affyGG: Computational protocols for genetical genomics with Affymetrix arrays. *Bioinformatics* **24**, 433–434.

Alter, O., *et al.* (2000). Singular value decomposition for genome-wide expression data processing and modeling. *Proc. Natl. Acad. Sci. USA* **97**, 10101–10106.

Badcock, J. C., *et al.* (2004). Speed of processing and individual differences in IQ in schizophrenia: General or specific cognitive deficits? *Cogn. Neuropsychiatry* **9**, 233–247.

Bailey, D. W. (1971). Recombinant-inbred strains. An aid to finding identity, linkage, and function of histocompatibility and other genes. *Transplantation* **11**, 325–327.

Baker, E. J., *et al.* (2009). Ontological discovery environment: A system for integrating gene-phenotype associations. *Genomics* **94**, 377–387.

Baldwin, N. E., *et al.* (2005). Computational, integrative, and comparative methods for the elucidation of genetic coexpression networks. *J. Biomed. Biotechnol.* **2005**, 172–180.

Barrett, T., *et al.* (2009). NCBI GEO: Archive for high-throughput functional genomic data. *Nucleic Acids Res.* **37**, D885–890.

Belknap, J. K., and Atkins, A. L. (2001). The replicability of QTLs for murine alcohol preference drinking behavior across eight independent studies. *Mamm. Genome* **12**, 893–899.

Belknap, J. K., *et al.* (1993). Quantitative trait loci (QTL) applications to substances of abuse: Physical dependence studies with nitrous oxide and ethanol in BXD mice. *Behav. Genet.* **23**, 213–222.

Benito, M., *et al.* (2004). Adjustment of systematic microarray data biases. *Bioinformatics* **20**, 105–114.

Bhave, S. V., *et al.* (2007). The PhenoGen informatics website: Tools for analyses of complex traits. *BMC Genet.* **8**, 59.

Bogue, M. A., and Grubb, S. C. (2004). The mouse phenome project. *Genetica* **122**, 71–74.

Bowers, B. J., *et al.* (1999). Decreased ethanol sensitivity and tolerance development in gamma-protein kinase C null mutant mice is dependent on genetic background. *Alcohol Clin. Exp. Res.* **23**, 387–397.

Bowers, B. J., *et al.* (2006). Microarray analysis identifies cerebellar genes sensitive to chronic ethanol treatment in PKCgamma mice. *Alcohol* **40**, 19–33.

Brazma, A., *et al.* (2001). Minimum information about a microarray experiment (MIAME)-toward standards for microarray data. *Nat. Genet.* **29**, 365–371.

Brem, R. B., *et al.* (2002). Genetic dissection of transcriptional regulation in budding yeast. *Science* **296**, 752–755.

Broman, K. (2009). A Guide to QTL Mapping with R/QTL. Springer, New York, NY.

Brown, P. O., and Botstein, D. (1999). Exploring the new world of the genome with DNA microarrays. *Nat. Genet.* **21**, 33–37.

Buck, K., *et al.* (1999). Quantitative trait loci affecting risk for pentobarbital withdrawal map near alcohol withdrawal loci on mouse chromosomes 1, 4, and 11. *Mamm. Genome* **10**, 431–437.

Buck, K. J., *et al.* (1997). Quantitative trait loci involved in genetic predisposition to acute alcohol withdrawal in mice. *J. Neurosci.* **17**, 3946–3955.

Buck, K. J., *et al.* (2002). Mapping murine loci for physical dependence on ethanol. *Psychopharmacology (Berl.)* **160**, 398–407.

Carr, L. G., *et al.* (2007). Identification of candidate genes for alcohol preference by expression profiling of congenic rat strains. *Alcohol. Clin. Exp. Res.* **31**, 1089–1098.

Chee, M., *et al.* (1996). Accessing genetic information with high-density DNA arrays. *Science* **274**, 610–614.

Chen, H., and Sharp, B. M. (2004). Content-rich biological network constructed by mining PubMed abstracts. *BMC Bioinformatics* **5**, 147.

Chen, J., *et al.* (2009). ToppGene suite for gene list enrichment analysis and candidate gene prioritization. *Nucleic Acids Res.* **37**, W305–311.

Chen, S., and Charness, M. E. (2008). Ethanol inhibits neuronal differentiation by disrupting activity-dependent neuroprotective protein signaling. *Proc. Natl. Acad. Sci. USA* **105**, 19962–19967.

Chesler, E. J., *et al.* (2004). WebQTL: Rapid exploratory analysis of gene expression and genetic networks for brain and behavior. *Nat. Neurosci.* **7**, 485–486.

Chesler, E. J., *et al.* (2003). Genetic correlates of gene expression in recombinant inbred strains: A relational model system to explore neurobehavioral phenotypes. *Neuroinformatics* **1**, 343–357.

Choi, D. S., *et al.* (2002). Conditional rescue of protein kinase C epsilon regulates ethanol preference and hypnotic sensitivity in adult mice. *J. Neurosci.* **22**, 9905–9911.

Churchill, G. A. (2002). Fundamentals of experimental design for cDNA microarrays. *Nat. Genet.* **32**(Suppl.), 490–495.

Churchill, G. A., *et al.* (2004). The collaborative cross, a community resource for the genetic analysis of complex traits. *Nat. Genet.* **36**, 1133–1137.

Crabbe, J. C. (2008). Review. Neurogenetic studies of alcohol addiction. *Philos. Trans. R Soc. Lond. B. Biol. Sci.* **363**, 3201–3211.

Crabbe, J. C., *et al.* (1999). Genetics of mouse behavior: Interactions with laboratory environment. *Science* **284**, 1670–1672.

Crabbe, J. C., *et al.* (2006). Alcohol-related genes: Contributions from studies with genetically engineered mice. *Addict. Biol.* **11**, 195–269.

Damerval, C., *et al.* (1994). Quantitative trait loci underlying gene product variation: A novel perspective for analyzing regulation of genome expression. *Genetics* **137**, 289–301.
Daniels, G. M., and Buck, K. J. (2002). Expression profiling identifies strain-specific changes associated with ethanol withdrawal in mice. *Genes Brain Behav.* **1**, 35–45.
Dennis, G., *et al.* (2003). DAVID: Database for annotation, visualization, and integrated discovery. *Genome Biol.* **4** (5), P3.
Diamond, I., and Gordon, A. S. (1997). Cellular and molecular neuroscience of alcoholism. *Physiol Rev.* **77**, 1–20.
Doss, S., *et al.* (2005). *cis*-Acting expression quantitative trait loci in mice. *Genome Res.* **15**, 681–691.
Edgar, R., *et al.* (2002). Gene expression omnibus: NCBI gene expression and hybridization array data repository. *Nucleic Acids Res.* **30**, 207–210.
Engel, S. R., *et al.* (1998). 5-HT3 receptor over-expression decreases ethanol self administration in transgenic mice. *Psychopharmacology (Berl.)* **140**, 243–248.
Fehr, C., *et al.* (2002). Congenic mapping of alcohol and pentobarbital withdrawal liability loci to a <1 centimorgan interval of murine chromosome 4: Identification of Mpdz as a candidate gene. *J. Neurosci.* **22**, 3730–3738.
Goldstein, D. B. (1973). Letter: Inherited differences in intensity of alcohol withdrawal reactions in mice. *Nature* **245**, 154–156.
Guo, A. Y., *et al.* (2009). ERGR: An ethanol-related gene resource. *Nucleic Acids Res.* **37**, D840–845.
Hill, S. Y., *et al.* (2008). Psychopathology in offspring from multiplex alcohol dependence families with and without parental alcohol dependence: A prospective study during childhood and adolescence. *Psychiatry Res.* **160**, 155–166.
Hodge, C. W., *et al.* (1999). Supersensitivity to allosteric GABA(A) receptor modulators and alcohol in mice lacking PKCepsilon. *Nat. Neurosci.* **2**, 997–1002.
Horvath, S., and Dong, J. (2008). Geometric interpretation of gene coexpression network analysis. *PLoS Comput. Biol.* **4**, e1000117.
Huang, D. W., *et al.* (2009). Systematic and integrative analysis of large gene lists using DAVID bioinformatics resources. *Nat. Protocols* **4**, 44–57.
Iwamoto, K., *et al.* (2004). Decreased expression of NEFH and PCP4/PEP19 in the prefrontal cortex of alcoholics. *Neurosci. Res.* **49**, 379–385.
Jansen, R. C., and Nap, J. P. (2001). Genetical genomics: The added value from segregation. *Trends Genet.* **17**, 388–391.
Jeanblanc, J., *et al.* (2009). Endogenous BDNF in the dorsolateral striatum gates alcohol drinking. *J. Neurosci.* **29**, 13494–13502.
Johnson, W. E., *et al.* (2007). Adjusting batch effects in microarray expression data using empirical Bayes methods. *Biostatistics* **8**, 118–127.
Junker, B. H., and Schreiber, F. (2008). Analysis of Biological Networks. Wiley, New York, NY.
Kang, H. M., *et al.* (2008). Accurate discovery of expression quantitative trait loci under confounding from spurious and genuine regulatory hotspots. *Genetics* **180**, 1909–1925.
Karpyak, V. M., *et al.* (2009). Sequence variations of the human MPDZ gene and association with alcoholism in subjects with European ancestry. *Alcohol. Clin. Exp. Res.* **33**, 712–721.
Kerns, R. T., and Miles, M. F. (2008). Microarray analysis of ethanol-induced changes in gene expression. *Methods Mol. Biol.* **447**, 395–410.
Kerns, R. T., *et al.* (2003). Application of the *S*-score algorithm for analysis of oligonucleotide microarrays. *Methods* **31**, 274–281.
Kerns, R. T., *et al.* (2005). Ethanol-responsive brain region expression networks: Implications for behavioral responses to acute ethanol in DBA/2J versus C57BL/6J mice. *J. Neurosci.* **25**, 2255–2266.
Kril, J. J., and Harper, C. G. (1989). Neuronal counts from four cortical regions of alcoholic brains. *Acta Neuropathol.* **79**, 200–204.

Leek, J. T., and Storey, J. D. (2007). Capturing heterogeneity in gene expression studies by surrogate variable analysis. *PLoS Genet.* **3**, 1724–1735.

Letwin, N. E., *et al.* (2006). Combined application of behavior genetics and microarray analysis to identify regional expression themes and gene–behavior associations. *J. Neurosci.* **26**, 5277–5287.

Lewandoski, M. (2001). Conditional control of gene expression in the mouse. *Nat. Rev. Genet.* **2**, 743–755.

Lewohl, J. M., *et al.* (2000). Gene expression in human alcoholism: Microarray analysis of frontal cortex. *Alcohol. Clin. Exp. Res.* **24**, 1873–1882.

Lewohl, J. M., *et al.* (2005). Expression of MBP, PLP, MAG, CNP, and GFAP in the human alcoholic brain. *Alcohol. Clin. Exp. Res.* **29**, 1698–1705.

Liu, B., *et al.* (2008). Gene network inference via structural equation modeling in genetical genomics experiments. *Genetics* **178**, 1763–1776.

Lyons, P. (2003). Advances in spotted microarray resources for expression profiling. *Brief. Funct. Genomic. Proteomic.* **2**, 21–30.

Mayfield, R. D., *et al.* (2002). Patterns of gene expression are altered in the frontal and motor cortices of human alcoholics. *J. Neurochem.* **81**, 802–813.

McBride, W. J., *et al.* (2010). Changes in gene expression in regions of the extended amygdala of alcohol-preferring rats after binge-like alcohol drinking. *Alcohol* **44**, 171–183.

McClearn, G. E., and Rodgers, D. A. (1959). Differences in alcohol preference among inbred strains of mice. *Q. J. Stud. Alcohol* **20**, 691–695.

Merikangas, K. R., *et al.* (1998). Familial transmission of substance use disorders. *Arch Gen. Psychiatry* **55**, 973–979.

Metten, P., *et al.* (1998a). Drug withdrawal convulsions and susceptibility to convulsants after short-term selective breeding for acute ethanol withdrawal. *Behav. Brain Res.* **95**, 113–122.

Metten, P., *et al.* (1998b). High genetic susceptibility to ethanol withdrawal predicts low ethanol consumption. *Mamm. Genome* **9**, 983–990.

Miles, M. F. (1995). Alcohol's effects on gene expression. *Alcohol Health Res. World* **19**, 237–243.

Miles, M. F. (2001). Microarrays: Lost in a storm of data? *Nat. Rev. Neurosci.* **2**, 441–443.

Mozhui, K., *et al.* (2008). Dissection of a QTL hotspot on mouse distal chromosome 1 that modulates neurobehavioral phenotypes and gene expression. *PLoS Genet.* **4**, e1000260.

Mulligan, M. K., *et al.* (2006). Toward understanding the genetics of alcohol drinking through transcriptome meta-analysis. *Proc. Natl. Acad. Sci. USA* **103**, 6368–6373.

Nestler, E. J., and Aghajanian, G. K. (1997). Molecular and cellular basis of addiction. *Science* **278**, 58–63.

Parkinson, H., *et al.* (2009). ArrayExpress update—From an archive of functional genomics experiments to the atlas of gene expression. *Nucleic Acids Res.* **37**, D868–872.

Peirce, J. L., *et al.* (2004). A new set of BXD recombinant inbred lines from advanced intercross populations in mice. *BMC Genet.* **5**, 7.

Peters, L. L., *et al.* (2007). The mouse as a model for human biology: A resource guide for complex trait analysis. *Nat. Rev. Genet.* **8**, 58–69.

Phillips, T. J., *et al.* (1994). Behavioral sensitization to drug stimulant effects in C57BL/6J and DBA/2J inbred mice. *Behav. Neurosci.* **108**, 789–803.

Prescott, C. A., *et al.* (1999). Sex differences in the sources of genetic liability to alcohol abuse and dependence in a population-based sample of US twins. *Alcohol Clin. Exp. Res.* **23**, 1136–1144.

Rimondini, R., *et al.* (2002). Long-lasting increase in voluntary ethanol consumption and transcriptional regulation in the rat brain after intermittent exposure to alcohol. *FASEB J.* **16**, 27–35.

Rockman, M. V., and Kruglyak, L. (2006). Genetics of global gene expression. *Nat. Rev. Genet.* **7**, 862–872.

Rodd, Z. A., *et al.* (2008). Differential gene expression in the nucleus accumbens with ethanol self-administration in inbred alcohol-preferring rats. *Pharmacol. Biochem. Behav.* **89**, 481–498.

Rosenbloom, M., *et al.* (2003). Using magnetic resonance imaging and diffusion tensor imaging to assess brain damage in alcoholics. *Alcohol Res. Health* **27**, 146–152.

Schadt, E. E., *et al.* (2005). An integrative genomics approach to infer causal associations between gene expression and disease. *Nat. Genet.* **37**, 710–717.

Schadt, E. E., *et al.* (2003). Genetics of gene expression surveyed in maize, mouse and man. *Nature* **422**, 297–302.

Schena, M., *et al.* (1995). Quantitative monitoring of gene expression patterns with a complementary DNA microarray. *Science* **270**, 467–470.

Schuckit, M. A. (1994). Low level of response to alcohol as a predictor of future alcoholism. *Am. J. Psychiatry* **151**, 184–189.

Scott, R. E., *et al.* (2005). P2P-R expression is genetically coregulated with components of the translation machinery and with PUM2, a translational repressor that associates with the P2P-R mRNA. *J. Cell Physiol.* **204**, 99–105.

Shirley, R. L., *et al.* (2004). *Mpdz* is a quantitative trait gene for drug withdrawal seizures. *Nat. Neurosci.* **7**, 699–700.

Smith, A. M., *et al.* (2006). Microarray analysis of the effects of a gamma-protein kinase C null mutation on gene expression in striatum: A role for transthyretin in mutant phenotypes. *Behav. Genet.* **36**, 869–881.

Sokolov, B. P. (2007). Oligodendroglial abnormalities in schizophrenia, mood disorders and substance abuse. Comorbidity, shared traits, or molecular phenocopies? *Int. J. Neuropsychopharmacol.* **10**, 547–555.

Stuart, J. M., *et al.* (2003). A gene-coexpression network for global discovery of conserved genetic modules. *Science* **302**, 249–255.

Tabakoff, B., *et al.* (2003). Selective breeding, quantitative trait locus analysis, and gene arrays identify candidate genes for complex drug-related behaviors. *J. Neurosci.* **23**, 4491–4498.

Thiele, T. E., *et al.* (1998). Ethanol consumption and resistance are inversely related to neuropeptide Y levels. *Nature* **396**, 366–369.

Treadwell, J. A., and Singh, S. M. (2004). Microarray analysis of mouse brain gene expression following acute ethanol treatment. *Neurochem. Res.* **29**, 357–369.

Tusher, V. G., *et al.* (2001). Significance analysis of microarrays applied to the ionizing radiation response. *Proc. Natl. Acad. Sci. USA* **98**, 5116–5121.

van Noort, V., *et al.* (2004). The yeast coexpression network has a small-world, scale-free architecture and can be explained by a simple model. *EMBO Rep.* **5**, 280–284.

Walter, N. A., *et al.* (2007). SNPs matter: Impact on detection of differential expression. *Nat. Methods* **4**, 679–680.

Xu, Y., *et al.* (2001). Comparison of global brain gene expression profiles between inbred long-sleep and inbred short-sleep mice by high-density gene array hybridization. *Alcohol Clin. Exp. Res.* **25**, 810–818.

Yang, R. J., *et al.* (2008). Variation in mouse basolateral amygdala volume is associated with differences in stress reactivity and fear learning. *Neuropsychopharmacology* **33**, 2595–2604.

Zachariou, V., *et al.* (2003). Essential role for RGS9 in opiate action. *Proc. Natl. Acad. Sci. USA* **100**, 13656–13661.

GENETIC VARIATION AND BRAIN GENE EXPRESSION IN RODENT MODELS OF ALCOHOLISM: IMPLICATIONS FOR MEDICATION DEVELOPMENT

Karl Björk*, Anita C. Hansson†, and Wolfgang H. Sommer†

*Translational Neuropharmacology, Center for Molecular Medicine, Karolinska Institutet, SE-171 77 Stockholm, Sweden

†Department of Psychopharmacology, Central Institute for Mental Health, D 68159 Mannheim, Germany

Much research on experimental animals that is aimed to decipher genetic factors involved in alcoholism has been devoted to either models of innate alcohol-related phenotypes or responses after acute alcohol challenge. Such focus has, however, limitations when it comes to the pathogenetic mechanism underlying alcohol addiction, because the progression into the disorder takes years and genetic

INTERNATIONAL REVIEW OF NEUROBIOLOGY, VOL. 91
DOI: 10.1016/S0074-7742(10)91005-2

as well as environmental factors may exert different influences along this trajectory. Animal models of the neuroadaptations involved in the development of dependence exist, but have been difficult to implement for genetic and genomics analysis. Consequently, currently available data have been difficult to reconcile with the human condition and could be misleading in predicting targets for medication development. This review will illustrate strengths and pitfalls of genomic approaches in rodent models of alcoholism and emphasize the need for convergent lines of evidence to improve the predictive value of such studies. Examples of a convergent research approach include validation studies for *Agt, Arrb2, Crhr1, Grin3a*, and *Npy*.

I. Alcohol Addiction, Genetic Susceptibility, and Neuroadaptation

Humans like to consume alcohol for various reasons, mostly because it changes emotional states, but also for its taste or its ability to alter palatability of food. Individuals differ in their response to alcohol, which may range from strong euphoria to outright aversive reactions in some people, but the majority of individuals experience and appreciate the mild anxiolytic or sedative effects of alcohol. These responses depend largely on individual constitutive factors. After some trial-and-error period, most people develop a stable pattern of consumption that may be modulated by various life events, but is generally characterized by control over the drinking situation. Unfortunately, some people cannot maintain controlled drinking. Over years they gradually develop an alcohol use disorder, in the most severe cases alcohol addiction. The disorder is characterized by compulsive, uncontrolled alcohol intake and chronic relapses that can occur even after years of abstinence. Alcohol consumption in this stage is mainly motivated by negative drives, such as anxiety, low mood or increased sensitivity to stress, and is thus fundamentally different from occasional, but controlled drinking that is generally positively motivated (Koob and Volkow, 2010). The risk to develop alcoholism is to a large proportion (50–60%) heritable, but shared by dozens or even hundreds of genes that interact with each other and with the environment, a pattern typical for most psychiatric disorders (Goldman *et al.*, 2005).

In contrast to most addictive substances, such as central stimulants or opioids, which have well-established effects on specific targets, alcohol affects a wide range of neurochemical systems (Spanagel, 2009). Alcohol has no high affinity receptor or binding site and the weak reactivity of the molecule results in low potency to produce its pharmacological effects. Nevertheless, a few specific binding sites have been identified for alcohol including the NMDA, $GABA_A$, 5-HT3, and nACh receptors as well as L-type calcium channels and GIRKs (for recent reviews see Harris *et al.*, 2008; Spanagel, 2009). Physiological relevant doses of alcohol, ranging from 5 to 100 mM blood and brain concentrations, alter the function of these targets and produce the acute neural effects that lead to the feeling of intoxication ranging depending on the dose, from disinhibition to sedation and even hypnosis. On the neural level, the primary effects of alcohol are followed by a wave of indirect effects involving a variety of neurotransmitter and neuropeptide systems—mainly monoamines, endogenous opioids, and endocannabinoids—which are thought to underlie the positively reinforcing effects. These are likely to be most relevant in early stages of alcohol use when alcohol associated cues are predictive of positively reinforcing alcohol effects, leading to what can be called "reward craving." A key factor in determining an individual's response to alcohol, including the feeling of positive effects or of being intoxicated, is genetic predisposition. Various candidate genes for these reactions have been put forward from human genetic studies (Ray and Hutchison, 2004; Schuckit *et al.*, 2009). However, it is still a matter of debate to what extent these traits and associated candidate genes contribute to an increased risk for developing alcoholism.

One example of how genetic factors may mediate individual responses to alcohol is a single-nucleotide polymorphism within the first exon of μ-opioid receptor OPRM1. The variant A118G affects binding of endogenous ligands and the G-allele is consistently associated with increased experience of euphorigenic effects of alcohol (Bond *et al.*, 1998; Kakko *et al.*, 2008; King *et al.*, 1997; Ray and Hutchison, 2004; Wand *et al.*, 2002). A functionally equivalent variant in *rhesus macaques* is associated with increased psychomotor stimulation by alcohol, increased alcohol preference, and increased frequency of alcohol consumption to the level is associated with in the rhOPRM1 gene (Barr *et al.*, 2008). Although the convergence of phenotypic responses is compelling, translation on the level of sequence variation may not be possible, in particular when it comes to assigning complex functional or behavioral effects to a single-point mutation that could be in linkage disequilibrium with an unknown number of other variants. One possibility for such a direct interrogation of distinct human genetic variants is to generate mice that carry human DNA sequences, sometimes referred to as humanized mice. In the case of OPRM1 A118G this was accomplished by replacing the endogenous mouse exon 1 with either allele of huOPRM1 A118G (Ramchandani *et al.*, 2010). The resulting mouse lines thus differ only in a single nucleotide. We demonstrated that G-allele carrying mice have a strongly enhanced alcohol-evoked release of dopamine in the nucleus accumbens compared to mice homozygote for huOPRM1 A-allele, thereby linking this individual polymorphism to a distinct neurochemical mechanism. These findings were paralleled by PET imaging of healthy human subjects demonstrating increased alcohol-induced displacement of the D2 ligand ^{11}C-raclopride in the ventral striatum of human G allele carriers compared to AA homozygotes (Ramchandani *et al.*, 2010). Together, these studies provide converging evidence across species that genetic variation at the OPRM1 locus modulates the acute responses to alcohol. The mouse model will be suitable to investigate whether excessive activation of brain reward systems by alcohol in OPRM1 118G carriers confers susceptibility for alcohol use disorders and to what extend this initial response to alcohol is changed once dependence has developed. In humans, being an OPRM1 118G carrier does not seem to predict increased risk for alcohol use disorders (Arias *et al.*, 2006). However, this conclusion is drawn from a meta-analysis of available data for A118G, and it is likely that such an approach pools genetically and pathophysiologically highly heterogeneous populations within the current diagnostic categories of alcoholism.

A critical point that deserves to be emphasized is that the responses to alcohol, both in regards to its pharmacological effects and to what comprises the motivational forces driving alcohol-related behaviors, change with repeated use. If excessive alcohol consumption progresses, brain stress systems involving various neuropetides (e.g., CRH, NK1, or NPY) are becoming engaged and increasingly sensitized, which in turn results in further escalation of alcohol use and high propensity to relapse, which is now driven by negative reinforcement or "relief" craving (Heilig *et al.*, 2009). Genetic factors may produce their effects at any stage of the addiction trajectory: initially by mediating direct drug effects, providing high or low motivation for drug taking, then through different adjustments to increasing or rapidly changing alcohol levels, and finally by building up powerful incentives to maintain an obviously aberrant, negative behavior. Most individuals that are exposed to alcohol, even those subjected to potentially harmful environmental factors, do not develop addiction. This suggests that the brain is equipped with adaptive systems providing resiliency toward such insults. However, in some instances adaptive responses or the breakdown of adaptive capacity in one or several brain systems may lead to the development of addiction (Koob, 2003).

A further important factor in the acute and chronic responses to alcohol is pharmacokinetics, because individuals differ widely in alcohol metabolism. The primary metabolite of alcohol is acetaldehyde, which when it accumulates can trigger a number of unpleasant vegetative reactions known as "flushing response." This reaction may act as a deterrent to further alcohol use. As an exception to the rule there is no major contribution of individual genes in complex psychiatric disorders, certain variants within alcohol metabolizing genes, that is, the alcohol/aldehyde dehydrogenase gene complex exert indeed strong protective effects on the risk of developing alcoholism (Edenberg, 2007), but do so likely at the expense of decreased defense against reactive metabolites and increased cancer liability (Brooks *et al.*, 2009; Sinclair, 2006).

II. Animal Models and the Trajectory of Alcohol Addiction

The purpose of an animal model of psychiatric disorders cannot be to perfectly copy the human condition, but to inform about (neuro)biological mechanisms underlying the disease process and to provide a tool for hypothesis testing, for example, if a molecular target might be useful for medication development. There is still some debate whether an animal model should cover as much of the symptomatology of a given psychiatric disorder as possible, for example, fulfill all diagnostic criteria of the Diagnostic and Statistical Manual for Mental Disorders (American Psychiatric Association, 1994) for alcoholism, or whether it is more useful to focus on a distinct core symptom that is recognized as a significant obstacle in treatment outcome. From a clinical perspective, the latter strategy appears to be more relevant. First, alcoholics, at least in early stages of their disease, present a highly heterogeneous, sometime diagnostically challenging clinical picture, which makes it difficult to identify common mechanisms. Second, the major problem in treating alcoholism is relapse and the compulsive mechanisms that may lead to it. Thus, the occurrence of relapse behavior or the intensity of craving may have little to do with tolerance to alcohol's pharmacological effects or the severity of the acute withdrawal reaction, two main diagnostic criteria for alcoholism. In fact, some of the most valid rodent models for medication development in alcoholism, that are, the alcohol deprivation effect or reinstatement of alcohol seeking, do not encompass any signs of tolerance or physical withdrawal (Egli, 2005; Sanchis-Segura and Spanagel, 2006).

Our understanding of the neurobiology of alcoholism comes largely from animal studies and is validated by at least two major developments: first, by the fact that all existing medications for this disorder are directly derived from animal experiments (Egli, 2005), and second the general concordance with an increasing amount of human brain imaging studies (Volkow *et al.*, 2003). Laboratory rodents can adequately model distinct characteristics of an addictive behavioral syndrome related to excessive drug intake, seeking and craving. These models have shown good predictive validity for identifying targets for novel pharmacological treatments for alcohol dependence (Egli, 2005). Consequently, these models are expected to be useful for genetic and genomic studies aimed to identify genes involved in the development of alcohol addiction.

Three classes of animal models can be recognized that have been considered for the study of genes and transcriptome responses in alcoholism. The first class addresses acute or short-term effects of alcohol and has been of key importance for identifying the pharmacological targets of alcohol and in studying its immediate behavioral consequences. The second class of models addresses genetically encoded vulnerability for the development of alcoholism and has been mainly used to study traits related to initial alcohol responses. Within this category fall some inbred mice lines with exceptional high innate alcohol preference as well as rodent lines that have been developed by selective breeding for high or low alcohol consumption. The third category is based on alcohol-induced neuroadaptations, relating to the fact that addiction inevitably requires exposure to the drug. Animals either have access to alcohol over long periods of time or are involuntarily exposed to sufficiently high alcohol levels to induce dependence, which may reflect various states in the progression of the disorder.

As illustrated in Fig. 1, the three types of models can be placed along the addiction trajectory, and accordingly could be considered as relatively separate entities that may or may not be connected by shared neurobiological mechanisms. In other words, genetic factors that play an important role in mediating

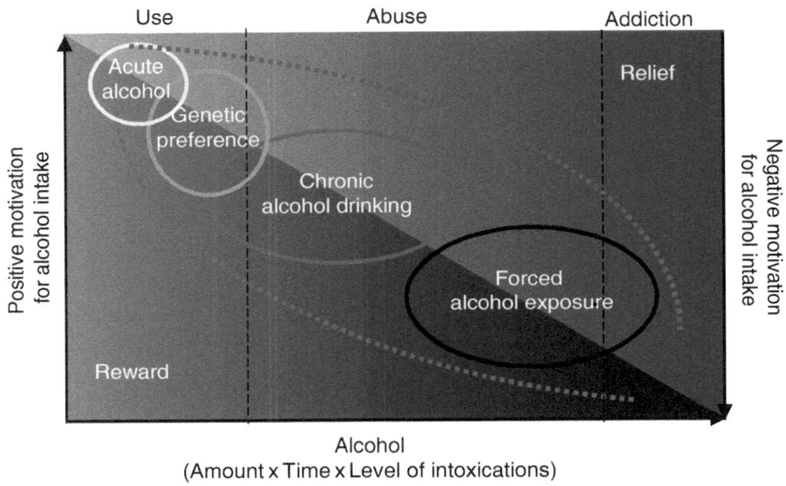

FIG. 1. Conceptual framework placing the common classes of animal models along the trajectory of addiction. Alcohol dependence progresses over time from initial, largely positively reinforcing, pleasurable alcohol effects (reward), to be maintained by relief from a negative emotional state (negative reinforcement). Genetic and environmental factors interact with each other and determine an individual's response to alcohol. Importantly, the influences of these factors vary along the trajectory of accumulated alcohol effects. Thus, an individual may experience alcohol consumption for a long time as positive, but finally show a rapid break down of the reward system (red dotted line). Alternatively, ones constitution may entail sensitized stress systems which could rapidly take-over the incentive for repeated alcohol use and entrap this person by powerful negative reinforcement already early in the addition process (blue dotted line). Animal models can only address distinct aspects of the addiction trajectory. Widely used genetic models and acute alcohol challenges cover the early, primarily positively reinforced phases of alcohol actions, but miss the neuroadaptive processes characterizing the progression into addiction. Chronic alcohol consumption and forced intoxication procedures have been shown to involve various degrees and aspects of negative reinforcement, but are generally difficult to utilize for candidate gene discovery (See color plate 2).

early alcohol responses, such as the opiate–dopamine cascade in positive reinforcement, may be completely disengaged when behavioral responses are uncoupled from normal motivational control at the final stages of addiction. Similarly, stress response genes, such as the corticotropin-releasing factor receptor subtype 1 (CRHR1), may have little weight for some of the initial positive emotional actions of alcohol. The distinction of the various models seems to be obvious, but to neglect these facts is problematic for studying the mechanisms underlying alcohol addiction and particularly for identifying therapeutic candidate targets. A wealth of candidate genes has been obtained using acute and genetic models but the relevance of many of these genes remains to be established. For long-term alcohol exposure models, extensive genomic and transciptome studies are lacking, but the emerging data suggest that these experiments can result in the identification of candidate genes that show greater consilience with human studies.

In the following we will briefly review some of the genetics and genomics findings from the various types of animal models. We will present our view on the strength and weaknesses of these models and their applications. Our main position is that clinically relevant genetic and genomic factors for the etiology of alcohol addiction can only be identified by attending to the drug-induced neuroplasticity and the formation of robust addiction-related behaviors, that is, by using animal models of the disease state. Trait models contribute importantly to the understanding of alcohol's various mechanisms of action, but they may be of limited use for finding mechanism involved in alcoholism, because none of the initial alcohol response traits is strongly associated to the human disorder. The best chance for finding consilience with human data and so to increase our knowledge about the human condition, results from animal studies searching for multiple lines of evidence that converge on a candidate mechanism.

III. Acute Pharmacological Responses to Alcohol

A. Transcriptome Studies of Acutely Alcohol-Responsive Genes

Genome-wide expression analysis of the acute responses to alcohol has been performed by several groups (Kerns *et al.*, 2005; Rulten *et al.*, 2006; Treadwell and Singh, 2004; Uddin *et al.*, 2005). Large numbers of genes affected by the alcohol challenge in a brain region-wise manner have been reported, but as to be expected from alcohol's complex pharmacology, no prominent mechanism could be unraveled. The most robust response to alcohol show gene clusters related to plasticity, myelination, neurogenesis, metal ion homeostasis, and cellular stress response. In the process, web-based tools were developed for browsing the alcohol-responsive transcriptome (e.g., at http://bioinfo.vipbg.vcu.edu/ERGR (Guo *et al.*, 2009)).

Without doubt, genome-wide expression experiments give important clues about mechanistic aspects of the acute alcohol response. The findings converge with in vitro and tissue culture methods for promoter analysis and point to a distinct, but large set of genes that could be regulated by alcohol through shared promoter elements, including the cAMP-response (CRE), activator protein-1 (AP-1), specificity protein-1 (SP-1), and the alcohol response (ARE) binding elements (recently reviewed by Pignataro *et al.*, 2009). Importantly, the mechanisms involved in the acute responses to low and high doses of alcohol are conserved across a wide range of species, including worms, flies, fish, and rodents (Davies *et al.*, 2003; Moore *et al.*, 1998; Peng *et al.*, 2009). The functional implication of these acutely responding genes for alcohol's various effects has only been proven for relatively few candidates and requires generally multiple lines of evidence dissecting the actions of involved genes. A recent study tested the importance of decreased *Kras* expression by alcohol in mutant mice showing either up- or down-regulation of *Kras* (Repunte-Canonigo *et al.*, 2010). Genetic interference with the expression of this small G-protein was associated with altered transcriptomic response to alcohol challenge involving distinct signal transduction pathways such as PI3K, NFKB, and Jak/Stat signaling. These findings are functionally validated by the demonstration that reduced *Kras* expression prevented the development of addiction-related behavior, that is, increased voluntary alcohol consumption after chronic intermittent alcohol vapor intoxication (see Section V.B.). Notably, *Kras* is also strongly upregulated in the brain of genetically high alcohol-preferring (HAP, see Section IV.A.) mice (Mulligan *et al.*, 2006). Given the pleiotropic actions of alcohol, an involvement of the RAS pathway is to be expected, but the convergence on distinct members of the RAS family and their interacting proteins may point to specific roles of these molecules for distinct alcohol responses. It will be interesting to see whether such observations will hold up as candidate genes for human alcohol consumption or dependence.

Furthermore, the induction of transcription factors such as *Fos* or *Egr1* by alcohol has been highly useful for delineating the brain sites that respond acutely to alcohol (reviewed in Vilpoux *et al.*, 2009). However, comparing maps of alcohol-sensitive regions from naïve- and alcohol-experienced rats clearly shows that this response is changing strongly with the development of dependence (Hansson *et al.*, 2008). Such information is lacking for nearly all acutely alcohol-responsive genes, but addressing these issues will importantly contribute to our understanding of how transcriptional control mechanisms participate in the disease process.

B. Models Derived by Reverse Genetic Design

Transgenic and knockout animal models have been utilized to a large extent for studying the role of individual candidate genes in the acute or early responses to alcohol. A recent review covering nearly 150 mouse mutants found that the majority of these lines show at least some alteration in alcohol consumption and other related phenotypes (Crabbe *et al.*, 2006). The results from these mutant lines emphasize the complexity of alcohol-related traits. Because the response to alcohol is intricately linked to a wide range of neurobiological systems, it will be difficult to prove alcohol-specific mechanisms in individual global gene knockouts. Conditional genetic interference will allow more specific perturbation, but the problem remains that alcohol responses are likely mediated by combined actions of many rather than a few genes. Furthermore, existing genetically engineered lines are so far rarely studied for neuroadaptations that are brought about by chronic alcohol use and that may be involved in the development of addiction. The importance of such studies is exemplified by knockout mice for the *Crhr1* gene. These mutants are not impaired in their acute responses to alcohol, but show dependence-related phenotypes after long-term exposure to the drug (Chu *et al.*, 2007; Sillaber *et al.*, 2002). In addition, the deletion of a single gene may have widespread consequences for the transcriptomic response to alcohol. For example, mice deficient for the gamma subunit of protein kinase C, *Prkcg*, do not develop tolerance to alcohol, and more than 100 genes react different to chronic, high-dose alcohol treatment between mutant and wild-type mice (Bowers *et al.*, 2006). The study identified a *Prkcg*-dependent gene, *Kcnk1*, which codes for a potassium channel and was related to resistance to tolerance.

In summary, experiments involving acute or short lasting alcohol challenges show wide spread genomic responses pointing to mechanisms of action. These responses change with repeated or chronic exposure to alcohol, but so far these dynamics have been little studied in the various experimental paradigms, and this information is particularly missing for the many transgenic lines showing altered initial alcohol responses. Another underdeveloped area is the validation of findings for the human condition. Investigations on human postmortem brains are not very helpful in this regard, because these studies cannot be designed in a manner that proves causality. A potential approach could be to study transcriptome responses of peripheral blood cells in humans after well-controlled alcohol challenge. Comparison of such data to corresponding blood and brain expression profiles from experimental animals could give important clues about translatable genomic responses to alcohol and may ultimately result in accessible biomarkers for diagnostic and monitoring of alcohol use disorders.

IV. Models of Genetic Susceptibility: Selected Rat Lines and Inbred Mouse Strains

A. Genetic Models Derived by Selective Breeding

A common strategy to establish the heritability of a distinct phenotype or disorder is the so called forward genetic approach. The term includes phenotype-driven selective breeding and selection of animal lines with innate phenotypic differences. Among the earliest attempts to develop a disease model was selective breeding for the amount of alcohol an animal is willing to consume under the assumption that a high preference for alcohol may be a contributing factor to developing alcohol addiction. The first of these experiments was successfully reported by Mardones of the University of Chile more then 60 years ago (Mardones *et al.*, 1953). Animals were selectively bred for their daily consumption of a 10% alcohol solution. Within a few generations there was virtually no overlap in consumption between animals of the high- and low-preferring lines. Ultimately these researchers found that variation within alcohol metabolizing enzymes, in particular variants of the aldehyde dehydrogenase *Aldh2*, predict the major proportion of variance in alcohol consumption (Quintanilla *et al.*, 2006). The validity of this experimental approach is clearly demonstrated by the well-established fact that in humans certain variants within the alcohol/aldehyde dehydrogenase complex have a clear protective effect on the risk for developing alcoholism (Edenberg, 2007).

Due to its simplicity, many laboratories around the world have tried to replicate Mardones' experiment and succeeded in establishing at least eight rat and one mouse line with high and low alcohol preference (recently summarized in Crabbe *et al.*, 2010). Genetically preferring rats will consume about 5–10 g/kg/day and mice up to 20 g/kg/day. The underlying idea is that the selection pressure applied gradually leads to enrichment of alleles promoting or preventing alcohol drinking in the high- and low-preferring lines, respectively. Genetic factors controlling the behavior should thus be more easy to identify. By and large, it has been proven difficult to link observed differences conclusively to alcohol-drinking behavior. Because behavioral traits are multifactorial controlled and genes have pleiotropic actions, the different alcohol-preferring lines show considerable phenotypic diversity. Some of the behavioral or neurochemical traits may in fact be co-segregated with alcohol preference. In line with our general concepts of human alcoholism, co-segregated traits may fall into the anxiety domain as observed in the P and sP rat lines (Colombo *et al.*, 1995; Hansson *et al.*, 2006; Stewart *et al.*, 1993), or alternatively could reflect increase impulsivity like in AA rats (Möller *et al.*, 1997). The major source of line differences, however, is random fixation of alleles that are neutral to the selective pressure. These are expected to segregate randomly, but in populations with a small number of breeders such alleles will either be fixed or lost with time. Hence, each of these lines and their respective controls represent a unique combination of selected and random traits. One way to partially control for the effects of random fixation is to generate multiple replicate lines for a selection experiment, and approach that has also been utilized in the breeding of some alcohol-preferring lines (i.e., HAD/LAD rat and HAP/LAP mouse lines).

1. *Transcriptome Analysis Illustrates Caveats of Selected Lines*

Comparing genome-wide expression patterns from the pairs of lines is a common approach to identify candidate genes whose differential expression is the result of allelic segregation. Various brain regions from alcohol-naïve animals have been subjected to this type of analysis resulting in identification of large numbers of differential expressed genes between the respective high and low alcohol-preferring lines (Arlinde *et al.*, 2004; Björk *et al.*, 2006; Ciccocioppo *et al.*, 2006; Edenberg *et al.*, 2005; Kimpel *et al.*, 2007; Mulligan *et al.*, 2006; Sommer *et al.*, 2001, 2006; Worst *et al.*, 2005). A somewhat surprising result of these studies was the notion of extremely low overlap in differential gene expression between the various models. A recent direct comparison of brain expression profiles from three alcohol-preferring rat lines gives a possible explanation for this observation by demonstrating that about one-third of the 6000 interrogated gene probes were significantly different expressed between P, HAD, and AA rats (Fig. 2; Matthaus *et al.*, 2009).

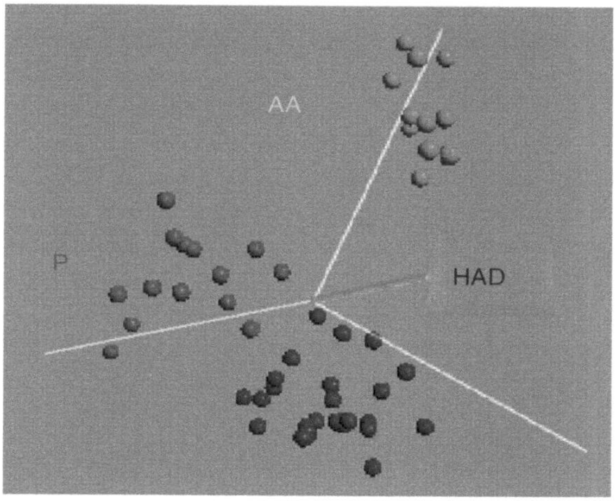

FIG. 2. Profound differences in brain gene expression between three selected lines for high alcohol preference. Principal component analysis was used to analyze microarray expression profiles from caudate–putamen or amygdala of AA, P, and HAD rats (Matthaus *et al.*, 2009). Each blob represents the summarized expression data from an individual microarray. Data are normalized for the effect of brain region. Analysis was done using Qlucore Omics Explorer (http://www.qlucore.com).

Because of the factors described above, the causal relationship between enrichment of alleles and increased alcohol consumption is often unclear, constituting a major hurdle in identifying candidate loci and genes for a specific trait as exemplified by studies on *Gsta4* (glutathione S-transferase alpha 4) in P/NP and AA/ANA rats. *Gsta4* was identified as differentially expressed in a high-throughput expression screen of several brain regions of P and NP rats (Liang *et al.*, 2004). Prompted by the observation that *Gsta4* is also located in proximity of a quantitative trait locus for alcohol preference in mice (Belknap and Atkins, 2001) the authors further investigated this gene and confirmed that alcohol-preferring P rats compared to non-preferring NP rats display a robust decrease in both *Gsta4* mRNA and protein levels. In contrast, alcohol-preferring AA rats show nearly twofold higher *Gsta4* brain expression compared to their low preferring genetic counterpart line (Björk *et al.*, 2006). P and ANA lines carry a different variant haplotype of the gene correlated with lower expression, suggesting that the differences are most likely mediated by cis-regulatory factors. The opposing correlations between alcohol preference and *Gsta4* expression observed in the two pairs of selected lines suggests that this gene is not likely involved in alcohol preference but was likely subject to random genetic fixation or may interact with the different genetic backgrounds of the founder lines by yet unknown mechanism (involving epistasis) pointing to the complexity and heterogeneous nature of alcohol preference.

Behavioral assessments of the three alcohol-preferring lines revealed also a potential pitfall of the genetic models, that is, a narrowed behavioral reaction norm. This term describes the amount of change in a quantitative phenotypic trait displayed in response to a certain amount of environmental challenge or pressure (Levine *et al.*, 1979). A line generated through the pressure of selective breeding may be "locked into" a particular phenotype, and thus will be less suita +ble for studies addressing changes in behavioral phenotype and gene expression patterns in relation to, for example, pharmacological or environmental challenge. Thus, for the P, HAD and AA lines it appears that that selection for high alcohol preference does not predict responsiveness in terms of alcohol-related behaviors. Only P rats, but not AA or HAD animals, responded with increased alcohol consumption to challenges such as alcohol deprivation or environmental stress (Vengeliene *et al.*, 2003).

2. *Gene Expression, Pharmacoresponse, and Medication Development*

Albeit selected lines have limitations when it comes to identifying genetic components intimately associated with alcohol-drinking behavior, they have become valuable tools for medication development (Egli, 2005). In spite of that alcohol addiction is a highly prevalent psychiatric disorder, only two medications, the glutamatergic modulator acamprosate and the opioid antagonist naltrexone, are internationally approved for the treatment. These drugs have provided proof-of-concept in large clinical studies, but also point to a need for identifying additional pharmacological treatments (Bouza *et al.*, 2004). Established differences in genomic and neurochemical profiles between the various lines may not be a disadvantage, but rather can be an asset for medication development. Thus, some lines are responsive to certain classes of drugs, but may be inadequate for testing other compounds as illustrated in the following example.

Alcohol, as mentioned earlier, physically interacts with and antagonizes the NMDA receptor. Accordingly, NMDA antagonists have been proposed for the treatment of alcoholism (Spanagel, 2009). Based on preclinical findings, neramexane, a noncompetitive NMDA receptor antagonist, has been deemed especially promising. However, clinical studies with alcohol-dependent patients have so far yielded negative results, probably due to insufficient dosage (Spanagel and Kiefer, 2008). Neramexane suppresses voluntary alcohol consumption in AA rats but not in another alcohol-preferring line, the msP line (Spanagel, 2009). In search of genetic factors underlying such a difference in pharmacoresponsiveness, microarray-derived brain expression profiles were compared between these lines and the gene for the NMDA receptor subtype 3 A, *Grin3a*, was found to be significantly upregulated in cortex of msP rats compared to AA rats. The increased expression is likely due to genetic variation at the *Grin3a* locus in the msP rats (KB, unpublished data). NMDA receptors composed of NR1–NR3A subunits exhibit a reduced sensitivity to channel blockers compared with NR1–NR2A receptors (Chatterton *et al.*, 2002). Thus, it is possible that the increased NR3A expression in msP rats results in an overall downregulation of NMDA signaling, rendering these rats insensitive to further suppression by neramexane. Markedly elevated expression levels of NR3A have also been found in the brains of psychiatric patients (Mueller and Meador-Woodruff, 2004), supporting the notion that NMDA receptor blockers can provide a therapeutic venue in alcohol-dependent patients only when sufficient doses of these drugs are applied.

The concept of personalized medicine, that is, achieving increased treatment response by the use of genotype directed treatment has received considerable attention recently. Studies like the one described above raises the interesting prospect that selected lines might be used to identify genes underlying the differential treatment response. Such a strategy would include obtaining a pharmacological profile for drugs relevant for clinical treatment of alcoholism in a panel of genetically selected alcohol-preferring rat lines. Lines that display a divergent response to a specific drug would then be screened for gene expression differences and genetic variation. Follow up on these findings would include going back to the clinic and conduct targeted genetic screens in patients where the selection of candidate genes is based on the rodent data.

3. *Candidate Gene Selection from Selected Lines Using Convergent Lines of Evidence*

Even though identification of candidate genes for alcohol preference is difficult and multiple lines of evidence are needed to link any of these findings with good confidence to the phenotype, such support has indeed been compiled for some genes, here exemplified by studies on β-arrestin 2 (*Arrb2*), neuropeptide Y (*Npy*) and angiotensinogen (*Agt*) .

Arrb2 was identified as a differentially expressed gene in a global expression screening of several brain regions of AA and ANA rats (Arlinde *et al.*, 2004). Further investigation of this gene was incited by studies showing that mouse lines selected for high alcohol preference display robustly upregulated *Arrb2* levels compared to their non-preferring counterparts (Mulligan *et al.*, 2006). The major function of the β-arrestins is to facilitate desensitization and internalization of G-protein coupled receptors following agonist stimulation (Björk *et al.*, 2010; Lefkowitz and Shenoy, 2005). *Arrb2* is a critical modulator of μ-opioid receptor signaling (Bohn *et al.*, 1999, 2003). Interestingly, alcohol-preferring AA rats exhibit increased *Arrb2* mRNA transcript and protein levels in several brain regions compared to ANA rats, and these differences appear to translate to protein levels (Björk *et al.*, 2008). To assess the functional importance of *Arrb2* for alcohol-related behaviors, mice lacking this gene were tested in a number of paradigms. Mutant mice display both reduced voluntary alcohol consumption and alcohol-evoked psychomotor stimulation, suggesting a role for *Arrb2* in mediating the rewarding effects of alcohol. There are currently no data regarding a role of ARRB2 in human alcoholism, but an involvement of β-arrestins in nicotine dependence is strongly suggested by human genetic studies (Sun *et al.*, 2008).

An extensive body of evidence from genetically modified animals exists supporting a role for *NPY* in alcohol preference (reviewed in (Thorsell *et al.*, 2006) and relapse behavior (Cippitelli *et al.*, 2010; Gilpin *et al.*, 2003)). Two of the selected rat lines, P and AA, show decreased *NPY* mRNA levels in several brain areas compared to their respective genetic counterparts (Arlinde *et al.*, 2004; Caberlotto *et al.*, 2001; Ehlers *et al.*, 1998). Expression of the peptide in the amygdala appears of special importance, not only for the trait anxiety shown by P rats, but also for alcohol consumption and relapse-like behaviors as demonstrated by specific interference with gene expression in this region using recombinant viral vectors (Thorsell *et al.*, 2007). As for *Arrb2*, the expression differences are likely due to *cis*-regulatory factors and genetic variants possibly leading to reduced expression have been identified (Spence *et al.*, 2005). These studies are of relevance for the human condition because the importance of variation in the NPY promoter for brain NPY expression and its association with stress responses and alcohol consumption was also demonstrated in primates, including *Rhesus macaques* and humans (Lindell *et al.*, 2010; Sommer *et al.*, 2010). Together, these findings suggest that variation at NPY locus interacts with the degree of stress loading raising the possibility that human NPY variation could potentially increase risk of alcohol dependence more so among individuals with especially traumatic life experiences and early or high cumulative levels of stress exposure. In support of this argument, the only reports of a link between NPY variation and alcohol dependence have studied individuals with late-onset alcoholism or samples highly represented by war veterans (Mottagui-Tabar *et al.*, 2005; Zhu *et al.*, 2003). These studies underline the importance of gene–environment interaction for the development of alcoholism and further studies in this direction are warranted.

A wider ranging approach for candidate gene identification was recently proposed by merging brain gene expression data with human genetic linkage data, as well as human postmortem tissue data and information on biological function (Rodd *et al.*, 2006). This convergent functional genomics approach combined the analysis of three animal paradigms based on the P and NP rats at base line condition and their response to treatments with alcohol. Microarray gene expression data from five key brain regions (frontal cortex, amygdala, caudate–putamen, nucleus accumbens, and hippocampus) were analyzed. The integration of multiple independent lines of evidence, each by itself lacking sufficient discriminatory power, led to the identification of high probability candidate genes, pathways, and mechanisms for alcoholism, among those highly interesting targets for potential pharmacological intervention with existing nonaddictive agents including ACE inhibitors for the angiotensin pathway, and mood stabilizers (lithium for PKC epsilon, lamotrigene and zonisamide for alpha-synuclein), or olanzapine for GABA-A1 receptor. The approach was validated using the angiotensin-converting enzyme (ACE) inhibitor lisinopril that markedly decreased alcohol intake in P rats (Rodd *et al.*, 2006). In line with the P rat, also other selected lines with high alcohol preference, that is, AA rats and HAP mice, show increased brain expression of the angiotensin precursor gene *Agt* (Saba *et al.*, 2006; Sommer *et al.*, 2006). A role for *Agt* as a positive modulator of alcohol consumption is further strengthened by work done in mutant mice with various types of genetic modifications in the angiotensin-signaling pathway (Maul *et al.*, 2001, 2005). Importantly, alcohol addicted rats show a long-lasting increase of brain *Agt* expression and respond to ACE inhibition with a more pronounced decrease in alcohol consumption compared to nonaddicted rats (Sommer *et al.*, 2007). Interaction of brain angiotensin and CRH systems has been suggested as an alcohol intake promoting mechanism (Sommer and Saavedra, 2008). These interactions may take place on a prewired background in the genetic models or may emerge as a dysregulation between various neurochemical systems during development of alcohol dependence.

B. Applying Inbred Mice Strains for the Investigation of Alcohol-Related Phenotypes

The use of mice in the search for alcohol-related genes has been different from rat. This is due to the availability of a large number of different inbred strains, which widely differ in their phenotypes and cover a rich reservoir of the natural genetic variance of this species. Inbred lines result from brother–sister mating for at least 20 generations and are genetically homogeneous, in other words, individuals of the same line are genetically identical. Provided the genotype is stable, these lines allow cumulative data collection over many generations. Great innate differences in alcohol preference between inbred strains of mice have been long known (McClearn and Rodgers, 1959). For example, the C57BL/6 consumes about 10 times more of a 10% alcohol solution compared to the DBA/2 line, and these differences in alcohol preference have remained stable over nearly 50 years (Wahlsten *et al.*, 2006). Consequently, much less emphasis has been given to selective breeding for alcohol preference/consumption.

Inbred mouse lines, in particular of the C57BL/6 and DBA/2 lines, have been intensely used to generate intercross populations (B6D2) for gene mapping using quantitative trait locus analysis. This approach correlates continuous measures of a distinct trait, for example, alcohol preference or withdrawal severity, with the information from allele specific markers spaced out across the genome. It is thereby possible to link the trait to one or several regions of the genome. With the availability of high-resolution genomic maps of the mouse genome and advanced methods to generate mapping populations, this approach becomes increasingly sensitive. Large scale re-sequencing of the mouse genome has produced high density maps of the genetic variation between more than 50 commonly used inbred mouse strains and identified over eight million genetic differences between them (http://mouse.perlegen.com/mouse/index.html), and thus allowed the construction of haplotype blocks of the mouse genome (Frazer *et al.*, 2007). Today, more than hundred QTLs for alcohol-related traits are recorded, again showing the complexity of the genetic control underlying these phenotypes. A database summarizing information on alcohol-related QTLs and related fine-mapping by congenic strains (mice that possess a short interval of DNA surrounding the QTL from a donor strain transferred onto the distinct genetic background of a recipient strain) is available at www.ohsu.edu/parc/data/qtl/by_phen.shtml

The major challenge of QTL mapping remains the identification of QTGs, the genes responsible for the trait. The difficulties arise from the fact that QTLs represent a relatively large genomic region that could harbor hundreds of genes. Thus far only one gene, *Mpdz*, implicated in protein–protein interactions and a known interaction partner of 5-HT_2 and $GABA_B$ receptors, has been conclusively linked to be the source of the effect of a QTL for alcohol withdrawal seizure intensity. The path to this discovery from the initial description of the QTL to mounting enough sufficient evidence for proving the case of the candidate took a substantial amount of time (Buck *et al.*, 1997; Shirley *et al.*, 2004). A second strong candidate QTG is *Scn4b* (voltage-gated sodium channel, subunit beta-4), which has been extracted out of a large meta-analysis of differential expression between alcohol-preferring mouse lines and fine mapping of a QTL using congenic lines for Chr9 (Mulligan *et al.*, 2006). It remains to be proven that *Scn4b* is the sole source of the effect of this QTL.

C. Genetical Genomics: Combining QTL Mapping and Genome Wide-Wide Expression Profiling as a Short Cut to Candidate Gene Identification?

Differentially expressed genes can be mapped to alcohol QTLs and those who pass this filter can be searched for polymorphisms. Using this strategy the carnitine transporter, *Slc22a4*, was found be linked to alcohol sensitivity (loss of righting response) and to contain several promoter polymorphisms that disrupt putative transcription factor binding sites (Maclaren *et al.*, 2006). Interestingly, the human ortholog maps to a genomic region associated with human alcohol sensitivity in linkage studies.

Moreover, gene expression can be seen as a quantitative trait by itself and consequently be mapped to a distinct position in the genome, thereby delineating loci that are involved in the control of expression of individual genes (this has been termed expression QTL or eQTL for distinction from QTLs obtained for a behavior, in this context referred to as bQTL). By overlaying a bQTL with eQTLs for genes differentially expressed in the mapping population one can considerably narrow down the number of genes that arises from expression profiling studies and that might be in control of the behavior. This integration of transcriptional profiling and linkage analysis has been termed "genetical genomics" (Jansen and Nap, 2001). Extensive data on brain expression difference between the C57BL/6 and DBA/2 mice and their derivative intercross or recombinant lines have been collected by several consortia and revealed potential regulatory networks involved in various alcohol-related phenotypes. These data are comprehensively and freely available to the community on the WebQTL site, http://www.genenetwork.org (Chesler *et al.*, 2005).

For alcohol-related phenotypes, this approach has been repeatedly used and refined to identify candidate genes in mice and rats (Hu *et al.*, 2008; Saba *et al.*, 2006; Tabakoff *et al.*, 2008, 2009). Data from this elegant set of studies are available for further mining at http://phenogen.uchsc.edu/PhenoGen/index.jsp (Hoffman *et al.*, 2010). One interesting conclusion from these experiments is that the genetic networks controlling alcohol action at low doses, that is, alcohol preference, seem to be completely different from the ones involved in tolerance to high doses of alcohol (Saba *et al.*, 2006). Furthermore, it appears that some of the candidate genes for alcohol preference in mice implicate the olfactory system (Tabakoff *et al.*, 2008). Support for this hypothesis comes from experiments demonstrating that stimulation of the rat tongue with a drop of aqueous alcohol in vivo alters expression profiles in the trigeminal ganglion, which in turn could trigger some of the gustatory or bodily responses to alcohol (Matsumoto *et al.*, 2006). It remains hitherto unclear if these findings are a species-specific characteristic of the rodent model or if they can be generalized to human alcohol consumption.

Recent advances in rat genome research allow extending the genetical genomics approach toward this species. The HxB/BxH panel of recombinant inbred rat strains is extensively characterized for cardiovascular and metabolic traits but has received little attention so far in the field of behavioral and psychiatric genetics. Alcohol-related traits are well distributed among the 25 or so recombinant inbred lines so that bQTLs for alcohol consumption could be determined in this population, the most robust one located on Chr1. The overlap of brain eQTLs with bQTLs for alcohol drinking produced a set of 12 candidate genes involved in GABA signaling, activation of dopaminergic neurons as well as energy metabolism and caloric intake (Tabakoff *et al.*, 2009). This study also asked whether these results are relevant for human alcohol consumption by testing association of about 850 gene-related SNPs with daily intake levels in two human populations. Significant association was found for some variants related to GABA genes. The study, thus, presents initial evidence for convergence of human and animal genetic data.

The low number of candidate genes derived from the above experiments points to an important problem of the genetical genomics approach: detection of highly specific candidate genes comes at the expense of sensitivity. The magnitude of this problem becomes evident when one compares the number of genes that are contained under a single QTL and that are actually differentially expressed. Such an experiment was reported recently for the highly significant alcohol preference QTL on Chr4 in P rats (Liang *et al.*, 2010). This QTL was isolated from its genomic background by producing reciprocal chromosome 4 congenic strains. In other words, the genomic region containing the QTL from the P line was transferred onto the NP background and vice versa. Any true difference in gene expression that is caused by locus specific genetic variation should carry along with this switch. Indeed, more than 70 cis-regulated, differentially expressed genes could be assigned to this QTL by this method, and these are proposed as strong candidates for affecting alcohol consumption. On the other hand, there is no hint of an alcohol preference QTL on Chr4 in the HxB/BxH panel of rat lines, and also with the AA model no overlap in differential expression can be found, except for a few genes such as *Snca* (α-synuclein), *Dgki* (diacylglycerolkinase iota), and *Npy* (Arlinde *et al.*, 2004; Sommer *et al.*, 2001, 2006; WHS, unpublished data). Thus, while the QTL on Chr4 seems to be highly important for drinking behavior in P rats, the same genes may not play a major role for controlling this behavior in the other rat models. This obviously poses a problem for generalizing findings from rodent genetic models to human alcohol consumption or alcohol use disorders and highlights the significant genetic heterogeneity underlying alcohol preference.

D. Common Themes From Transcriptome Studies in Genetic Models of Alcoholism

Despite differences in experimental design and the difficulties in linking a specific candidate gene to the high-drinking phenotype, a recurring theme in examining potential transcriptome differences between selected lines under alcohol-naïve conditions are two major functional groups, neuroplasticity and metabolism. Nearly all studies point to few, brain region-specific signal transduction pathways (e.g., cAMP/CREB/PKA/MAPK, Jak/Stat, NF-κB, AKT) or plasticity markers (neuropeptides, synaptic proteins) and a wide range of differences in metabolic pathways. Differential expression of signaling molecules under alcohol-naïve conditions seems to predict some alcohol-responsive genes. Thus, most of the responsive genes after an acute alcohol challenge of C57BL/6 mice show constitutively different levels between alcohol-naïve C57BL/6 and DBA/2 animals (Rulten *et al.*, 2006). Key pathways identified in this study are also differentially expressed between AA and ANA rats, and are functionally affected by an acute alcohol challenge as demonstrated by different phosphorylation of AKT, GSK3β, and ERK1/2 in these rat lines (Neznanova *et al.*, 2009; Sommer *et al.*, 2006). On the other hand, high alcohol preference does not predict general responsiveness to the drug. In fact, alcohol-avoiding DBA/2 mice show a stronger transcriptome response in the prefrontal cortex as well as stronger locomotor activation upon acute alcohol challenge relative to high-drinking C57BL/6 mice (Crabbe *et al.*, 1982; Kerns *et al.*, 2005).

Observed alterations in metabolic gene expression are only partially attributable to variances in the ADH/ALDH complex. Alcohol avoidance seems to emerge very rapidly during selective breeding, which in fact would support a major gene effect on this behavior. On the other hand, expression of *Aldh2* is higher in DBA/2 compared to C57BL/6 mice (Bhave *et al.*, 2006), which is not consistent with the general notion of a protective effect of this enzyme against increased alcohol consumption (Edenberg, 2007). An alternative possibility to be considered is that metabolic adaptation to higher alcohol levels may be an essential co-selected trait when selective pressure toward high alcohol preference is applied. Support for this hypothesis comes from two independent lines of evidence. Human brain imaging data demonstrate that low doses of alcohol decrease brain glucose metabolism in human subjects to a much larger degree than expected from observed cognitive and behavioral symptoms (Volkow *et al.*, 2006). Upon 0.5 g/kg alcohol, which leaves subjects alert and only moderately intoxicated, whole-brain glucose metabolism could be reduced by 30%, which is close to levels encountered by doses of anesthetics that induce unconsciousness. To substitute this energy deficit, brain metabolism has to be able to shift toward alternative substrates, such as acetate which can enter the brain through active transport mechanisms (Cruz *et al.*, 2005; Pawlosky *et al.*, 2010; Waniewski and Martin, 1998). Furthermore, organisms that consume large amounts of fermented food seem to be enzymatically well equipped to ensure that the ingested alcohol does not interfere with their survival. An extreme case was recently reported where the pentailed treeshrew (*Ptilocercus lowii*) in its natural habitat frequently consumes alcohol doses that would intoxicate humans without showing signs of intoxication (Wiens *et al.*, 2008). Taken together, it appears that coping with high alcohol levels requires metabolic flexibility that likely reaches beyond tolerance mediated by rapid metabolism through the ADH/ALDH complex, but may be obtained through a wide range of tuning mechanism to secure sufficient ATP production for proper brain functioning.

E. Consilience Between Candidate Genes Found in Genetic Animal Models and for Human Alcoholism

With the growing number of candidate genes for which sufficient evidence for their involvement in acute or early alcohol responses is mounted, the question arises to what extend these findings are relevant for the human condition. In other words, are the genes that mediate innate alcohol-related behaviors in mice the same as those that comprise risk factors for human alcoholism?

Some examples for such overlap can indeed be found. MPDZ, identified as a QTG for withdrawal seizure intensity in mice (Shirley *et al.*, 2004), was associated with alcohol-related phenotypes in two human populations, though not with alcohol withdrawal seizures (Karpyak *et al.*, 2010; Tabakoff *et al.*, 2009). Likewise, in an attempt to find overlapping confirmed QTLs for specific alcohol-related behaviors in mice, that is, alcohol preference and alcohol withdrawal severity, and any alcohol-related phenotype in humans, including DSM-IV defined alcoholism, it was realized this is not an easy task (Ehlers *et al.*, 2010). In the end, two cases of overlap were found. A human QTL on distal chromosome 1 for alcohol dependence overlaps with the mouse QTL for alcohol withdrawal on distal chromosome 4 that was conclusively linked to *Mpdz*. In the second case, a human QTL for alcohol dependence severity and withdrawal on distal chromosome 15 maps on the region of mouse Chr9, where a broad QTL for alcohol preference was found. Thus, the paired phenotypes are only loosely matched. The authors conclude that exact phenotypic matching might not be necessary or optimal for finding genes that affect alcohol-related behaviors. From the human disease perspective, this is frustrating news because it implies that either existing data are not sufficiently comparable or the applied models do not adequately reflect the clinical phenotype. As often, truth lies in both arguments and an extensive discussion of this subject including what might be necessary to reach better consilience between clinical and animal research fields on alcoholism can be found in a recent special issue of Addiction Biology (Crabbe, 2010).

On the animal side, there has been a strong bias toward easy assessable phenotypes, that is, alcohol preference, tolerance or withdrawal severity, but judging by the positions expressed in the consilience papers we still lack a good understanding of how these animal responses can be translated into human alcohol behaviors. To illustrate the point, we will briefly go over some issues raised in this debate regarding these phenotypes. Preference drinking in rodents relates to alcohol's reinforcing effects, but this test lacks any measure of the motivational component of this behavior (Leeman *et al.*, 2010; Stephens *et al.*, 2010). It is possible to measure reinforcement and reward sensitivity quite precisely in animal models, but for these tests genetic or transcriptome data are hitherto lacking. As detailed above it is still a matter of debate to what extend increased alcohol preference in humans may contribute to the risk of developing alcoholism (Arias *et al.*, 2006). Also a low level of response to alcohol intoxication has some degree of heritability in humans and seems predicative for developing alcohol addiction (Schuckit *et al.*, 2009). In animals this genetic relationship is often assessed by the loss of righting reflex (LORR), in which hypnotic doses of alcohol are injected and time is measured until the animal regains the ability to right itself. Plenty of genetic and expression data have been obtained for this test, but its validity for human low level of response to alcohol is not demonstrated (Crabbe *et al.*, 2010). Finally, seeking relief from acute withdrawal symptoms by renewed alcohol intake is one diagnostic criterion for alcoholism (DSM-IV). A hallmark feature of alcohol withdrawal is hyperexcitability, which can be relatively easy measured in animals as withdrawal-induced seizures. Although the genetic determinants of withdrawal seizure susceptibility are well characterized, it is uncertain how this susceptibility connects to the development of an addictive process (Heilig *et al.*, 2010).

In summary, most of the innate responses in commonly employed animal models miss the aspect of alcohol and behaviorally induced neuroadaptations that underlie the motivational shift occurring during the progression into addiction. Therefore, many of these efforts might fail to detect disease relevant genetic mechanisms. The high alcohol consuming phenotype has been viewed as an important methodological advantage for pharmacological studies, and it has been repeatedly argued that these models possess good predictive validity based on the fact that they have played an important role in the development of available medications for alcoholism. Such an argument carries the risk of being circular and given the low efficacy of current treatments, the suitability of a particular animal model for testing a particular drug should be continuously reevaluated. Further development of genetic models should reach beyond stable intake patterns, which likely reflect controlled drinking. Changing environmental conditions may provide a challenge for animals with high alcohol preference that may drive them to override their internal control mechanisms preventing intoxication. It has been shown that scheduling alcohol access to short periods at the beginning of their active (circadian dark) phase can make rodents to engage in more harmful drinking patterns ("too fast – too much"). Selection for this type of drinking was successfully demonstrated recently and these mice consistently reach blood alcohol levels >1 g/L (Crabbe *et al.*, 2009; Rhodes *et al.*, 2007). It will be interesting to see what genes control those safety signals and the propensity of an animal to ignore them. Apart from high alcohol intake, other traits predictive of an increased addiction risk should be subjected to behavioral genetics analysis. Such traits may include impulsive choices, impaired extinction of conditioned alcohol responses or increased sensitivity for reinstatement of alcohol-seeking behavior. Future research will show to what extent new models will contribute to increased consillience of human and animal data and to a better understanding of alcohol addiction.

V. Modeling Addiction-Related Neuroadaptations by Long-Term Alcohol Access or Forced Intoxication

The studies reviewed above have mainly focused on the acquisition of alcohol drinking or the maintenance of relatively constant and controlled drinking behaviors. This work has led to the identification of genetic determinants of alcohol's positive reinforcement processes. The emergence of uncontrolled, compulsive alcohol consumption and seeking behaviors typically seen in addicted individuals is likely driven by negative motivational forces, probably triggered as an opponent process by acute, protracted, or conditioned withdrawal. The allostatic dysregulation that is hypothesized to underlie this development requires long-lasting neuroadaptations of the reinforcement systems (Koob and Le Moal, 2001).

Until recently, the study of long-term neuroadaptations in alcohol addiction has been limited, in part because of lack of suitable animal models. A major obstacle of experimental alcohol research is that even in high-preferring lines voluntary alcohol drinking rarely reaches blood or brain alcohol concentrations (BACs) required for dependence induction (> 1 to 2 g/L). Pharmacological effects to modulate this level of intake may be of limited use for modeling treatment of human alcoholism (Egli, 2005; Heilig and Egli, 2005). In order to drive the type of stable plasticity associated with or underlying dependence, robust changes in multiple transmitter systems are likely to be required. However, glutamate signaling is thought to be of particular importance and can likely only be provided by high enough alcohol concentrations that lead to consecutive withdrawal reactions.

Large amounts of alcohol are indeed consumed voluntarily also by nonselected laboratory rodents under certain experimental conditions. For example, by scheduling access to alcohol drinking to the beginning of the circadian dark phase, rats can overcome the internal control signals that limit their intake and thus engage in excessive drinking (Rhodes *et al.*, 2005). In a very different approach, rats will increase their alcohol consumption in time when given access over many months and may actually double or triple their daily intake (Holter *et al.*, 1998; Wolffgramm *et al.*, 2000). Also, alternate access to 20% alcohol and water every other day has successfully been used to increase voluntary alcohol consumption in a matter of weeks (Steensland *et al.*, 2007; Wise, 1975). The excessive intake observed under all these procedures can be selectively reduced by acamprosate or naltrexone, providing validation for their use in medication development. However, their genetic or transcriptome foundation has yet to be established.

Two well-established models for studying aspects of relapse behavior are the reinstatement of alcohol seeking and the alcohol deprivation effect (Bachteler *et al.*, 2005; Katner *et al.*, 1999; Sanchis-Segura and Spanagel, 2006; Shaham *et al.*, 2003). In the former, animals are trained to respond for alcohol in self-administer chambers. Then, no reinforcement is given during the self-administration sessions until operant responding is extinguished. Following extinction, a priming dose of alcohol, stress or conditioned stimuli can reinstate alcohol-seeking behavior. Although the obtained BACs during alcohol self-administration are below 1 g/L, the procedure clearly induces long-term neuroadaptation within addiction-related brain circuits. However, reinstatement of alcohol seeking can be elicited both in alcohol dependent and in nondependent animals, and thus, the specificity of the underlying processes for alcohol addiction is unclear. Nevertheless, the blockade of CRHR1 receptors prevents stress-induced reinstatement of alcohol seeking only in dependent animals, demonstrating that the mechanisms leading to this behavior in dependent and nondependent animals are different (Gehlert *et al.*, 2007). Various pharmacological targets interfere with reinstatement behavior, but so far experiments to analyze the genetic factors or transcriptomic responses are lacking.

The term alcohol deprivation effect (ADE) refers to a phenomenon that is observable across many species including mice, rats, monkeys, and humans (Spanagel, 2009). Long-term access to alcohol, when interspersed by short periods where access is prevented (deprivation) will result in temporarily excessive alcohol consumption. In normal, unselected male rats, daily intake may increase from normal 1–2 g/kg alcohol to above 6 g/kg after the deprivation. Importantly, repeating the pattern of access and deprivation progressively elicits a compulsive-like alcohol seeking. The uncontrolled and compulsive component of this drinking behavior is demonstrated by taste adulteration of the alcohol solution with quinine, a highly bitter substance that usually produces strong taste aversion in rats (Spanagel and Holter, 2000; Vengeliene *et al.*, 2009). The ADE is widely used in preclinical medication development and the results for many compounds have been systematically reviewed recently by Spanagel (2009).

A. Neuroadaptations Involved in Compulsive-Like Behaviors After Long-Term Alcohol Access

Long-term alcohol access with repeated deprivation periods has substantial effects on the striatal and amygdala transcriptome (Matthaus *et al.*, 2009). From the overall expression changes several networks could be identified related to heme synthesis, immunoregulation, and cytoskeleton. Whether these alterations reflect compensatory mechanisms to the long-term alcohol intake or are involved in the addiction-related behavioral responses remains to be established. Another finding from this experiment was upregulation of striatal dopamine D3 receptors and the demonstration of their functional involvement in the behavioral control of the ADE by specific antagonists (Vengeliene *et al.*, 2006). These data are highly relevant for human alcoholism because a functional polymorphism in the human DRD3 gene (Ser9Gly) affects P300 event-related potentials, an EEG based endophenotype consistently associated with alcoholism (Mulert *et al.*, 2006). Of even greater importance seems to be the fact that the chronic alcohol intake altered expression in a set of genes that was predicative for finding susceptibility genes for human alcoholism in a genome-wide association study (Treutlein *et al.*, 2009). Out of fifteen SNPs, that could be confirmed in the replication study, three where put forward by the animal experiment. These three SNPs were in CDH13 (cadhedrin 13), ADH1C (alcohol dehydrogenase 1C), and GATA4 (GATA binding protein 4). Thus, combining animal and human data in such a convergent approach can indeed result in increased consilience for our knowledge about human alcoholism.

A new study in the long-term access/repeated deprivation model focused on alterations in glutamate signaling in the striatum associated with development of compulsive-like drinking behavior (Vengeliene *et al.*, 2010). Expression of many glutamatergic genes in the dorsal striatum was altered by the long-term alcohol access. Importantly, about a third could be reverted to baseline by treatment with a glycine receptor antagonist that showed long-lasting anti-relapse properties in this model. Further data mining points to an important role of *Adcyap1* (adenylate cyclase activating polypeptide 1, also known as PACAP) in mediating this effect. These findings fit well with the proposed role of striatal circuits in habit formation and that their dysfunction may underlie compulsive drug-seeking behavior (Vanderschuren and Everitt, 2005).

Another study testing the effects of 1-year access to alcohol (Sake) was conducted in Fisher rats. However, because in this study rats did not have a choice between alcohol and nonalcohol solutions, it is unlikely that similar neuroadaptive response should have taken place as in the ADE model. Gene expression profiling was combined with proteomics analysis. No overlap with the above reviewed studies was reported, but a strong downregulation of NADH dehydrogenase (ubiquinone) Fe-S protein 1, *Ndufs1*, was found pointing again to altered mitochondrial function in the brain of alcohol-drinking rats.

B. Forced Alcohol Exposure, Sensitization of Stress Systems, and the Post-Dependent State

The major disadvantage of the long-term alcohol access model may be its high logistic requirements, which are difficult to afford for most laboratories. Alternatively, forced alcohol exposure can produce over relatively short periods of time a behavioral profile consistent with alcohol addiction. Forced intoxication can be administered via different routes, by offering alcohol-containing liquid diet as the sole source of food, by intragastric alcohol gavages, by breathing alcohol vapor or by systemic injection (Majchrowicz, 1975; Rogers *et al.*, 1979). For a long time, chronic alcohol administration via these routes was known to produce robust withdrawal reactions (Becker, 2000), but only transient motivation to self-administer alcohol (Roberts *et al.*, 2000). Thus, in contrast to the clinical situation, increased alcohol preference as well as most of the induced neuroplasticity was fading away along with the recovery from the acute withdrawal reaction. The missing piece in these experimental setups was consideration of the temporal pattern of alcohol exposure. Even in heavy drinkers, the brain is not bathed constantly in high alcohol concentrations, but has to cope with ever changing alcohol levels, from rapidly rising back to zero, thereby going repeatedly through cycles of intoxication and various degrees of withdrawal. Once the element of intermittent exposure patterns was introduced into the routes of chronic administration, researchers began to see behavioral consequences that persist long beyond completion of physical withdrawal (Lopez and Becker, 2005; O'Dell *et al.*, 2004; Rimondini *et al.*, 2002). Progressively increasing activation of glutamatergic transmission over repeated withdrawals is likely to provide the signal for this plasticity (De Witte *et al.*, 2003).

One of the most reliable and versatile tools for precise control of brain alcohol exposure is vapor exposure. Using this approach it was shown that prolonged brain alcohol exposure to levels commonly occurring in human alcoholics (~150–250 mg/dL or 35–55 mM) leads to behavioral consequences that seem to be relevant for alcoholism (Rimondini *et al.*, 2002). Prolonged duration (Rimondini *et al.*, 2003) and an intermittent pattern of exposure (O'Dell *et al.*, 2004; Rimondini *et al.*, 2002), two features that mimic the exposure profile in clinical alcoholism, appear critical for induction of the key behavioral consequences that persist long beyond completion of physical withdrawal: (1) escalation of subsequent voluntary alcohol intake, measured using both two-bottle free-choice drinking (Griffin III *et al.*, 2009; Lopez and Becker, 2005; Rimondini *et al.*, 2002) and operant responding for alcohol (Roberts *et al.*, 2000); (2) sensitization of behavioral stress responses (Breese *et al.*, 2005b; Overstreet *et al.*, 2002; Sommer *et al.*, 2008; Valdez *et al.*, 2002, 2003, 2004). The resulting behavioral syndrome has been termed the "post-dependent state" (Heilig and Koob, 2007; Heilig *et al.*, 2009). This term is used to describe the sum of the behavioral and neuroadaptive consequences that are induced as an animal becomes dependent on alcohol, and that remain for extended periods of time thereafter even in its absence. Examples of such long-lasting neuroadaptations after intermittent alcohol vapor exposure from our own work are summarized in Fig. 3.

Other methods that lead to repeated cycles of intoxication and withdrawal exist, such as for example through forced liquid diet or intragastric alcohol gavages. These have been recently reevaluated for inducing a post-dependent state (Braconi *et al.*, 2010; Gilpin *et al.*, 2009). Although perhaps less potent and less easy to control, the procedures can induce similar sets of behavioral consequences (for review see Breese *et al.*, 2005a).

Although vapor exposure by itself and the rapidly changing alcohol levels are a demanding challenge to the animals, the procedure is safe and we found no overt health problems in post-dependent rats that had been abstinent for 3 weeks. Besides numerous unchanged blood parameters, also basal corticosterone levels and alcohol metabolism were normal. (Rimondini *et al.*, 2002, 2008). Alcohol vapor exposure has been criticized to be a highly non-physiological method of

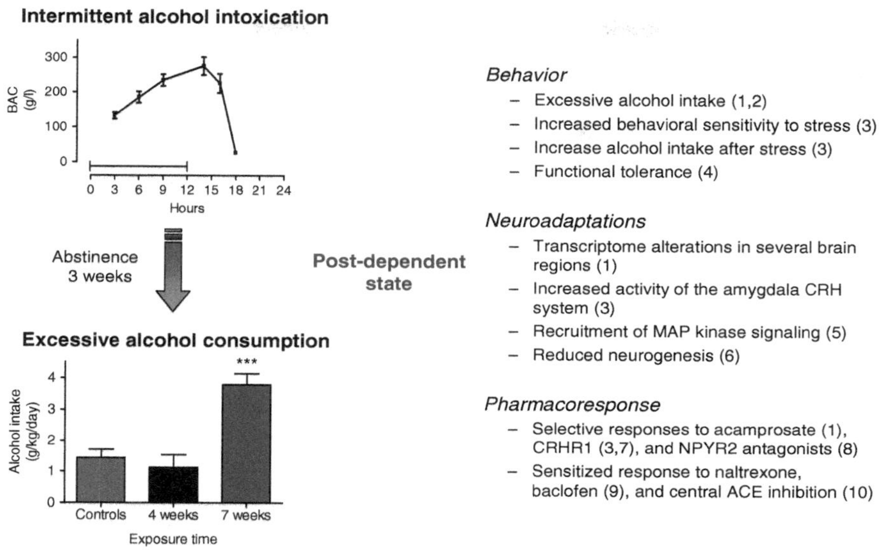

FIG. 3. Intermittent alcohol intoxication leads to the development of a post-dependent state. Left: The top graph shows blood alcohol concentrations in rat obtained during one daily cycle of alcohol vapor exposure and withdrawal. The bottom graph shows voluntary alcohol consumption in a two-bottle free-choice, continuous access paradigm (6% alcohol in water vs. water only) after 4 and 7 weeks of daily intermittent exposure cycles. Seven weeks exposed animals develop a persistent phenotype of excessive voluntary alcohol consumption and increased behavioral sensitivity to stress (post-dependent state). Right: Summary of our observations in post-dependent rats. Numbers in parenthesis refer to the following publications: (1) Rimondini *et al.*, 2002; (2) Rimondini *et al.*, 2003; (3) Sommer *et al.*, 2008; (4) Rimondini *et al.*, 2008; (5) Hansson *et al.*, 2008; (6) Hansson *et al.*, 2010; (7) Gehlert *et al.*, 2007; (8) Rimondini *et al.*, 2005; (9) unpublished results; (10) Sommer *et al.*, 2007. $^{***}p < .001$.

alcohol delivery. In fact, alcohol inhalers (AWOL™) have been introduced for human consumption, which clearly demonstrates that, given the choice, humans voluntarily use alcohol vaporization as a method of intoxication. Scientific reports on the health consequences of this consumption pattern in humans are still lacking. Exposed animals show no apparent pathology in respiratory function. Olfactory and gustatory systems have not been investigated so far; however, since vapor exposed animals do not differ in preference for sweet and bitter taste from controls, basic function of these systems seems to be intact. To this point, the only known long-term pathology in post-dependent rats is secondary osteoporosis (Torricelli *et al.*, 2007), a condition that is commonly seen also in alcoholic patients (Kanis *et al.*, 2005).

A key neuroadaptation in post-dependent animals and widely replicated finding is the recruitment of the CRH system within the amygdala brain region

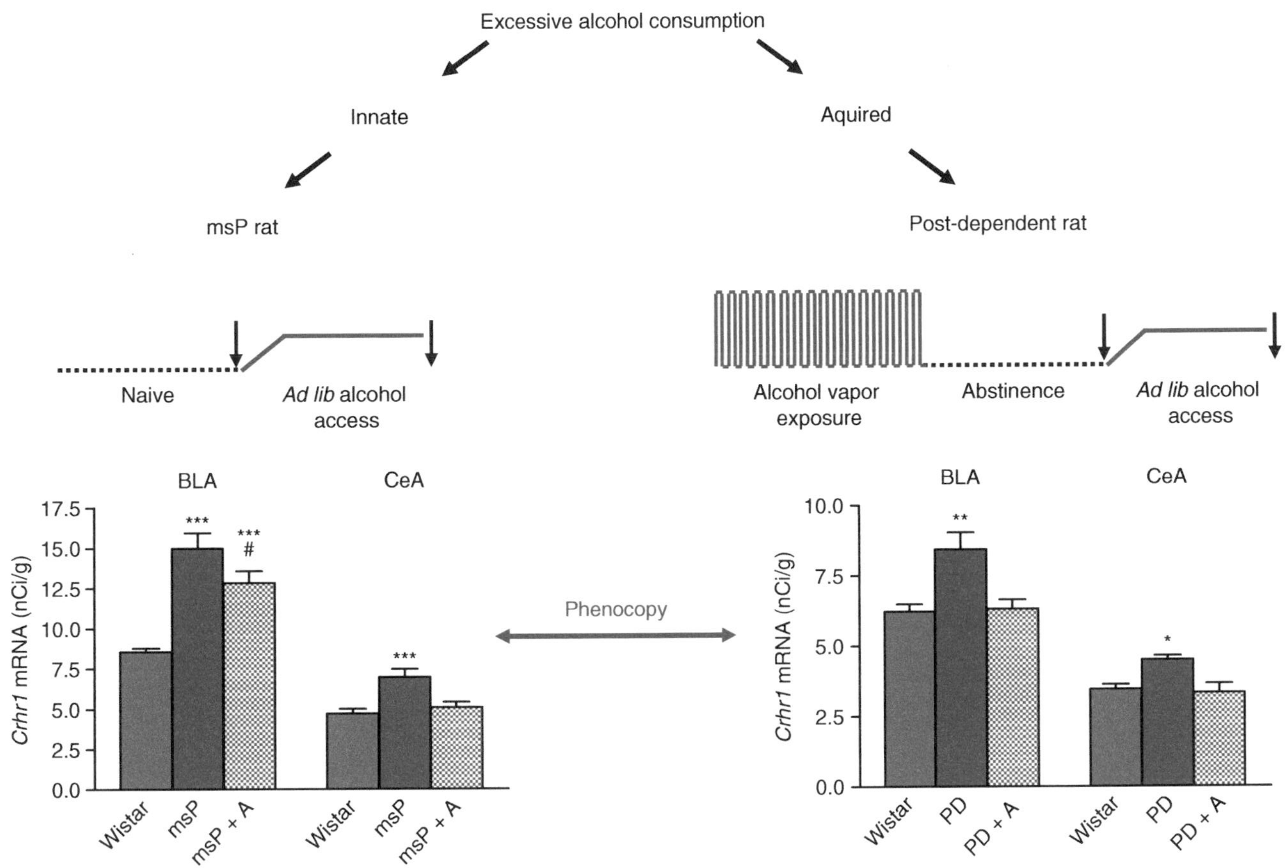

Excessive alcohol consumption
Innate
Aquired
msP rat
Post-dependent rat
Naive
Ad lib alcohol access
Alcohol vapor exposure
Abstinence
Ad lib alcohol access
BLA
CeA
Crhr1 mRNA (nCi/g)
Wistar
msP
msP + A
Phenocopy
PD
PD + A

(Heilig and Koob, 2007), which is critically involved in mediating dependence-induced increased behavioral stress responses, elevated alcohol self-administration and alcohol-seeking behavior during both acute and protracted abstinence (Funk *et al.*, 2006; Gehlert *et al.*, 2007; Sommer *et al.*, 2008; Valdez *et al.*, 2003).

Interestingly, robust *Crhr1* upregulation was also demonstrated in the genetically selected, alcohol-preferring Marchigian Sardinian Preferring (msP) rat and could be linked to increased stress reactivity, excessive alcohol self-administration, and increased propensity for relapse-like behavior in this line (Hansson *et al.*, 2006). An intriguing observation is that both post-dependent and msP rats given *ad lib* access to alcohol will reverse increased amygdala *Crhr1* transcripts down to control levels, suggesting that alcohol intake may act as a functional CRHR1 antagonists in these animals. Thus, msP animals seem to share similar features with post-dependent animals and may therefore represent a corresponding behavioral *phenocopy* of post-dependent rats (Fig. 4). It has to be pointed out that msP rats only share distinct neurochemical features with post-dependent animals. In fact, *Crhr1* expression in msP rat is upregulated across many brain regions (Hansson *et al.*, 2006), while in post-dependent animals increased expression of this gene is restricted to the amygdala (Sommer *et al.*, 2008). In other words, investigations in msP cannot replace studies in the more laborious neuroadaptation model but seem to provide an excellent opportunity for functional validation of specific targets. Such convergent use of genetic and neuroadaptation models appears as a useful path to increased consilience with human data. Indeed, genetic variation at the *Crhr1* locus as a susceptibility factor for excessive alcohol drinking might have parallels in higher species, including rhesus macaques (Barr *et al.*, 2009) and humans (Chen *et al.*, 2010; Nelson *et al.*, 2010; Treutlein *et al.*, 2006).

FIG. 4. Convergent evidence from genetic and neuroadaptation rat models for a role of amygdala *Crhr1* expression in excessive alcohol consumption. Increased *Crhr1* mRNA levels are clearly seen within the basolateral (BLA) and central (CeA) amygdala brain regions in both post-dependent and msP rats by means of in situ hybridization. Upon *ad lib* access to alcohol a downregulation of *Crhr1* transcripts toward control levels occur within these amygdala regions in animals of both models, suggesting that alcohol is voluntary consumed to counteract the overactivity of the CRH system in this region. Right: Post-dependent rats (Wistar) were generated by intermittent alcohol vapor exposure for 7 weeks (rolling line) and sacrificed (black arrows) either after 3 weeks of abstinence (PD) or after additional 3 weeks of *ad lib* access to alcohol (PD+A) in a continuous access, two-bottle free-choice paradigm (6% alcohol vs. water). During this period PD+A rats consumed 3.5–4 g/kg/day alcohol. Alcohol-naïve Wistar rats were used as controls. Left: Genetically preferring msP rats were sacrificed either alcohol naïve (msP) or after *ad lib* access to alcohol (msP+A, 10% alcohol vs. water) for 15 day during which they consumed 7.1–7.7 g/kg/day alcohol. Nonselected, alcohol-naïve Wistar rats were used as controls. Bar graphs showing quantitative *Crhr1* expression levels (mean ± SEM, $n = 7$ to 8) from in situ hybridization on amygdala sections. Statistical analysis was performed by region-wise one-way ANOVA ($^{*}p < 0.05$, $^{**}p < 0.01$, $^{***}p < 0.001$ vs. Wistar controls; $^{\#}p < 0.05$ msP+E vs. msP). Data have been partly reported previously (Hansson *et al.*, 2006, 2007b; Sommer *et al.*, 2008).

C. Transcriptome Studies of the Post-Dependent State

While the post-dependent state model appears highly versatile and is increasingly utilized in pharmacological experiments, only a few studies have assessed transcriptome-wide responses associated with the neuroadaptations involved in the addiction-related behaviors. One major problem in the design of such studies is to choose the relevant time points. Experimental data obtained during alcohol intoxication, in early withdrawal or after prolonged abstinence will likely point to very different groups of genes. Yet, among the sparse reports using the intermittent alcohol vapor intoxication there is surprising overlap.

To capture long-term neuroadaptations we chose a three week abstinence period. At this point effects of acute withdrawal have passed and most of the expression changes should have stabilized. Also, because the behavioral symptoms seen in the post-dependent state are subtle, no grave alterations in gene expression are to be expected either potentially being below the sensitivity threshold of the microarray platform. Nevertheless, our initial study found dysregulation of about 30 genes in the medial prefrontal cortex or amygdala persisting 3 weeks into abstinence (Rimondini *et al.*, 2002). Although the evidence from this microarray experiment was modest, several findings were pointing to pharmacological targets and have been functionally validated in subsequent studies, for example, the endocannabinoid system, MAP kinases pathways, and glutamate transporter (Hansson *et al.*, 2007a, 2008; Karlsson, 2009). An extended analysis points to the prefrontal projection neurons as the major place of long-term plasticity with profoundly decreased expression of several cAMP/PKA/CREB regulated transcription factors (e.g., *Egr1, Fos, Junb, Nr4a1*) and functional important effector genes including *Bdnf, Homer1*, and *Scn4b* (WHS, unpublished data). Moreover, the transcription factors show a blunted response to alcohol challenge (Hansson *et al.*, 2008). Interestingly, some of these genes seem to be downregulated in the medial prefrontal cortex of intermittently alcohol vapor exposed animals already during intoxication (*Egr1, Nr4a1, Homer1)* and in early withdrawal (*Egr1, Bdnf*) (Melendez *et al.*, 2006; Repunte-Canonigo *et al.*, 2007). Together, these data are highly consistent and point to the possibility that the cAMP/PKA/CREB pathway in the medial prefrontal cortex has lost, at least in part, the ability to generate an adaptive response in the dependent state. This would have important functional implications likely to be contributing to the addiction-related behaviors.

The same network of transcription factors, that is, *Fos, Jun, Junb* and *Nr4a1* and *Nr4a3,* was found to be upregulated in the nucleus accumbens of P rats that had either continuous or intermittent access to alcohol for 2 months and were sacrificed 15 hours after the last alcohol access period (Bell *et al.*, 2009). These rats may have reached blood alcohol levels up to 0.8 g/L, which is likely not sufficient to induce neuroadaptations of the type found in the studies reviewed above using intermittent alcohol vapor exposure. The results from these different models are obviously difficult to compare, but the finding of correlated expression in this regulatory network could point to opposing activity states in the two regions after long-term exposure to alcohol. Increased activity in the nucleus accumbens is produced by the pharmacological effects of alcohol, but higher doses may also involve neurotoxic actions, for which the mPFC seems to be particularly vulnerable to, and hence result in decreased activity of this region. On the behavioral level, activation of accumbal and inhibition of prefrontal activity may both lead to increased alcohol intake, but while the former seems to reflect adaptation within the homeostatic borders of normal or controlled function, the latter appears to be dysregulation leading to loss of control.

Although post-dependent animals express addiction-related phenotypes such as excessive voluntary alcohol consumption and increased anxiety responses, the model is not informative about predisposing factors for these behavioral responses. Instead post-dependent animals can inform how repeated alcohol intoxication and withdrawal impact on systems that mediate these phenotypic responses. Thus, alterations in gene expression are not necessarily found within the genes mediating the initial response to alcohol, but reflect a pathological process that has recruited additional pathways which in turn set the stage for aberrant response on the systems level which become evident as addictive behaviors. Thus, in post-dependent animals we find a mix of pharmacological effects of alcohol including tolerance and withdrawal reactions as well as neurodegenerative and compensatory responses to alcohol. These vary in a brain region specific manner and together comprise the pathological phenotype. In contrast, long-term access models such as repeated ADE exposure are likely to be driven by the reinforcing effects of alcohol and resulting in a rewiring of brain circuits and formation of habitual responses. For example, sensitized stress responses are found in post-dependent, but not in repeated ADE exposed animals. Thus, both types of neuroadatation models result in different behavioral syndromes associated with distinct transcriptomic profiles, but both reflect important aspects of human alcoholism.

D. Attempts to Combine Genetic and Neuroadaptation Models for Studying Gene-Environment Interactions

After identifying genetic factors underlying alcohol responses and neuroadaptive mechanisms involved in the development of addictive behaviors, it appears to be a logical continuation of these lines of research to ask, how these two main categories of causal factors interact and to what extend these models are suitable to study such interactions? There seem to be severe limitations to such an approach resulting from the complexity of each type of model. For example, we mentioned that some of the genetic lines have a severely narrowed behavioral reaction norm and thus only a limited ability to react to challenge, as exemplified by the observation that AA and HAD rats do not express an ADE (Vengeliene *et al.*, 2003). In addition, we found that both AA and ANA rats show resistance to the intermittent alcohol vapor exposure procedure; specifically they do not alter their pattern of alcohol consumption (Hoffman *et al.*, 2003; Sommer *et al.*, 2005). In our experience, this is unusual because using the intermittent vapor exposure paradigm we robustly induce post-dependent excessive alcohol drinking in a variety of rat lines, including Wistar, Sprague-Dawley, Fisher, and a genetically modified rat line (Sommer *et al.*, 2007). If the findings in AA and ANA rats do indeed represent innate resistance to developing a post-dependent state, it would be highly interesting to identify involved genetic factors.

Other genetic lines might be more useful to study the genetic determinants underlying the development or maintenance of a post-dependent state. Withdrawal seizure prone and withdrawal seizure resistant (WSP/WSR) mouse lines were selectively bred for high or low alcohol withdrawal severity following chronic ethanol vapor exposure (Kosobud and Crabbe, 1986). A recent study examined 15 common inbred mouse lines after 3 days of intermittent alcohol vapor intoxication for withdrawal severity (Metten *et al.*, 2010). In line with previous results, the severity of handling induced convulsions during alcohol withdrawal is strongly determined by genotype, but the genetic factors seem to act independent of the intoxication paradigm, that is, the same genes are likely to be involved in withdrawal severity regardless whether intoxication was chronic, intermittent, or acute. Given the logistic efforts necessary for maintaining the intermittent exposure paradigm over several weeks, there is justified hesitation to embark on selective breeding for susceptibility or resilience to developing a post-dependent state.

In summary, neuroadaptations that occur after a prolonged history of alcohol access or exposure seem to persist long into abstinence, some of those probably for the lifetime of the individual, and reflect important aspects of human alcoholism. Several interesting candidate mechanism have been put forward by these models, most notably a widespread dysregulation of glutamatergic genes in the repeated ADE model and the sensitization of brain stress and fear systems in the post-dependent state.

VI. Conclusions

The last decade has seen a widespread use of high-throughput methods for transcriptome analysis to study brain function, and a wealth of genome-wide expression data from various models has been accumulated by alcohol researchers. We have considered commonly used rodent models in alcoholism research from the perspective of their potential place on the disease trajectory. Such a view emphasizes mechanistic differences in the observed responses to alcohol as they relate to alcoholism. Taking into account the dynamics of the addiction process, which are mainly driven by time-dependent brain exposure to alcohol, led us to critically reconsider the potential knowledge gain that could be obtained from the study of initial responses or relatively brief exposure to alcohol in terms of the disorder and its treatment. In fact, the large number of genetic and genomic investigations based on experimental paradigms mostly reflective of acute alcohol effects or its controlled use has resulted in surprisingly few candidate genes with potential implications for addiction treatment. Instead, the research focus should be shifted toward alcohol-induced long-term neuroadaptations, where initial transcriptome studies have already put forward interesting and verifiable candidate mechanisms. The emerging transcriptome studies from these models point to alterations in distinct vulnerable brain regions and circuits, such as stress response and fear circuits, rather than widespread dysregulation. Thus, future genomics analysis may focus on affected networks and subpopulations on the basis of detailed neuroanatomical, pharmacological, and in vivo neuroimaging mappings. Most importantly, evidence needs to be collected from multiple experimental lines including different types of animal models as well as data from human observations in order to reach increased consilience about what are the genetic and transcriptional mechanism underlying the clinical condition of alcohol addiction and how to apply this knowledge to better satisfy the largely unmet medical needs.

References

American Psychiatric Association (1994). Diagnostic and Statistical Manual of Mental Disorders 4th Ed. American Psychiatric Press, Washington, DC.

Arias, A., Feinn, R., and Kranzler, H. R. (2006). Association of an Asn40Asp (A118G) polymorphism in the mu-opioid receptor gene with substance dependence: A meta-analysis. *Drug Alcohol Depend* **83**, 262–268.

Arlinde, C., Sommer, W., Bjork, K., Reimers, M., Hyytia, P., Kiianmaa, K., and Heilig, M. (2004). A cluster of differentially expressed signal transduction genes identified by microarray analysis in a rat genetic model of alcoholism. *Pharmacogenomics J.* **4**, 208–218.

Bachteler, D., Economidou, D., Danysz, W., Ciccocioppo, R., and Spanagel, R. (2005). The effects of acamprosate and neramexane on cue-induced reinstatement of ethanol-seeking behavior in rat. *Neuropsychopharmacology* **30**, 1104–1110.

Barr, C. S., Dvoskin, R. L., Gupte, M., Sommer, W., Sun, H., Schwandt, M. L., Lindell, S. G., Kasckow, J. W., Suomi, S. J., Goldman, D., Higley, J. D., and Heilig, M. (2009). Functional CRH variation increases stress-induced alcohol consumption in primates. *Proc. Natl. Acad. Sci. USA* **106**, 14593–14598.

Barr, C. S., Dvoskin, R. L., Yuan, Q., Lipsky, R. H., Gupte, M., Hu, X., Zhou, Z., Schwandt, M. L., Lindell, S. G., McKee, M., Becker, M. L., Kling, M. A., Gold, P. W., Higley, D., Heilig, M., Suomi, S. J., and Goldman, D. (2008). CRH haplotype as a factor influencing cerebrospinal fluid levels of corticotropin-releasing hormone, hypothalamic-pituitary-adrenal axis activity, temperament, and alcohol consumption in rhesus macaques. *Arch. Gen. Psychiatry.* **65**, 934–944.

Becker, H. C. (2000). Animal models of alcohol withdrawal. *Alcohol Res. Health* **24**, 105–113.

Belknap, J. K., and Atkins, A. L. (2001). The replicability of QTLs for murine alcohol preference drinking behavior across eight independent studies. *Mamm. Genome* **12**, 893–899.

Bell, R. L., Kimpel, M. W., McClintick, J. N., Strother, W. N., Carr, L. G., Liang, T., Rodd, Z. A., Mayfield, R. D., Edenberg, H. J., and McBride, W. J. (2009). Gene expression changes in the nucleus accumbens of alcohol-preferring rats following chronic ethanol consumption. *Pharmacol. Biochem. Behav.* **94**, 131–147.

Bhave, S. V., Hoffman, P. L., Lassen, N., Vasiliou, V., Saba, L., Deitrich, R. A., and Tabakoff, B. (2006). Gene array profiles of alcohol and aldehyde metabolizing enzymes in brains of C57BL/6 and DBA/2 mice. *Alcohol Clin. Exp. Res.* **30**, 1659–1669.

Björk, K., Rimondini, R., Hansson, A. C., Terasmaa, A., Hyytia, P., Heilig, M., and Sommer, W. H. (2008). Modulation of voluntary ethanol consumption by beta-arrestin 2. *FASEB J.* **22**, 2552–2560.

Björk, K., Saarikoski, S. T., Arlinde, C., Kovanen, L., Osei-Hyiaman, D., Ubaldi, M., Reimers, M., Hyytia, P., Heilig, M., and Sommer, W. H. (2006). Glutathione-S-transferase expression in the brain: Possible role in ethanol preference and longevity. *FASEB J.* **20**, 1826–1835.

Björk, K., Sjögren, B., and Svenningsson, P. (2010). Regulation of serotonin receptor function in the nervous system by lipid rafts and adaptor proteins. *Exp. Cell Res.* **316**, 1351–1356.

Bohn, L. M., Gainetdinov, R. R., Sotnikova, T. D., Medvedev, I. O., Lefkowitz, R. J., Dykstra, L. A., and Caron, M. G. (2003). Enhanced rewarding properties of morphine, but not cocaine, in beta (arrestin)-2 knock-out mice. *J. Neurosci.* **23**, 10265–10273.

Bohn, L. M., Lefkowitz, R. J., Gainetdinov, R. R., Peppel, K., Caron, M. G., Lin, and F. T. (1999). Enhanced morphine analgesia in mice lacking beta-arrestin 2. *Science* **286**, 2495–2498.

Bond, C., LaForge, K. S., Tian, M. T., Melia, D., Zhang, S. W., Borg, L., Gong, J. H., Schluger, J., Strong, J. A., Leal, S. M., Tischfield, J. A., Kreek, M. J., and Yu, L. (1998). Single-nucleotide polymorphism in the human mu opioid receptor gene alters beta-endorphin binding and activity: Possible implications for opiate addiction. *Proc. Natl. Acad. Sci. USA* **95**, 9608–9613.

Bouza, C., Angeles, M., Munoz, A., and Amate, J. M. (2004). Efficacy and safety of naltrexone and acamprosate in the treatment of alcohol dependence: A systematic review. *Addiction* **99**, 811–828.

Bowers, B. J., Radcliffe, R. A., Smith, A. M., Miyamoto-Ditmon, J., and Wehner, J. M. (2006). Microarray analysis identifies cerebellar genes sensitive to chronic ethanol treatment in PKCgamma mice. *Alcohol* **40**, 19–33.

Braconi, S., Sidhpura, N., Aujla, H., Martin-Fardon, R., Weiss, F., and Ciccocioppo, R. (2010). Revisiting intragastric ethanol intubation as a dependence induction method for studies of ethanol reward and motivation in rats. *Alcohol Clin. Exp. Res.* **34**, 538–544.

Breese, G. R., Overstreet, D. H., and Knapp, D. J. (2005a). Conceptual framework for the etiology of alcoholism: A "kindling"/stress hypothesis. *Psychopharmacology (Berl.)* **178**, 367–380.

Breese, G. R., Overstreet, D. H., Knapp, D. J., and Navarro, M. (2005b). Prior multiple ethanol withdrawals enhance stress-induced anxiety-like behavior: Inhibition by CRF1- and benzodiazepine-receptor antagonists and a 5-HT1a-receptor agonist. *Neuropsychopharmacology* **30**, 1662–1669.

Brooks, P. J., Enoch, M. A., Goldman, D., Li, T. K., and Yokoyama, A. (2009). The alcohol flushing response: An unrecognized risk factor for esophageal cancer from alcohol consumption. *PLoS Med.* **6**, e50.

Buck, K. J., Metten, P., Belknap, J. K., and Crabbe, J. C. (1997). Quantitative trait loci involved in genetic predisposition to acute alcohol withdrawal in mice. *J. Neurosci.* **17**, 3946–3955.

Caberlotto, L., Thorsell, A., Rimondini, R., Sommer, W., Hyytia, P., and Heilig, M. (2001). Differential expression of NPY and its receptors in alcohol-preferring AA and alcohol-avoiding ANA rats. *Alcohol. Clin. Exp. Res.* **25**, 1564–1569.

Chatterton, M. L., Lofland, J. H., Shechter, A., Curtice, W. S., Hu, X. H., Lenow, J., Smullens, S. N., Nash, D. B., and Silberstein, S. D. (2002). Reliability and validity of the migraine therapy assessment questionnaire. *Headache* **42**, 1006–1015.

Chen, A. C., Manz, N., Tang, Y., Rangaswamy, M., Almasy, L., Kuperman, S., Nurnberger, J. Jr., O'Connor, S. J., Edenberg, H. J., Schuckit, M. A., Tischfield, J., Foroud, T., Bierut, L. J., Rohrbaugh, J., Rice, J. P., Goate, A., Hesselbrock, V., and Porjesz, B. (2010). Single-nucleotide polymorphisms in corticotropin releasing hormone receptor 1 gene (CRHR1) are associated with quantitative trait of event-related potential and alcohol dependence. *Alcohol. Clin. Exp. Res.* **34**, 988–996.

Chesler, E. J., Lu, L., Shou, S., Qu, Y., Gu, J., Wang, J., Hsu, H. C., Mountz, J. D., Baldwin, N. E., Langston, M. A., Threadgill, D. W., Manly, K. F., and Williams, R. W. (2005). Complex trait analysis of gene expression uncovers polygenic and pleiotropic networks that modulate nervous system function. *Nat. Genet.* **37**, 233–242.

Chu, K., Koob, G. F., Cole, M., Zorrilla, E. P., and Roberts, A. J. (2007). Dependence-induced increases in ethanol self-administration in mice are blocked by the CRF(1) receptor antagonist antalarmin and by CRF(1) receptor knockout. *Pharmacol. Biochem. Behav.* **86**, 813–821.

Ciccocioppo, R., Economidou, D., Cippitelli, A., Cucculelli, M., Ubaldi, M., Soverchia, L., Lourdusamy, A., and Massi, M. (2006). Genetically selected Marchigian Sardinian alcohol-preferring (msP) rats: An animal model to study the neurobiology of alcoholism. *Addict. Biol.* **11**, 339–355.

Cippitelli, A., Damadzic, R., Hansson, A. C., Singley, E., Sommer, W. H., Eskay, R., Thorsell, A., and Heilig, M. (2010). Neuropeptide Y (NPY) suppresses yohimbine-induced reinstatement of alcohol seeking. *Psychopharmacology (Berl.)* **208**, 417–426.

Colombo, G., Agabio, R., Lobina, C., Reali, R., Zocchi, A., Fadda, F., and Gessa, G. L. (1995). Sardinian alcohol-preferring rats: A genetic animal model of anxiety. *Physiol. Behav.* **57**, 1181–1185.

Crabbe, J. C. (2010). Consilience of rodent and human phenotypes relevant for alcohol dependence. *Addict. Biol.* **15**, 103–108.

Crabbe, J. C., Bell, R. L., and Ehlers, C. L. (2010). Human and laboratory rodent low response to alcohol: Is better consilience possible? *Addict. Biol.* **15**, 125–144.

Crabbe, J. C. Jr., Johnson, N. A., Gray, D. K., Kosobud, A., and Young, E. R. (1982). Biphasic effects of ethanol on open-field activity: Sensitivity and tolerance in C57BL/6N and DBA/2N mice. *J. Comp. Physiol. Psychol.* **96**, 440–451.

Crabbe, J. C., Metten, P., Rhodes, J. S., Yu, C. H., Brown, L. L., Phillips, T. J., and Finn, D. A. (2009). A line of mice selected for high blood ethanol concentrations shows drinking in the dark to intoxication. *Biol. Psychiatry* **65**, 662–670.

Crabbe, J. C., Phillips, T. J., Harris, R. A., Arends, M. A., and Koob, G. F. (2006). Alcohol-related genes: Contributions from studies with genetically engineered mice. *Addict. Biol.* **11**, 195–269.

Cruz, N. F., Lasater, A., Zielke, H. R., and Dienel, G. A. (2005). Activation of astrocytes in brain of conscious rats during acoustic stimulation: Acetate utilization in working brain. *J. Neurochem.* **92**, 934–947.

Davies, A. G., Pierce-Shimomura, J. T., Kim, H., VanHoven, M. K., Thiele, T. R., Bonci, A., Bargmann, C. I., and McIntire, S. L. (2003). A central role of the BK potassium channel in behavioral responses to ethanol in *C. elegans*. *Cell* **115**, 655–666.

De Witte, P., Pinto, E., Ansseau, M., and Verbanck, P. (2003). Alcohol and withdrawal: From animal research to clinical issues. *Neurosci. Biobehav. Rev.* **27**, 189–197.

Edenberg, H. J. (2007). The genetics of alcohol metabolism: Role of alcohol dehydrogenase and aldehyde dehydrogenase variants. *Alcohol Res. Health* **30**, 5–13.

Edenberg, H. J., Strother, W. N., McClintick, J. N., Tian, H., Stephens, M., Jerome, R. E., Lumeng, L., Li, T. K., and McBride, W. J. (2005). Gene expression in the hippocampus of inbred alcohol-preferring and -nonpreferring rats. *Genes Brain Behav.* **4**, 20–30.

Egli, M. (2005). Can experimental paradigms and animal models be used to discover clinically effective medications for alcoholism? *Addict. Biol.* **10**, 309–319.

Ehlers, C. L., Li, T. K., Lumeng, L., Hwang, B. H., Somes, C., Jimenez, P., and Mathé, A. A. (1998). Neuropeptide Y levels in ethanol-naive, alcohol-preferring, and nonpreferring rats and in Wistar rats after ethanol exposure. *Alcohol. Clin. Exp. Res.* **22**, 1778–1782.

Ehlers, C. L., Walter, N. A., Dick, D. M., Buck, K. J., and Crabbe, J. C. (2010). A comparison of selected quantitative trait loci associated with alcohol use phenotypes in humans and mouse models. *Addict. Biol.* **15**, 185–199.

Frazer, K. A., Eskin, E., Kang, H. M., Bogue, M. A., Hinds, D. A., Beilharz, E. J., Gupta, R. V., Montgomery, J., Morenzoni, M. M., Nilsen, G. B., Pethiyagoda, C. L., Stuve, L. L., Johnson, F. M., Daly, M. J., Wade, C. M., and Cox, D. R. (2007). A sequence-based variation map of 8.27 million SNPs in inbred mouse strains. *Nature* **448**, 1050–1053.

Funk, C. K., O'Dell, L. E., Crawford, E. F., and Koob, G. F. (2006). Corticotropin-releasing factor within the central nucleus of the amygdala mediates enhanced ethanol self-administration in withdrawn, ethanol-dependent rats. *J. Neurosci.* **26**, 11324–11332.

Gehlert, D. R., Cippitelli, A., Thorsell, A., Le, A. D., Hipskind, P. A., Hamdouchi, C., Lu, J., Hembre, E. J., Cramer, J., Song, M., McKinzie, D., Morin, M., Ciccocioppo, R., and Heilig, M. (2007). 3-(4-Chloro-2-morpholin-4-yl-thiazol-5-yl)-8-(1-ethylpropyl)-2,6-dimethyl- imidazo [1,2-b]pyridazine: A novel brain-penetrant, orally available corticotropin-releasing factor receptor 1 antagonist with efficacy in animal models of alcoholism. *J. Neurosci.* **27**, 2718–2726.

Gilpin, N. W., Smith, A. D., Cole, M., Weiss, F., Koob, G. F., and Richardson, H. N. (2009). Operant behavior and alcohol levels in blood and brain of alcohol-dependent rats. *Alcohol. Clin. Exp. Res.* **33**, 2113–2123.

Gilpin, N. W., Stewart, R. B., Murphy, J. M., Li, T. K., and Badia-Elder, N. E. (2003). Neuropeptide Y reduces oral ethanol intake in alcohol-preferring (P) rats following a period of imposed ethanol abstinence. *Alcohol. Clin. Exp. Res.* **27**, 787–794.

Goldman, D., Oroszi, G., and Ducci, F. (2005). The genetics of addictions: Uncovering the genes. *Nat. Rev. Genet.* **6**, 521–532.

Griffin, W. C. III, Lopez, M. F., Yanke, A. B., Middaugh, L. D., and Becker, H. C. (2009). Repeated cycles of chronic intermittent ethanol exposure in mice increases voluntary ethanol drinking and ethanol concentrations in the nucleus accumbens. *Psychopharmacology (Berl.)* **201**, 569–580.

Guo, A. Y., Webb, B. T., Miles, M. F., Zimmerman, M. P., Kendler, K. S., and Zhao, Z. (2009). ERGR: An ethanol-related gene resource. *Nucleic Acids Res.* **37**, D840–D845.

Hansson, A. C., Bermudez-Silva, F. J., Malinen, H., Hyytia, P., Sanchez-Vera, I., Rimondini, R., Rodriguez de, F. F., Kunos, G., Sommer, W. H., and Heilig, M. (2007b). Genetic impairment of frontocortical endocannabinoid degradation and high alcohol preference. *Neuropsychopharmacology* **32**, 117–126.

Hansson, A. C., Cippitelli, A., Sommer, W., Ciccocioppo, R., and Heilig, M. (2007b). Region-specific down regulation of *Crhr1* gene expression in alcohol preferring msP rats following *ad lib* access to alcohol. *Addict. Biol.* **12**, 30–34.

Hansson, A. C., Cippitelli, A., Sommer, W. H., Fedeli, A., Bjork, K., Soverchia, L., Terasmaa, A., Massi, M., Heilig, M., and Ciccocioppo, R. (2006). Variation at the rat *Crhr1* locus and sensitivity to relapse into alcohol seeking induced by environmental stress. *Proc. Natl. Acad. Sci. USA* **103**, 15236–15241.

Hansson, A. C., Nixon, K., Rimondini, R., Damadzic, R., Sommer, W. H., Eskay, R., Crews, F. T., and Heilig, M. (2010). Long-term suppression of forebrain neurogenesis and loss of neuronal progenitor cells following prolonged alcohol dependence in rats. *Int. J. Neuropsychopharmacol.* **13(5)**, 583–593.

Hansson, A. C., Rimondini, R., Neznanova, O., Sommer, W. H., and Heilig, M. (2008). Neuroplasticity in brain reward circuitry following a history of ethanol dependence. *Eur. J. Neurosci.* **27**, 1912–1922.

Harris, R. A., Trudell, J. R., and Mihic, S. J. (2008). Ethanol's molecular targets. *Sci. Signal.* **1**, re7.

Heilig, M., and Egli, M. (2005). Models for alcohol dependence: A clinical perspective. *Drug Discov. Today Dis. Models* **2**, 313–318.

Heilig, M., Egli, M., Crabbe, J. C., and Becker, H. C. (2010). Acute withdrawal, protracted abstinence and negative affect in alcoholism: Are they linked? *Addict. Biol.* **15**, 169–184.

Heilig, M., and Koob, G. F. (2007). A key role for corticotropin-releasing factor in alcohol dependence. *Trends Neurosci.* **30**, 399–406.

Heilig, M., Thorsell, A., Sommer, W. H., Hansson, A. C., Ramchandani, V. A., George, D. T., Hommer, D., and Barr, C. S. (2009). Translating the neuroscience of alcoholism into clinical treatments: From blocking the buzz to curing the blues. *Neurosci. Biobehav. Rev.* doi:10.1016/j.neubiorev.2009.11.018.

Hoffman, P. L., Bennett, B., Saba, L. M., Bhave, S. V., Carosone-Link, P. J., Hornbaker, C. K., Kechris, K. J., Williams, R. W., and Tabakoff, B. (2010). Using the phenoGen website for "In Silico" analysis of morphine-induced analgesia: Identifying candidate genes.

Hoffman, P. L., Miles, M., Edenberg, H. J., Sommer, W., Tabakoff, B., Wehner And J. M., and Lewohl, J. (2003). Gene expression in brain: A window on ethanol dependence, neuroadaptation, and preference. *Alcohol. Clin. Exp. Res.* **27**, 155–168.

Holter, S. M., Engelmann, M., Kirschke, C., Liebsch, G., Landgraf, R., and Spanagel, R. (1998). Long-term ethanol self-administration with repeated ethanol deprivation episodes changes ethanol drinking pattern and increases anxiety-related behaviour during ethanol deprivation in rats. *Behav. Pharmacol.* **9**, 41–48.

Hu, W., Saba, L., Kechris, K., Bhave, S. V., Hoffman, P. L., and Tabakoff, B. (2008). Genomic insights into acute alcohol tolerance. *J. Pharmacol. Exp. Ther.* **326**, 792–800.

Jansen, R. C., and Nap, J. P. (2001). Genetical genomics: The added value from segregation. *Trends Genet.* **17**, 388–391.

Kakko, J., von, W. J., Svanborg, K. D., Lidstrom, J., Barr, C. S., and Heilig, M. (2008). Mood and neuroendocrine response to a chemical stressor, metyrapone, in buprenorphine-maintained heroin dependence. *Biol. Psychiatry* **63**, 172–177.

Kanis, J. A., Johansson, H., Johnell, O., Oden, A., De Laet, C., Eisman, J. A., Pols, H., and Tenenhouse, A. (2005). Alcohol intake as a risk factor for fracture. *Osteoporos. Int.* **16**, 737–742.

Karlsson, R. M. (2009). Neuropeptide Y (NPY) and Glutamate Transporter (GLAST) in Behavioral Models of Psychiatric Disorders. Karolinska Institute, Stockholm, Sweden.

Karpyak, V. M., Biernacka, J. M., Weg, M. W., Stevens, S. R., Cunningham, J. M., Mrazek, D. A., and Black, J. L. (2010). Interaction of SLC6A4 and DRD2 polymorphisms is associated with a history of delirium tremens. *Addict. Biol.* **15**, 23–34.

Katner, S. N., Magalong, J. G., and Weiss, F. (1999). Reinstatement of alcohol-seeking behavior by drug-associated discriminative stimuli after prolonged extinction in the rat. *Neuropsychopharmacology* **20**, 471–479.

Kerns, R. T., Ravindranathan, A., Hassan, S., Cage, M. P., York, T., Sikela, J. M., Williams, R. W., and Miles, M. F. (2005). Ethanol-responsive brain region expression networks: Implications for

behavioral responses to acute ethanol in DBA/2J versus C57BL/6J mice. *J. Neurosci.* **25**, 2255–2266.
Kimpel, M. W., Strother, W. N., McClintick, J. N., Carr, L. G., Liang, T., Edenberg, H. J., and McBride, W. J. (2007). Functional gene expression differences between inbred alcohol-preferring and -non-preferring rats in five brain regions. *Alcohol* **41**, 95–132.
King, A. C., Volpicelli, J. R., Frazer, A., and O'Brien, C. P. (1997). Effect of naltrexone on subjective alcohol response in subjects at high and low risk for future alcohol dependence. *Psychopharmacology (Berl.)* **129**, 15–22.
Koob, G. F. (2003). Alcoholism: Allostasis and beyond. *Alcohol. Clin. Exp. Res.* **27**, 232–243.
Koob, G. F., and Le Moal, M. (2001). Drug addiction, dysregulation of reward, and allostasis. *Neuropsychopharmacology* **24**, 97–129.
Koob, G. F., and Volkow, N. D. (2010). Neurocircuitry of addiction. *Neuropsychopharmacology* **35**, 217–238.
Kosobud, A., and Crabbe, J. C. (1986). Ethanol withdrawal in mice bred to be genetically prone or resistant to ethanol withdrawal seizures. *J. Pharmacol. Exp. Ther.* **238**, 170–177.
Leeman, R. F., Heilig, M., Cunningham, C. L., Stephens, D. N., Duka, T., and O'Malley, S. S. (2010). Ethanol consumption: How should we measure it? Achieving consilience between human and animal phenotypes. *Addict. Biol.* **15**, 109–124.
Lefkowitz, R. J., and Shenoy, S. K. (2005). Transduction of receptor signals by beta-arrestins. *Science* **308**, 512–517.
Levine, L., Grossfield, J., and Rockwell, R. F. (1979). Functional relationships between genotypes and environments in behavior. Effects of different kinds of early social experience on interstrain fighting in male mice. *J. Hered.* **70**, 317–320.
Liang, T., Habegger, K., Spence, J. P., Foroud, T., Ellison, J. A., Lumeng, L., Li, T. K., and Carr, L. G. (2004). Glutathione S-transferase 8–8 expression is lower in alcohol-preferring than in alcohol-nonpreferring rats. *Alcohol. Clin. Exp. Res.* **28**, 1622–1628.
Liang, T., Kimpel, M. W., McClintick, J. N., Skillman, A. R., McCall, K., Edenberg, H. J., and Carr, L. G. (2010). Candidate genes for alcohol preference identified by expression profiling in alcohol-preferring and -nonpreferring reciprocal congenic rats. *Genome Biol.* **11**, R11.
Lindell, S. G., Schwandt, M. L., Sun, H., Sparenborg, J. D., Bjork, K., Kasckow, J. W., Sommer, W. H., Goldman, D., Higley, J. D., Suomi, S. J., Heilig, M., and Barr, C. S. (2010). Functional NPY variation as a factor in stress resilience and alcohol consumption in rhesus macaques. *Arch. Gen. Psychiatry* **67**, 423–431.
Lopez, M. F., and Becker, H. C. (2005). Effect of pattern and number of chronic ethanol exposures on subsequent voluntary ethanol intake in C57BL/6J mice. *Psychopharmacology (Berl.)* **181**, 688–696.
Maclaren, E. J., Bennett, B., Johnson, T. E., and Sikela, J. M. (2006). Expression profiling identifies novel candidate genes for ethanol sensitivity QTLs. *Mamm. Genome* **17**, 147–156.
Majchrowicz, E. (1975). Induction of physical dependence upon ethanol and the associated behavioral changes in rats. *Psychopharmacologia* **43**, 245–254.
Mardones, J., Segovia, N., and Hederra, A. (1953). Heredity of experimental alcohol preference in rats. II. Coefficient of heredity. *Q. J. Stud. Alcohol* **14**, 1–2.
Matsumoto, I., Abe, K., and Arai, S. (2006). Molecular logic of alcohol and taste. *Nihon Arukoru Yakubutsu Igakkai Zasshi* **41**, 431–444.
Matthaus, F., Smith, V. A., Fogtman, A., Sommer, W. H., Leonardi-Essmann, F., Lourdusamy, A., Reimers, M. A., Spanagel, R., and Gebicke-Haerter, P. J. (2009). Interactive molecular networks obtained by computer-aided conversion of microarray data from brains of alcohol-drinking rats. *Pharmacopsychiatry* **42**(Suppl. 1), S118–S128.
Maul, B., Krause, W., Pankow, K., Becker, M., Gembardt, F., Alenina, N., Walther, T., Bader, M., and Siems, W. E. (2005). Central angiotensin II controls alcohol consumption via its AT1 receptor. *FASEB J.* **19**, 1474–1481.

Maul, B., Siems, W. E., Hoehe, M. R., Grecksch, G., Bader, M., and Walther, T. (2001). Alcohol consumption is controlled by angiotensin II. *FASEB J.* **15**, 1640–1642.

McClearn, G. E., and Rodgers, D. A. (1959). Differences in alcohol preference among inbred strain of mice. *Q. J. Stud. Alcohol.* **20**, 691–695.

Melendez, R. I., Kalivas, P. W., McGinty, G. F., and Becker, H. C. (2006). Differential gene expression alterations induced by chronic ethanol exposure and withdrawal in C57BL/6J mice. *Alcohol. Clin. Exp. Res.* **12**, 2017–2025.

Metten, P., Sorensen, M. L., Cameron, A. J., Yu, C.-H., and Crabbe, J. C. (2010). Withdrawal severity after chronic intermittent ethanol in inbred mouse strains. *Alcohol. Clin. Exp. Res.* **34**, 1–14.

Möller, C., Wiklund, L., Thorsell, A., Hyytia, P., and Heilig, M. (1997). Decreased measures of experimental anxiety in rats bred for high alcohol preference. *Alcohol. Clin. Exp. Res.* **21**, 656–660.

Moore, M. S., DeZazzo, J., Luk, A. Y., Tully, T., Singh, C. M., and Heberlein, U. (1998). Ethanol intoxication in Drosophila: Genetic and pharmacological evidence for regulation by the cAMP signaling pathway. *Cell* **93**, 997–1007.

Mottagui-Tabar, S., Prince, J. A., Wahlestedt, C., Zhu, G., Goldman, D., and Heilig, M. (2005). A novel single nucleotide polymorphism of the neuropeptide Y (NPY) gene associated with alcohol dependence. *Alcohol. Clin. Exp. Res.* **29**, 702–707.

Mueller, H. T., and Meador-Woodruff, J. H. (2004). NR3A NMDA receptor subunit mRNA expression in schizophrenia, depression and bipolar disorder. *Schizophr. Res.* **71**, 361–370.

Mulert, C., Juckel, G., Giegling, I., Pogarell, O., Leicht, G., Karch, S., Mavrogiorgou, P., Moller, H. J., Hegerl, U., and Rujescu, D. (2006). A Ser9Gly polymorphism in the dopamine D3 receptor gene (DRD3) and event-related P300 potentials. *Neuropsychopharmacology* **31**, 1335–1344.

Mulligan, M. K., Ponomarev, I., Hitzemann, R. J., Belknap, J. K., Tabakoff, B., Harris, R. A., Crabbe, J. C., Blednov, Y. A., Grahame, N. J., Phillips, T. J., Finn, D. A., Hoffman, P. L., Iyer, V. R., Koob, G. F., and Bergeson, S. E. (2006). Toward understanding the genetics of alcohol drinking through transcriptome meta-analysis. *Proc. Natl. Acad. Sci. USA* **103**, 6368–6373.

Nelson, E. C., Agrawal, A., Pergadia, M. L., Wang, J. C., Whitfield, J. B., Saccone, F. S., Kern, J., Grant, J. D., Schrage, A. J., Rice, J. P., Montgomery, G. W., Heath, A. C., Goate, A. M., Martin, N. G., and Madden, P. A. (2010). H2 haplotype at chromosome 17q21.31 protects against childhood sexual abuse-associated risk for alcohol consumption and dependence. *Addict. Biol.* **15**, 1–11.

Neznanova, O., Bjork, K., Rimondini, R., Hansson, A. C., Hyytia, P., Heilig, M., and Sommer, W. H. (2009). Acute ethanol challenge inhibits glycogen synthase kinase-3beta in the rat prefrontal cortex. *Int. J. Neuropsychopharmacol.* **12**, 275–280.

O'Dell, L. E., Roberts, A. J., Smith, R. T., and Koob, G. F. (2004). Enhanced alcohol self-administration after intermittent versus continuous alcohol vapor exposure. *Alcohol. Clin. Exp. Res.* **28**, 1676–1682.

Overstreet, D. H., Knapp, D. J., and Breese, G. R. (2002). Accentuated decrease in social interaction in rats subjected to repeated ethanol withdrawals. *Alcohol. Clin. Exp. Res.* **26**, 1259–1268.

Pawlosky, R. J., Kashiwaya, Y., Srivastava, S., King, M. T., Crutchfield, C., Volkow, N., Kunos, G., Li, T. K., and Veech, R. L. (2010). Alterations in brain glucose utilization accompanying elevations in blood ethanol and acetate concentrations in the rat. *Alcohol. Clin. Exp. Res.* **34**, 375–381.

Peng, J., Wagle, M., Mueller, T., Mathur, P., Lockwood, B. L., Bretaud, S., and Guo, S. (2009). Ethanol-modulated camouflage response screen in zebrafish uncovers a novel role for cAMP and extracellular signal-regulated kinase signaling in behavioral sensitivity to ethanol. *J. Neurosci.* **29**, 8408–8418.

Pignataro, L., Varodayan, F. P., Tannenholz, L. E., and Harrison, N. L. (2009). The regulation of neuronal gene expression by alcohol. *Pharmacol. Ther.* **124**, 324–335.

Quintanilla, M. E., Israel, Y., Sapag, A., and Tampier, L. (2006). The UChA and UChB rat lines: Metabolic and genetic differences influencing ethanol intake. *Addict. Biol.* **11**, 310–323.

Ramchandani, V. A., Umhau, J., Pavon, F. J., Ruiz-Velasco, V., Margas, W., Sun, H., Damadzic, R., Eskay, R., Schoor, M., Thorsell, A., Schwandt, M. L., Sommer, W. H., George, D. T., Parsons, L. H., Herscovitch, P., Hommer, D., and Heilig, M. (2010). A genetic determinant of the striatal dopamine response to alcohol in men. *Mol. Psychiatry.* doi:10.1038/mp.2010.56.

Ray, L. A., and Hutchison, K. E. (2004). A polymorphism of the mu-opioid receptor gene (OPRM1) and sensitivity to the effects of alcohol in humans. *Alcohol. Clin. Exp. Res.* **28**, 1789–1795.

Repunte-Canonigo, V., Lutjens, R., van der Stap, L. D., and Sanna, P. P. (2007). Increased expression of protein kinase A inhibitor alpha (PKI-alpha) and decreased PKA-regulated genes in chronic intermittent alcohol exposure. *Brain Res.* **1138**, 48–56.

Repunte-Canonigo, V., van der Stap, L. D., Chen, J., Sabino, V., Wagner, U., Zorrilla, E. P., Schumann, G., Roberts, A. J., and Sanna, P. P. (2010). Genome-wide gene expression analysis identifies K-ras as a regulator of alcohol intake. *Brain Res.* **1339**, 11–17.

Rhodes, J. S., Best, K., Belknap, J. K., Finn, D. A., and Crabbe, J. C. (2005). Evaluation of a simple model of ethanol drinking to intoxication in C57BL/6J mice. *Physiol. Behav.* **84**, 53–63.

Rhodes, J. S., Ford, M. M., Yu, C. H., Brown, L. L., Finn, D. A., Garland, T. Jr., and Crabbe, J. C. (2007). Mouse inbred strain differences in ethanol drinking to intoxication. *Genes Brain Behav.* **6**, 1–18.

Rimondini, R., Arlinde, C., Sommer, W., and Heilig, M. (2002). Long-lasting increase in voluntary ethanol consumption and transcriptional regulation in the rat brain after intermittent exposure to alcohol. *FASEB J.* **16**, 27–35.

Rimondini, R., Sommer, W., and Heilig, M. (2003). A temporal threshold for induction of persistent alcohol preference: Behavioral evidence in a rat model of intermittent intoxication. *J. Stud. Alcohol* **64**, 445–449.

Rimondini, R., Sommer, W. H., Dall'olio, R., and Heilig, M. (2008). Long-lasting tolerance to alcohol following a history of dependence. *Addict. Biol.* **13**, 26–30.

Rimondini, R., Thorsell, A., and Heilig, M. (2005). Suppression of ethanol self-administration by the neuropeptide Y (NPY) Y2 receptor antagonist BIIE0246: Evidence for sensitization in rats with a history of dependence. *Neurosci. Lett.* **375**, 129–133.

Roberts, A. J., Heyser, C. J., Cole, M., Griffin, P., and Koob, G. F. (2000). Excessive ethanol drinking following a history of dependence: Animal model of allostasis. *Neuropsychopharmacology* **22**, 581–594.

Rodd, Z. A.,Bertsch, B. A., Strother, W. N., Le-Niculescu, H., Balaraman, Y., Hayden, E., Jerome, R. E., Lumeng, L., Nurnberger, J. I. Jr., Edenberg, H. J., McBride, W. J., and Niculescu, A. B. (2006). Candidate genes, pathways and mechanisms for alcoholism: An expanded convergent functional genomics approach. *Pharmacogenomics J.* **6**, 1–3.

Rogers, J., Wiener, S. G., and Bloom, F. E. (1979). Long-term ethanol administration methods for rats: Advantages of inhalation over intubation or liquid diets. *Behav. Neural Biol.* **27**, 466–486.

Rulten, S. L., Ripley, T. L., Hunt, C. L., Stephens, D. N., and Mayne, L. V. (2006). Sp1 and NFkappaB pathways are regulated in brain in response to acute and chronic ethanol. *Genes Brain Behav.* **5**, 257–273.

Saba, L., Bhave, S. V., Grahame, N., Bice, P., Lapadat, R., Belknap, J., Hoffman, P. L., and Tabakoff, B. (2006). Candidate genes and their regulatory elements: Alcohol preference and tolerance. *Mamm. Genome* **17**, 669–688.

Sanchis-Segura, C., and Spanagel, R. (2006). Behavioural assessment of drug reinforcement and addictive features in rodents: An overview. *Addict. Biol.* **11**, 2–38.

Schuckit, M. A., Smith, T. L., Danko, G. P., Trim, R., Bucholz, K. K., Edenberg, H. J., Hesselbrock, V., Kramer, J. J., and Dick, D. M. (2009). An evaluation of the full level of response to alcohol model of heavy drinking and problems in COGA offspring. *J. Stud. Alcohol Drugs* **70**, 436–445.

Shaham, Y., Shalev, U., Lu, L., de Wit, H., and Stewart, J. (2003). The reinstatement model of drug relapse: History, methodology and major findings. *Psychopharmacology (Berl.)* **168**, 3–20.

Shirley, R. L., Walter, N. A., Reilly, M. T., Fehr, C., and Buck, K. J. (2004). Mpdz is a quantitative trait gene for drug withdrawal seizures. *Nat. Neurosci.* **7**, 699–700.

Sillaber, I., Rammes, G., Zimmermann, S., Mahal, B., Zieglgansberger, W., Wurst, W., Holsboer, F., and Spanagel, R. (2002). Enhanced and delayed stress-induced alcohol drinking in mice lacking functional CRH1 receptors. *Science* **296**, 931–933.

Sinclair, D. (2006). Deleterious pleiotropic effects of the atypical aldehyde dehydrogenase 2 (ALDH2) allele: Comment on Luo *et al.*, 2005. *Biochem. Genet.* **44**, 385–390.

Sommer, W., Arlinde, C., Caberlotto, L., Thorsell, A., Hyytia, P., and Heilig, M. (2001). Differential expression of diacylglycerol kinase iota and L18A mRNAs in the brains of alcohol-preferring AA and alcohol-avoiding ANA rats. *Mol. Psychiatry.* **6**, 103–108.

Sommer, W., Arlinde, C., and Heilig, M. (2005). The search for candidate genes of alcoholism: Evidence from expression profiling studies. *Addict. Biol.* **10**, 71–79.

Sommer, W., Hyytia, P., and Kiianmaa, K. (2006). The alcohol-preferring AA and alcohol-avoiding ANA rats: Neurobiology of the regulation of alcohol drinking. *Addict. Biol.* **11**, 289–309.

Sommer, W. H., Lidström, J., Sun, H., Passer, D., Eskay, R. L., Parker, S., Witt, S. H., Zimmermann, U., Nieratschker, V., Rietschel, M., Margulies, M., Palkovits, M., Laucht, M., and Heilig, M. (2010). Human NPY promoter variation rs16147 as a moderator of prefrontal NPY gene expression and negative affect. *Hum. Mutat.* doi:10.1002/humu.21299.

Sommer, W. H., Rimondini, R., Hansson, A. C., Hipskind, P. A., Gehlert, D. R., Barr, C. S., and Heilig, M. (2008). Upregulation of voluntary alcohol intake, behavioral sensitivity to stress, and Amygdala crhr1 expression following a history of dependence. *Biol. Psychiatry* **63**, 139–145.

Sommer, W. H., Rimondini, R., Marquitz, M., Lidstrom, J., Siems, W. E., Bader, M., and Heilig, M. (2007). Plasticity and impact of the central renin-angiotensin system during development of ethanol dependence. *J. Mol. Med.* **85**, 1089–1097.

Sommer, W. H., and Saavedra, J. M. (2008). Targeting brain angiotensin and corticotrophin-releasing hormone systems interaction for the treatment of mood and alcohol use disorders. *J. Mol. Med.* **86**, 723–728.

Spanagel, R. (2009). Alcoholism: A systems approach from molecular physiology to addictive behavior. *Physiol. Rev.* **89**, 649–705.

Spanagel, R., and Holter, S. M. (2000). Pharmacological validation of a new animal model of alcoholism. *J. Neural Transm. (Budapest)* **107**, 669–680.

Spanagel, R., and Kiefer, F. (2008). Drugs for relapse prevention of alcoholism: Ten years of progress. *Trends Pharmacol. Sci.* **29**, 109–115.

Spence, J. P., Liang, T., Habegger, K., and Carr, L. G. (2005). Effect of polymorphism on expression of the neuropeptide Y gene in inbred alcohol-preferring and -nonpreferring rats. *Neuroscience* **131**, 871–876.

Steensland, P., Simms, J. A., Holgate, J., Richards, J. K., and Bartlett, S. E. (2007). Varenicline, an alpha4beta2 nicotinic acetylcholine receptor partial agonist, selectively decreases ethanol consumption and seeking. *Proc. Natl. Acad. Sci. USA* **104**, 12518–12523.

Stephens, D. N., Duka, T., Crombag, H. S., Cunningham, C. L., Heilig, M., and Crabbe, J. C. (2010). Reward sensitivity: Issues of measurement, and achieving consilience between human and animal phenotypes. *Addict. Biol.* **15**, 145–168.

Stewart, R. B., Gatto, G. J., Lumeng, L., Li, T. K., and Murphy, J. M. (1993). Comparison of alcohol-preferring (P) and nonpreferring (NP) rats on tests of anxiety and for the anxiolytic effects of ethanol. *Alcohol* **10**, 1–10.

Sun, D., Ma, J. Z., Payne, T. J., and Li, M. D. (2008). Beta-arrestins 1 and 2 are associated with nicotine dependence in European American smokers. *Mol. Psychiatry* **13**, 398–406.

Tabakoff, B., Saba, L., Kechris, K., Hu, W., Bhave, S. V., Finn, D. A., Grahame, N. J., and Hoffman, P. L. (2008). The genomic determinants of alcohol preference in mice. *Mamm. Genome* **19**, 352–365.

Tabakoff, B., *et al.* (2009). Genetical genomic determinants of alcohol consumption in rats and humans. *BMC Biol.* **7**, 70.

Thorsell, A., Karlsson, R. M., and Heilig, M. (2006). NPY in alcoholism and psychiatric disorders. *EXS.* **95**, 183–192.

Thorsell, A., Repunte-Canonigo, V., O'Dell, L. E., Chen, S. A., King, A. R., Lekic, D., Koob, G. F., and Sanna, P. P. (2007). Viral vector-induced amygdala NPY overexpression reverses increased alcohol intake caused by repeated deprivations in Wistar rats. *Brain* **130**, 1330–1337.

Torricelli, P., Fini, M., Giavaresi, G., Borsari, V., Rimondini, L., Rimondini, R., Carrassi, A., and Giardino, R. (2007). Intermittent exposure to ethanol vapor affects osteoblast behaviour more severely than estrogen deficiency does in vitro study on rat osteoblasts. *Toxicology* **237**, 168–176.

Treadwell, J. A., and Singh, S. M. (2004). Microarray analysis of mouse brain gene expression following acute ethanol treatment. *Neurochem. Res.* **29**, 357–369.

Treutlein, J., *et al.* (2009). Genome-wide association study of alcohol dependence. *Arch. Gen. Psychiatry* **66**, 773–784.

Treutlein, J., Kissling, C., Frank, J., Wiemann, S., Dong, L., Depner, M., Saam, C., Lascorz, J., Soyka, M., Preuss, U. W., Rujescu, D., Skowronek, M. H., Rietschel, M., Spanagel, R., Heinz, A., Laucht, M., Mann, K., and Schumann, G. (2006). Genetic association of the human corticotropin releasing hormone receptor 1 (CRHR1) with binge drinking and alcohol intake patterns in two independent samples. *Mol. Psychiatry* **11**, 594–602.

Uddin, R. K., Treadwell, J. A., and Singh, S. M. (2005). Towards unraveling ethanol-specific neurometabolomics based on ethanol responsive genes in vivo. *Neurochem. Res.* **30**, 1179–1190.

Valdez, G. R., Roberts, A. J., Chan, K., Davis, H., Brennan, M., Zorrilla, E. P., and Koob, G. F. (2002). Increased ethanol self-administration and anxiety-like behavior during acute ethanol withdrawal and protracted abstinence: Regulation by corticotropin-releasing factor. *Alcohol. Clin. Exp. Res.* **26**, 1494–1501.

Valdez, G. R., Sabino, V., and Koob, G. F. (2004). Increased anxiety-like behavior and ethanol self-administration in dependent rats: Reversal via corticotropin-releasing factor-2 receptor activation. *Alcohol. Clin. Exp. Res.* **28**, 865–872.

Valdez, G. R., Zorrilla, E. P., Roberts, A. J., and Koob, G. F. (2003). Antagonism of corticotropin-releasing factor attenuates the enhanced responsiveness to stress observed during protracted ethanol abstinence. *Alcohol* **29**, 55–60.

Vanderschuren, L. J., and Everitt, B. J. (2005). Behavioral and neural mechanisms of compulsive drug seeking. *Eur. J. Pharmacol.* **526**, 77–88.

Vengeliene, V., Celerier, E., Chaskiel, L., Penzo, F., and Spanagel, R. (2009). Compulsive alcohol drinking in rodents. *Addict. Biol.* **14**, 384–396.

Vengeliene, V., Leonardi-Essmann, F., Perreau-Lenz, S., Gebicke-Haerter, P., Drescher, K., Gross, G., and Spanagel, R. (2006). The dopamine D3 receptor plays an essential role in alcohol-seeking and relapse. *FASEB J.* **20**, 2223–2233.

Vengeliene, V., Leonardi-Essmann, F., Sommer, W. H., and Spanagel, R. (2010). Glycine transporter-1 blockade leads to persistently reduced relapse-like alcohol drinking in rats. *Biol. Psychiatry.* doi:10.1016/j.biopsych.2010.05.029.

Vengeliene, V., Siegmund, S., Singer, M. V., Sinclair, J. D., Li, T. K., and Spanagel, R. (2003). A comparative study on alcohol-preferring rat lines: Effects of deprivation and stress phases on voluntary alcohol intake. *Alcohol. Clin. Exp. Res.* **27**, 1048–1054.

Vilpoux, C., Warnault, V., Pierrefiche, O., Daoust, M., and Naassila, M. (2009). Ethanol-sensitive brain regions in rat and mouse: A cartographic review, using immediate early gene expression. *Alcohol. Clin. Exp. Res.* **33**, 945–969.

Volkow, N. D., Fowler, J. S., and Wang, G. J. (2003). The addicted human brain: Insights from imaging studies. *J. Clin. Invest.* **111**, 1444–1451.

Volkow, N. D., Wang, G. J., Franceschi, D., Fowler, J. S., Thanos, P. P., Maynard, L., Gatley, S. J., Wong, C., Veech, R. L., Kunos, G., and Kai, L. T. (2006). Low doses of alcohol substantially decrease glucose metabolism in the human brain. *Neuroimage* **29**, 295–301.

Wahlsten, D., Bachmanov, A., Finn, D. A., and Crabbe, J. C. (2006). Stability of inbred mouse strain differences in behavior and brain size between laboratories and across decades. *Proc. Natl. Acad. Sci. USA* **103**, 16364–16369.

Wand, G. S., McCaul, M., Yang, X., Reynolds, J., Gotjen, D., Lee, S., and Ali, A. (2002). The mu-opioid receptor gene polymorphism (A118G) alters HPA axis activation induced by opioid receptor blockade. *Neuropsychopharmacology* **26**, 106–114.

Waniewski, R. A., and Martin, D. L. (1998). Preferential utilization of acetate by astrocytes is attributable to transport. *J. Neurosci.* **18**, 5225–5233.

Wiens, F., Zitzmann, A., Lachance, M. A., Yegles, M., Pragst, F., Wurst, F. M., von, H. D., Guan, S. L., and Spanagel, R. (2008). Chronic intake of fermented floral nectar by wild treeshrews. *Proc. Natl. Acad. Sci. USA* **105**, 10426–10431.

Wise, R. A. (1975). Maximization of ethanol intake in the rat. [Review] [51 refs]. *Adv. Exp. Med. Biol.* **59**, 279–294.

Wolffgramm, J., Galli, G., Thimm, F., and Heyne, A. (2000). Animal models of addiction: models for therapeutic strategies? *J. Neural Transm. (Budapest)* **107**, 649–668.

Worst, T. J., Tan, J. C., Robertson, D. J., Freeman, W. M., Hyytia, P., Kiianmaa, K., and Vrana, K. E. (2005). Transcriptome analysis of frontal cortex in alcohol-preferring and nonpreferring rats. *J. Neurosci. Res.* **80**, 529–538.

Zhu, G., Pollak, L., Mottagui-Tabar, S., Wahlestedt, C., Taubman, J., Virkkunen, M., Goldman, D., and Heilig, M. (2003). NPY leu7pro and alcohol dependence in Finnish and Swedish populations. *Alcohol. Clin. Exp. Res.* **27**, 19–24.

IDENTIFYING QUANTITATIVE TRAIT LOCI (QTLs) AND GENES (QTGs) FOR ALCOHOL-RELATED PHENOTYPES IN MICE

Lauren C. Milner and Kari J. Buck

Department of Behavioral Neuroscience, VA Medical Center and Oregon Health & Science University, Portland, OR 97239, USA

Alcoholism is a complex clinical disorder with genetic and environmental contributions. Although no animal model duplicates alcoholism, models for specific factors, such as the withdrawal syndrome, are useful to identify potential genetic determinants of liability in humans. Murine models have been invaluable to identify quantitative trait loci (QTLs) that influence a variety of alcohol responses. However, the QTL regions are typically large, at least initially, and contain numerous genes,

INTERNATIONAL REVIEW OF
NEUROBIOLOGY, VOL. 91
DOI: 10.1016/S0074-7742(10)91006-4

making identification of the causal quantitative trait gene(s) (QTGs) challenging. Here, we present QTG identification strategies currently used in the field of alcohol genetics and discuss relevance to alcoholic human populations.

I. Introduction

A. Alcohol Abuse and Alcoholism

Alcohol (ethanol) dependence and abuse disorders are prevalent in American society, with >18 million Americans (8.5% of the population 18 years of age and older) estimated to meet the diagnostic criteria of alcohol abuse or dependence (Grant *et al.*, 2004), and lifetime prevalences for these disorders up to 17% (Regier *et al.*, 1990). The evident genetic contribution to alcoholism is supported by adoption studies that demonstrate an increased risk for severe alcohol-related problems in children of alcoholics who were adopted out, even if raised without knowledge of their biological parents' substance use (Cadoret *et al.*, 1980; Schuckit *et al.*, 1972). Several large twin and nuclear family studies published in the 1990s substantiate the conclusion that alcoholism is >50% heritable (Goldman, 1993; Goldman and Bergen, 1998; Prescott and Kendler, 1999; Reich *et al.*, 1999).

Despite the sizable heritability component contributing to these disorders, the development of alcoholism is determined by multiple genetic and environmental contributions, and specific genetic determinants of alcohol abuse and alcoholism remain largely unknown. To date, only a few genes have consistently demonstrated a role in alcoholism or associated phenotypes (i.e., discrete behavioral symptoms or biological measures associated with the disorder of interest also referred to as endophenotypes) in human studies. These include isoforms of the alcohol-metabolizing enzymes *ADH1, ADH4*, and *ALDH2*, and neurotransmitter receptor subunits *GABRB1, GABRA2*, and *CHRM2* (reviewed in Dick and Foroud, 2003; Edenberg and Foroud, 2006; Kuo *et al.*, 2008, 2009). Despite significant advances in candidate gene classification for alcohol-related traits in human populations, initial identification of these genes has relied heavily on *a priori* knowledge of physiological responses to alcohol and neurotransmitter systems previously associated with this drug (Zhu and Zhou, 2007). Systematic approaches to gene discovery outside of these pathways are critical if novel genes and mechanisms underlying alcohol dependence are to be discovered and translated to improved treatment and prevention strategies in clinical populations. However, the many practical and ethical limitations involved with identification of novel genes in human subjects necessitate the use of animal models to examine various genetic and physiological aspects associated with this disorder.

Quantitative trait locus (QTL) mapping in animal models is an approach that has become common in the field of alcohol genetics research to determine chromosomal locations harboring genes(s) that influence complex traits, such as alcohol-related behaviors. This chapter will focus on recent QTL fine-mapping and candidate gene-testing strategies in murine models to identify novel and discrete genetic influences on alcohol withdrawal behaviors, as well as potential brain regions in which particular genes may be exerting their influence on these alcohol phenotypes. The human relevance of these types of studies depends upon using robust animal models (genotypic and phenotypic) to examine alcohol-related behaviors and the high degree of homology between the human and mammalian animal model genomes at the gene and linkage levels, discussed in Section VII (Peters *et al.*, 2007; Spanagel, 2003).

B. Alcohol Physiological Dependence and Associated Withdrawal

In addition to the initial effects of alcohol, the continued use of alcohol depends on multiple factors that profoundly affect the pattern of alcohol use and the likelihood that alcohol use will, for some individuals, progress from casual use to alcoholism. Among these is the capacity of alcohol to produce physical dependence. When alcohol is eliminated, and its depressant effects disappear, characteristic symptoms of central nervous system (CNS) hyperexcitability first wax and then wane, including tremor, autonomic nervous system overactivity, and, in extreme cases, convulsions, which can be lethal (McKeon *et al.*, 2008). These withdrawal symptoms define a preexisting state of physical dependence on the drug. Physiological dependence and associated withdrawal episodes are thought to constitute a motivational force that perpetuates alcohol use and abuse, and may contribute to relapse (Little *et al.*, 2005).

Although many symptoms are associated with alcohol withdrawal in affected populations, CNS hyperexcitability associated with alcohol withdrawal has allowed for the creation of a robust phenotype that has allowed great strides for gene identification associated with alcohol withdrawal in mouse models.

C. Mouse Models of Alcohol Withdrawal

A comprehensive understanding of genetic variation in humans and informative animal models is crucial to establish relationships between genotype and biological function (Collins *et al.*, 2003). Although no animal model exactly duplicates clinically defined alcoholism, robust animal models for specific factors (e.g., acute sensitivity, excessive drinking, tolerance, physical dependence, and associated withdrawal) are useful to identify potential genetic determinants of liability in humans (Uhl *et al.*, 2008). It is important to note that genetic studies using mouse models have been undertaken for a variety of alcohol-related behaviors, many of which are currently being used to identify chromosomal regions associated with these behaviors. Here, we primarily focus on the significant strides toward discerning specific genes contributing to alcohol withdrawal behaviors as a means of providing one example of current methods used to fine-map alcohol-related QTLs.

Alcohol withdrawal convulsions occur in all species studied, including humans (Friedman, 1980), and provide a quantitative index of ethanol withdrawal severity in mice (Goldstein and Pal, 1971). In a series of studies in the 1970s, Dr. Dora Goldstein developed a system for inducing physical dependence on alcohol in mice by administering ethanol vapor continuously to animals confined in an inhalation chamber. She also described and quantified the characteristic handling-induced convulsion (HIC) displayed by withdrawing mice (Goldstein and Pal, 1971). An HIC is assessed by lifting a mouse gently by the tail, spinning it in a 180° arc, and measuring the convulsion response, the severity of response being rated by the experimenter on a scale from 0 (no convulsion) to 7 (spontaneous convulsion). HICs are assessed hourly between 2 and 12 h after administration of a sedative-hypnotic dose of alcohol. The HIC phenotype is easily measured and quantified, allowing for a standard measure of withdrawal severity that can be reliably measured across mouse models and alcohol research laboratories. HIC severity after inhalation is dose and duration dependent (Goldstein, 1972), and there is a clear genetic contribution to individual differences in this trait (Goldstein, 1973). Crabbe *et al.* (1983a) subsequently showed that there were substantial genetically determined differences among inbred mouse strains and that these differences were pharmacodynamic rather than pharmacokinetic in nature. The establishment of this withdrawal phenotype as a quantifiable trait with clear genetic influence has allowed for the development of specific breeding and molecular strategies to identify discrete genetic influences underlying withdrawal severity in mice.

II. Quantitative Trait Locus Mapping in Mice

A. Definition of a QTL

A quantitative trait is one that has measurable phenotypic variation within a population owing to underlying variability in genetic and/or environmental influences. A QTL is a genetic locus in which allelic variation affects variation in the observed phenotype (Grisel, 2000). Generally, quantitative traits are multifactorial, meaning they are influenced by several polymorphic genes and environmental conditions. To map a QTL, its influence on a trait must be detected amid considerable "noise" from other QTLs and non-genetic sources of individual variation. This has been made feasible through the implementation of technologies to identify genetic polymorphisms throughout the genome and the development of statistical methods to map QTLs from specific genetic marker and phenotypic (i.e., trait) data (Doerge, 2002). The identification of the chromosomal regions where marker allelic and phenotype variation covary implicates the presence of a QTL. Each QTL identifies the genomic location of a gene or genes (referred to as quantitative trait genes or QTGs) affecting the trait of interest. The power of this approach was demonstrated first in plants and later in rodents (Crabbe and Belknap, 1993; Plomin *et al.*, 1991; Stuber, 1995), and has been used widely to identify genetic contributions to a wide variety of complex phenotypes (Almasy and Blangero, 2009; Hunter and Crawford, 2008; Johnson *et al.*, 1992; Kliebenstein, 2009; Roff, 2007; Rothschild *et al.*, 2007).

B. QTLs for Alcohol-Related Traits

Over the past 15 years, numerous QTLs for alcohol-related behaviors have been detected in various mouse crosses with logarithm of odds (LOD) scores that achieve "highly significant" status ($p < 10^{-4}$; Abiola *et al.*, 2003). This impressive accomplishment has contributed greatly to the now widely held view that the genetic influence on most behavioral phenotypes involves the actions of multiple gene products, each with a moderate to small effect (typically responsible for <5% of the observed phenotypic variation; Flint *et al.*, 2005). Mapping populations that have been successfully used to detect and map alcohol response QTLs in mice include F2 intercross populations and recombinant inbred (RI) strains, short- and long-term selected lines, backcrosses, congenic and consomic strains, and heterogeneous stocks (HSs). Strategies that increase the number of genetic recombinations within a mapping population provide higher mapping resolution, but require a larger number of animals to achieve the statistical power required for small-effect QTL detection. For alcohol-related traits, initial detection of QTLs has most often employed RI and F2 populations, which can typically map QTLs to ~10–50 centimorgan (cM) regions, corresponding to ~20–100 megabases (Mb) in the mouse genome. For reviews on QTL mapping population development and discovery for alcohol-related traits, see Crabbe *et al.* (1998, 1999a) Grisel, (2000), Phillips *et al.* (2002), and Spence *et al.* (2009).

Populations derived from the C57BL/6J (B6) and DBA/2J (D2) inbred strains have been widely examined for alcohol-related behaviors due to their phenotypic divergence in a variety of alcohol responses and the high degree of genetic polymorphism among mice derived from these two progenitor strains (Crabbe, 2002). Although crosses derived from these two strains have proven to be extremely useful to detect significant QTLs for alcohol withdrawal (see Section II.C), tolerance (Bennett *et al.*, 2007; Crabbe *et al.*, 1994a; Kirstein *et al.*, 2002), sensitization (Phillips *et al.*, 1995, 1998a), conditioned taste aversion (Risinger and Cunningham, 1998), and preference (Belknap *et al.*, 1997; Bice *et al.*, 2006, 2009; Phillips *et al.*, 1998b; Rodriguez *et al.*, 1995; Tabakoff *et al.*, 2008), they interrogate only about 20% of the total genetic variance in mice (Roberts *et al.*, 2007). However, studies employing more genetically rich mapping populations can interrogate 90% or more of the mouse genome (Churchill *et al.*, 2004).

C. Identification of Alcohol Withdrawal QTLs

The well-documented difference between the B6 and D2 inbred mouse strains in susceptibility to withdrawal after chronic alcohol exposure (Crabbe *et al.*, 1983b; Kakihana, 1979), or after a single sedative-hypnotic dose of alcohol (Metten *et al.*, 1998), has provided an excellent starting point for dissecting genetic influences involved in physical/physiological dependence on alcohol and to study how common allele variants influence genetic predisposition to physical dependence on alcohol. Initial genome-wide evaluations of chronic and acute alcohol withdrawal in populations derived from the D2 and B6 progenitor strains (including RI, F2, and short-term selectively bred lines for alcohol withdrawal severity) identified significant QTLs that determine genetic predisposition to alcohol withdrawal on chromosomes 1, 4, 11, and 19 (Buck *et al.*, 1997, 2002). This approach using more than one mapping population for identification and verification of QTLs takes into consideration the consequences of type I (identification of a false QTL) and type II (failure to detect an actual QTL) errors (Belknap *et al.*, 1996), and, in our experience, the ability to discriminate the QTL of interest on different genetic backgrounds increases the likelihood that detection and fine-mapping in other genetic populations will be successful.

Each of these QTL intervals harbors a gene (or genes) responsible for the QTL's phenotypic effect on the trait of interest. Although more than one gene can be the basis for a QTL, this is largely mitigated by fine-mapping strategies, which can dissociate closely linked QTLs and narrow QTL regions to sizes amenable to individual gene identification (Darvasi, 1997, 1998).

III. High-Resolution QTL Mapping

A. Toward Candidate Gene Identification

Despite the success of QTL mapping approaches to detect chromosomal regions of interest, QTLs are typically coarsely mapped, with hundreds of potential candidate genes within the QTL region. Effective strategies to identify causal genes within a large QTL region are lacking, with the successes largely relying upon *a priori* knowledge of a gene's involvement in biological systems associated with the trait of interest (e.g., the identification of genes encoding for receptors within the dopaminergic system). Although QTL mapping to genes/gene systems already implicated in alcohol-related behaviors is generally parsimonious, many genes underlying alcohol-related QTLs are not unambiguously linked to neurotransmitter systems associated with this drug. For these cases, a number of strategies have been proposed to reduce the number of potential candidate genes within the QTL, including examination of polymorphisms between strains used for mapping, and *in vitro* examination of gene function (Abiola *et al.*, 2003). Additionally, two *in vivo* strategies have been employed to further narrow the QTL interval to a size amenable to QTG identification. The first utilizes marker-assisted breeding strategies to fine-map the QTL to regions amenable to gene identification, and has been used in our laboratory to identify genes associated with alcohol withdrawal severity. The second examines gene expression patterns in previously established QTLs of interest to determine potential candidate genes, a method recently used to examine genetic contributions to alcohol preference in mice.

B. Interval-Specific Congenic Strains

One of the most powerful strategies to precisely map a QTL employs the development of interval-specific congenic strain (ISCS) models (Darvasi, 1997). This approach can be both time and cost efficient, particularly when a traditional congenic or consomic strain already exists, which is frequently the case (Bennett, 2000). Congenic and consomic strains are marker-assisted backcross breeding strategies to isolate a homologous chromosome (consomic) or chromosomal segment (congenic) from one inbred strain onto the genetic background of another. After 10 generations of backcrossing, the congenic strain has effectively isolated the QTL of interest by full introgression from a "donor" strain onto a "recipient" (or background) strain, allowing researchers to study QTL effects without background genomic influences from the donor strain (Rogner and

Avner, 2003). However, in most standard congenic strains, the introgressed interval is large (10–35 cM or more) and may contain hundreds of genes (Nadeau and Frankel, 2000).

ISCSs are one approach to further refine the QTL of interest and identify discrete genes that are contributing to the phenotype under examination (Darvasi, 1997). ISCSs are developed by backcrossing a standard congenic (or consomic) strain to background strain mice to create N_1 generation mice. These mice are backcrossed to create N_2 mice that exhibit distinct regions of progenitor allelic recombination within the starting congenic interval. The recombinant N_2 mice are genotyped across the QTL to determine precise boundaries of the individuals' allelic recombination within the region. Mice that contain the QTL of interest are backcrossed to produce multiple offspring carrying the same introgressed region (referred to as an interval-specific congenic line or ISCL), which carry one copy of the donor QTL and one copy of the background QTL. ISCL mice can then be intercrossed to create inbred strains with a homozygous donor region (i.e., a finished ISCS). A panel of ISCSs can be used to isolate a much smaller interval of interest (as small as 1 Mb) within the starting QTL region (Kozell *et al.*, 2008; Shirley *et al.*, 2004), thus facilitating QTG candidate identification and analyses (Denmark and Buck, 2008; Shirley *et al.*, 2004).

In addition to facilitating fine-mapping, congenic strains (particularly ISCS) are useful to confirm QTLs (Bennett *et al.*, 2008; Fehr *et al.*, 2002; Ferraro *et al.*, 2005; Lorenzana *et al.*, 2007; Ma *et al.*, 2009), as well as to examine potential pleiotropic effects of QTLs (Kim *et al.*, 2009a; Kumar *et al.*, 2010; Shirley *et al.*, 2004), to assess interactions among QTLs (Burgio *et al.*, 2007; Moisan, 2007), and to assess neural circuitry and mechanisms by which a QTL may affect behavior (Chen *et al.*, 2008; Reilly *et al.*, 2008; Rogers *et al.*, 2008). The utility of these models to both identify specific genes potentially underlying the phenotype of interest while simultaneously examining multiple behavioral and physiological characteristics of the QTL makes ISCLs/ISCSs an especially attractive model to study alcohol-related traits.

C. High-Resolution Mapping of Alcohol Withdrawal QTLs Using Congenic Models

Fine-mapping of an alcohol withdrawal QTL mapped to mouse chromosome 4 (*Alcw2*) used a traditional D2.B6 congenic strain (D2.B6-*D4Mit142*) as the point of departure to develop a series of ICSL and ISCS mice (Fehr *et al.*, 2002, Shirley *et al.*, 2004). The development and phenotypic analyses of six ISCLs with distinct introgressed intervals narrowed the QTL to a 1.8 Mb interval (Fehr *et al.*, 2002; Shirley *et al.*, 2004), which was key toward the identification of a single gene within this interval with the coding and expression differences indicative of a potential QTG (Section IV.C). A separate panel of ISCSs was developed and tested in order to fine-map QTLs on distal chromosome 1 that affect withdrawal after acute and chronic ethanol exposure (*Alcw1* and *Alcdp1*, respectively) to a 1.1–1.7 Mb interval (Buck *et al.*, 1997, 2002). Notably, QTLs for a variety of alcohol-related behaviors are also detected on distal chromosome 1, including alcohol-conditioned taste aversion (Risinger and Cunningham, 1998), alcohol preference/consumption (Tarantino *et al.*, 1998), alcohol-induced locomotor activation (Demarest *et al.*, 1999), and alcohol-induced hypothermia (Crabbe *et al.*, 1994b). This convergence of alcohol-related QTLs makes it tempting to speculate that the gene(s) underlying the influence of *Alcdp1*/*Alcw1* on withdrawal may have pleiotropic effects on other behavioral responses to alcohol, and shows the cumulative power of QTL mapping to detect multiple potential effects of what could be the same gene. A more definitive confirmation of this hypothesis will require additional work (e.g., testing a small donor segment ISCS) for behavioral responses to ethanol to determine whether the congenic versus background differences are congruent with those observed for ethanol withdrawal severity (Kozell *et al.*, 2008). The possibility that *Alcdp1*/*Alcw1* plays an important role in such diverse responses to alcohol makes it an important target for further investigations. However, it is important to note that this region of chromosome 1 is extremely gene dense, containing at least 40 genes (Denmark and Buck, 2008), many of which exhibit allelic differences in sequence and/or expression that may underlie or contribute to the QTL effects on withdrawal as well as other alcohol responses.

In addition to facilitating fine-mapping, traditional congenic and ISCS models can be extremely useful to rigorously assess QTG candidates for genotype-dependent differential expression on a uniform genetic background, and to begin identifying brain regions that may be involved in mediating the QTLs' phenotypic effects, potential pleiotropic effects of a QTL, and relevant neurotransmitter systems by which the gene(s) is exerting a phenotypic effect (see Section IV.C for details).

D. High-Resolution Mapping Using Heterogeneous Stock Animals

A second approach used for QTL fine-mapping employs HS animals (Mott and Flint, 2002; Johannesson *et al.*, 2009). HS populations provide a method to map QTLs with high precision by taking advantage of both the genetic heterogeneity produced by intercrosses of multiple inbred strains and the dense polymorphism maps available for these strains (Wade *et al.*, 2002). Originally derived from four or eight progenitor mouse strains, HS mice are pseudorandomly mated for each generation to increase recombination events between strains, thus increasing individual genetic variation and, within 50 generations, showing recombination intervals with an average distance of <2 cM across the genome (Valdar *et al.*, 2006). The small boundaries identified using HS populations are made possible by polymorphisms present in haplotypes (sets of allelic variants for a gene or set of genes underlying a QTL that are transmitted together). As recombination events become denser across the genome, haplotype polymorphisms between progenitor strains are mapped through generations of HS progeny to determine and reduce the boundaries of QTLs for traits of interest.

Currently, four mouse HS populations are available, viz. Hs/Ibg (Erwin *et al.*, 1980), HS-NPT (Valdar *et al.*, 2006), HS4, and HS-CC populations (Hitzemann *et al.*, 2008). To date, three of these populations have been used to confirm, fine-map, and test reciprocal effects of QTLs associated with alcohol-induced locomotor sensitization, preference, and withdrawal (Demarest *et al.*, 2001; Hitzemann *et al.*, 2009; Malmanger *et al.*, 2006). Discrete candidate genes have yet to be confirmed, although some have been proposed for the ethanol preference (Malmanger *et al.*, 2006).

IV. QTG Identification

A. Challenges and Successful Strategies

Identification of the QTGs underlying the phenotypic effects of QTLs is a challenge in the translation of preclinical research (Flint *et al.*, 2005). Successful strategies have most often relied upon evidence from several sources, and many have used congenic strains to their advantage (Belknap *et al.*, 2001; Cicila *et al.*, 2001; Liang *et al.*, 2007; Shirley *et al.*, 2004). Approximately 20 genes are accepted as QTGs for complex traits in mammals, and in most cases were identified using a stepwise process that eliminated all other possibilities, involving positional cloning (fine-mapping), comparing the sequences and expression levels of the genes in the progenitor strains or congenic versus background strains, and, when feasible, *in vivo* confirmation. We and others have discussed various sources of proof that support the conclusion that a candidate gene is a QTG (e.g., Belknap *et al.*, 2001; Phillips *et al.*, 2002; Flint *et al.*, 2005).

B. Genomic Resources to Facilitate QTG Identification

Within the QTL interval, variation must exist that is relevant to the trait of interest, for example, in QTG sequence and/or expression, although differences need not be within a coding region (Yalcin *et al.*, 2004). Recent advances in genomic resources for mice (and other species) have greatly facilitated QTG identification. These include physical maps which allow identification of all the sequences (and genes) within a QTL interval (see Phillips *et al.*, 2003). Other resources are useful to determine what brain region or regions may be most impacted by a QTG candidate. For example, the Allen Brain Atlas (www.brain-map.org) and Unigene (www.ncbi.nlm.nih.gov/entrez/query) Web sites provide data on the regional expression patterns of genes in mice and other species, including humans. Although these resources are pioneers in attempting to provide public access to gene expression patterns, the sheer volume of expression data limits access to information on different inbred strains. For example, the Allen Brain Atlas shows regional expression patterns using one representative C57BL/6J male mouse, so between-strain expression studies still often necessitate that microarray and quantitative expression analyses be performed within the laboratory. However, as public databases provide increased expression information across strains (as noted below for single nucleotide polymorphism or SNP databases), the analysis of gene expression patterns between murine models will be made increasingly less onerous and more efficient.

Resources are also increasingly available for many mouse models to interrogate naturally occurring gene sequence variation, including SNPs and insertion/deletions. The Sanger Center Mouse Genomes Project recently released genome variation data using next-generation sequencing (NGS) for 17 mouse strains (129P2, 129S1/SvImJ, 129S5, A/J, AKR/J, BALB/cJ, C3H/HeJ, C57BL/6NJ, CAST/EiJ, CBA/J, DBA/2J, LP/J, NOD/ShiLtJ, NZO/HiLtJ, PWK/PhJ, Spretus/EiJ, and WSB/EiJ; www.sanger.ac.uk/resources/mouse/genomes/). Most of these strains now have at least 20× coverage, and *de novo* assemblies to detect large insertion/deletions and other sequence variations will soon be completed. However, for murine models used less commonly in research and for novel animal models, sequence data are far from complete or lacking entirely. In such cases, next-generation sequencing has proven to be a powerful approach to identify "cryptic" sequence variation within a QTL interval and inform QTG identification (Walter *et al.*, 2009).

In recent years, the integration of QTL and gene expression data has proven to be an efficient strategy to progress toward QTG identification. Combined use of QTL and expression analyses and the development of resources to identify relevant gene expression networks (e.g., GeneNet) allows for the identification of candidate genes without necessitating the production of novel mouse strains, although the sheer number of expression differences exhibited by experimental strains may still necessitate fine-mapping. For the most appropriate comparisons (e.g., ISCS vs. non-congenic littermates) or brain regions, expression data may be lacking and must be generated *de novo*. Furthermore, it is critical that the deleterious effects of allelic sequence variation on hybridization-based gene expression data (including microarray data) be considered and minimized (see Walter *et al.*, 2007, 2009). These have been used effectively to integrate QTL and gene expression data (Section IV.D).

C. Identification of *Mpdz* as a QTG for Alcohol Withdrawal

As outlined in Section III.C, ISCS mice were developed to narrow *Alcw2*, an alcohol withdrawal QTL on mouse chromosome 4, to a 1.8 Mb interval. Using public and private sequencing databases (e.g., GenBank and Ensembl), five known and predicted genes are identified within this QTL interval. Among these five genes, only one, *Mpdz*, has been found to exhibit genotype-dependent differences between chromosome 4 congenic and background strain mice in gene expression or coding sequence (structural differences) (Shirley *et al.*, 2004). *Mpdz* and its human homolog (*MPDZ*) encode the multi-PDZ domain protein (MPDZ, also called MUPP1; Simpson *et al.*, 1999; Ullmer *et al.*, 1998). As a member of the PDZ protein family, MPDZ is thought to act as a scaffolding protein that alters the rate and/or fidelity of signal transduction mediated by proteins with which it physically associates (Nourry *et al.*, 2003; Sheng and Sala, 2001). MPDZ's known interactions with serotonin 5-HT_{2C} and gamma-aminobutyric acid receptor B ($GABA_B$) receptors are of interest to *Mpdz*'s role in alcohol withdrawal, as both of these receptor systems are implicated in alcohol withdrawal (Addolorato *et al.*, 2000, 2002; Knapp *et al.*, 2004) as well as other CNS hyperexcitability convulsion phenotypes (Isaac, 2005; Sperk *et al.*, 2004) in clinical and preclinical models. The link between MPDZ and these receptors has been previously confirmed *in vitro* (Balasubramanian *et al.*, 2007; Becamel *et al.*, 2004) and our finding that response to a 5-HT_{2C} receptor ligand is reduced in chromosome 4 congenic mice compared to background strain mice (Reilly *et al.*, 2008) is consistent with the hypothesis that this chromosome 4 QTL influences 5-HT_{2C} receptor function *in vivo*. However, future studies will be needed in order to rigorously test the role of MPDZ in regulating 5-HT_{2C} receptor and the function of other proteins with which MPDZ physically associates in the brain.

Mapping of this QTL and the development of congenic animal models has also allowed the identification of brain regions that exhibit neural activation associated with alcohol withdrawal that is QTL genotype dependent (Chen *et al.*, 2008). Using c-Fos as a high-resolution histological marker of neuronal activation associated with alcohol withdrawal, chromosome 4 congenic and background strain mice differ in neuronal activation associated with alcohol withdrawal in an extended basal ganglia circuit, including the substantia nigra (SN) and subthalamic nucleus (STN). Subsequent electrolytic and chemical lesion analyses confirmed the role of the caudolateral SNr in ethanol withdrawal, with SNr-lesioned mice showing significantly reduced withdrawal compared to sham mice following acute or repeated exposure to ethanol (Chen *et al.*, 2008, and unpublished results). 5-HT_{2C} receptors are distributed widely throughout the brain, with dense labeling observed in the basal ganglia, especially the subregions of the SN, where they are believed to primarily influence GABA-containing neurons (Leysen, 2004). The overlapping of distribution of 5-HT_{2C} receptors and MPDZ protein in this brain region make the SN (Pompeiano *et al.*, 1994) an interesting region in which to study *Mpdz*'s influence on ethanol withdrawal. Lentiviral-mediated knockdown of endogenous MPDZ in SN (discussed in Section V.C) is currently being undertaken in our laboratory in order to clarify the role of *Mpdz*'s influence on neuronal activity in this region during alcohol withdrawal.

D. Integration of QTL and Microarray Data for QTG Identification

The integration of QTL and microarray data has been used extensively to identify QTG candidates for a variety of alcohol response phenotypes including withdrawal, preference drinking, and several measures of acute sensitivity (Hu *et al.*, 2008; Liang *et al.*, 2010; Tabakoff *et al.*, 2008, 2009). In addition to our work identifying *Mpdz* as a QTG for alcohol withdrawal, we have integrated QTL and gene expression data in order to progress toward QTG identification for additional alcohol withdrawal QTLs (Denmark and Buck, 2008; Kozell *et al.*, 2009).

Saba *et al.* (2006) examined expression differences in naïve brains of mouse lines genetically selected for high or low alcohol tolerance (HAFT/LAFT lines, defined by high or low acute functional tolerance to a single dose of alcohol) and identified 8 and 22 differentially expressed genes underlying previously identified QTLs for alcohol preference and tolerance behaviors, respectively. Pathway analyses for these genes suggested a role for genes encoding for transcriptional and protein binding elements for the development of alcohol tolerance, with genes encoding for protein kinases and intracellular signal transduction implicated in alcohol preference. Tabakoff and colleagues (2008) compiled regional expression and phenotype data from 20 standard inbred mouse strains, 20 BXD RI strains, and replicate lines selectively bred for high or low preference for alcohol drinking (HAP and LAP, respectively) to identify genes that may underlie alcohol preference QTLs. Six QTG candidates were identified using this strategy, with most being expressed in the hippocampus and olfactory regions, suggesting that both chemosensory and memory processes may play an important role in the expression of the alcohol preference phenotype in mice. A meta-analysis compiled expression data from multiple mouse models, including three selected lines for alcohol preference (two being HAP/LAP replicate lines), six inbred strains, and a chromosome 9 congenic strain, and identified 20 QTG candidates within an alcohol preference QTL on mouse chromosome 9 (Mulligan *et al.*, 2006). One gene, *Scn4b*, exhibited differential expression in multiple mapping populations, making it a promising QTG candidate for alcohol preference. *Scn4b* encodes for the β_4 subunit of the voltage-gated sodium channel and is highly expressed in limbic regions of the mouse brain, including the nucleus accumbens, amygdala, and ventral striatum, which are regions implicated in preference (Bell *et al.*, 2009; Lesscher *et al.*, 2009; Sommer *et al.*, 2006). These findings are currently being followed up by examining the impact of shRNA-mediated reduction of *Scn4b* expression in the brain on preference (Dr. Robert Hitzemann, personal communication).

Although the integration of expression and QTL analyses provides a promising method to identify QTG candidates, these analyses are still limited in that many differentially expressed genes may be present within a large QTL, inflating the false negative rate of identification and requiring much work to identify the causative gene or genes. In some cases, fine-mapping may yield only one QTG candidate (Shirley *et al.*, 2004). However, most often, more than one high-quality QTG candidate is identified even after fine-mapping of a QTL to less than 2 Mb (Denmark and Buck, 2008). Nonetheless, in both scenarios, narrowing the QTL interval of interest dramatically reduces the number of QTG candidates that remain within the fine-mapped interval, making fine-mapping a crucial step in progressing from QTL to QTG identification.

V. *In Vivo* QTG Analyses

Examination of specific genes *in vivo* is necessary before the gene can be confirmed as a QTG for a behavior of interest. Although *in vitro* analyses provide extremely useful information outlining basic functional and mechanistic components of the potential QTG, cellular models cannot demonstrate complex behavioral and/or physiological responses, such as those resulting from exposure to alcohol. In order for a behavioral QTG to be confirmed as influencing an alcohol-related behavioral trait of interest, mammalian models must be developed which can alter the gene of interest to examine relevant biological and behavioral effects in a representative mammalian system. Three strategies employed for confirmation of alcohol-related QTGs are given below.

A. Targeted Gene Knockout Models

Knockout mice have revolutionized our ability to test hypotheses about the function of genes *in vivo*, which has provided insights into the nature of human diseases and other traits. Knockout mouse models are used to assess a gene's involvement in a trait of interest by nullifying the function of that gene in an otherwise functioning animal. To date, thousands of knockout mouse models have been completed or are being developed (see www.knockoutmouse.org and www.credrivermice.org), many of which have been tested for alcohol response phenotypes (Crabbe *et al.*, 2006). After a QTG (or a high-quality QTG candidate) has been identified, analyses utilizing an existing knockout model or models are common approaches to confirm and assess potential QTG effects *in vivo* (Boyce-Rustay *et al.*, 2006; Kozell *et al.*, 2009; Morice *et al.*, 2010; Thanos *et al.*, 2005). Our laboratory recently used a knockout mouse model to provide evidence for a compelling QTG (*Kcnj9*) candidate underlying an alcohol withdrawal QTL on mouse chromosome 1. *Kcnj9* has been previously linked to morphine behaviors, including morphine withdrawal (Cruz *et al.*, 2008; Smith *et al.*, 2008), and initial results using *Kcnj9* knockout mice in our laboratory have displayed a genotype-dependent withdrawal response to alcohol and other sedative-hypnotic drugs (Kozell *et al.*, 2009). Knockout models are also useful for identifying potential pleiotropic effects of a gene of interest, whereby a single gene or genetic variant mediates phenotypic responses to multiple traits (Blednov *et al.*, 2005; Solecki *et al.*, 2008; Suzuki *et al.*, 2009), and for the assessment of potential mechanisms associated with a gene's effects *in vivo* (Fox *et al.*, 2007; Lobo and Harris, 2008).

Despite the clear strength of the knockout approach, it has limitations, one of which is particularly relevant to QTG identification. Embryonic stem (ES) cell lines used to generate targeted knockout models are most commonly derived from the 129 strain (see Simpson *et al.*, 1997), although some successes have been reported using ES cells derived from B6 strain mice (Seong *et al.*, 2004). The 129 strain of mice can be challenging to work with because of the divergent genotypes (and physiological phenotypes) present in many of the 129 substrains (Threadgill *et al.*, 1997). Additionally, this strain may not exhibit phenotypes relevant/sufficient to study the behavioral trait of interest. Therefore, after gene targeting in the ES cells of the 129 strain, knockout mouse models are often developed onto a different background strain in order to observe relevant phenotypic effects of the mutation. However, due to recombination restriction, the region immediately flanking the null allele will inevitably contain ES-derived genes, so phenotypic differences between knockout and background strain mice may be due to the target (e.g., QTG candidate) gene, a 129-derived passenger gene or genes, or both (Su *et al.*, 2009). This can be addressed, at least in part, by examining the other linked QTG candidates for differential expression and/or sequence variation between knockout and wild-type littermates (Kozell *et al.*, 2009) or by comparison of the knockout strain with a co-developed 129.B6 co-isogenic strain spanning the QTL of interest (Crusio, 2004). The major drawback to these solutions, however, is that simultaneous development of the multiple control lines required can be both time and cost prohibitive for the examination of a single QTG. Additional approaches to target specific candidate genes linked to alcohol-related behaviors include transgenic overexpressing and rescue models and RNAi-mediated gene knockdown.

B. Transgenic Overexpression and Rescue

Transgenesis is commonly used to study gene functions in biological processes or to devise animal models of human diseases. A limitation of the method using standard constructs is the uncertainty regarding the expression of the transgene because host sequences surrounding the place of transgene integration can modify the expression pattern, potentially causing it to be ectopic, weak, or undetectable. This problem is largely mitigated in transgenesis using bacterial artificial chromosomes (BACs). BACs are circular plasmid DNA molecules that are hosted in *E. coli* and can accommodate large genomic DNA inserts up to 300 kb (Shizuya *et al.*, 1992). BAC vectors carry all sequences needed for autonomous replication and copy number control. A major advantage of BAC transgenesis is that it is likely to ensure position-independent, copy-number–dependent, and optimal levels of expression of the transgene, provided all regulatory sequences needed for the establishment and maintenance of the expression domain are located within the BAC (Giraldo and Montoliu, 2001). Thus, BAC transgenes can circumvent most position effects observed with standard constructs and are preferred when regulatory sequences of a gene are not known.

The BAC transgene is microinjected into blastocysts of a recipient animal model. In addition to transgene expression, one must also consider expression of the endogenous allele if it exists in the recipient. However, in many cases, the degree of transgene expression can overcome background strain phenotypes, providing an *in vivo* link between the transgene expressed and behaviors of interest. For example, our laboratory has recently developed an *Mpdz* BAC transgenic model; a BAC clone harboring *Mpdz* as the only full-length gene was introduced into B6D2-derived mice, and then backcrossed to D2 strain mice. Preliminary analyses indicate that increased *Mpdz* expression is associated with reduced alcohol withdrawal severity (Milner and Buck, unpublished results), consistent with our previous work identifying *Mpdz* as a QTG candidate for alcohol withdrawal convulsions (Shirley *et al.*, 2004).

Another strategy combines the method of transgenic overexpression with knockout models for a particular gene of interest in order to examine restoration of a genetic function, a method called "transgenic rescue." The main strength of this approach is that by observing a "rescued" phenotype when the QTG is restored to the knockout model, the role of the QTG for the behavior of interest can be confirmed without the creation of multiple control lines. This strategy has been used to confirm genes for a wide variety of behavioral phenotypes, including alcohol-related behaviors such as alcohol sensitivity (Findlay *et al.*, 2002) and sedation (Rothenfluh *et al.*, 2006). Choi *et al.* (2002) used transgenic "rescue" methodology to restore protein kinase C-ε (PKCε) function in specific brain regions (basal forebrain, amygdala, and cerebellum) of PKCε knockout mice. The restoration of PKCε in these regions effectively restored alcohol intake and sensitivity behaviors altered in the PKCε knockout mice, both confirming this gene's role in alcohol behaviors and identifying relevant brain regions through which this gene influences alcohol intake.

Despite successes in the development of genetically manipulated mouse models to examine discrete genetic influences on complex traits, a number of limitations still exist for these types of methods. For example, in both knockout and transgenic models, the genetic mutation is introduced at the blastocyst stage, so the possibility of developmental compensation resulting from the mutation can influence phenotypes observed in the model (Crabbe *et al.*, 2006). Novel mouse models developed to temporally (or physically) restrict candidate gene expression are being rapidly developed (Mallo, 2006; Stuber, 2010; Sun *et al.*, 2007), although they have not yet, to our knowledge, been employed in the field of alcohol behavior research. Viral methods can also be incredibly useful to dissociate target gene effects in the adult to avoid developmental compensation confounds, and are being increasingly employed by alcohol researchers to study the effects of alcohol-related QTGs.

C. Viral-Mediated RNA Interference (RNAi)

A common RNAi strategy is to employ virally mediated short hairpin loop RNA (shRNA) or short-interfering RNA (siRNA) to disrupt gene transcription and silence expression of a gene of interest *in vivo* (Pushparaj *et al.*, 2008). However, these do not cross the blood–brain barrier efficiently and stereotaxic injection into the brain is required. Once in the brain, it binds to an endogenous silencing complex (RNA-induced silencing complex, or RISC) to prevent translation of endogenous mRNA from the gene of interest into a functional protein (Lu *et al.*, 2005), thereby reducing endogenous protein levels of the gene of interest in a targeted (i.e., region-specific) manner. One commonly employed method of administration of RNAi is achieved through the use of lentiviral (Singer and Verma, 2008) or adenoviral (Kuninger *et al.*, 2004) vectors. Unlike siRNA, which needs to be repeatedly infused and can show poor cell uptake if administered alone (Kim *et al.*, 2009b), these vectors can provide a method of RNAi transfection that has been shown to stably reduce targeted gene expression levels for a least 1 month post-transfection (Van den Haute *et al.*, 2003). The viral infection spreads very little from the site of injection, so altered gene expression can be limited to a precise location. This is beneficial when target gene knockdown in a discrete brain region is necessary, but is limiting if it is not known if one brain region determines the target gene effect on phenotype or, if it exists, its precise location. New approaches may allow peripheral administration to achieve widespread RNAi in the brain. Viral vectors can be pseudotyped with rabies viral glycoprotein (RVG) to confer tropism for the CNS (Cronin *et al.*, 2005), potentially allowing the vectors to efficiently cross the blood–brain barrier after peripheral injection and target cells in the CNS for RNAi delivery (Kumar and Clarke, 2007).

This approach is being increasingly used to assess target gene effects on behavior, including the role of brain-derived neurotrophic factor (BDNF) in the dorsolateral striatum (Jeanblanc *et al.*, 2009), μ-opioid receptors in the ventral tegmental area (Lasek *et al.*, 2007), and $\alpha 4$ subunit–containing $GABA_A$ receptors in the nucleus accumbens (Rewal *et al.*, 2009) on alcohol consumption and preference behaviors in mice. Our laboratory is currently using a lentiviral-mediated shRNA construct targeted against *Mpdz* to examine this QTG's influence on ethanol withdrawal in brain regions implicated in alcohol withdrawal HIC severity (Section IV.C).

Viral-mediated overexpression of cDNA encoding a QTG is also a widely used strategy to examine the phenotypic effects of discrete genes. Viral-mediated overexpression has been used to confirm the role of nucleus accumbal $5HT_{1B}$ receptors (Hoplight *et al.*, 2006) and amygdalar neuropeptide Y (NPY) (Primeaux *et al.*, 2006) on alcohol consumption and self-administration, respectively, in rats. One striking example of this methodology in alcohol research used adeno-associated viral (AAV) transfection of the *Homer2* to restore alcohol consumption and sensitization phenotypes to *Homer2* knockout mice (Szumlinski *et al.*, 2005) and increased drinking phenotypes in B6 mice when administered directly into the nucleus accumbens (Szumlinski *et al.*, 2008).

The ability of RNAi to directly manipulate QTG expression levels in discrete brain regions, as well as its ability to circumvent potential developmental confounds present in knockout and transgenic models, makes RNAi an especially attractive new method for use in the field of alcohol genetics. However, all studies using these *in vivo* models must also take other factors (such as the epigenetic and environmental factors noted below) into consideration before the phenotypic effects of the QTG can be confirmed.

VI. Epistatic and Environmental Considerations in QTL Mapping

An important consideration that must be taken into account when examining a discrete genetic influence on an alcohol-related trait is the likely possibility that the QTL's influence on trait expression is modified by other genetic and environmental influences. In QTL mapping, interactions between genes (called epistasis) can significantly alter the phenotype of interest. When investigating the influence of a particular QTL or gene on a specific inbred strain background, epistatic influence can often obscure genes with small effect sizes (Crabbe, 2002), highlighting the importance of using multiple inbred strain (or inbred strain–derived) models to examine potential QTGs. However, researchers are currently developing methods to identify regions of epistatic interaction even before a candidate gene has been identified (Phillips *et al.*, 2002). For example, one study performed a genome-wide scan on an F_2 intercross derived from B6 and 129 substrains discovered modifying loci for the primary QTLs identified on chromsomes 4 and 7, as well as significant interactions between suggestive QTLs identified in this study (Bachmanov *et al.*, 2002). Fernández *et al.* (2000) used a multiple regression model to identify an epistatic interaction between loci on mouse chromosomes 2 and 3 that individually accounted for $\sim$8% of phenotypic variation in alcohol preference in their model. Although epistasis has the potential to obscure relevant, small-effect QTLs for alcohol-related behaviors, developing strategies to identify these types of interactions also have the possibility of providing additional information about relevant loci outside the primary region of interest that may influence phenotypic expression and ultimately provide a more accurate picture of a QTL's effect across heterogeneous populations.

Another important consideration influencing phenotypic variation is environmental influence. Gene (or QTL) $\times$ environment interactions can have a potentially negative effect in complex trait research, as inconsistencies between testing environments can influence behavioral responses in these strains despite experimenter attempts for consistency (Crabbe *et al.*, 1999b) and can be a confounding factor during initial discovery or replication of genetic findings when using specific mouse models (Wahlsten, 2001). However, laboratory environment influences do not generally obscure large phenotypic differences between strains (Wahlsten *et al.*, 2003), suggesting that the QTLs with substantial effect sizes discovered for alcohol-related traits are not significantly confounded by small variations in testing environment. Additionally, as gene × environment interactions occur frequently in human populations for all behavioral traits, often with significant behavioral outcomes (Caspi *et al.*, 2003; Laucht *et al.*, 2009), researchers are beginning to use rodent models to explicitly examine these types of interactions. The influence of stress on alcohol-related behaviors has been well documented for many mammals (Campbell *et al.*, 2009) and much work is currently being done using rodent and primate models to examine the relationship between stress exposure and discrete genetic factors to influence alcohol behaviors (reviewed in Stacey *et al.*, 2009). Sher *et al.* (2010) outline two additional environmental factors (early alcohol exposure and social environment) that may be studied using different genetic rodent models to dissect genetic and environmental influences. Although the potential for gene × environmental interaction can be confounding, the ability to manipulate and regulate genetic and environmental factors associated with alcohol use disorders allows preclinical researchers an additional avenue to examine the complexity of genetic and environmental influence on alcohol behaviors in order to more accurately model the human condition and provide more accurate mechanistic/risk information relating to discrete QTGs associated with alcohol behaviors in rodents.

VII. Human Relevance of QTGs Identified in Mice

The human relevance of mouse QTG data depends on using robust animal models and on the high degree of homology between human and mouse genomes at the gene level (Gregory *et al.*, 2002). Moreover, QTG identification is likely to identify genes for regulatory or rate-limiting proteins that are important therapeutic targets for populations affected by alcohol use disorders (Korstanje and Paigen, 2002).

The QTGs identified for alcohol responses in mice can also play an important role in human subjects. One of the most striking examples of overlap is the distal region of chromosome 1, where alcohol traits have been identified in both human (Bierut *et al.*, 2004; Nurnberger *et al.*, 2001; Reich *et al.*, 1998) and mouse (Buck *et al.*, 2002; Denmark and Buck, 2008; Kozell *et al.*, 2008) models. Ehlers *et al.* (2010) recently reported on this region, showing that multiple QTLs for human alcohol and tobacco dependence are syntenic with mouse QTLs for alcohol and pentobarbital withdrawal severity, suggesting that high-quality

QTGs identified using mouse models may be directly relevant to substance abuse risk in human populations.

MPDZ is present in humans and may also contribute to risk for alcoholism in some clinical populations. At least three QTLs on human chromosome 9 are detected in mapping studies for alcoholism and associated endophenotypes (Bergen *et al.*, 2003; Daw *et al.*, 2003; Long *et al.*, 1998; Ma *et al.*, 2003; Saunders *et al.*, 2009; Williams *et al.*, 2005), although most are still suggestive associations and mapped to relatively large regions. At least two of these studies identify markers on chromosome 9p22-p21 associated with alcoholism (Long *et al.*, 1998; Williams *et al.*, 2005), making *MPDZ* a viable candidate gene for these loci detected in human studies. Furthermore, recent studies in human populations have identified *MPDZ* as a potential candidate gene associated with alcohol consumption behaviors (Tabakoff *et al.*, 2009) and risk of dependence (Karpyak *et al.*, 2009), although association with withdrawal symptoms was not observed. These human association studies provide persuasive evidence that specific genes associated with alcohol behaviors in rodent populations may be directly relevant to alcohol risk factors in clinical populations.

We view the purpose of QTL research in mice as the identification of genetic targets for humans. In some cases the QTG will be the same in mouse and man (Mogil *et al.*, 2003). In other cases QTL research will identify a relevant gene network. A comprehensive understanding of genetic variation, both in humans and animal models, is crucial to establish relationships between genotype and biological function (Collins *et al.*, 2003).

VIII. General Conclusions

QTL mapping using rodent populations has created an abundance of genetic data with which to identify specific genetic risk factors associated with alcohol use disorders and other alcohol-related behaviors. Although rodent mapping populations have been used by alcohol researchers for some time, the large size of many QTL regions makes the task of identifying individual genetic contributions challenging. However, strategies developed to fine-map these QTLs using ISCS and HS models, as well as substantial advances in the ability to access shared genetic resources, is beginning to make candidate gene identification more plausible. Additionally, the advancement of molecular strategies to isolate and examine discrete genes in a controlled environment allows genetic alcohol researchers to both confirm a candidate gene's involvement in the behavior of interest and examine the mechanisms by which the candidate gene exerts its effect. Although clear limitations to this type of research still exist, QTL mapping in rodent populations serves as an exceptionally useful tool for the examination of genetic contributions to human alcohol use disorders.

Acknowledgments

The authors would like to thank Dr. Laura Kozell, Nikki Walter, and journal editors Drs. Matthew Reilly and David Lovinger for their helpful comments and suggestions on this manuscript. Supported by grants AA011114, AA010760, AA017031, AA07468, DA05228, and the VA Medical Center.

References

Abiola, O., Angel, J. M., Avner, P., Bachmanov, A. A., Belknap, J. K., Bennett, B., Blankenhorn, E. P., Blizard, D. A., Bolivar, V., Brockmann, G. A., Buck, K. J., Bureau, J. F., Casley, W. L., Chesler, E. J., Cheverud, J. M., Churchill, G. A., Cook, M., Crabbe, J. C., Crusio, W. E., Darvasi, A., de Haan, G., Dermant, P., Doerge, R. W., Elliot, R. W., Farber, C. R., Flaherty, L., Flint, J., Gershenfeld, H., Gibson, J. P., Gu, J., Gu, W., Himmelbauer, H., Hitzemann, R., Hsu, H. C., Hunter, K., Iraqi, F. F., Jansen, R. C., Johnson, T. E., Jones, B. C., Kempermann, G., Lammert, F., Lu, L., Manly, K. F., Matthews, D. B., Medrano, J. F., Mehrabian, M., Mittlemann, G., Mock, B. A., Mogil, J. S., Montagutelli, X., Morahan, G., Mountz, J. D., Nagase, H., Nowakowski, R. S., O'Hara, B. F., Osadchuk, A. V., Paigen, B., Palmer, A. A., Peirce, J. L., Pomp, D., Rosemann, M., Rosen, G. D., Schalkwyk, L. C., Seltzer, Z., Settle, S., Shimomura, K., Shou, S., Sikela, J. M., Siracusa, L. D., Spearow, J. L., Teuscher, C., Threadgill, D. W., Toth, L. A., Toye, A. A., Vadasz, C., Van Zant, G., Wakeland, E., Williams, R. W., Zhang, H. G., and Zou, F. (2003). The nature and identification of quantitative trait loci: A community's view. *Nat. Rev. Genet.* **4**, 911–916.

Addolorato, G., Caputo, F., Capristo, E., Colombo, G., Gessa, G. L., and Gasbarrini, G. (2000). Ability of baclofen in reducing alcohol craving and intake: II. Preliminary clinical evidence. *Alcohol. Clin. Exp. Res.* **24**, 67–71.

Addolorato, G., Caputo, F., Capristo, E., Domenicali, M., Bernardi, M., Janiri, L., Agabio, R., Colombo, G., Gessa, G. L., and Gasbarrini, G. (2002). Baclofen efficacy in reducing alcohol craving and intake: A preliminary double-blind randomized controlled study. *Alcohol Alcohol.* **37**, 504–508.

Almasy, L., and Blangero, J. (2009). Human QTL linkage mapping. *Genetica* **136**, 333–340.

Bachmanov, A. A., Reed, D. R., Li, X., Li, S., Beauchamp, G. K., and Tordoff, M. G. (2002). Voluntary ethanol consumption by mice: Genome-wide analysis of quantitative trait loci and their interactions in a C57BL/6ByJ x 129P3/J F2 intercross. *Genome Res.* **12**, 1257–1268.

Balasubramanian, S., Fam, S. R., and Hall, R. A. (2007). $GABA_B$ receptor association with the PDZ scaffold Mupp1 alters receptor stability and function. *J. Biol. Chem.* **282**, 4162–4171.

Becamel, C., Gavarini, S., Chanrion, B., Alonso, G., Galeotti, N., Dumuis, A., Bockaert, J., and Marin, P. (2004). The serotonin 5-HT_{2A} and 5-HT_{2C} receptors interact with specific sets of PDZ proteins. *J. Biol. Chem.* **279**, 20257–20266.

Belknap, J. K., Hitzemann, R., Crabbe, J. C., Phillips, T. J., Buck, K. J., and Williams, R. W. (2001). QTL analysis and genomewide mutagenesis in mice: Complementary genetic approaches to the dissection of complex traits. *Behav. Genet.* **31**, 5–15.

Belknap, J. K., Mitchell, S. R., O'Toole, L. A., Helms, M. L., and Crabbe, J. C. (1996). Type I and type II error rates for quantitative trait loci (QTL) mapping studies using recombinant inbred mouse strains. *Behav. Genet.* **26**, 149–160.

Belknap, J. K., Richards, S. P., O'Toole, L. A., Helms, M. L., and Phillips, T. J. (1997). Short-term selective breeding as a tool for QTL mapping: Ethanol preference drinking in mice. *Behav. Genet.* **27**, 55–66.

Bell, R. L., Kimpel, M. W., McClintick, J. N., Strother, W. N., Carr, L. G., Liang, T., Rodd, Z. A., Mayfield, R. D., Edenberg, H. J., and McBride, W. J. (2009). Gene expression changes in the nucleus accumbens of alcohol-preferring rats following chronic ethanol consumption. *Pharmacol. Biochem. Behav.* **94**, 131–147.

Bennett, B. (2000). Congenic strains developed for alcohol- and drug-related phenotypes. *Pharmacol. Biochem. Behav.* **67**, 671–681.

Bennett, B., Carosone-Link, P., Beeson, M., Gordon, L., Phares-Zook, N., and Johnson, T. E. (2008). Genetic dissection of quantitative trait locus for ethanol sensitivity in long- and short-sleep mice. *Genes Brain Behav.* **7**, 659–668.

Bennett, B., Downing, C., Carosone-Link, P., Ponicsan, H., Ruf, C., and Johnson, T. E. (2007). Quantitative trait locus mapping for acute functional tolerance to ethanol in the L × S recombinant inbred panel. *Alcohol. Clin. Exp. Res.* **31**, 200–208.

Bergen, A. W., Yang, X. R., Bai, Y., Beerman, M. B., Goldstein, A. M., and Goldin, L. R. (2003). Genomic regions linked to alcohol consumption in the Framingham Heart Study. *BMC Genet.* **4** (Suppl. 1), S101.

Bice, P. J., Foroud, T., Carr, L. G., Zhang, L., Liu, L., Grahame, N. J., Lumeng, L., Li, T. K., and Belknap, J. K. (2006). Identification of QTLs influencing alcohol preference in the high alcohol preferring (HAP) and low alcohol preferring (LAP) mouse lines. *Behav. Genet.* **36**, 248–260.

Bice, P., Valdar, W., Zhang, L., Liu, L., Lai, D., Grahame, N., Flint, J., Li, T. K., Lumeng, L., and Foroud, T. (2009). Genomewide SNP screen to detect quantitative trait loci for alcohol preference in the high alcohol preferring and low alcohol preferring mice. *Alcohol. Clin. Exp. Res.* **33**, 531–537.

Bierut, L. J., Rice, J. P., Goate, A., Hinrichs, A. L., Saccone, N. L., Foroud, T., Edenberg, H. J., Cloninger, C. R., Begleiter, H., Conneally, P. M., Crowe, R. R., Hesselbrock, V., Li, T. K., Nurnberger, J. I., Jr., Porjesz, B., Schuckit, M. A., and Reich, T. (2004). A genomic scan for habitual smoking in families of alcoholics: Common and specific genetic factors in substance dependence. *Am. J. Med. Genet. A* **124**, 19–27.

Blednov, Y. A., Bergeson, S. E., Walker, D., Ferreira, V. M., Kuziel, W. A., and Harris, R. A. (2005). Perturbation of chemokine networks by gene deletion alters the reinforcing actions of ethanol. *Behav. Brain Res.* **165**, 110–125.

Boyce-Rustay, J. M., Wiedholz, L. M., Millstein, R. A., Carroll, J., Murphy, D. L., Daws, L. C., and Holmes, A. (2006). Ethanol-related behaviors in serotonin transporter knockout mice. *Alcohol. Clin. Exp. Res.* **30**, 1957–1965.

Buck, K. J., Metten, P., Belknap, J. K., and Crabbe, J. C. (1997). Quantitative trait loci involved in genetic predisposition to acute alcohol withdrawal in mice. *J. Neurosci.* **17**, 3946–3955.

Buck, K. J., Rademacher, B. S., Metten, P., and Crabbe, J. C. (2002). Mapping murine loci for physical dependence on ethanol. *Psychopharmacology (Berl.)* **160**, 398–407.

Burgio, G., Szatanik, M., Guenet, J. L., Arnau, M. R., Panthier, J. J., and Montagutelli, X. (2007). Interspecific recombinant congenic strains between C57BL/6 and mice of the *Mus spretus* species: A powerful tool to dissect genetic control of complex traits. *Genetics* **177**, 2321–2333.

Cadoret, R. J., Cain, C. A., and Grove, W. M. (1980). Development of alcoholism in adoptees raised apart from alcoholic biologic relatives. *Arch. Gen. Psychiatry* **37**, 561–563.

Campbell, J. C., Szumlinski, K. K., and Kippin, T. E. (2009). Contribution of early environmental stress to alcoholism vulnerability. *Alcohol* **43**, 547–554.

Caspi, A., Sugden, K., Moffitt, T. E., Taylor, A., Craig, I. W., Harrington, H., McClay, J., Mill, J., Martin, J., Braithwaite, A., and Poulton, R. (2003). Influence of life stress on depression: Moderation by a polymorphism in the *5-HTT* gene. *Science* **301**, 386–389.

Chen, G., Kozell, L. B., Hitzemann, R., and Buck, K. J. (2008). Involvement of the limbic basal ganglia in ethanol withdrawal convulsivity in mice is influenced by a chromosome 4 locus. *J. Neurosci.* **28**, 9840–9849.

Choi, D. S., Wang, D., Dadgar, J., Chang, W. S., and Messing, R. O. (2002). Conditional rescue of protein kinase C epsilon regulates ethanol preference and hypnotic sensitivity in adult mice. *J. Neurosci.* **22**, 9905–9911.

Churchill, G. A., Airey, D. C., Allayee, H., Angel, J. M., Attie, A. D., *et al.* (2004). The Collaborative Cross, a community resource for the genetic analysis of complex traits. *Nat. Genet.* **36**, 1133–1137.

Cicila, G. T., Garrett, M. R., Lee, S. J., Liu, J., Dene, H., and Rapp, J. P. (2001). High-resolution mapping of the blood pressure QTL on chromosome 7 using Dahl rat congenic strains. *Genomics* **72**, 51–60.

Collins, L. J., Poole, A. M., and Penny, D. (2003). Using ancestral sequences to uncover potential gene homologues. *Appl. Bioinformatics* **2**, S85–95.

Crabbe, J. C. (2002). Alcohol and genetics: New models. *Am. J. Med. Genet.* **114**, 969–974.

Crabbe, J. C., and Belknap, J. K. (1993). Behavior genetic analyses of drug withdrawal. *Alcohol Alcohol. Suppl.* **2**, 477–482.

Crabbe, J. C., Belknap, J. K., Buck, K. J., and Metten, P. (1994a). Use of recombinant inbred strains for studying genetic determinants of responses to alcohol. *Alcohol Alcohol. Suppl.* **2**, 67–71.

Crabbe, J. C., Belknap, J. K., Metten, P., Grisel, J. E., and Buck, K. J. (1998). Quantitative trait loci: Mapping drug and alcohol-related genes. *Adv. Pharmacol.* **42**, 1033–1037.

Crabbe, J. C., Belknap, J. K., Mitchell, S. R., and Crawshaw, L. I. (1994b). Quantitative trait loci mapping of genes that influence the sensitivity and tolerance to ethanol-induced hypothermia in BXD recombinant inbred mice. *J. Pharmacol. Exp. Ther.* **269**, 184–192.

Crabbe, J. C., Kosobud, A., and Young, E. R. (1983a). Genetic selection for ethanol withdrawal severity: Differences in replicate mouse lines. *Life Sci.* **33**, 955–962.

Crabbe, J. C., Phillips, T. J., Buck, K. J., Cunningham, C. L., and Belknap, J. K. (1999a). Identifying genes for alcohol and drug sensitivity: Recent progress and future directions. *Trends Neurosci.* **22**, 173–179.

Crabbe, J. C., Phillips, T. J., Harris, R. A., Arends, M. A., and Koob, G. F. (2006). Alcohol-related genes: Contributions from studies with genetically engineered mice. *Addict. Biol.* **11**, 195–269.

Crabbe, J. C., Wahlsten, D., and Dudek, B. C. (1999b). Genetics of mouse behavior: Interactions with laboratory environment. *Science* **284**, 1670–1672.

Crabbe, J. C., Jr., Young, E. R., and Kosobud, A. (1983b). Genetic correlations with ethanol withdrawal severity. *Pharmacol. Biochem. Behav.* **18**(Suppl. 1), 541–547.

Cronin, J., Zhang, X. Y., and Reiser, J. (2005). Altering the tropism of lentiviral vectors through pseudotyping. *Curr. Gene. Ther.* **5**, 387–398.

Crusio, W. E. (2004). Flanking gene and genetic background problems in genetically manipulated mice. *Biol. Psychiatry* **56**, 381–385.

Cruz, H. G., Berton, F., Sollini, M., Blanchet, C., Pravetoni, M., Wickman, K., and Luscher, C. (2008). Absence and rescue of morphine withdrawal in GIRK/Kir3 knock-out mice. *J. Neurosci.* **28**, 4069–4077.

Darvasi, A. (1997). Interval-specific congenic strains (ISCS): An experimental design for mapping a QTL into a 1-centimorgan interval. *Mamm. Genome.* **8**, 163–167.

Darvasi, A. (1998). Experimental strategies for the genetic dissection of complex traits in animal models. *Nat. Genet.* **18**, 19–24.

Daw, E. W., Morrison, J., Zhou, X., and Thomas, D. C. (2003). Genetic Analysis Workshop 13: Simulated longitudinal data on families for a system of oligogenic traits. *BMC Genet.* **4** (Suppl. 1), S3.

Demarest, K., Koyner, J., McCaughran, J., Jr., Cipp, L., and Hitzemann, R. (2001). Further characterization and high-resolution mapping of quantitative trait loci for ethanol-induced locomotor activity. *Behav. Genet.* **31**, 79–91.

Demarest, K., McCaughran, J., Jr., Mahjubi, E., Cipp, L., and Hitzemann, R. (1999). Identification of an acute ethanol response quantitative trait locus on mouse chromosome 2. *J. Neurosci.* **19**, 549–561.

Denmark, D. L., and Buck, K. J. (2008). Molecular analyses and identification of promising candidate genes for loci on mouse chromosome 1 affecting alcohol physical dependence and associated withdrawal. *Genes Brain Behav.* **7**, 599–608.

Dick, D. M., and Foroud, T. (2003). Candidate genes for alcohol dependence: A review of genetic evidence from human studies. *Alcohol. Clin. Exp. Res.* **27**, 868–879.

Doerge, R. W. (2002). Mapping and analysis of quantitative trait loci in experimental populations. *Nat. Rev. Genet.* **3**, 43–52.

Edenberg, H. J., and Foroud, T. (2006). The genetics of alcoholism: Identifying specific genes through family studies. *Addict. Biol.* **11**, 386–396.

Ehlers, C. L., Walter, N. A., Dick, D. M., Buck, K. J., and Crabbe, J. C. (2010). A comparison of selected quantitative trait loci associated with alcohol use phenotypes in humans and mouse models. *Addict. Biol.* **15**, 185–199.

Erwin, V. G., McClearn, G. E., and Kuse, A. R. (1980). Interrelationships of alcohol consumption, actions of alcohol, and biochemical traits. *Pharmacol. Biochem. Behav.* **13**(Suppl. 1), 297–302.

Fehr, C., Shirley, R. L., Belknap, J. K., Crabbe, J. C., and Buck, K. J. (2002). Congenic mapping of alcohol and pentobarbital withdrawal liability loci to a <1 centimorgan interval of murine chromosome 4: Identification of *Mpdz* as a candidate gene. *J. Neurosci.* **22**, 3730–3738.

Fernández, J. R., Tarantino, L. M., Hofer, S. M., Vogler, G. P., and McClearn, G. E. (2000). Epistatic quantitative trait loci for alcohol preference in mice. *Behav. Genet.* **30**, 431–437.

Ferraro, T. N., Golden, G. T., Smith, G. G., Martin, J. F., Schwebel, C. L., Doyle, G. A., Buono, R. J., and Berrettini, W. H. (2005). Confirmation of a major QTL influencing oral morphine intake in C57 and DBA mice using reciprocal congenic strains. *Neuropsychopharmacology* **30**, 742–746.

Findlay, G. S., Wick, M. J., Mascia, M. P., Wallace, D., Miller, G. W., Harris, R. A., and Blednov, Y. A. (2002). Transgenic expression of a mutant glycine receptor decreases alcohol sensitivity of mice. *J. Pharmacol. Exp. Ther.* **300**, 526–534.

Flint, J., Valdar, W., Shifman, S., and Mott, R. (2005). Strategies for mapping and cloning quantitative trait genes in rodents. *Nat. Rev. Genet.* **6**, 271–286.

Fox, M. A., Jensen, C. L., Gallagher, P. S., and Murphy, D. L. (2007). Receptor mediation of exaggerated responses to serotonin-enhancing drugs in serotonin transporter (SERT)-deficient mice. *Neuropharmacology* **53**, 643–656.

Friedman, H. J. (1980). Assessment of physical dependence on and withdrawal from ethanol in animals. *In* Alcohol Tolerance and Dependence (H. Rigter and J. C. Crabbe, Eds.), pp. 93–121. Elsevier, North-Holland, Amsterdam.

Giraldo, P., and Montoliu, L. (2001). Size matters: Use of YACs, BACs and PACs in transgenic animals. *Transgenic Res.* **10**, 83–103.

Goldman, D. (1993). Recent developments in alcoholism: Genetic transmission. *Recent Dev. Alcohol.* **11**, 231–248.

Goldman, D., and Bergen, A. (1998). General and specific inheritance of substance abuse and alcoholism. *Arch. Gen. Psychiatry* **55**, 964–965.

Goldstein, D. B. (1972). Relationship of alcohol dose to intensity of withdrawal signs in mice. *J. Pharmacol. Exp. Ther.* **180**, 203–215.

Goldstein, D. B. (1973). Quantitative study of alcohol withdrawal signs in mice. *Ann. NY Acad. Sci.* **215**, 218–223.

Goldstein, D. B., and Pal, N. (1971). Alcohol dependence produced in mice by inhalation of ethanol: Grading the withdrawal reaction. *Science* **172**, 288–290.

Grant, B. F., Stinson, F. S., Dawson, D. A., Chou, S. P., Ruan, W. J., and Pickering, R. P. (2004). Co-occurrence of 12-month alcohol and drug use disorders and personality disorders in the United States: Results from the National Epidemiologic Survey on Alcohol and Related Conditions. *Arch. Gen. Psychiatry* **61**, 361–368.

Gregory, S. G., Sekhon, M., Schein, J., Zhao, S., Osoegawa, K., Scott, C. E., Evans, R. S., Burridge, P. W., Cox, T. V., Fox, C. A., Hutton, R. D., Mullenger, I. R., Phillips, K. J., Smith, J., Stalker, J., Threadgold, G. J., Birney, E., Wylie, K., Chinwalla, A., Wallis, J., Hillier, L., Carter, J., Gaige, T., Jaeger, S., Kremitzki, C., Layman, D., Maas, J., McGrane, R., Mead, K., Walker, R., Jones, S., Smith, M., Asano, J., Bosdet, I., Chan, S., Chittaranjan, S., Chiu, R., Fjell, C., Fuhrmann, D., Girn, N., Gray, C., Guin, R., Hsiao, L., Krzywinski, M., Kutsche, R., Lee, S. S., Mathewson, C., McLeavy, C., Messervier, S., Ness, S., Pandoh, P., Prabhu, A. L., Saeedi, P., Smailus, D., Spence, L., Stott, J., Taylor, S., Terpstra, W., Tsai, M., Tsai, M., Vardy, J., Wye, N., Yang, G., Shatsman, S., Ayodeji, B., Geer, K., Tsegaye, G., Osoegawa, K., Shvartsbeyn, A., Gebregeorgis, E., Krol, M., Russell, D., Overton, L., Malek, J. A., Holmes, M., Heaney, M., Shetty, J., Feldblyum, T., Nierman, W. C., Catanese, J. J., Hubbard, T., Waterston, R. H., Rogers, J., de Jong, P. J., Fraser, C. M., Marra, M., McPherson, J. D., and Bentley, D. R. (2002). A physical map of the mouse genome. *Nature* **418**, 743–750.

Grisel, J. E. (2000). Quantitative trait locus analysis. *Alcohol Res. Health* **24**, 169–174.

Hitzemann, R., Belknap, J. K., and McWeeney, S. K. (2008). Quantitative trait locus analysis: Multiple cross and heterogeneous stock mapping. *Alcohol Res. Health* **31**, 261–265.

Hitzemann, R., Edmunds, S., Wu, W., Malmanger, B., Walter, N., Belknap, J., Darakjian, P., and McWeeney, S. (2009). Detection of reciprocal quantitative trait loci for acute ethanol withdrawal and ethanol consumption in heterogeneous stock mice. *Psychopharmacology (Berl.)* **203**, 713–722.

Hoplight, B. J., Sandygren, N. A., and Neumaier, J. F. (2006). Increased expression of 5-HT_{1B} receptors in rat nucleus accumbens via virally mediated gene transfer increases voluntary alcohol consumption. *Alcohol* **38**, 73–79.

Hu, W., Saba, L., Kechris, K., Bhave, S. V., Hoffman, P. L., and Tabakoff, B. (2008). Genomic insights into acute alcohol tolerance. *J. Pharmacol. Exp. Ther.* **326**, 792–800.

Hunter, K. W., and Crawford, N. P. (2008). The future of mouse QTL mapping to diagnose disease in mice in the age of whole-genome association studies. *Annu. Rev. Genet.* **42**, 131–141.

Isaac, M. (2005). Serotonergic 5-HT_{2C} receptors as a potential therapeutic target for the design antiepileptic drugs. *Curr. Top. Med. Chem.* **5**, 59–67.

Jeanblanc, J., He, D. Y., Carnicella, S., Kharazia, V., Janak, P. H., and Ron, D. (2009). Endogenous BDNF in the dorsolateral striatum gates alcohol drinking. *J. Neurosci.* **29**, 13494–13502.

Johannesson, M., Lopez-Aumatell, R., Stridh, P., Diez, M., Tuncel, J., Blazquez, G., Martinez-Membrives, E., Canete, T., Vicens-Costa, E., Graham, D., Copley, R. R., Hernandez-Pliego, P., Beyeen, A. D., Ockinger, J., Fernandez-Santamaria, C., Gulko, P. S., Brenner, M., Tobena, A., Guitart-Masip, M., Gimenez-Llort, L., Dominiczak, A., Holmdahl, R., Gauguier, D., Olsson, T., Mott, R., Valdar, W., Redei, E. E., Fernandez-Teruel, A., and Flint, J. (2009). A resource for the simultaneous high-resolution mapping of multiple quantitative trait loci in rats: The NIH heterogeneous stock. *Genome Res.* **19**, 150–158.

Johnson, T. E., DeFries, J. C., and Markel, P. D. (1992). Mapping quantitative trait loci for behavioral traits in the mouse. *Behav. Genet.* **22**, 635–653.

Kakihana, R. (1979). Alcohol intoxication and withdrawal in inbred strains of mice: Behavioral and endocrine studies. *Behav. Neural Biol.* **26**, 97–105.

Karpyak, V. M., Kim, J. H., Biernacka, J. M., Wieben, E. D., Mrazek, D. A., Black, J. L., and Choi, D. S. (2009). Sequence variations of the human *MPDZ* gene and association with alcoholism in subjects with European ancestry. *Alcohol. Clin. Exp. Res.* **33**, 712–721.

Kim, S., Zhang, S., Choi, K. H., Reister, R., Do, C., Baykiz, A. F., and Gershenfeld, H. K. (2009a). An E3 ubiquitin ligase, Really Interesting New Gene (*RING*) Finger 41, is a candidate gene for anxiety-like behavior and beta-carboline-induced seizures. *Biol. Psychiatry* **65**, 425–431.

Kim, S. S., Garg, H., Joshi, A., and Manjunath, N. (2009b). Strategies for targeted nonviral delivery of siRNAs *in vivo*. *Trends Mol. Med.* **15**, 491–500.

Kirstein, S. L., Davidson, K. L., Ehringer, M. A., Sikela, J. M., Erwin, V. G., and Tabakoff, B. (2002). Quantitative trait loci affecting initial sensitivity and acute functional tolerance to ethanol-induced ataxia and brain cAMP signaling in BXD recombinant inbred mice. *J. Pharmacol. Exp. Ther.* **302**, 1238–1245.

Kliebenstein, D. (2009). Quantitative genomics: Analyzing intraspecific variation using global gene expression polymorphisms or eQTLs. *Annu. Rev. Plant Biol.* **60**, 93–114.

Knapp, D. J., Overstreet, D. H., Moy, S. S., and Breese, G. R. (2004). SB242084, flumazenil, and CRA1000 block ethanol withdrawal-induced anxiety in rats. *Alcohol* **32**, 101–111.

Korstanje, R., and Paigen, B. (2002). From QTL to gene: The harvest begins. *Nat. Genet.* **31**, 235–236.

Kozell, L., Belknap, J. K., Hofstetter, J. R., Mayeda, A., and Buck, K. J. (2008). Mapping a locus for alcohol physical dependence and associated withdrawal to a 1.1 Mb interval of mouse chromosome 1 syntenic with human chromosome 1q23.2-23.3. *Genes Brain Behav.* **7**, 560–567.

Kozell, L. B., Walter, N. A., Milner, L. C., Wickman, K., and Buck, K. J. (2009). Mapping a barbiturate withdrawal locus to a 0.44 Mb interval and analysis of a novel null mutant identify a role for *Kcnj9* (GIRK3) in withdrawal from pentobarbital, zolpidem, and ethanol. *J. Neurosci.* **29**, 11662–11673.

Kumar, K. G., DiCarlo, L. M., Volaufova, J., Zuberi, A. R., and Richards, B. K. (2010). Increased physical activity cosegregates with higher intake of carbohydrate and total calories in a subcongenic mouse strain. *Mamm. Genome* **21**, 52–63.

Kumar, L. D., and Clarke, A. R. (2007). Gene manipulation through the use of small interfering RNA (siRNA): From *in vitro* to *in vivo* applications. *Adv. Drug Deliv. Rev.* **59**, 87–100.

Kuninger, D., Stauffer, D., Eftekhari, S., Wilson, E., Thayer, M., and Rotwein, P. (2004). Gene disruption by regulated short interfering RNA expression, using a two-adenovirus system. *Hum. Gene. Ther.* **15**, 1287–1292.

Kuo, P. H., Kalsi, G., Prescott, C. A., Hodgkinson, C. A., Goldman, D., Alexander, J., van den Oord, E. J., Chen, X., Sullivan, P. F., Patterson, D. G., Walsh, D., Kendler, K. S., and Riley, B. P. (2009). Associations of glutamate decarboxylase genes with initial sensitivity and age-at-onset of alcohol dependence in the Irish Affected Sib Pair Study of Alcohol Dependence. *Drug Alcohol Depend.* **101**, 80–87.

Kuo, P. H., Kalsi, G., Prescott, C. A., Hodgkinson, C. A., Goldman, D., van den Oord, E. J., Alexander, J., Jiang, C., Sullivan, P. F., Patterson, D. G., Walsh, D., Kendler, K. S., and Riley, B. P. (2008). Association of *ADH* and *ALDH* genes with alcohol dependence in the Irish Affected Sib Pair Study of Alcohol Dependence (IASPSAD) sample. *Alcohol. Clin. Exp. Res.* **32**, 785–795.

Lasek, A. W., Janak, P. H., He, L., Whistler, J. L., and Heberlein, U. (2007). Downregulation of mu opioid receptor by RNA interference in the ventral tegmental area reduces ethanol consumption in mice. *Genes Brain Behav.* **6**, 728–735.

Laucht, M., Treutlein, J., Schmid, B., Blomeyer, D., Becker, K., Buchmann, A. F., Schmidt, M. H., Esser, G., Jennen-Steinmetz, C., Rietschel, M., Zimmermann, U. S., and Banaschewski, T. (2009). Impact of psychosocial adversity on alcohol intake in young adults: Moderation by the LL genotype of the serotonin transporter polymorphism. *Biol. Psychiatry* **66**, 102–109.

Lesscher, H. M., Wallace, M. J., Zeng, L., Wang, V., Deitchman, J. K., McMahon, T., Messing, R. O., and Newton, P. M. (2009). Amygdala protein kinase C epsilon controls alcohol consumption. *Genes Brain Behav.* **8**, 493–499.

Leysen, J. E. (2004). 5-HT2 receptors. *Curr. Drug Targets CNS Neurol. Disord.* **3**, 11–26.

Liang, Y., Jansen, M., Aronow, B., Geiger, H., and Van Zant, G. (2007). The quantitative trait gene latexin influences the size of the hematopoietic stem cell population in mice. *Nat. Genet.* **39**, 178–188.

Liang, T., Kimpel, M. W., McClintick, J. N., Skillman, A. R., McCall, K., Edenberg, H. J., and Carr, L. G. (2010). Candidate genes for alcohol preference identified by expression profiling in alcohol-preferring and -nonpreferring reciprocal congenic rats. *Genome Biol.* **11**, R11.

Little, H. J., Stephens, D. N., Ripley, T. L., Borlikova, G., Duka, T., Schubert, M., Albrecht, D., Becker, H. C., Lopez, M. F., Weiss, F., Drummond, C., Peoples, M., and Cunningham, C. (2005). Alcohol withdrawal and conditioning. *Alcohol. Clin. Exp. Res.* **29**, 453–464.

Lobo, I. A., and Harris, R. A. (2008). $GABA_A$ receptors and alcohol. *Pharmacol. Biochem. Behav.* **90**, 90–94.

Long, J. C., Knowler, W. C., Hanson, R. L., Robin, R. W., Urbanek, M., Moore, E., Bennett, P. H., and Goldman, D. (1998). Evidence for genetic linkage to alcohol dependence on chromosomes 4 and 11 from an autosome-wide scan in an American Indian population. *Am. J. Med. Genet.* **81**, 216–221.

Lorenzana, A., Chancer, Z., and Schauwecker, P. E. (2007). A quantitative trait locus on chromosome 18 is a critical determinant of excitotoxic cell death susceptibility. *Eur. J. Neurosci.* **25**, 1998–2008.

Lu, P. Y., Xie, F., and Woodle, M. C. (2005). *In vivo* application of RNA interference: From functional genomics to therapeutics. *Adv. Genet.* **54**, 117–142.

Ma, J. Z., Zhang, D., Dupont, R. T., Dockter, M., Elston, R. C., and Li, M. D. (2003). Mapping susceptibility loci for alcohol consumption using number of grams of alcohol consumed per day as a phenotype measure. *BMC Genet.* **4**(Suppl. 1), S104.

Ma, Y., Miller, J. C., Crandall, H., Larsen, E. T., Dunn, D. M., Weiss, R. B., Subramanian, M., Weis, J. H., Zachary, J. F., Teuscher, C., and Weis, J. J. (2009). Interval-specific congenic lines reveal quantitative trait loci with penetrant lyme arthritis phenotypes on chromosomes 5, 11, and 12. *Infect. Immun.* **77**, 3302–3311.

Mallo, M. (2006). Controlled gene activation and inactivation in the mouse. *Front. Biosci.* **11**, 313–327.

Malmanger, B., Lawler, M., Coulombe, S., Murray, R., Cooper, S., Polyakov, Y., Belknap, J., and Hitzemann, R. (2006). Further studies on using multiple-cross mapping (MCM) to map quantitative trait loci. *Mamm. Genome.* **17**, 1193–1204.

McKeon, A., Frye, M. A., and Delanty, N. (2008). The alcohol withdrawal syndrome. *J. Neurol. Neurosurg. Psychiatry* **79**, 854–862.

Metten, P., Belknap, J. K., and Crabbe, J. C. (1998). Drug withdrawal convulsions and susceptibility to convulsants after short-term selective breeding for acute ethanol withdrawal. *Behav. Brain Res.* **95**, 113–122.

Mogil, J. S., Wilson, S. G., Chesler, E. J., Rankin, A. L., Nemmani, K. V., Lariviere, W. R., Groce, M. K., Wallace, M. R., Kaplan, L., Staud, R., Ness, T. J., Glover, T. L., Stankova, M., Mayorov, A., Hruby, V. J., Grisel, J. E., and Fillingim, R. B. (2003). The melanocortin-1 receptor gene mediates female-specific mechanisms of analgesia in mice and humans. *Proc. Natl. Acad. Sci. USA* **100**, 4867–4872.

Moisan, M. P. (2007). From QTL detection to gene identification. In Neurobehavioral Genetics: Methods and Applications (B. C. Jones and P. Mormède, Eds.), 2nd Ed., pp. 55–71. Taylor & Francis Group, Boca Raton, FL.

Morice, E., Denis, C., Giros, B., and Nosten-Bertrand, M. (2010). Evidence of long-term expression of behavioral sensitization to both cocaine and ethanol in dopamine transporter knockout mice. *Psychopharmacology (Berl.)* **208**, 57–66.

Mott, R., and Flint, J. (2002). Simultaneous detection and fine mapping of quantitative trait loci in mice using heterogeneous stocks. *Genetics* **160**, 1609–1618.

Mulligan, M. K., Ponomarev, I., Hitzemann, R. J., Belknap, J. K., Tabakoff, B., Harris, R. A., Crabbe, J. C., Blednov, Y. A., Grahame, N. J., Phillips, T. J., Finn, D. A., Hoffman, P. L., Iyer, V. R., Koob, G. F., and Bergeson, S. E. (2006). Toward understanding the genetics of alcohol drinking through transcriptome meta-analysis. *Proc. Natl. Acad. Sci. USA* **103**, 6368–6373.

Nadeau, J. H., and Frankel, W. N. (2000). The roads from phenotypic variation to gene discovery: Mutagenesis versus QTLs. *Nat. Genet.* **25**, 381–384.

Nourry, C., Grant, S. G., and Borg, J. P. (2003). PDZ domain proteins: Plug and play! *Sci. STKE* **2003**, RE7.

Nurnberger, J. I., Jr., Foroud, T., Flury, L., Su, J., Meyer, E. T., Hu, K., Crowe, R., Edenberg, H., Goate, A., Bierut, L., Reich, T., Schuckit, M., and Reich, W. (2001). Evidence for a locus on chromosome 1 that influences vulnerability to alcoholism and affective disorder. *Am. J. Psychiatry* **158**, 718–724.

Peters, L. L., Robledo, R. F., Bult, C. J., Churchill, G. A., Paigen, B. J., and Svenson, K. L. (2007). The mouse as a model for human biology: A resource guide for complex trait analysis. *Nat. Rev. Genet.* **8**, 58–69.

Phillips, T. J., Belknap, J. K., Buck, K. J., and Cunningham, C. L. (1998b). Genes on mouse chromosomes 2 and 9 determine variation in ethanol consumption. *Mamm. Genome.* **9**, 936–941.

Phillips, T. J., Belknap, J. K., Hitzemann, R. J., Buck, K. J., Cunningham, C. L., and Crabbe, J. C. (2002). Harnessing the mouse to unravel the genetics of human disease. *Genes Brain Behav.* **1**, 14–26.

Phillips, T. J., Belknap, J. K., Hitzemann, R. J., Buck, K. J., Cunningham, C. L., and Crabbe, J. C. (2003). Internet resources for genomic, bioinformatics, and medical genetics information. *Curr. Protoc. Neurosci.* Appendix 5, Appendix 5A.

Phillips, T. J., Huson, M., Gwiazdon, C., Burkhart-Kasch, S., and Shen, E. H. (1995). Effects of acute and repeated ethanol exposures on the locomotor activity of BXD recombinant inbred mice. *Alcohol. Clin. Exp. Res.* **19**, 269–278.

Phillips, T. J., Huson, M. G., and McKinnon, C. S. (1998a). Localization of genes mediating acute and sensitized locomotor responses to cocaine in BXD/Ty recombinant inbred mice. *J. Neurosci.* **18**, 3023–3034.

Plomin, R., McClearn, G. E., Gora-Maslak, G., and Neiderhiser, J. M. (1991). Use of recombinant inbred strains to detect quantitative trait loci associated with behavior. *Behav. Genet.* **21**, 99–116.

Pompeiano, M., Palacios, J. M., and Mengod, G. (1994). Distribution of the serotonin 5-HT_2 receptor family mRNAs: Comparison between 5-HT_{2A} and 5-HT_{2C} receptors. *Brain Res. Mol. Brain Res.* **23**, 163–178.

Prescott, C. A., and Kendler, K. S. (1999). Genetic and environmental contributions to alcohol abuse and dependence in a population-based sample of male twins. *Am. J. Psychiatry* **156**, 34–40.

Primeaux, S. D., Wilson, S. P., Bray, G. A., York, D. A., and Wilson, M. A. (2006). Overexpression of neuropeptide Y in the central nucleus of the amygdala decreases ethanol self-administration in "anxious" rats. *Alcohol. Clin. Exp. Res.* **30**, 791–801.

Pushparaj, P. N., Aarthi, J. J., Manikandan, J., and Kumar, S. D. (2008). siRNA, miRNA, and shRNA: *In vivo* applications. *J. Dent. Res.* **87**, 992–1003.

Regier, D. A., Farmer, M. E., Rae, D. S., Locke, B. Z., Keith, S. J., Judd, L. L., and Goodwin, F. K. (1990). Comorbidity of mental disorders with alcohol and other drug abuse. Results from the Epidemiologic Catchment Area (ECA) study. *JAMA* **264**, 2511–2518.

Reich, T., Edenberg, H. J., Goate, A., Williams, J. T., Rice, J. P., Van Eerdewegh, P., Foroud, T., Hesselbrock, V., Schuckit, M. A., Bucholz, K., Porjesz, B., Li, T. K., Conneally, P. M., Nurnberger, J. I., Jr., Tischfield, J. A., Crowe, R. R., Cloninger, C. R., Wu, W., Shears, S.,

Carr, K., Crose, C., Willig, C., and Begleiter, H. (1998). Genome-wide search for genes affecting the risk for alcohol dependence. *Am. J. Med. Genet.* **81**, 207–215.

Reich, T., Hinrichs, A., Culverhouse, R., and Bierut, L. (1999). Genetic studies of alcoholism and substance dependence. *Am. J. Hum. Genet.* **65**, 599–605.

Reilly, M. T., Milner, L. C., Shirley, R. L., Crabbe, J. C., and Buck, K. J. (2008). 5-HT_{2C} and $GABA_B$ receptors influence handling-induced convulsion severity in chromosome 4 congenic and DBA/2J background strain mice. *Brain Res.* **1198**, 124–131.

Rewal, M., Jurd, R., Gill, T. M., He, D. Y., Ron, D., and Janak, P. H. (2009). Alpha4-containing $GABA_A$ receptors in the nucleus accumbens mediate moderate intake of alcohol. *J. Neurosci.* **29**, 543–549.

Risinger, F. O., and Cunningham, C. L. (1998). Ethanol-induced conditioned taste aversion in BXD recombinant inbred mice. *Alcohol. Clin. Exp. Res.* **22**, 1234–1244.

Roberts, A., Pardo-Manuel de Villena, F., Wang, W., McMillan, L., and Threadgill, D. W. (2007). The polymorphism architecture of mouse genetic resources elucidated using genome-wide resequencing data: Implications for QTL discovery and systems genetics. *Mamm. Genome* **18**, 473–481.

Rodriguez, L. A., Plomin, R., Blizard, D. A., Jones, B. C., and McClearn, G. E. (1995). Alcohol acceptance, preference, and sensitivity in mice: II. Quantitative trait loci mapping analysis using BXD recombinant inbred strains. *Alcohol. Clin. Exp. Res.* **19**, 367–373.

Roff, D. A. (2007). A centennial celebration for quantitative genetics. *Evolution* **61**, 1017–1032.

Rogers, S. W., Weis, J. J., Ma, Y., Teuscher, C., and Gahring, L. C. (2008). Mouse chromosome 11 harbors genetic determinants of hippocampal strain-specific nicotinic receptor expression. *Hippocampus* **18**, 750–757.

Rogner, U. C., and Avner, P. (2003). Congenic mice: Cutting tools for complex immune disorders. *Nat. Rev. Immunol.* **3**, 243–252.

Rothenfluh, A., Threlkeld, R. J., Bainton, R. J., Tsai, L. T., Lasek, A. W., and Heberlein, U. (2006). Distinct behavioral responses to ethanol are regulated by alternate RhoGAP18B isoforms. *Cell* **127**, 199–211.

Rothschild, M. F., Hu, Z. L., and Jiang, Z. (2007). Advances in QTL mapping in pigs. *Int. J. Biol. Sci.* **3**, 192–197.

Saba, L., Bhave, S. V., Grahame, N., Bice, P., Lapadat, R., Belknap, J., Hoffman, P. L., and Tabakoff, B. (2006). Candidate genes and their regulatory elements: Alcohol preference and tolerance. *Mamm. Genome* **17**, 669–688.

Saunders, E. F., Zhang, P., Copeland, J. N., McLnnis, M. G., and Zollner, S. (2009). Suggestive linkage at 9p22 in bipolar disorder weighted by alcohol abuse. *Am. J. Med. Genet. B. Neuropsychiatr. Genet.* **150**, 1133–1138.

Schuckit, M. A., Goodwin, D. A., and Winokur, G. (1972). A study of alcoholism in half siblings. *Am. J. Psychiatry* **128**, 1132–1136.

Seong, E., Saunders, T. L., Stewart, C. L., and Burmeister, M. (2004). To knockout in 129 or in C57BL/6: That is the question. *Trends Genet.* **20**, 59–62.

Sheng, M., and Sala, C. (2001). PDZ domains and the organization of supramolecular complexes. *Annu. Rev. Neurosci.* **24**, 1–29.

Sher, K. J., Dick, D. M., Crabbe, J. C., Hutchison, K. E., O'Malley, S. S., and Heath, A. C. (2010). Consilient research approaches in studying gene × environment interactions in alcohol research. *Addict. Biol.* **15**, 200–216.

Shirley, R. L., Walter, N. A., Reilly, M. T., Fehr, C., and Buck, K. J. (2004). *Mpdz* is a quantitative trait gene for drug withdrawal seizures. *Nat. Neurosci.* **7**, 699–700.

Shizuya, H., Birren, B., Kim, U. J., Mancino, V., Slepak, T., Tachiiri, Y., and Simon, M. (1992). Cloning and stable maintenance of 300-kilobase-pair fragments of human DNA in *Escherichia coli* using an F-factor-based vector. *Proc. Natl. Acad. Sci. USA* **89**, 8794–8797.

Simpson, E. M., Linder, C. C., Sargent, E. E., Davisson, M. T., Mobraaten, L. E., and Sharp, J. J. (1997). Genetic variation among 129 substrains and its importance for targeted mutagenesis in mice. *Nat. Genet.* **16**, 19–27.

Simpson, E. H., Suffolk, R., and Jackson, I. J. (1999). Identification, sequence, and mapping of the mouse multiple PDZ domain protein gene, *Mpdz*. *Genomics* **59**, 102–104.

Singer, O., and Verma, I. M. (2008). Applications of lentiviral vectors for shRNA delivery and transgenesis. *Curr. Gene Ther.* **8**, 483–488.

Smith, S. B., Marker, C. L., Perry, C., Liao, G., Sotocinal, S. G., Austin, J. S., Melmed, K., Clark, J. D., Peltz, G., Wickman, K., and Mogil, J. S. (2008). Quantitative trait locus and computational mapping identifies *Kcnj9* (GIRK3) as a candidate gene affecting analgesia from multiple drug classes. *Pharmacogenet. Genomics* **18**, 231–241.

Solecki, W., Krowka, T., Kubik, J., Kaczmarek, L., and Przewlocki, R. (2008). Increased analgesic tolerance to acute morphine in fosB knock-out mice: A gender study. *Pharmacol. Biochem. Behav.* **90**, 512–516.

Sommer, W., Hyytia, P., and Kiianmaa, K. (2006). The alcohol-preferring AA and alcohol-avoiding ANA rats: Neurobiology of the regulation of alcohol drinking. *Addict. Biol.* **11**, 289–309.

Spanagel, R. (2003). Alcohol addiction research: From animal models to clinics. *Best. Pract. Res. Clin. Gastroenterol.* **17**, 507–518.

Spence, J. P., Liang, T., Liu, L., Johnson, P. L., Foroud, T., Carr, L. G., and Shekhar, A. (2009). From QTL to candidate gene: A genetic approach to alcoholism research. *Curr. Drug Abuse Rev.* **2**, 127–134.

Sperk, G., Furtinger, S., Schwarzer, C., and Pirker, S. (2004). GABA and its receptors in epilepsy. *Adv. Exp. Med. Biol.* **548**, 92–103.

Stacey, D., Clarke, T. K., and Schumann, G. (2009). The genetics of alcoholism. *Curr. Psychiatry Rep.* **11**, 364–369.

Stuber, C. W. (1995). Mapping and manipulating quantitative traits in maize. *Trends Genet.* **11**, 477–481.

Stuber, G. D. (2010). Dissecting the neural circuitry of addiction and psychiatric disease with optogenetics. *Neuropsychopharmacology* **35**, 341–342.

Su, Z., Wang, X., Tsaih, S. W., Zhang, A., Cox, A., Sheehan, S., and Paigen, B. (2009). Genetic basis of HDL variation in 129/SvImJ and C57BL/6J mice: Importance of testing candidate genes in targeted mutant mice. *J. Lipid Res.* **50**, 116–125.

Sun, Y., Chen, X., and Xiao, D. (2007). Tetracycline-inducible expression systems: New strategies and practices in the transgenic mouse modeling. *Acta Biochim. Biophys. Sin. (Shanghai)* **39**, 235–246.

Suzuki, G., Harper, K. M., Hiramoto, T., Sawamura, T., Lee, M., Kang, G., Tanigaki, K., Buell, M., Geyer, M. A., Trimble, W. S., Agatsuma, S., and Hiroi, N. (2009). *Sept5* deficiency exerts pleiotropic influence on affective behaviors and cognitive functions in mice. *Hum. Mol. Genet.* **18**, 1652–1660.

Szumlinski, K. K., Ary, A. W., Lominac, K. D., Klugmann, M., and Kippin, T. E. (2008). Accumbens *Homer2* overexpression facilitates alcohol-induced neuroplasticity in C57BL/6J mice. *Neuropsychopharmacology* **33**, 1365–1378.

Szumlinski, K. K., Lominac, K. D., Kleschen, M. J., Oleson, E. B., Dehoff, M. H., Schwarz, M. K., Seeburg, P. H., Worley, P. F., and Kalivas, P. W. (2005). Behavioral and neurochemical phenotyping of *Homer1* mutant mice: Possible relevance to schizophrenia. *Genes Brain Behav.* **4**, 273–288.

Tabakoff, B., Saba, L., Kechris, K., Hu, W., Bhave, S. V., Finn, D. A., Grahame, N. J., and Hoffman, P. L. (2008). The genomic determinants of alcohol preference in mice. *Mamm. Genome* **19**, 352–365.

Tabakoff, B., Saba, L., Printz, M., Flodman, P., Hodgkinson, C., Goldman, D., Koob, G., Richardson, H. N., Kechris, K., Bell, R. L., Hubner, N., Heinig, M., Pravenec, M., Mangion, J., Legault, L., Dongier, M., Conigrave, K. M., Whitfield, J. B., Saunders, J., Grant, B., and Hoffman, P. L. (2009). Genetical genomic determinants of alcohol consumption in rats and humans. *BMC Biol.* **7**, 70.

Tarantino, L. M., McClearn, G. E., Rodriguez, L. A., and Plomin, R. (1998). Confirmation of quantitative trait loci for alcohol preference in mice. *Alcohol. Clin. Exp. Res.* **22**, 1099–1105.

Thanos, P. K., Dimitrakakis, E. S., Rice, O., Gifford, A., and Volkow, N. D. (2005). Ethanol self-administration and ethanol conditioned place preference are reduced in mice lacking cannabinoid CB1 receptors. *Behav. Brain Res.* **164**, 206–213.

Threadgill, D. W., Yee, D., Matin, A., Nadeau, J. H., and Magnuson, T. (1997). Genealogy of the 129 inbred strains: 129/SvJ is a contaminated inbred strain. *Mamm. Genome* **8**, 390–393.

Uhl, G. R., Drgon, T., Johnson, C., Fatusin, O. O., Liu, Q. R., Contoreggi, C., Li, C. Y., Buck, K., and Crabbe, J. (2008). "Higher order" addiction molecular genetics: Convergent data from genome-wide association in humans and mice. *Biochem. Pharmacol.* **75**, 98–111.

Ullmer, C., Schmuck, K., Figge, A., Lubbert, H. (1998). Cloning and characterization of *MUPP1*, a novel PDZ domain protein. *FEBS Lett.* **424**, 63–68.

Valdar, W., Solberg, L. C., Gauguier, D., Burnett, S., Klenerman, P., Cookson, W. O., Taylor, M. S., Rawlins, J. N., Mott, R., and Flint, J. (2006). Genome-wide genetic association of complex traits in heterogeneous stock mice. *Nat. Genet.* **38**, 879–887.

Van den Haute, C., Eggermont, K., Nuttin, B., Debyser, Z., and Baekelandt, V. (2003). Lentiviral vector-mediated delivery of short hairpin RNA results in persistent knockdown of gene expression in mouse brain. *Hum. Gene. Ther.* **14**, 1799–1807.

Wade, C. M., Kulbokas, E. J., 3rd, Kirby, A. W., Zody, M. C., Mullikin, J. C., Lander, E. S., Lindblad-Toh, K., and Daly, M. J. (2002). The mosaic structure of variation in the laboratory mouse genome. *Nature* **420**, 574–578.

Wahlsten, D. (2001). Standardizing tests of mouse behavior: Reasons, recommendations, and reality. *Physiol. Behav.* **73**, 695–704.

Wahlsten, D., Metten, P., Phillips, T. J., Boehm, S. L., 2nd, Burkhart-Kasch, S., Dorow, J., Doerksen, S., Downing, C., Fogarty, J., Rodd-Henricks, K., Hen, R., McKinnon, C. S., Merrill, C. M., Nolte, C., Schalomon, M., Schlumbohm, J. P., Sibert, J. R., Wenger, C. D., Dudek, B. C., and Crabbe, J. C. (2003). Different data from different labs: Lessons from studies of gene–environment interaction. *J. Neurobiol.* **54**, 283–311.

Walter, N. A., Bottomly, D., Laderas, T., Mooney, M. A., Darakjian, P., Searles, R. P., Harrington, C. A., McWeeney, S. K., Hitzemann, R., and Buck, K. J. (2009). High throughput sequencing in mice: A platform comparison identifies a preponderance of cryptic SNPs. *BMC Genomics* **10**, 379.

Walter, N. A., McWeeney, S. K., Peters, S. T., Belknap, J. K., Hitzemann, R., and Buck, K. J. (2007). SNPs matter: Impact on detection of differential expression. *Nat. Methods* **4**, 679–680.

Williams, A. H., Brown, W. M., and Langefeld, C. D. (2005). Multilocus and interaction-based genome scan for alcoholism risk factors in Caucasian Americans: The COGA study. *BMC Genet.* **6**(Suppl. 1), S37.

Yalcin, B., Fullerton, J., Miller, S., Keays, D. A., Brady, S., Bhomra, A., Jefferson, A., Volpi, E., Copley, R. R., Flint, J., and Mott, R. (2004). Unexpected complexity in the haplotypes of commonly used inbred strains of laboratory mice. *Proc. Natl. Acad. Sci. USA* **101**, 9734–9739.

Zhu, M., and Zhao, S. (2007). Candidate gene identification approach: Progress and challenges. *Int. J. Biol. Sci.* **3**, 420–427.

GLUTAMATE PLASTICITY IN THE DRUNKEN AMYGDALA: THE MAKING OF AN ANXIOUS SYNAPSE

Brian A. McCool,[*,†,‡] Daniel T. Christian,[*] Marvin R. Diaz,[‡] and Anna K. Läck[*]

[*]Department of Physiology & Pharmcology, Wake Forest University School of Medicine, Winston-Salem, NC 27157, USA

[†]Translational Center for the Neurobehavioral Study of Alcohol, Wake Forest University School of Medicine, Winston-Salem, NC 27157, USA

[‡]Neuroscience Training Program, Wake Forest University School of Medicine, Winston-Salem, NC 27157, USA

Plasticity at glutamatergic synapses is believed to be the cellular correlate of learning and memory. Classic fear conditioning, for example, is dependent upon NMDA-type glutamate receptor activation in the lateral/basolateral amygdala followed by increased synaptic expression of AMPA-type glutamate receptors. This review provides an extensive comparison between the initiation and expression of glutamatergic plasticity during learning/memory and glutamatergic alterations associated with chronic ethanol exposure and withdrawal. The parallels between these neuro-adaptive processes suggest that long-term ethanol exposure might "chemically condition" amygdala-dependent fear/anxiety via the increased function of pre- and post-synaptic glutamate signaling.

INTERNATIONAL REVIEW OF
NEUROBIOLOGY, VOL. 91
DOI: 10.1016/S0074-7742(10)91007-6

I. Introduction

The enduring behavioral changes wrought by chronic alcohol abuse and alcoholism are among the most challenging obstacles for the development of effective therapies. These behaviors are currently believed to represent a shift from the initially rewarding effects of ethanol to an avoidance of the negative consequences during alcohol withdrawal following long-term abuse (Koob, 2003). One pervasive manifestation of chronic drug abuse in general and alcohol dependence in particular is the robust and persistent anxiety experienced during withdrawal and throughout abstinence. Long-abstinent alcoholics frequently cite anxiety as a significant contributing factor to relapse (Brown *et al.*, 1990; Noone *et al.*, 1999; Schneider *et al.*, 2001; Thevos *et al.*, 1991). Independent measures of anxiety in recently abstinent alcoholics suggest this negatively reinforcing emotion persists weeks to months after the acute withdrawal (Cohn *et al.*, 2003) and can predict relapse rates in some populations (Roberts *et al.*, 1999). Cause–effect relationships between alcohol dependence, relapse, and anxiety are nonetheless difficult to study in humans (Kushner *et al.*, 2000; Merikangas *et al.*, 1998; Rodgers *et al.*, 2000). However, animal models clearly suggest that withdrawal-associated increases in anxiety-like behavior persist well after the acute withdrawal syndrome (Santucci *et al.*, 2008) and contribute to subsequent increases in ethanol drinking (Valdez *et al.*, 2002). While antecedent anxiety disorders contribute to ethanol drinking and abuse and are likely to make important contributions to the transition from use to dependence, we will focus specifically upon dependence-related anxiety and the excitatory synaptic mechanisms that potentially help regulate this persistent increase in anxiety following chronic ethanol exposure.

Enduring changes in behavior, whether initiated by environmental cues or by chronic drug and alcohol exposure, are most certainly dictated by long-term changes in neurotransmitter signaling and neuronal excitability. Recent evidence in animal models of cocaine (Borgland *et al.*, 2004; Goussakov *et al.*, 2006; Ungless *et al.*, 2001) and opiate (Glass *et al.*, 2005) exposure support the hypothesis that abused drugs can engage many of the cellular mechanisms that are believed to govern long-term changes in synaptic plasticity, particularly within reward- and anxiety-related neural circuits. This neurobiological "allostasis" likely reflects the subversion of synaptic and intrinsic mechanisms that normally serve functions such as learning and memory. This chapter will attempt to consolidate recent findings that suggest chronic ethanol exposure, like cocaine and opiates, can specifically upregulate glutamatergic synaptic function in ways that parallel activity-related changes in synaptic efficacy.

What then are the neuroanatomical substrates that govern anxiety-like behavior in general? There have been many excellent reviews discussing the neuroanatomical contributions to learned fear behaviors as well as innate anxiety-like responses to more naturally aversive environments (Charney and Deutch, 1996;

Davidson, 2002; Davis, 2006; Mathew *et al.*, 2008; Phelps and LeDoux, 2005). Critical to both learned fear-like and innate anxiety-like behaviors is the amygdala. As part of the limbic system, the amygdala serves as a central "hub" for information dealing with emotional behaviors and therefore sends/receives information from a number of brain regions. Highly processed sensory and cognitive information flows into the amygdala from the cortex and thalamus as well as the hippocampus (LeDoux, 1996). The amygdala subsequently projects to the lateral hypothalamus to regulate the HPA axis responses to learned fear (Gray *et al.*, 1989) and to the parabrachial nucleus and dorsal motor nucleus of the vagus, which modulate the autonomic and physical components of the conditioned fear (Davis, 1992). In addition, the amygdala modulates innate anxiety-related behaviors via projections to forebrain regions such as the prefrontal cortex, nucleus accumbens, and bed nucleus of the stria terminalis, which have been associated with risk assessment (Jinks and McGregor, 1997; Simpson *et al.*, 2001), reward (Prado-Alcala and Wise, 1984), and unconditioned fear (Walker and Davis, 1997), respectively. Hence the central contributions by the amygdala to both conditioned fear responses and unconditioned anxiety-like behavior make it a likely candidate when looking for neural adaptations related to chronic ethanol exposure/withdrawal-associated increases in anxiety.

The amygdala itself contains several distinct nuclei, which are all associated with specific efferent and afferent projections that serve to modulate conditioned and unconditioned behaviors. The lateral and basolateral nuclei (BLA) are the primary "input" regions of the amygdala. We refer to these regions together as "BLA" to emphasize their cytoarchitectural similarities and their extensively overlapping efferent/afferent arrangements. These nuclei represent the initial "stop" for cognitive and sensory information flowing into the anxiety circuit. The BLA integrates information from sensory and executive cortices, hippocampus, and sensory thalamus and sends glutamatergic efferents to the central nucleus of the amygdala (CeA), the bed nucleus of the stria terminalis, and the nucleus accumbens (De Olmos *et al.*, 1985). Ninety to ninety-five percent of BLA neurons are large, pyramidal-shaped glutamatergic projection neurons and 5–10% small, nonpyramidal GABAergic interneurons (McDonald, 1982). The information processing in the BLA is dictated by local inhibitory GABAergic inputs and excitatory glutamatergic synaptic inputs arising both locally and from distant brain regions. The delicate balance between excitation and inhibition in the BLA ultimately influences the activity of the principal glutamatergic projection neurons.

Neuroadaptations within the BLA due to chronic ethanol and withdrawal would significantly impact the flow of information both within the amygdala and along the amygdala efferent pathways and ultimately influence the expression of anxiety/fear-related behaviors. For example, classic behavioral pharmacology experiments have shown that the BLA plays a central role in the acquisition of learned-fear behaviors (Blair *et al.*, 2001; Fanselow and LeDoux, 1999; Maren, 2005) and can also modulate the expression of innate anxiety-like behaviors (McCool and Chappell, 2007; Sajdyk and Shekhar, 1997). Both behavior measures are dependent upon BLA neurotransmitter systems and can be modulated by direct manipulation of glutamatergic (Fanselow and Kim, 1994; Sajdyk and Shekhar, 1997), GABAergic (Muller *et al.*, 1997; Sanders and Shekhar, 1995), catecholaminergic (Gonzalez *et al.*, 1996; Guarraci *et al.*, 1999), and neuropeptide (Gutman *et al.*, 2008; Sajdyk *et al.*, 1999) systems. Consistent with this, we have also recently shown that microinjection of the AMPA receptor antagonist DNQX into the BLA can alleviate withdrawal-associated increases in anxiety-like behavior as they are represented in the light/dark transition test (Lack *et al.*, 2007). Previous work has also found increased c-fos expression in the BLA following multiple withdrawals from an ethanol liquid diet (Borlikova *et al.*, 2006). Although these findings do not directly implicate glutamatergic dysregulation during ethanol withdrawal, the data do suggest a central role of the lateral/basolateral amygdala in the regulation of withdrawal-associated anxiety. We therefore began our investigations with examining glutamatergic alterations following chronic ethanol and withdrawal. Our rationale centered upon the important role this system plays in both learned and unconditioned amygdala-dependent behaviors.

II. Use-Dependent Synaptic Plasticity

Synaptic plasticity has been generally defined as a modification in synaptic efficacy resulting from some salient behavioral experience or overt experimental manipulation. This plasticity is currently believed to be the neurobiological mechanism that underlies long-term behavioral changes that manifest as learning and memory (Sigurdsson *et al.*, 2007) or, in the context of the amygdala, Pavlovian fear conditioning (McKernan and Shinnick-Gallagher, 1997; Rogan *et al.*, 1997). The physiological manifestation of synaptic plasticity has a direction (increase/decrease or potentiation/depression of activity) and a time course (lasting minutes to days or short-term/long-term plasticity) that ultimately defines its character. While short-term changes in synaptic efficacy are clearly important for signal integration and processing, longer-lasting changes in synaptic function are arguably more relevant for enduring changes in behavior.

The hippocampus has provided the most detailed insights into the neurobiological mechanisms governing long-term synaptic plasticity at glutamatergic synapses. Hippocampal long-term potentiation (LTP) has been suggested as a possible mechanism underlying learning and memory formation. At the Schaffer collateral CA3–CA1 glutamatergic synapse, activity-dependent *N*-methyl-D-aspartate (NMDA) receptor activation initiates plastic changes in glutamatergic signaling via calcium influx and initiation of downstream second messenger systems. NMDA receptor-dependent activation of calcium–calmodulin dependent protein kinase II (CamKII) has been identified as a critical step in this process (Barria and Malinow, 2005) with additional important roles played by other second messenger systems (cAMP) and protein kinases (PKC and src-family kinases). Initiation of NMDA-independent forms of LTP at Schaffer collateral synapses likewise involves increases in intracellular calcium via voltage-gated calcium channels and subsequent activation of calcium-dependent kinases and signaling pathways (Jilla Sabeti, 2007; Stefani *et al.*, 1998). These Schaffer collateral/CA1 initiation mechanisms, particularly calcium influx and calcium-dependent signaling processes, parallel similar initiation mechanisms at other hippocampal glutamatergic inputs (Colino and Malenka, 1993) and at glutamate synapses in many other brain regions (Berretta *et al.*, 2008; Feldman *et al.*, 1999) including the amygdala (see below). These similarities underscore the fundamental importance of calcium-dependent signaling pathways during the initiation of synaptic plasticity.

However, the mechanisms responsible for the expression of LTP at hippocampal Shaffer collateral–CA1 synapses have been the subject of some debate. It is now generally accepted that LTP expression is the result of increased contributions by α-amino-3-hydroxy-5-methyl-4-isoxazolepropionic (AMPA) receptors at the affected synapses. Calcium-dependent protein kinases (CamKII, PKC) or other calcium-activated signaling pathways, presumably activated by the influx of calcium through NMDA receptors, appear to phosphorylate AMPA receptors (Barria *et al.*, 1997; Hayashi *et al.*, 2000; Lee *et al.*, 2009). This not only can directly increase receptor function (Derkach *et al.*, 1999) but has also been shown to increase receptor trafficking from intracellular pools to dendritic membranes (Kerchner and Nicoll, 2008). As with initiation, AMPAR-dependent expression of synaptic plasticity is characteristic of LTP in other brain regions (Kakegawa and Yuzaki, 2005; Yu *et al.*, 2008).

In contrast to the postsynaptic initiation/expression at Schaffer collateral–CA1 synapses, the hippocampus also provides an example of presynaptic initiation and expression of synaptic plasticity. At the mossy fiber–CA3 glutamatergic synapses, use-dependent LTP is initiated by activation of calcium-permeable kainate-type glutamate receptors (Schmitz *et al.*, 2003). This is believed to initiate calcium-dependent signaling in the presynaptic terminal and enhance neurotransmitter release by increasing quantal content (Xiang *et al.*, 1994) and/or the number of release sites (Tong *et al.*, 1996). Presynaptic initiation/expression of use-dependent synaptic plasticity is also found in other brain regions like the lateral (Tsvetkov *et al.*, 2002) and central amygdala (Samson and Pare, 2005) and cerebellum (Bender *et al.*, 2009). Thus, synaptic plasticity can involve either pre or postsynaptic neurobiological mechanisms depending upon the specific synapse involved.

For fear conditioning, the acquisition phase of behavioral training was ultimately shown to involve NMDA-dependent (Fanselow and Kim, 1994), long-lasting plastic changes in glutamatergic synaptic transmission in the lateral/basolateral amygdala (McKernan and Shinnick-Gallagher, 1997; Rogan *et al.*, 1997). These LTP-like changes in lateral/basolateral amygdala neurophysiology were among the first to provide a direct link between synaptic plasticity in a specific brain region and long-term changes in behavior. LTP at thalamic- and cortical–lateral/basolateral amygdala synapses can utilize either NMDA-dependent activity (Drephal *et al.*, 2006) or voltage-gated calcium channel signaling (Bauer *et al.*, 2002; Chapman and Bellavance, 1992) depending upon the precise stimulation parameters and recording conditions. Both "initiation" pathways result in the activation of CamKII (Chapman *et al.*, 2003) and subsequent downstream effectors (Huang and Kandel, 1998). These molecular pathways are very similar to those found in the hippocampus. These calcium-dependent signals are believed to increase AMPA receptor trafficking to the plasma membrane (Yu *et al.*, 2008), which ultimately regulates the postsynaptic expression of plasticity in these studies. In contrast to these postsynaptic pathways, recent evidence also suggests that LTP at the cortico-amygdala glutamatergic synapses can be attributed to presynaptic increases in neurotransmitter release (Humeau *et al.*, 2003; Tsvetkov *et al.*, 2002) and cAMP-dependent increases in presynaptic calcium channel function (Fourcaudot *et al.*, 2008, 2009). Thus, use-dependent plasticity in the lateral/basolateral amygdala shares many of the same characteristics as hippocampal plasticity, including the input-specific mechanisms (both pre and postsynaptic) related to LTP initiation and expression.

A. Amygdala Use-Dependent Synaptic Plasticity and Ethanol

Direct measures of the acute effects of ethanol on amygdala-dependent behaviors are complicated by the amnestic and sedative properties of ethanol. Despite this, doses of acute ethanol that suppress the acquisition of fear-conditioned behaviors in rats can be separated from higher doses that suppress motor function (Sonner *et al.*, 1998), thus segregating a potential confound for measures of both fear-potentiated startle and anxiety-like behavior. Similar findings have been reported in mice (Gould, 2003) and importantly, in humans (Moberg and Curtin, 2009). Given the critical role of the lateral/basolateral amygdala in the acquisition of fear-conditioned and anxiety-related behaviors, these findings suggest that the well-established learning/memory-impairing effects of acute ethanol likewise disrupt amygdala-dependent fear learning.

The modulation of amygdala-dependent behaviors has also been studied in the context of chronic exposure and withdrawal. A history of binge drinking in humans can impair auditory fear conditioning (Stephens *et al.*, 2005), a behavioral learning process that is dependent upon an intact amygdala (Bechara *et al.*, 1995). This is paralleled by findings that withdrawal from chronic ethanol exposure can disrupt the acquisition of auditory fear conditioning in rats (Stephens *et al.*, 2001) and can impair contextual fear conditioning in mice (Celerier *et al.*, 2000). Since the acquisition of contextual fear conditioning depends upon NMDA-dependent processes in the basolateral amygdala (Fanselow and Kim, 1994), these findings together imply that chronic ethanol and withdrawal must influence amygdala neural plasticity and subsequently impair the acquisition of fear conditioning. In fact, repeated withdrawal from a chronic ethanol liquid diet attenuates calcium-dependent postsynaptic LTP induction at rat cortical–lateral amygdala and Schaffer collateral–CA1 glutamatergic synapses (Roberto *et al.*, 2002; Stephens *et al.*, 2005). This might be interpreted as a simple "inhibition" of plasticity initiation or expression in this brain region following chronic ethanol and withdrawal. However, withdrawal from chronic ethanol increases anxiety-like behavior; and glutamate signaling in the lateral/basolateral amygdala plays a critical role in this withdrawal-associated anxiety (Lack *et al.*, 2007). Both seem inconsistent with any direct suppression of glutamate signaling. These inconsistencies set the stage for our recent attempts to directly measure the glutamate signaling pathways responsible for the initiation and expression of amygdala LTP following chronic ethanol exposure and withdrawal.

III. Lateral/Basolateral Amygdala, Chronic Ethanol, and Glutamatergic Mechanisms Associated with Plasticity Initiation

A. NMDA Receptors and Chronic Alcohol/Withdrawal

To understand the inhibitory effect of chronic ethanol/withdrawal on LTP initiation/expression in the amygdala, we began by exploring functional alterations of NMDA-type glutamate receptors. NMDA receptors are heteromeric receptors whose subunits include the NR1 as well as NR2A-D subunits; the NR1 subunit is necessary for the formation of a functional channel whereas NR2 subunits confer native pharmacological and biophysical properties to the receptor complex. All NMDA NR1–NR2 subunit combinations are inhibited by acute ethanol; however, ethanol appears to be most potent at NMDA receptors containing combinations of NR1 paired with either NR2A or 2B (Masood *et al.*, 1994). Parallels to these heterologous system studies are evident in whole-cell

NMDA-gated currents expressed by cultured neurons (Lovinger, 1995; Lovinger *et al.*, 1989; Popp *et al.*, 1999) and by acutely isolated neurons (Criswell *et al.*, 2003; Grover *et al.*, 1998). In addition to the acute ethanol inhibition of NMDA receptors, chronic ethanol exposure upregulates native NMDA receptor function in many different brain regions (Grover *et al.*, 1998; Gulya *et al.*, 1991). This functional increase is not only sometimes associated with increased subunit mRNA and protein expression (Follesa and Ticku, 1995; Snell *et al.*, 1996) but may also be related to increased trafficking of NMDA receptors from intracellular pools to the cell surface (Carpenter-Hyland *et al.*, 2004). Given the central role of NMDA receptors in the initiation of synaptic plasticity, it seemed reasonable to hypothesize that the ethanol-related effects on amygdala LTP might involve alterations in NMDA receptor function.

We recently examined the impact of chronic intermittent ethanol vapor inhalation on rat NMDA receptor synaptic function in principal neurons found in the lateral/basolateral amygdala. This paradigm exposes rats to ethanol vapor 12 h/day for 10 consecutive days; and blood ethanol levels at the end of the exposure are typically >200 mg/dL. Glutamatergic synaptic currents were measured using whole-cell patch-clamp methods and electrical stimulation of "local" afferents. We found that electrically evoked NMDA-mediated synaptic currents were increased in BLA slices prepared from intoxicated animals sacrificed immediately following the last ethanol exposure (CIE) as well as animals sacrificed after 24 h of withdrawal (WD) (Lack *et al.*, 2007) relative to control (air-exposed) animals. Both CIE and WD caused a left- and upward shift in an NMDAR-mediated stimulus–response relationship indicating greater synaptic activation of these receptors at a given electrical stimulus. These findings suggest either that NMDAR-mediated synaptic currents were more sensitive to electrically evoked glutamate release or that NMDAR function/expression at the synapse was increased by chronic ethanol and withdrawal. The latter interpretation was supported by another set of experiments examining the relative contribution of NMDA- and AMPA-mediated synaptic currents. In this study, the amplitude of an NMDAR-mediated synaptic current was normalized to the amplitude of an AMPA receptor-mediated synaptic current. This NMDA:AMPA ratio was increased in BLA neurons recorded from both CIE and WD slices. Since AMPA receptor function is increased in animals following both chronic ethanol and withdrawal (see below), increases in the NMDA:AMPA ratio would be due to a relatively greater contribution by NMDA receptors. These findings are consistent with the hypothesis that chronic ethanol and withdrawal enhance synaptic NMDA receptor expression/function in the BLA. Additional studies on acutely isolated BLA neurons also found an increase in NMDAR-mediated current density following chronic exposure to an ethanol-containing liquid diet (Floyd *et al.*, 2003). This increased function was associated with decreased calcium-dependent desensitization and increased sensitivity to the NR2B-selective allosteric inhibitor ifenprodil, but did not change acute ethanol sensitivity. Levels of BLA NR1 subunit mRNA were altered in these animals, but expression level and cellular distribution of NR2A and NR2B mRNAs remained unchanged. We also found that the chronic ethanol exposure did not alter NR1, NR2A, or NR2B peptide levels (unpublished observations). Together the synaptic and whole-cell measures suggest a substantial upregulation in the function of BLA NMDA receptors following chronic ethanol exposure. Although the molecular mechanism responsible for this effect remains to be clearly established, the absence of any substantive modulation of NMDAR subunit expression suggest that nongenomic mechanisms, such as increased trafficking of NMDA receptors to the cell surface (Carpenter-Hyland *et al.*, 2004), are likely to be engaged by chronic ethanol exposure in the BLA.

Chronic ethanol also alters NMDA receptors in the neighboring central amygdaloid nucleus. While many of these changes are similar to those we have identified in the BLA, some are distinct. For example, chronic ethanol inhalation increased the ifenprodil sensitivity of NMDAR-mediated synaptic responses in central amygdala neurons (Roberto *et al.*, 2004b), similar to our results in acutely isolated neurons (Floyd *et al.*, 2003). However, unlike BLA NMDA receptors, the synaptic function of central amygdala NMDA receptors is unchanged by chronic ethanol exposure (Roberto *et al.*, 2004b, 2006) whereas the acute ethanol sensitivity of NMDA-mediated synaptic currents is increased (Roberto *et al.*, 2004b, 2006). Even the molecular alterations of BLA and central amygdala NMDA receptors in response to chronic ethanol/withdrawal share some characteristics but differ in others. Similar to the BLA, chronic ethanol inhalation not only increases NR1 subunit mRNA levels in the central amygdala but also increases NR2B mRNA levels and NR2A/2B subunit peptide expression (Roberto *et al.*, 2006). These effects appear transient as NR1- and NR2-subunit mRNA expression is significantly decreased compared to control levels following 1 week of withdrawal from chronic alcohol and return to control levels 2 weeks after the last ethanol exposure. Similarly, protein levels for these subunits also return to control levels, but much sooner—at 1 week of withdrawal (Roberto *et al.*, 2006). It is important to note that methodological differences may influence the unique chronic ethanol/withdrawal-dependent alterations in central and lateral/basolateral amygdala NMDA receptors. For example, a more modest chronic exposure to an ethanol-containing liquid diet caused only minimal effects to NR2A subunit mRNA and peptide expression and did not alter NR1 or NR2B mRNA or protein expression (Lack *et al.*, 2005). This suggests that exposure duration or intensity might dramatically influence NMDA receptor alterations. Since NMDA receptors initiate plastic changes in synaptic activity in the amygdala, we subsequently examined the influence of chronic ethanol/withdrawal on neurobiological mechanisms involved with the expression of synaptic plasticity.

B. Kainate Receptor-Dependent Glutamatergic Plasticity in the Amygdala

Kainate receptors are tetrameric cation channels made up of five possible subunits with $GluR_{5-7}$ needed for functional channels, as well as KA_{1-2} (reviewed in Braga *et al.*, 2004). Similar to AMPA receptors, kainate receptors exhibit rapid activation and desensitization kinetics (reviewed in Pinheiro and Mulle, 2006). The receptors can be subject to posttranscriptional RNA editing at a Q/R site in the membrane re-entrant domain that directly modulates calcium permeability of the receptor (Burnashev *et al.*, 1995). Kainate receptor subunits are found throughout the nervous system, including the cortex, striatum, hippocampus, cerebellum, retina, spinal cord, dorsal root ganglia, and amygdala (reviewed in Huettner, 2003). These receptors share many molecular and functional characteristics with AMPA receptors, yet form their own unique receptor assemblies (Puchalski *et al.*, 1994). The historical precedent grouping AMPA and kainate glutamate receptors together was based largely on the absence of any pharmacological tools to functionally segregate native channels. However, it is worth noting that, unlike AMPA receptors (see below), kainate receptors expressed in both heterologous and native systems are inhibited by relatively low concentrations of acute ethanol (Costa *et al.*, 2000; Dildy-Mayfield and Harris, 1992; Valenzuela *et al.*, 1998; Weiner *et al.*, 1999). This suggests a potentially unique contribution of kainate receptors to glutamatergic alterations in response to chronic ethanol and withdrawal.

Over the past decade a number of receptor-selective compounds have been developed that differentiate between AMPA and kainate receptors (reviewed in Jane *et al.*, 2009). The compound GYKI 53655 is a noncompetitive antagonist of the AMPA receptors and has allowed direct measurement of native kainate receptor-mediated currents in the absence of a contaminating AMPA-mediated component (Paternain *et al.*, 1995). Conversely, UBP 296, a selective and competitive kainate receptor antagonist, allowed investigators to confirm that isolated GYKI53655-insensitive currents were in fact kainate receptor mediated (Andrasfalvy and Magee, 2004). This pharmacology has recently allowed the characterization of neurobiological processes mediated specifically by kainate receptors. In the lateral/basolateral amygdala, for example, evidence from a number of studies indicates that GluR5-containing kainate receptors mediate excitatory synaptic responses that can be directly measured in principal neurons (Li *et al.*, 2001; Li and Rogawski, 1998; Rogawski *et al.*, 2001). Kainate receptor-mediated responses can represent up to 30% of the overall glutamatergic synaptic response measured in a principal BLA neuron (Li and Rogawski, 1998). Trains of electrical stimuli can increase the magnitude of this synaptic kainate current up to 190% suggesting that kainate receptors may be most active under conditions of "strong synaptic drive"

(Li and Rogawski, 1998). These same studies showed that kainate receptor-mediated postsynaptic currents can be evoked from cortical inputs arriving via the external capsule but appear to be absent from other glutamatergic inputs. We recently showed that kainate receptor-mediated BLA synaptic responses are inhibited both by the competitive antagonist UBP296 and by acute ethanol (Lack *et al.*, 2008). In fact, acute ethanol was more potent at inhibiting kainate receptor-mediated synaptic currents than at NMDA-mediated synaptic responses in the same neurons.

Kainate receptor-dependent synaptic plasticity in the amygdala appears to have a much different biological profile from that expressed in the hippocampus (see above). BLA kainate receptors mediate a form of heterosynaptic plasticity (Li *et al.*, 2001) where receptor activation at one synapse can cause plastic changes in other adjacent glutamatergic synapses on the same neuron. This decidedly "postsynaptic" plasticity can be initiated by both low-frequency electrical stimulation (1 Hz, 15 min) and the kainate receptor-selective agonist (RS)-2-amino-3(3-hydroxy-5-tert-butylisoxazol-4-yl)propanoic acid (ATPA). Using field excitatory postsynaptic potentials, we were recently able to replicate the findings that application of kainate receptor agonist ATPA leads to long-term increases in synaptic strength in the BLA (Lack *et al.*, 2008). Acute ethanol blocked the increased synaptic strength seen with ATPA administration. Confirmation that this effect was kainate receptor mediated was demonstrated when we were able to inhibit ATPA-LTP with the specific kainate receptor antagonist, UBP 296. Since activation of BLA kainate receptors increases anxiety-like behaviors (Lack *et al.*, 2008), this research suggests that the inhibitory effects of acute ethanol on kainate receptors may contribute to the anxiolytic behaviors elicited by ethanol intoxication.

The acute ethanol sensitivity of BLA kainate receptors and their ability to initiate plastic changes in glutamatergic synaptic transmission suggest parallels with NMDA receptors in this brain region. However, the effects of chronic ethanol or withdrawal on kainate receptors have not been extensively studied. Chronic ethanol/withdrawal does not affect protein expression for kainate receptor subunits in cultured cortical neurons (Chandler *et al.*, 1999) but enhances both subunit protein and kainate receptor function in cultured hippocampal neurons (Carta *et al.*, 2002). Supporting this latter study, chronic cocaine exposure increases kainate receptor mRNA and protein in both rats and monkeys (Ghasemzadeh *et al.*, 1996; Hemby *et al.*, 2005). Given these findings and the prominent role for kainate receptors in postsynaptic glutamatergic responses in the BLA, we recently examined the effects of chronic ethanol and withdrawal on kainate receptor neurophysiology in this brain region. Using the same ethanol exposure paradigm employed for NMDA receptors, we found that chronic ethanol exposure enhanced BLA kainate receptor-mediated synaptic responses relative to control neurons (Lack *et al.*, 2009). Unlike NMDA receptors, this effect was entirely reversed after 24 h of withdrawal from the exposure. In this same study, we subsequently showed that chronic ethanol and withdrawal significantly reduced ATPA-induced synaptic plasticity. These data provide remarkable parallels with our findings with NMDA receptors. Namely chronic ethanol and/or withdrawal enhances the function of both NMDA and kainate receptors—receptors that both initiate forms of synaptic plasticity in the lateral/basolateral amygdala—but paradoxically reduces glutamatergic synaptic plasticity in this same brain region.

IV. Chronic Alcohol, Withdrawal, and the Expression of Synaptic Plasticity

A. Amygdala AMPA Receptors

AMPA-type ionotropic glutamate receptors mediate the vast majority of excitatory neurotransmission in the nervous system. These receptors are composed of four closely related subunits, GluR1–4, that form tetrameric assemblies to produce functional channels (Rosenmund *et al.*, 1998). GluR1–4 subunits are all expressed within the rat BLA (Farb *et al.*, 1995; McDonald, 1994). The specific subunit composition of the native channel dictates ion permeability. Although all AMPARs are nonselectively permeable to cations like sodium and potassium, subsets of these channels, particularly those lacking the GluR2 subunit, are also permeable to calcium (Bochet *et al.*, 1994). GluR2-less channels are in fact found in BLA interneurons (McDonald, 1996), where they mediate synaptic Ca^{2+} signaling and initiate plastic changes in GABAergic function (Szinyei *et al.*, 2007). In BLA projection neurons, AMPA receptors are responsible for the expression of use-dependent plasticity at glutamatergic synapses (Humeau *et al.*, 2007; Yu *et al.*, 2008). Thus, BLA AMPA receptors mediate a complex array of neurophysiological functions.

AMPA-type glutamate receptors seem to be relatively insensitive to modulation by acute ethanol. For example, 80 mM ethanol did not alter synaptic AMPA-mediated currents measured in hippocampal dentate granule neurons in either rats or monkeys despite robust ethanol inhibition of NMDA-mediated synaptic currents in these same neurons (Ariwodola *et al.*, 2003). AMPA-mediated synaptic responses from locus ceruleus (Nieber *et al.*, 1998) and spinal motoneurons (Ziskind-Conhaim *et al.*, 2003) are likewise insensitive to the direct, postsynaptic effects of acute ethanol. This rather modest sensitivity of AMPA receptors to intoxicating concentrations of acute ethanol has also been noted in whole-cell currents from cultured cortical neurons (Lovinger, 1993), acutely isolated medial septum/diagonal band neurons (Frye and Fincher, 2000), and AMPA-gated whole-cell currents recorded from nucleus accumbens neurons in brain slices (Nie *et al.*, 1994). This latter finding contrasts with earlier work in the nucleus accumbens showing robust ethanol sensitivity of AMPA-mediated excitatory postsynaptic potentials (Nie *et al.*, 1993). This effect was reversed with the opioid antagonist naloxone suggesting an indirect ethanol modulation of the excitatory postsynaptic potential (EPSP) in this instance. However, it should be noted that 60 mM acute ethanol modestly inhibits (20%) AMPA-mediated synaptic responses in the central amygdala (Roberto *et al.*, 2004b). Similarly, acute ethanol robustly inhibits AMPA-gated whole-cell currents recorded from neonatal (P4) but not juvenile (P26) CA3 hippocampal neurons (Mameli *et al.*, 2005). These findings suggest that intoxicating concentrations of acute ethanol produce relatively modest effects on AMPA receptor function in most cases but that the magnitude of this modulation may be developmentally regulated and brain region dependent.

There has been very limited interest in the effects of chronic ethanol on AMPA-type glutamate receptors. Chronic exposure does seem upregulate AMPA receptor subunit mRNA levels in hippocampus (Bruckner *et al.*, 1997), AMPAR subunit protein expression in primary cortical cultures (Chandler *et al.*, 1999), AMPA receptor-dependent calcium signaling in cerebellar Purkinje neurons (Netzeband *et al.*, 1999), and AMPA receptor binding in cortical membranes (Haugbol *et al.*, 2005). However, AMPA receptor synaptic function following chronic ethanol exposure has not been extensively studied.

Our laboratory recently found that AMPA receptor synaptic function was significantly upregulated in the BLA during withdrawal from chronic ethanol inhalation. We showed specifically that the amplitude of AMPA-mediated, tetrodotoxin (TTX)-insensitive spontaneous glutamatergic events (mEPSCs) recorded from BLA projection neurons was potentiated in slices from 24-h withdrawal animals, but not from slices prepared from CIE animals (Lack *et al.*, 2007). This increased synaptic function was also associated with an increase in the decay kinetics of AMPA-mediated mEPSCs. These findings are consistent with altered BLA postsynaptic AMPA receptor synaptic function. Although the mechanism for the increase mEPSC amplitude is unclear at present, preliminary data suggest that neither chronic ethanol nor withdrawal increases AMPA receptor subunit peptide levels in the BLA (unpublished observations). Despite this, withdrawal from chronic benzodiazepine treatment increases AMPA-mediated whole-cell current density (Song *et al.*, 2007), increases mEPSC amplitude (Van Sickle *et al.*, 2004), and reduces hippocampal LTP expression (Shen *et al.*, 2009). These findings are very similar to the effects of chronic ethanol/withdrawal on the BLA. Importantly, chronic benzodiazepine treatment increases the delivery of GluR1-containing AMPA receptors to hippocampal glutamatergic synapses (Das *et al.*, 2008). This is similar to the increased delivery of GluR1 subunits to the synaptic compartment following synaptic plasticity in the barrel cortex, which is associated with faster mEPSC decay (Clem and Barth, 2006). Together these data suggest that the altered postsynaptic properties of BLA AMPA-mediated mEPSCs following withdrawal from chronic ethanol might involve increased trafficking of AMPA receptors to the cell surface.

B. BLA Extracellular Recordings

The implication of increased AMPA-receptor signaling is that BLA neurons will be more responsive to glutamatergic inputs during chronic ethanol/withdrawal. In support of this, we recently found that chronic ethanol/withdrawal increased extracellular excitatory postsynaptic potentials (field EPSPs) measured in BLA brain slices by electrical stimulation of cortical inputs (Lack *et al.*, 2009). This increase was evident across a range of electrical stimulation intensities and occurred in the absence of any significant presynaptic effect of chronic ethanol or withdrawal at these particular glutamatergic synapses (see below). These data suggest that chronic ethanol and withdrawal increase the responses of BLA neurons to electrical stimulation via increased excitability of individual neurons and/or increased AMPA receptor-mediated synaptic responses. Since BLA fEPSPs are entirely sensitive to the AMPA receptor antagonist 6-cyano-7-nitroquinoxaline-2,3-dione (CNQX) (unpublished observations), it is difficult to differentiate between these possible mechanisms. However, chronic ethanol increases the extracellular population EPSP recorded in CA1 hippocampus (Whittington and Little, 1990) suggesting this may be a common outcome. Regardless, increased neuronal responsiveness during chronic ethanol/withdrawal would certainly interfere with traditional postsynaptic measures of LTP expression.

C. Presynaptic Glutamatergic Function

In addition to the altered postsynaptic characteristics seen during withdrawal from chronic ethanol, we also found that the glutamatergic synapses themselves were altered by the exposure. CIE and WD increased the frequency of mEPSCs recorded from BLA projection neurons (Lack *et al.*, 2007). This is difficult to interpret in the face of the increased mEPSC amplitude found in that same study. However, there were two pieces of evidence that supported the hypothesis that chronic ethanol/withdrawal altered presynaptic glutamatergic function in this brain region. First, we found increases in the probability of glutamate release when measured by pairs of locally applied electrical stimuli. Second, we found an increased frequency of spontaneous EPSCs in CIE slices relative to control, but not when tetrodotoxin was present in the extracellular bath. These findings are consistent with the idea that glutamatergic release is significantly elevated in the BLA during both CIE and withdrawal, albeit via distinct, TTX-sensitive and -resistant mechanisms. Interestingly, chronic ethanol treatment also enhances presynaptic glutamate release (Zhu *et al.*, 2007) and increases extracellular glutamate levels (Roberto *et al.*, 2004b) in the neighboring central amygdala.

The precise mechanisms governing these presynaptic alterations are not clear. However, chronic ethanol increases the number of glutamatergic terminals in the shell of the nucleus accumbens (Zhou *et al.*, 2006). Alternatively, acute ethanol can depress glutamate release via inhibition of voltage-gated calcium channels in some brain regions (Mameli *et al.*, 2005; Zhu *et al.*, 2007). It is possible then that adaptations increasing presynaptic voltage-gated calcium channels might underlie enhanced glutamate release during chronic ethanol/withdrawal. It should be noted that presynaptic effects of chronic ethanol are not consistently observed, even in the BLA. Our laboratory has shown that presynaptic release probability at cortical glutamatergic inputs seems insensitive to chronic ethanol/withdrawal (Lack *et al.*, 2009). Given that "local" stimulation within the BLA itself reveals clear increases in glutamatergic release probability (Lack *et al.*, 2007), these data suggest that presynaptic effects of chronic ethanol/withdrawal may be input specific. Regardless, the increased glutamatergic presynaptic function at "local" synapses, similar to enhanced postsynaptic function of AMPA receptors, parallels those mechanisms responsible for the expression of use-dependent synaptic plasticity at hippocampal mossy fiber terminals.

V. Ethanol-Dependent Plasticity of Additional Amygdala Neurotransmitter Systems

This review has focused primarily upon the plasticity of glutamatergic ionotropic receptors following chronic ethanol and withdrawal. However, we must emphasize that there are many additional signaling pathways which can contribute to increased neuronal output from the BLA. Along with the pre and postsynaptic glutamatergic alterations described here, these additional pathways could also help regulate the enhanced anxiety-related behaviors evident during withdrawal. For example, increased glutamate release from some synaptic inputs might potentially enhance signaling by G-protein-coupled metabotropic glutamate receptors. In fact, chronic ethanol/withdrawal enhances mGluR-dependent phosphoinositide hydrolysis in the hippocampus (Valles *et al.*, 1995) and intracellular calcium levels in cultured Purkinje neurons (Netzeband *et al.*, 2002).

Since Group I mGluR activation in the BLA enhances the acquisition of fear-conditioned behaviors (Rudy and Matus-Amat, 2009), increased mGluR signaling during withdrawal might provide an additional glutamatergic influence to tip the balance toward enhanced BLA output and increased anxiety-like behavior.

In addition to excitatory neurotransmission, BLA output is regulated by a balance between excitatory and inhibitory synaptic and intrinsic activity. Like many brain regions, fast synaptic inhibition in this brain region is governed by GABAergic mechanisms. Chronic ethanol exposure alters BLA $GABA_A$ receptor expression and function (Anderson *et al.*, 2007; Floyd *et al.*, 2004; McCool *et al.*, 2003; Papadeas *et al.*, 2001). Recent work also suggests that GABAergic synaptic responses evoked by electrical stimulation of the lateral external capsule are substantially reduced by withdrawal from chronic ethanol and diazepam treatments as well as repeated restraint stress (Isoardi *et al.*, 2007). Although the effects of chronic ethanol/withdrawal on BLA GABAergic synaptic transmission remain to be fully characterized, chronic ethanol enhances both GABA presynaptic release and extracellular GABA levels in the neighboring central amygdala (Roberto *et al.*, 2004a). This CeA upregulation is contrasted with diminished GABAergic synaptic/extrasynaptic function in the hippocampus (Cagetti *et al.*, 2003; Liang *et al.*, 2006, 2007) and reduced expression of $GABA_A$ subunit mRNAs/proteins in several different brain regions (reviewed in Kumar *et al.*, 2009). Thus, GABAergic neuroadaptations to chronic ethanol/withdrawal in the BLA are likely to be complex and potentially specific to this brain region.

In addition to the glutamate and GABA systems, BLA catecholamine and neuropeptide neurotransmitters also play an important regulatory role during chronic ethanol and withdrawal. For example, D1 receptor activation enhances GABAergic synaptic activity arising from local feedback neurons (Kroner *et al.*, 2005) while providing simultaneous suppression of GABAergic activity arising feed-forward inhibitory synapses (Marowsky *et al.*, 2005). Similarly, BLA mu-opioid receptors suppress GABAergic but not glutamatergic synaptic function (Finnegan *et al.*, 2006). These data suggest that increased binding density of BLA D1-like receptors (Sari *et al.*, 2006) and mu-opioid receptors (Djouma and Lawrence, 2002) following chronic ethanol drinking may dramatically alter GABAergic synaptic signaling in this brain region. However, this remains to be more directly established.

Finally, chronic ethanol/withdrawal may modulate the intrinsic membrane properties of lateral/basolateral amygdala neurons. Oddly, single-unit recordings of lateral amygdala neurons in awake animals have consistently shown *reduced* firing rates during ethanol withdrawal (Feng and Faingold, 2008; Feng *et al.*, 2007). This seems inconsistent with the emerging picture of an "excitable" BLA following chronic ethanol. However, interpretation of *in vivo* BLA data is complicated by the robust tonic activity and high intrinsic firing rates of GABAergic interneurons relative to glutamatergic principal neurons (Rainnie *et al.*, 1993;

Washburn and Moises, 1992). These characteristics suggest that GABAergic interneurons might be over-represented in the population of cells sampled during *in vivo* unit recordings. Similarly, glutamatergic projection neurons increase firing in response to emotionally salient (Chang *et al.*, 2005; Hobin *et al.*, 2003; Pelletier *et al.*, 2005) but not emotionally neutral environmental stimuli (Cromwell *et al.*, 2005). Since most *in vivo* BLA studies to date employ habituated behaviors during the unit recording, they would also tend to under-represent contributions by glutamatergic projection neuron firing during ethanol withdrawal. Regardless, the precise effect of chronic ethanol exposure on the excitability of projection neurons and local-circuit feed-forward and feedback interneurons remains to be characterized. Ethanol withdrawal reduces the firing threshold and increases spontaneous firing rates of periaqueductal gray neurons (Yang *et al.*, 2002). Potential cellular mechanisms regulating increased excitability might include increased voltage-gated calcium channel function (Whittington and Little, 1993) or reduced function of calcium-dependent potassium channels or nonselective hyperpolarization activated cation (I_h) channels (Hopf *et al.*, 2007) that serve to re-polarize or delay recovery from the after-hyperpolarization following an action potential, respectively. While future studies will focus on the effects of chronic ethanol and withdrawal on intrinsic membrane properties and conductances, interactions between BLA neuron excitability and the synaptic contributions outlined in the previous sections would ultimately govern the increased BLA-dependent anxiety-like behavior expressed during ethanol withdrawal.

VI. Conclusions and Future Directions

Our work and that of others have shown that chronic ethanol exposure and withdrawal can enhance function of glutamatergic receptors responsible for both the initiation and expression of BLA synaptic plasticity. The increased NMDA and kainate receptor function both speak to the diversity of LTP initiation-related signaling processes co-opted by a model of chronic alcohol exposure. Similarly, the enhanced AMPA receptor function, neuronal responsiveness, and presynaptic glutamatergic function that is evident following chronic ethanol/withdrawal suggest that mechanisms responsible for the expression of synaptic plasticity are likewise engaged. It is worth noting that these expression-related mechanisms are generally more resistant to the acute effects of ethanol suggesting that these alterations result from other cellular processes rather than a direct compensatory change to ethanol-related altered function. Our findings suggest that, rather than a simple inhibition of the mechanisms governing the use-dependent initiation or expression of glutamatergic plasticity in the lateral/basolateral amygdala, chronic

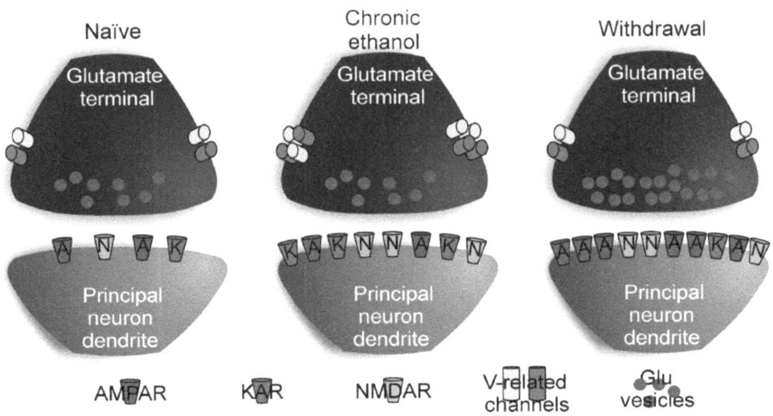

FIG. 1. Model describing glutamatergic alterations measured in lateral/basolateral amygdala principal neurons following chronic ethanol exposure and withdrawal. Presynaptic (purple) and postsynaptic (green) specializations represent a "generic" glutamatergic synapse. Synaptic responses mediated by postsynaptic AMPA (red)- , kainate (blue)-, and NMDA (yellow)-type glutamate receptors can be measured in ethanol-naive (left) neurons (Lack *et al.*, 2007, 2008, 2009). Chronic ethanol exposure (middle) enhances NMDA and kainate receptor synaptic function (here represented as increased numbers of receptors found at the postsynaptic specialization) and increases the probability of glutamate release measured by paired stimuli and the frequency of spontaneous AMPA-mediated synaptic events. Since this effect is sensitive to tetrodotoxin, we presume that presynaptic alterations involve voltage- or calcium-dependent processes. During withdrawal (right), kainate receptor-mediated synaptic responses return to control levels whereas AMPA receptor-mediated responses are markedly increased, possibly due to increased numbers of receptors at the synapse. Like chronic ethanol, withdrawal also increases presynaptic function but these changes become resistant to tetrodotoxin—here represented as increased numbers of vesicles but could also involve increases in the probability of release or quantal content at single synapses or increases in the total number of release sites. See text for details.

ethanol and withdrawal appear to occlude experimental activation of these processes by engaging these very same mechanisms.

A model representing ethanol/withdrawal-related changes in glutamatergic function is shown in Fig. 1. In the chronic ethanol/intoxicated condition, we found increased postsynaptic function of both NMDA and kainate receptors (Lack *et al.*, 2007, 2009). It is not clear yet if this increase can be found across all the various glutamatergic inputs arriving at the principal neurons from which we made these measures or if these postsynaptic alterations are selectively expressed at specific synaptic inputs. We also found that chronic ethanol may increase the frequency/amplitude of these AMPA-mediated events via the enhanced function of tetrodotoxin-sensitive mechanisms. Specifically, tetrodotoxin, a potent blocker of voltage-gated sodium channels, appeared to inhibit the increased amplitude/frequency of spontaneous EPSCs recorded from CIE neurons. This suggests that action potential-dependent processes help mediate some effects on glutamatergic neurotransmission immediately following chronic ethanol exposure. This is represented in the model by increased numbers of voltage-gated channels (Brodie and Sampson, 1990; McMahon *et al.*, 2000; Watson and Little, 1999) and might also be manifested as decreased expression of ethanol-sensitive potassium channels (Mulholland *et al.*, 2009; Pietrzykowski *et al.*, 2004; Sun *et al.*, 2008) that serve to re-polarize or hyperpolarize the presynaptic compartment and thus reduce or control presynaptic neurotransmitter release. We have not included this later possibility in the model for simplicity, but both mechanisms are plausible in that they would ultimately produce a tetrodotoxin-sensitive increase in spontaneous EPSC amplitude/frequency in the chronic ethanol-exposed animals.

In the withdrawal rats, postsynaptic alterations differ considerably between specific receptor families. The increased function of NMDA receptors measured immediately following chronic exposure is maintained at least 24 h after the exposure whereas kainate receptor function rapidly returns to control levels during this same period. In neither instance have we found any substantial changes in subunit protein levels (unpublished data). This suggests that any other mechanisms, like receptor trafficking or localization, must also be modified by chronic exposure/withdrawal in a receptor-dependent fashion. We also found that, unlike glutamatergic synapses measured during chronic ethanol exposure, presynaptic effects of withdrawal shift to a form that is resistant to tetrodotoxin and therefore independent of action potentials. Although the precise mechanism for this is unclear, increased presynaptic function can be associated with a change in release probability at individual synapses via an increase in the number of release sites (Bender *et al.*, 2009; Tong *et al.*, 1996), increased quantal size of individual vesicles (Bellingham *et al.*, 1998; Wilson *et al.*, 2005), or some combination of these mechanisms. Distinguishing between these possible mechanisms is an important future direction. However, increased release probability during withdrawal was evident at "local" glutamate synapses (Lack *et al.*, 2007) but not cortical synapses arriving via the external capsule (Lack *et al.*, 2009). Thus, like chronic ethanol, the presynaptic alterations evident during withdrawal may be input specific.

Future studies are likely to focus on the precise molecular mechanisms which govern both pre and postsynaptic functional alterations resulting from chronic ethanol exposure and withdrawal. Additional aspects to be explored relate to the duration of exposure needed to achieve a particular adaptation, the durability of the glutamatergic alterations we and others have described to date, and potential contributions by other neurotransmitter systems. Regardless, our evidence suggests that chronic ethanol exposure and withdrawal "chemically condition" the lateral/basolateral amygdala using glutamatergic mechanisms that closely parallel those engaged by Pavlovian fear conditioning.

The enduring behavioral changes resulting from environmental cue-dependent synaptic plasticity in this brain region likewise parallel the enhanced anxiety-like behaviors that are characteristic of ethanol dependence. It seems reasonable to propose that the cellular mechanisms governing the extinction of learned fear may also provide attractive therapeutic targets for ethanol withdrawal-associated anxiety.

Acknowledgments

This work was supported by NIH/NIAAA grants R01 AA014445 (BAM), P01 AA017056, T32 AA007565 (DTC), F31 AA017576 (MRD), and F31 AA016442 (AKL).

References

Anderson, N. J., Daunais, J. B., Friedman, D. P., Grant, K. A., and McCool, B. A. (2007). Long-term ethanol self-administration by the nonhuman primate, *Macaca fascicularis*, decreases the benzodiazepine sensitivity of amygdala GABA(A) receptors. *Alcohol. Clin. Exp. Res.* **31**, 1061–1070.

Andrasfalvy, B. K., and Magee, J. C. (2004). Changes in AMPA receptor currents following LTP induction on rat CA1 pyramidal neurones. *J. Physiol.* **559**, 543–554.

Ariwodola, O. J., Crowder, T. L., Grant, K. A., Daunais, J. B., Friedman, D. P., and Weiner, J. L. (2003). Ethanol modulation of excitatory and inhibitory synaptic transmission in rat and monkey dentate granule neurons. *Alcohol. Clin. Exp. Res.* **27**, 1632–1639.

Barria, A., and Malinow, R. (2005). NMDA receptor subunit composition controls synaptic plasticity by regulating binding to CaMKII. *Neuron* **48**, 289–301.

Barria, A., Muller, D., Derkach, V., Griffith, L. C., and Soderling, T. R. (1997). Regulatory phosphorylation of AMPA-type glutamate receptors by CaM-KII during long-term potentiation. *Science* **276**, 2042–2045.

Bauer, E. P., Schafe, G. E., and LeDoux, J. E. (2002). NMDA receptors and L-type voltage-gated calcium channels contribute to long-term potentiation and different components of fear memory formation in the lateral amygdala. *J. Neurosci.* **22**, 5239–5249.

Bechara, A., Tranel, D., Damasio, H., Adolphs, R., Rockland, C., and Damasio, A. R. (1995). Double dissociation of conditioning and declarative knowledge relative to the amygdala and hippocampus in humans. *Science* **269**, 1115–1118.

Bellingham, M. C., Lim, R., and Walmsley, B. (1998). Developmental changes in EPSC quantal size and quantal content at a central glutamatergic synapse in rat. *J. Physiol.* **511**(Pt 3), 861–869.

Bender, V. A., Pugh, J. R., and Jahr, C. E. (2009). Presynaptically expressed long-term potentiation increases multivesicular release at parallel fiber synapses. *J. Neurosci.* **29**, 10974–10978.

Berretta, N., Nistico, R., Bernardi, G., and Mercuri, N. B. (2008). Synaptic plasticity in the basal ganglia: A similar code for physiological and pathological conditions. *Prog. Neurobiol.* **84**, 343–362.

Blair, H. T., Schafe, G. E., Bauer, E. P., Rodrigues, S. M., and LeDoux, J. E. (2001). Synaptic plasticity in the lateral amygdala: A cellular hypothesis of fear conditioning. *Learn. Mem.* **8**, 229–242.

Bochet, P., Audinat, E., Lambolez, B., Crepel, F., Rossier, J., *et al.* (1994). Subunit composition at the single-cell level explains functional properties of a glutamate-gated channel. *Neuron* **12**, 383–388.

Borgland, S. L., Malenka, R. C., and Bonci, A. (2004). Acute and chronic cocaine-induced potentiation of synaptic strength in the ventral tegmental area: Electrophysiological and behavioral correlates in individual rats. *J. Neurosci.* **24**, 7482–7490.

Borlikova, G. G., Le Merrer, J., and Stephens, D. N. (2006). Previous experience of ethanol withdrawal increases withdrawal-induced c-fos expression in limbic areas, but not withdrawal-induced anxiety and prevents withdrawal-induced elevations in plasma corticosterone. *Psychopharmacology (Berl)* **185**, 188–200.

Braga, M. F., Aroniadou-Anderjaska, V., and Li, H. (2004). The physiological role of kainate receptors in the amygdala. *Mol. Neurobiol.* **30**, 127–141.

Brodie, C., and Sampson, S. R. (1990). Effects of ethanol on voltage-sensitive Na-channels in cultured skeletal muscle: Up-regulation as a result of chronic treatment. *J. Pharmacol. Exp. Ther.* **255**, 1195–1201.

Brown, S. A., Vik, P. W., McQuaid, J. R., Patterson, T. L., Irwin, M. R., and Grant, I. (1990). Severity of psychosocial stress and outcome of alcoholism treatment. *J. Abnorm. Psychol.* **99**, 344–348.

Bruckner, M. K., Rossner, S., and Arendt, T. (1997). Differential changes in the expression of AMPA receptors genes in rat brain after chronic exposure to ethanol: An *in situ* hybridization study. *J. Hirnforsch.* **38**, 369–376.

Burnashev, N., Zhou, Z., Neher, E., and Sakmann, B. (1995). Fractional calcium currents through recombinant GluR channels of the NMDA, AMPA and kainate receptor subtypes. *J. Physiol.* **485** (Pt 2), 403–418.

Cagetti, E., Liang, J., Spigelman, I., and Olsen, R. W. (2003). Withdrawal from chronic intermittent ethanol treatment changes subunit composition, reduces synaptic function, and decreases behavioral responses to positive allosteric modulators of $GABA_A$ receptors. *Mol. Pharmacol.* **63**, 53–64.

Carpenter-Hyland, E. P., Woodward, J. J., and Chandler, L. J. (2004). Chronic ethanol induces synaptic but not extrasynaptic targeting of NMDA receptors. *J. Neurosci.* **24**, 7859–7868.

Carta, M., Olivera, D. S., Dettmer, T. S., and Valenzuela, C. F. (2002). Ethanol withdrawal upregulates kainate receptors in cultured rat hippocampal neurons. *Neurosci. Lett.* **327**, 128–132.

Celerier, A., Ognard, R., Decorte, L., and Beracochea, D. (2000). Deficits of spatial and non-spatial memory and of auditory fear conditioning following anterior thalamic lesions in mice: Comparison with chronic alcohol consumption. *Eur. J. Neurosci.* **12**, 2575–2584.

Chandler, L. J., Norwood, D., and Sutton, G. (1999). Chronic ethanol upregulates NMDA and AMPA, but not kainate receptor subunit proteins in rat primary cortical cultures. *Alcohol. Clin. Exp. Res.* **23**, 363–370.

Chang, C. H., Liang, K. C., and Yen, C. T. (2005). Inhibitory avoidance learning altered ensemble activity of amygdaloid neurons in rats. *Eur. J. Neurosci.* **21**, 210–218.

Chapman, P. F., and Bellavance, L. L. (1992). Induction of long-term potentiation in the basolateral amygdala does not depend on NMDA receptor activation. *Synapse* **11**, 310–318.

Chapman, P. F., Ramsay, M. F., Krezel, W., and Knevett, S. G. (2003). Synaptic plasticity in the amygdala: Comparisons with hippocampus. *Ann. N. Y. Acad. Sci.* **985**, 114–124.

Charney, D. S., and Deutch, A. (1996). A functional neuroanatomy of anxiety and fear: Implications for the pathophysiology and treatment of anxiety disorders. *Crit. Rev. Neurobiol.* **10**, 419–446.

Clem, R. L., and Barth, A. (2006). Pathway-specific trafficking of native AMPARs by *in vivo* experience. *Neuron* **49**, 663–670.

Cohn, T. J., Foster, J. H., and Peters, T. J. (2003). Sequential studies of sleep disturbance and quality of life in abstaining alcoholics. *Addict. Biol.* **8**, 455–462.

Colino, A., and Malenka, R. C. (1993). Mechanisms underlying induction of long-term potentiation in rat medial and lateral perforant paths *in vitro*. *J. Neurophysiol.* **69**, 1150–1159.

Costa, E. T., Soto, E. E., Cardoso, R. A., Olivera, D. S., and Valenzuela, C. F. (2000). Acute effects of ethanol on kainate receptors in cultured hippocampal neurons. *Alcohol. Clin. Exp. Res.* **24**, 220–225.

Criswell, H. E., Ming, Z., Griffith, B. L., and Breese, G. R. (2003). Comparison of effect of ethanol on *N*-methyl-D-aspartate- and GABA-gated currents from acutely dissociated neurons: Absence of regional differences in sensitivity to ethanol. *J. Pharmacol. Exp. Ther.* **304**, 192–199.

Cromwell, H. C., Anstrom, K., Azarov, A., and Woodward, D. J. (2005). Auditory inhibitory gating in the amygdala: Single-unit analysis in the behaving rat. *Brain Res.* **1043**, 12–23.

Das, P., Lilly, S. M., Zerda, R., Gunning, W. T. 3rd, Alvarez, F. J., and Tietz, E. I. (2008). Increased AMPA receptor GluR1 subunit incorporation in rat hippocampal CA1 synapses during benzodiazepine withdrawal. *J. Comp. Neurol.* **511**, 832–846.

Davidson, R. J. (2002). Anxiety and affective style: Role of prefrontal cortex and amygdala. *Biol. Psychiatry* **51**, 68–80.

Davis, M. (1992). The role of the amygdala in fear and anxiety. *Annu. Rev. Neurosci.* **15**, 353–375.

Davis, M. (2006). Neural systems involved in fear and anxiety measured with fear-potentiated startle. *Am. Psychol.* **61**, 741–756.

De Olmos, J., Alheid, G. F., and Beltamino, C. A. (1985). Amygdala. In The Rat Nervous System, (G. Paxinos Ed.), pp. 223–335. Academic Press, Sydney.

Derkach, V., Barria, A., and Soderling, T. R. (1999). Ca2+/calmodulin-kinase II enhances channel conductance of alpha-amino-3-hydroxy-5-methyl-4-isoxazolepropionate type glutamate receptors. *Proc. Natl. Acad. Sci. U S A.* **96**, 3269–3274.

Dildy-Mayfield, J. E., and Harris, R. A. (1992). Comparison of ethanol sensitivity of rat brain kainate, DL-alpha-amino-3-hydroxy-5-methyl-4-isoxalone proprionic acid and *N*-methyl-D-aspartate receptors expressed in *Xenopus* oocytes. *J. Pharmacol. Exp. Ther.* **262**, 487–494.

Djouma, E., and Lawrence, A. J. (2002). The effect of chronic ethanol consumption and withdrawal on mu-opioid and dopamine D(1) and D(2) receptor density in Fawn-Hooded rat brain. *J. Pharmacol. Exp. Ther.* **302**, 551–559.

Drephal, C., Schubert, M., and Albrecht, D. (2006). Input-specific long-term potentiation in the rat lateral amygdala of horizontal slices. *Neurobiol. Learn. Mem.* **85**, 272–282.

Fanselow, M. S., and Kim, J. J. (1994). Acquisition of contextual Pavlovian fear conditioning is blocked by application of an NMDA receptor antagonist D,L-2-amino-5- phosphonovaleric acid to the basolateral amygdala. *Behav. Neurosci.* **108**, 210–212.

Fanselow, M. S., and LeDoux, J. E. (1999). Why we think plasticity underlying Pavlovian fear conditioning occurs in the basolateral amygdala. *Neuron* **23**, 229–232.

Farb, C. R., Aoki, C., and Ledoux, J. E. (1995). Differential localization of NMDA and AMPA receptor subunits in the lateral and basal nuclei of the amygdala: A light and electron microscopic study. *J. Comp. Neurol.* **362**, 86–108.

Feldman, D. E., Nicoll, R. A., and Malenka, R. C. (1999). Synaptic plasticity at thalamocortical synapses in developing rat somatosensory cortex: LTP, LTD, and silent synapses. *J. Neurobiol.* **41**, 92–101.

Feng, H. J., and Faingold, C. L. (2008). The effects of chronic ethanol administration on amygdala neuronal firing and ethanol withdrawal seizures. *Neuropharmacology* **55**, 648–653.

Feng, H. J., Yang, L., and Faingold, C. L. (2007). Role of the amygdala in ethanol withdrawal seizures. *Brain Res.* **1141**, 65–73.

Finnegan, T. F., Chen, S. R., and Pan, H. L. (2006). Mu opioid receptor activation inhibits GABAergic inputs to basolateral amygdala neurons through Kv1.1/1.2 channels. *J. Neurophysiol.* **95**, 2032–2041.

Floyd, D. W., Friedman, D. P., Daunais, J. B., Pierre, P. J., Grant, K. A., and McCool, B. A. (2004). Long-term ethanol self-administration by cynomolgus macaques alters the pharmacology and expression of $GABA_A$ receptors in basolateral amygdala. *J. Pharmacol. Exp. Ther.* **311**, 1071–1079.

Floyd, D. W., Jung, K. Y., and McCool, B. A. (2003). Chronic ethanol ingestion facilitates *N*-methyl-D-aspartate receptor function and expression in rat lateral/basolateral amygdala neurons. *J. Pharmacol. Exp. Ther.* **307**, 1020–1029.

Follesa, P., and Ticku, M. K. (1995). Chronic ethanol treatment differentially regulates NMDA receptor subunit mRNA expression in rat brain. *Brain Res. Mol. Brain Res.* **29**, 99–106.

Fourcaudot, E., Gambino, F., Casassus, G., Poulain, B., Humeau, Y., and Luthi, A. (2009). L-type voltage-dependent Ca(2+) channels mediate expression of presynaptic LTP in amygdala. *Nat. Neurosci.* **12**, 1093–1095.

Fourcaudot, E., Gambino, F., Humeau, Y., Casassus, G., Shaban, H., Poulain, B., and Luthi, A. (2008). cAMP/PKA signaling and RIM1alpha mediate presynaptic LTP in the lateral amygdala. *Proc. Natl. Acad. Sci. U S A.* **105**, 15130–15135.

Frye, G. D., and Fincher, A. (2000). Sustained ethanol inhibition of native AMPA receptors on medial septum/diagonal band (MS/DB) neurons. *Br. J. Pharmacol.* **129**, 87–94.

Ghasemzadeh, M. B., Sharma, S., Surmeier, D. J., Eberwine, J. H., and Chesselet, M. F. (1996). Multiplicity of glutamate receptor subunits in single striatal neurons: An RNA amplification study. *Mol. Pharmacol.* **49**, 852–859.

Glass, M. J., Kruzich, P. J., Colago, E. E., Kreek, M. J., and Pickel, V. M. (2005). Increased AMPA GluR1 receptor subunit labeling on the plasma membrane of dendrites in the basolateral amygdala of rats self-administering morphine. *Synapse* **58**, 1–12.

Gonzalez, L. E., Andrews, N., and File, S. E. (1996). 5-HT1A and benzodiazepine receptors in the basolateral amygdala modulate anxiety in the social interaction test, but not in the elevated plus-maze. *Brain Res.* **732**, 145–153.

Gould, T. J. (2003). Ethanol disrupts fear conditioning in C57BL/6 mice. *J. Psychopharmacol.* **17**, 77–81.

Goussakov, I., Chartoff, E. H., Tsvetkov, E., Gerety, L. P., Meloni, E. G., Carlezon, W. A. Jr., and Bolshakov, V. Y. (2006). LTP in the lateral amygdala during cocaine withdrawal. *Eur. J. Neurosci.* **23**, 239–250.

Gray, T. S., Carney, M. E., and Magnuson, D. J. (1989). Direct projections from the central amygdaloid nucleus to the hypothalamic paraventricular nucleus: Possible role in stress-induced adrenocorticotropin release. *Neuroendocrinology* **50**, 433–446.

Grover, C. A., Wallace, K. A., Lindberg, S. A., and Frye, G. D. (1998). Ethanol inhibition of NMDA currents in acutely dissociated medial septum/diagonal band neurons from ethanol dependent rats. *Brain Res.* **782**, 43–52.

Guarraci, F. A., Frohardt, R. J., and Kapp, B. S. (1999). Amygdaloid D1 dopamine receptor involvement in Pavlovian fear conditioning. *Brain Res.* **827**, 28–40.

Gulya, K., Grant, K. A., Valverius, P., Hoffman, P. L., and Tabakoff, B. (1991). Brain regional specificity and time-course of changes in the NMDA receptor-ionophore complex during ethanol withdrawal. *Brain Res.* **547**, 129–134.

Gutman, A. R., Yang, Y., Ressler, K. J., and Davis, M. (2008). The role of neuropeptide Y in the expression and extinction of fear-potentiated startle. *J. Neurosci.* **28**, 12682–12690.

Haugbol, S. R., Ebert, B., and Ulrichsen, J. (2005). Upregulation of glutamate receptor subtypes during alcohol withdrawal in rats. *Alcohol Alcohol.* **40**, 89–95.

Hayashi, Y., Shi, S. H., Esteban, J. A., Piccini, A., Poncer, J. C., and Malinow, R. (2000). Driving AMPA receptors into synapses by LTP and CaMKII: Requirement for GluR1 and PDZ domain interaction. *Science* **287**, 2262–2267.

Hemby, S. E., Tang, W., Muly, E. C., Kuhar, M. J., Howell, L., and Mash, D. C. (2005). Cocaine-induced alterations in nucleus accumbens ionotropic glutamate receptor subunits in human and non-human primates. *J. Neurochem.* **95**, 1785–1793.

Hobin, J. A., Goosens, K. A., and Maren, S. (2003). Context-dependent neuronal activity in the lateral amygdala represents fear memories after extinction. *J. Neurosci.* **23**, 8410–8416.

Hopf, F. W., Martin, M., Chen, B. T., Bowers, M. S., Mohamedi, M. M., and Bonci, A. (2007). Withdrawal from intermittent ethanol exposure increases probability of burst firing in VTA neurons *in vitro*. *J. Neurophysiol.* **98**, 2297–2310.

Huang, Y. Y., and Kandel, E. R. (1998). Postsynaptic induction and PKA-dependent expression of LTP in the lateral amygdala. *Neuron* **21**, 169–178.

Huettner, J. E. (2003). Kainate receptors and synaptic transmission. *Prog. Neurobiol.* **70**, 387–407.

Humeau, Y., Reisel, D., Johnson, A. W., Borchardt, T., Jensen, V., *et al.* (2007). A pathway-specific function for different AMPA receptor subunits in amygdala long-term potentiation and fear conditioning. *J. Neurosci.* **27**, 10947–10956.

Humeau, Y., Shaban, H., Bissiere, S., and Luthi, A. (2003). Presynaptic induction of heterosynaptic associative plasticity in the mammalian brain. *Nature* **426**, 841–845.

Isoardi, N. A., Bertotto, M. E., Martijena, I. D., Molina, V. A., and Carrer, H. F. (2007). Lack of feedback inhibition on rat basolateral amygdala following stress or withdrawal from sedative-hypnotic drugs. *Eur. J. Neurosci.* **26**(4), 1036–1044.

Jane, D. E., Lodge, D., and Collingridge, G. L. (2009). Kainate receptors: Pharmacology, function and therapeutic potential. *Neuropharmacology* **56**, 90–113.

Jilla Sabeti, T. E. N., Purdy, R. H., Gruol, D. L. (2007). Steroid pregnenolone sulfate enhances NMDA-receptor-independent long-term potentiation at hippocampal CA1 synapses: Role for L-type calcium channels and sigma-receptors. *Hippocampus* **17**, 349–369.

Jinks, A. L., and McGregor, I. S. (1997). Modulation of anxiety-related behaviours following lesions of the prelimbic or infralimbic cortex in the rat. *Brain Res.* **772**, 181–190.

Kakegawa, W., and Yuzaki, M. (2005). A mechanism underlying AMPA receptor trafficking during cerebellar long-term potentiation. *Proc. Natl. Acad. Sci. U S A.* **102**, 17846–17851.

Kerchner, G. A., and Nicoll, R. A. (2008). Silent synapses and the emergence of a postsynaptic mechanism for LTP. *Nat. Rev. Neurosci.* **9**, 813–825.

Koob, G. F. (2003). Alcoholism: Allostasis and beyond. *Alcohol. Clin. Exp. Res.* **27**, 232–243.

Kroner, S., Rosenkranz, J. A., Grace, A. A., and Barrionuevo, G. (2005). Dopamine modulates excitability of basolateral amygdala neurons *in vitro*. *J. Neurophysiol.* **93**, 1598–1610.

Kumar, S., Porcu, P., Werner, D. F., Matthews, D. B., Diaz-Granados, J. L., Helfand, R. S., and Morrow, A. L. (2009). The role of GABA(A) receptors in the acute and chronic effects of ethanol: A decade of progress. *Psychopharmacology (Berl)* **205**, 529–564.

Kushner, M. G., Abrams, K., and Borchardt, C. (2000). The relationship between anxiety disorders and alcohol use disorders: A review of major perspectives and findings. *Clin. Psychol. Rev.* **20**, 149–171.

Lack, A. K., Ariwodola, O. J., Chappell, A. M., Weiner, J. L., and McCool, B. A. (2008). Ethanol inhibition of kainate receptor-mediated excitatory neurotransmission in the rat basolateral nucleus of the amygdala. *Neuropharmacology* **55**, 661–668.

Lack, A. K., Christian, D. T., Diaz, M. R., and McCool, B. A. (2009). Chronic ethanol and withdrawal effects on kainate receptor-mediated excitatory neurotransmission in the rat basolateral amygdala. *Alcohol* **43**, 25–33.

Lack, A. K., Diaz, M. R., Chappell, A., DuBois, D. W., and McCool, B. A. (2007). Chronic ethanol and withdrawal differentially modulate pre- and postsynaptic function at glutamatergic synapses in rat basolateral amygdala. *J. Neurophysiol.* **98**, 3185–3196.

Lack, A. K., Floyd, D. W., and McCool, B. A. (2005). Chronic ethanol ingestion modulates proanxiety factors expressed in rat central amygdala. *Alcohol* **36**, 83–90.

LeDoux, J. (1996). Emotional networks and motor control: A fearful view. *Prog. Brain Res.* **107**, 437–446.

Lee, H. K., Takamiya, K., He, K., Song, L., and Huganir, R. L. (2009). Specific roles of AMPA receptor subunit GluR1 (GluA1) phosphorylation sites in regulating synaptic plasticity in the CA1 region of hippocampus. *J. Neurophysiol.* **103**, 479–489.

Li, H., Chen, A., Xing, G., Wei, M. L., and Rogawski, M. A. (2001). Kainate receptor-mediated heterosynaptic facilitation in the amygdala. *Nat. Neurosci.* **4**, 612–620.

Li, H., and Rogawski, M. A. (1998). GluR5 kainate receptor mediated synaptic transmission in rat basolateral amygdala *in vitro*. *Neuropharmacology* **37**, 1279–1286.

Liang, J., Suryanarayanan, A., Abriam, A., Snyder, B., Olsen, R. W., and Spigelman, I. (2007). Mechanisms of reversible GABAA receptor plasticity after ethanol intoxication. *J. Neurosci.* **27**, 12367–12377.

Liang, J., Zhang, N., Cagetti, E., Houser, C. R., Olsen, R. W., and Spigelman, I. (2006). Chronic intermittent ethanol-induced switch of ethanol actions from extrasynaptic to synaptic hippocampal $GABA_A$ receptors. *J. Neurosci.* **26**, 1749–1758.

Lovinger, D. M. (1993). High ethanol sensitivity of recombinant AMPA-type glutamate receptors expressed in mammalian cells. *Neurosci. Lett.* **159**, 83–87.

Lovinger, D. M. (1995). Developmental decrease in ethanol inhibition of *N*-methyl-D-aspartate receptors in rat neocortical neurons: Relation to the actions of ifenprodil. *J. Pharmacol. Exp. Ther.* **274**, 164–172.

Lovinger, D. M., White, G., and Weight, F. F. (1989). Ethanol inhibits NMDA-activated ion current in hippocampal neurons. *Science* **243**, 1721–1724.

Mameli, M., Zamudio, P. A., Carta, M., and Valenzuela, C. F. (2005). Developmentally regulated actions of alcohol on hippocampal glutamatergic transmission. *J. Neurosci.* **25**, 8027–8036.

Maren, S. (2005). Synaptic mechanisms of associative memory in the amygdala. *Neuron* **47**, 783–786.

Marowsky, A., Yanagawa, Y., Obata, K., and Vogt, K. E. (2005). A specialized subclass of interneurons mediates dopaminergic facilitation of amygdala function. *Neuron* **48**, 1025–1037.

Masood, K., Wu, C., Brauneis, U., and Weight, F. F. (1994). Differential ethanol sensitivity of recombinant *N*-methyl-D-aspartate receptor subunits. *Mol. Pharmacol.* **45**, 324–329.

Mathew, S. J., Price, R. B., and Charney, D. S. (2008). Recent advances in the neurobiology of anxiety disorders: Implications for novel therapeutics. *Am. J. Med. Genet. C. Semin. Med. Genet.* **148**, 89–98.

McCool, B. A., and Chappell, A. (2007). Strychnine and taurine modulation of amygdala-associated anxiety-like behavior is "state" dependent. *Behav. Brain Res.* **178**, 70–81.

McCool, B. A., Frye, G. D., Pulido, M. D., and Botting, S. K. (2003). Effects of chronic ethanol consumption on rat GABA(A) and strychnine-sensitive glycine receptors expressed by lateral/basolateral amygdala neurons. *Brain Res.* **963**, 165–177.

McDonald, A. J. (1982). Neurons of the lateral and basolateral amygdaloid nuclei: A Golgi study in the rat. *J. Comp. Neurol.* **212**, 293–312.

McDonald, A. J. (1994). Neuronal localization of glutamate receptor subunits in the basolateral amygdala. *Neuroreport* **6**, 13–16.

McDonald, A. J. (1996). Localization of AMPA glutamate receptor subunits in subpopulations of non-pyramidal neurons in the rat basolateral amygdala. *Neurosci. Lett.* **208**, 175–178.

McKernan, M. G., and Shinnick-Gallagher, P. (1997). Fear conditioning induces a lasting potentiation of synaptic currents *in vitro*. *Nature* **390**, 607–611.

McMahon, T., Andersen, R., Metten, P., Crabbe, J. C., and Messing, R. O. (2000). Protein kinase C epsilon mediates up-regulation of N-type calcium channels by ethanol. *Mol. Pharmacol.* **57**, 53–58.

Merikangas, K. R., Stevens, D. E., Fenton, B., Stolar, M., O'Malley, S., Woods, S. W., and Risch, N. (1998). Co-morbidity and familial aggregation of alcoholism and anxiety disorders. *Psychol. Med.* **28**, 773–788.

Moberg, C. A., and Curtin, J. J. (2009). Alcohol selectively reduces anxiety but not fear: Startle response during unpredictable versus predictable threat. *J. Abnorm. Psychol.* **118**, 335–347.

Mulholland, P. J., Hopf, F. W., Bukiya, A. N., Martin, G. E., Liu, J. *et al.* (2009). Sizing up ethanol-induced plasticity: The role of small and large conductance calcium-activated potassium channels. *Alcohol. Clin. Exp. Res.* **33**, 1125–1135.

Muller, J., Corodimas, K. P., Fridel, Z., and LeDoux, J. E. (1997). Functional inactivation of the lateral and basal nuclei of the amygdala by muscimol infusion prevents fear conditioning to an explicit conditioned stimulus and to contextual stimuli. *Behav. Neurosci.* **111**, 683–691.

Netzeband, J. G., Schneeloch, J. R., Trotter, C., Caguioa-Aquino, J. N., and Gruol, D. L. (2002). Chronic ethanol treatment and withdrawal alter ACPD-evoked calcium signals in developing Purkinje neurons. *Alcohol. Clin. Exp. Res.* **26**, 386–393.

Netzeband, J. G., Trotter, C., Caguioa, J. N., and Gruol, D. L. (1999). Chronic ethanol exposure enhances AMPA-elicited Ca2+ signals in the somatic and dendritic regions of cerebellar Purkinje neurons. *Neurochem. Int.* **35**, 163–174.

Nie, Z., Madamba, S. G., and Siggins, G. R. (1994). Ethanol inhibits glutamatergic neurotransmission in nucleus accumbens neurons by multiple mechanisms. *J. Pharmacol. Exp. Ther.* **271**, 1566–1573.

Nie, Z., Yuan, X., Madamba, S. G., and Siggins, G. R. (1993). Ethanol decreases glutamatergic synaptic transmission in rat nucleus accumbens *in vitro*: Naloxone reversal. *J. Pharmacol. Exp. Ther.* **266**, 1705–1712.

Nieber, K., Poelchen, W., Sieler, D., and Illes, P. (1998). Inhibition by ethanol of excitatory amino acid receptors in rat locus coeruleus neurons *in vitro*. *Naunyn Schmiedebergs Arch. Pharmacol.* **357**, 299–308.

Noone, M., Dua, J., and Markham, R. (1999). Stress, cognitive factors, and coping resources as predictors of relapse in alcoholics. *Addict. Behav.* **24**, 687–693.

Papadeas, S., Grobin, A. C., and Morrow, A. L. (2001). Chronic ethanol consumption differentially alters GABA(A) receptor alpha1 and alpha4 subunit peptide expression and GABA(A) receptor-mediated 36 Cl(-) uptake in mesocorticolimbic regions of rat brain. *Alcohol. Clin. Exp. Res.* **25**, 1270–1275.

Paternain, A. V., Morales, M., and Lerma, J. (1995). Selective antagonism of AMPA receptors unmasks kainate receptor-mediated responses in hippocampal neurons. *Neuron* **14**, 185–189.

Pelletier, J. G., Likhtik, E., Filali, M., and Pare, D. (2005). Lasting increases in basolateral amygdala activity after emotional arousal: Implications for facilitated consolidation of emotional memories. *Learn. Mem.* **12**, 96–102.

Phelps, E. A., and LeDoux, J. E. (2005). Contributions of the amygdala to emotion processing: From animal models to human behavior. *Neuron* **48**, 175–187.

Pietrzykowski, A. Z., Martin, G. E., Puig, S. I., Knott, T. K., Lemos, J. R., and Treistman, S. N. (2004). Alcohol tolerance in large-conductance, calcium-activated potassium channels of CNS terminals is intrinsic and includes two components: Decreased ethanol potentiation and decreased channel density. *J. Neurosci.* **24**, 8322–8332.

Pinheiro, P., and Mulle, C. (2006). Kainate receptors. *Cell Tissue Res.* **326**, 457–482.

Popp, R. L., Lickteig, R. L., and Lovinger, D. M. (1999). Factors that enhance ethanol inhibition of *N*-methyl-D-aspartate receptors in cerebellar granule cells. *J. Pharmacol. Exp. Ther.* **289**, 1564–1574.

Prado-Alcala, R., and Wise, R. A. (1984). Brain stimulation reward and dopamine terminal fields. I. Caudate-putamen, nucleus accumbens and amygdala. *Brain Res.* **297**, 265–273.

Puchalski, R. B., Louis, J. C., Brose, N., Traynelis, S. F., Egebjerg, J., *et al.* (1994). Selective RNA editing and subunit assembly of native glutamate receptors. *Neuron* **13**, 131–147.

Rainnie, D. G., Asprodini, E. K., and Shinnick-Gallagher, P. (1993). Intracellular recordings from morphologically identified neurons of the basolateral amygdala. *J Neurophysiol.* **69**, 1350–1362.

Roberto, M., Bajo, M., Crawford, E., Madamba, S. G., and Siggins, G. R. (2006). Chronic ethanol exposure and protracted abstinence alter NMDA receptors in central amygdala. *Neuropsychopharmacology* **31**, 988–996.

Roberto, M., Madamba, S. G., Stouffer, D. G., Parsons, L. H., and Siggins, G. R. (2004a) Increased GABA release in the central amygdala of ethanol-dependent rats, J Neurosci **24**, 10159–10166.

Roberto, M., Nelson, T. E., Ur, C. L., and Gruol, D. L. (2002). Long-term potentiation in the rat hippocampus is reversibly depressed by chronic intermittent ethanol exposure. *J. Neurophysiol.* **87**, 2385–2397.

Roberto, M., Schweitzer, P., Madamba, S. G., Stouffer, D. G., Parsons, L. H., and Siggins, G. R. (2004b) Acute and chronic ethanol alter glutamatergic transmission in rat central amygdala: An *in vitro* and *in vivo* analysis. *J. Neurosci.* **24**, 1594–1603.

Roberts, M. C., Emsley, R. A., Pienaar, W. P., and Stein, D. J. (1999). Anxiety disorders among abstinent alcohol-dependent patients. *Psychiatr. Serv.* **50**, 1359–1361.

Rodgers, B., Korten, A. E., Jorm, A. F., Jacomb, P. A., Christensen, H., and Henderson, A. S. (2000). Non-linear relationships in associations of depression and anxiety with alcohol use. *Psychol. Med.* **30**, 421–432.

Rogan, M. T., Staubli, U. V., and LeDoux, J. E. (1997). Fear conditioning induces associative long-term potentiation in the amygdala. *Nature* **390**, 604–607.

Rogawski, M. A., Kurzman, P. S., Yamaguchi, S. I., and Li, H. (2001). Role of AMPA and GluR5 kainate receptors in the development and expression of amygdala kindling in the mouse. *Neuropharmacology* **40**, 28–35.

Rosenmund, C., Stern-Bach, Y., and Stevens, C. F. (1998). The tetrameric structure of a glutamate receptor channel. *Science* **280**, 1596–1599.

Rudy, J. W., and Matus-Amat, P. (2009). DHPG activation of group 1 mGluRs in BLA enhances fear conditioning. *Learn. Mem.* **16**, 421–425.

Sajdyk, T. J., Schober, D. A., Gehlert, D. R., and Shekhar, A. (1999). Role of corticotropin-releasing factor and urocortin within the basolateral amygdala of rats in anxiety and panic responses. *Behav. Brain Res.* **100**, 207–215.

Sajdyk, T. J., and Shekhar, A. (1997). Excitatory amino acid receptors in the basolateral amygdala regulate anxiety responses in the social interaction test. *Brain Res.* **764**, 262–264.

Samson, R. D., and Pare, D. (2005). Activity-dependent synaptic plasticity in the central nucleus of the amygdala. *J. Neurosci.* **25**, 1847–1855.

Sanders, S. K., and Shekhar, A. (1995). Anxiolytic effects of chlordiazepoxide blocked by injection of $GABA_A$ and benzodiazepine receptor antagonists in the region of the anterior basolateral amygdala of rats. *Biol. Psychiatry* **37**, 473–476.

Santucci, A. C., Cortes, C., Bettica, A., and Cortes, F. (2008). Chronic ethanol consumption in rats produces residual increases in anxiety 4 months after withdrawal. *Behav. Brain Res.* **188**, 24–31.

Sari, Y., Bell, R. L., and Zhou, F. C. (2006). Effects of chronic alcohol and repeated deprivations on dopamine D1 and D2 receptor levels in the extended amygdala of inbred alcohol-preferring rats. *Alcohol. Clin. Exp. Res.* **30**, 46–56.

Schmitz, D., Mellor, J., Breustedt, J., and Nicoll, R. A. (2003). Presynaptic kainate receptors impart an associative property to hippocampal mossy fiber long-term potentiation. *Nat. Neurosci.* **6**, 1058–1063.

Schneider, U., Altmann, A., Baumann, M., Bernzen, J., Bertz, B., *et al.* (2001). Comorbid anxiety and affective disorder in alcohol-dependent patients seeking treatment: The first multicentre study in Germany. *Alcohol Alcohol.* **36**, 219–223.

Shen, G., Mohamed, M. S., Das, P., and Tietz, E. I. (2009). Positive allosteric activation of GABAA receptors bi-directionally modulates hippocampal glutamate plasticity and behaviour. *Biochem. Soc. Trans.* **37**, 1394–1398.

Sigurdsson, T., Doyère, V., Cain, C. K., and LeDoux, J. E. (2007). Long-term potentiation in the amygdala: A cellular mechanism of fear learning and memory. *Neuropharmacology* **52**, 215–227.

Simpson, J. R. Jr., Drevets, W. C., Snyder, A. Z., Gusnard, D. A., and Raichle, M. E. (2001). Emotion-induced changes in human medial prefrontal cortex: II. During anticipatory anxiety. *Proc. Natl. Acad. Sci. U S A.* **98**, 688–693.

Snell, L. D., Nunley, K. R., Lickteig, R. L., Browning, M. D., Tabakoff, B., and Hoffman, P. L. (1996). Regional and subunit specific changes in NMDA receptor mRNA and immunoreactivity in mouse brain following chronic ethanol ingestion. *Brain Res. Mol. Brain Res.* **40**, 71–78.

Song, J. *et al.*, Shen, G., Greenfield, L. J. Jr., and Tietz, E. I. (2007). Benzodiazepine withdrawal-induced glutamatergic plasticity involves upregulation of GluR1-containing AMPA receptors in hippocampal CA1 neurons. *J Pharmacol Exp Ther.* **322**(2), 569–581.

Sonner, J. M., Li, J., Kandel, L., and Eger, E. I. 2nd (1998). Ethanol concentrations approaching minimum alveolar anesthetic concentration are required to suppress learning in a fear-potentiated startle paradigm in rats. *Anesth. Analg.* **87**, 1398–1403.

Stefani, A., Chen, Q., Flores-Hernandez, J., Jiao, Y., Reiner, A., and Surmeier, D. J. (1998). Physiological and molecular properties of AMPA/Kainate receptors expressed by striatal medium spiny neurons. *Dev. Neurosci.* **20**, 242–252.

Stephens, D. N., Brown, G., Duka, T., and Ripley, T. L. (2001). Impaired fear conditioning but enhanced seizure sensitivity in rats given repeated experience of withdrawal from alcohol. *Eur. J. Neurosci.* **14**, 2023–2031.

Stephens, D. N., Ripley, T. L., Borlikova, G., Schubert, M., Albrecht, D., Hogarth, L., and Duka, T. (2005). Repeated ethanol exposure and withdrawal impairs human fear conditioning and depresses long-term potentiation in rat amygdala and hippocampus. *Biol. Psychiatry* **58**, 392–400.

Sun, H., Zhao, H., Sharpe, G. M., Arrick, D. M., and Mayhan, W. G. (2008). Influence of chronic alcohol consumption on inward rectifier potassium channels in cerebral arterioles. *Microvasc Res.* **75**, 367–372.

Szinyei, C., Narayanan, R. T., and Pape, H. C. (2007). Plasticity of inhibitory synaptic network interactions in the lateral amygdala upon fear conditioning in mice. *Eur. J. Neurosci.* **25**, 1205–1211.

Thevos, A. K., Johnston, A. L., Latham, P. K., Randall, C. L., Adinoff, B., and Malcolm, R. (1991). Symptoms of anxiety in inpatient alcoholics with and without DSM-III-R anxiety diagnoses. *Alcohol. Clin. Exp. Res.* **15**, 102–105.

Tong, G., Malenka, R. C., and Nicoll, R. A. (1996). Long-term potentiation in cultures of single hippocampal granule cells: A presynaptic form of plasticity. *Neuron* **16**, 1147–1157.

Tsvetkov, E., Carlezon, W. A., Benes, F. M., Kandel, E. R., and Bolshakov, V. Y. (2002). Fear conditioning occludes LTP-induced presynaptic enhancement of synaptic transmission in the cortical pathway to the lateral amygdala. *Neuron* **34**, 289–300.

Ungless, M. A., Whistler, J. L., Malenka, R. C., and Bonci, A. (2001). Single cocaine exposure *in vivo* induces long-term potentiation in dopamine neurons. *Nature* **411**, 583–587.

Valdez, G. R., Roberts, A. J., Chan, K., Davis, H., Brennan, M., Zorrilla, E. P., and Koob, G. F. (2002). Increased ethanol self-administration and anxiety-like behavior during acute ethanol withdrawal and protracted abstinence: Regulation by corticotropin-releasing factor. *Alcohol. Clin. Exp. Res.* **26**, 1494–1501.

Valenzuela, C. F., Bhave, S., Hoffman, P., and Harris, R. A. (1998). Acute effects of ethanol on pharmacologically isolated kainate receptors in cerebellar granule neurons: Comparison with NMDA and AMPA receptors. *J. Neurochem.* **71**, 1777–1780.

Valles, S., Felipo, V., Montoliu, C., and Guerri, C. (1995). Alcohol exposure during brain development reduces 3H-MK-801 binding and enhances metabotropic-glutamate receptor-stimulated phosphoinositide hydrolysis in rat hippocampus. *Life Sci.* **56**, 1373–1383.

Van Sickle, B. J., Xiang, K., and Tietz, E. I. (2004). Transient plasticity of hippocampal CA1 neuron glutamate receptors contributes to benzodiazepine withdrawal-anxiety. *Neuropsychopharmacology* **29**, 1994–2006.

Walker, D. L., and Davis, M. (1997). Double dissociation between the involvement of the bed nucleus of the stria terminalis and the central nucleus of the amygdala in startle increases produced by conditioned versus unconditioned fear. *J. Neurosci.* **17**, 9375–9383.

Washburn, M. S., and Moises, H. C. (1992). Electrophysiological and morphological properties of rat basolateral amygdaloid neurons *in vitro*. *J. Neurosci.* **12**, 4066–4079.

Watson, W. P., and Little, H. J. (1999). Correlation between increases in dihydropyridine binding *in vivo* and behavioural signs of ethanol withdrawal in mice. *Alcohol Alcohol.* **34**, 35–42.

Weiner, J. L., Dunwiddie, T. V., and Valenzuela, C. F. (1999). Ethanol inhibition of synaptically evoked kainate responses in rat hippocampal CA3 pyramidal neurons. *Mol. Pharmacol.* **56**, 85–90.

Whittington, M. A., and Little, H. J. (1990). Patterns of changes in field potentials in the isolated hippocampal slice on withdrawal from chronic ethanol treatment of mice *in vivo*. *Brain Res.* **523**, 237–244.

Whittington, M. A., and Little, H. J. (1993). Changes in voltage-operated calcium channels modify ethanol withdrawal hyperexcitability in mouse hippocampal slices. *Exp. Physiol.* **78**, 347–370.

Wilson, N. R., Kang, J., Hueske, E. V., Leung, T., Varoqui, H., *et al.* (2005). Presynaptic regulation of quantal size by the vesicular glutamate transporter VGLUT1. *J. Neurosci.* **25**, 6221–6234.

Xiang, Z., Greenwood, A. C., Kairiss, E. W., and Brown, T. H. (1994). Quantal mechanism of long-term potentiation in hippocampal mossy-fiber synapses, *J. Neurophysiol.* **71**, 2552–2556.

Yang, L., Long, C., Evans, M. S., and Faingold, C. (2002). Ethanol withdrawal results in aberrant membrane properties and synaptic responses in periaqueductal gray neurons associated with seizure susceptibility. *Brain Res.* **957**, 99–108.

Yu, S. Y., Wu, D. C., Liu, L., Ge, Y., and Wang, Y. T. (2008). Role of AMPA receptor trafficking in NMDA receptor-dependent synaptic plasticity in the rat lateral amygdala. *J. Neurochem.* **106**, 889–899.

Zhou, F. C., Sahr, R. N., Sari, Y., and Behbahani, K. (2006). Glutamate and dopamine synaptic terminals in extended amygdala after 14-week chronic alcohol drinking in inbred alcohol-preferring rats. *Alcohol* **39**, 39–49.

Zhu, W., Bie, B., and Pan, Z. Z. (2007). Involvement of non-NMDA glutamate receptors in central amygdala in synaptic actions of ethanol and ethanol-induced reward behavior. *J Neurosci* **27**, 289–298.

Ziskind-Conhaim, L., Gao, B. X., and Hinckley, C. (2003). Ethanol dual modulatory actions on spontaneous postsynaptic currents in spinal motoneurons. *J. Neurophysiol.* **89**, 806–813.

ETHANOL ACTION ON DOPAMINERGIC NEURONS IN THE VENTRAL TEGMENTAL AREA: INTERACTION WITH INTRINSIC ION CHANNELS AND NEUROTRANSMITTER INPUTS

Hitoshi Morikawa and Richard A. Morrisett

Waggoner Center for Alcohol and Addiction Research, University of Texas at Austin, Austin, TX 78712, USA

The dopaminergic system originating in the midbrain ventral tegmental area (VTA) has been extensively studied over the past decades as a critical neural substrate involved in the development of alcoholism and addiction to other drugs of abuse. Accumulating evidence indicates that ethanol modulates the functional output of this system by directly affecting the firing activity of VTA dopamine neurons, whereas withdrawal from chronic ethanol exposure leads to a reduction in the functional output of these neurons. This chapter will provide an update on the mechanistic investigations of the acute ethanol action on dopamine neuron activity and the neuroadaptations/plasticities in the VTA produced by previous ethanol experience.

INTERNATIONAL REVIEW OF NEUROBIOLOGY, VOL. 91
DOI: 10.1016/S0074-7742(10)91008-8

I. Introduction

In addressing the neurobiological mechanisms underlying alcohol drinking and the development of alcohol addiction (alcoholism), one can broadly ask the following two types of questions: (1) Why do we drink and why do we keep drinking? (2) How do we get drunk/intoxicated? For the latter question, studying the action of ethanol in the cerebellum, for example, would shed light on the neurobiology of motor incoordination, which causes potentially life-threatening impairment, as occurs when driving under the influence of alcohol. However, depending on the dosage and rate of consumption, ethanol affects diverse brain structures subserving different behavioral and cognitive functions, leading to a wide array of drunken/intoxicated states ranging from sedation and impaired memory to euphoria, which could drive alcohol drinking, especially during the initial stage. Conversely, withdrawal from chronic ethanol consumption results in a "negative affective state" characterized by dysphoria, anhedonia, and depression, which persists even after somatic withdrawal symptoms have subsided (Koob, 2009; Trevisan *et al.*, 1998). Thus one might simply argue that part of the answer to the first question can be found in that to the second question ("We drink because we like the state of being drunk and we keep drinking because we want to avoid the affective symptoms of withdrawal"). But how so in neurobiological terms? In this review chapter, we will focus on the mesolimbic dopamine (DA) system, which comprises DA neurons in the midbrain ventral tegmental area (VTA) and their projection targets (the nucleus accumbens (NAc) and other limbic structures), in order to provide insights into the critical components of these two interrelated questions.

II. The Mesolimbic DA System and Alcohol Drinking

Multiple lines of evidence implicate the mesolimbic DA system in the behavioral actions of ethanol. For example, neurotoxin lesioning of DA terminals in the NAc has been shown to suppress the acquisition of ethanol-taking behaviors, measured with a two-bottle choice or self-administration paradigms (Ikemoto *et al.*, 1997; Rassnick *et al.*, 1993). A number of studies have demonstrated that DA receptor blockade, either systemically or locally in the NAc, suppresses ethanol consumption/self-administration (Dyr *et al.*, 1993; Gonzales *et al.*, 2004; Hodge *et al.*, 1997; Myers and Robinson, 1999; Pfeffer and Samson, 1986, 1988; Rassnick *et al.*, 1992; Samson *et al.*, 1993). Furthermore, intra-VTA infusion of the DA D2 receptor agonist quinpirole, which would inhibit the firing activity of DA neurons, has been shown to attenuate ethanol consumption/self-administration (Hodge *et al.*, 1993; Nowak *et al.*, 2000). Interestingly, expression of ethanol-induced conditioned place preference is also decreased by inhibition of DA neurons via intra-VTA infusion of a gamma-aminobutyric acid B receptor ($GABA_B$ receptor) agonist, suggesting the involvement of DA neurons in the memory of ethanol-associated environmental stimuli (Bechtholt and Cunningham, 2005). More direct evidence for the role of VTA DA neurons in ethanol reinforcement comes from a series of studies demonstrating self-administration of ethanol directly into the VTA, which is blocked by co-administration of the D2 agonist quinpirole (Gatto *et al.*, 1994; Rodd *et al.*, 2005c, 2004; Rodd-Henricks *et al.*, 2003). Finally, a human study has reported that pharmacological blockade of DA synthesis can reduce ethanol-induced euphoria and behavioral stimulation (Ahlenius *et al.*, 1973).

It is well established that systemic administration of ethanol, either passively or by self-administration, produces an increase in DA levels in the NAc (Imperato and Di Chiara, 1986; Weiss *et al.*, 1993). This effect is thought to be caused by ethanol action in the VTA and not because of DA uptake inhibition in the NAc (Budygin *et al.*, 2001; Yim and Gonzales, 2000). Indeed, administration of ethanol directly into the VTA leads to DA release in the NAc (Ding *et al.*, 2009a; but see Ericson *et al.*, 2003). Consistent with these observations, it has been shown that ethanol stimulates the firing activity of VTA DA neurons both *in vivo* (Foddai *et al.*, 2004; Gessa *et al.*, 1985; Mereu *et al.*, 1984) and *in vitro* (Brodie *et al.*, 1999b, 1990). Accordingly, it is widely believed that ethanol-induced DA neuron stimulation and the resulting DA release in the NAc provide the neural substrate mediating the reinforcing and rewarding actions ethanol, that is, learning of the behaviors and environmental stimuli associated with ethanol consumption and, perhaps, perception of an aspect of the ethanol-induced interoceptive state ("euphoria").

In the following sections, we will first describe the regulation and function of VTA DA neurons and DA output in general. Then, we will discuss the neural mechanisms underlying ethanol actions on DA neuron activity and the neuroadaptations/plasticities in DA neurons produced by previous ethanol exposure. It should be noted that ethanol also produces various effects in the NAc and other DA projection areas that play important roles in the behavioral actions of ethanol. The readers of this chapter should refer to some recent reviews on this subject (Ron and Jurd, 2005; Zhang *et al.*, 2006).

III. Intrinsic and Synaptic Mechanisms Controlling DA Neuron Activity

A. Two Modes of DA Neuron Firing and Their Functions

DA neurons display two distinct modes of action potential (AP) firing *in vivo*: (1) tonic, regular/irregular single-spike activity at 1–5 Hz, and (2) phasic bursts comprising 2–10 APs at 10–30 Hz, which are often followed by a depression of activity (Fig. 1A) (Grace and Bunney, 1984a,b; Hyland *et al.*, 2002; Overton and Clark, 1997). In contrast, DA neurons show uniform pacemaker-like firing (0.5–5 Hz) in an *in vitro* brain slice preparation or even in an acutely dissociated preparation, indicating that it reflects the intrinsic activity of these neurons removed from the influence of active synaptic inputs (Grace and Onn, 1989; Hainsworth *et al.*, 1991; Lacey, 1993; Puopolo *et al.*, 2007). However, recent studies demonstrate that phasic burst firing can be reproduced in brain slices by

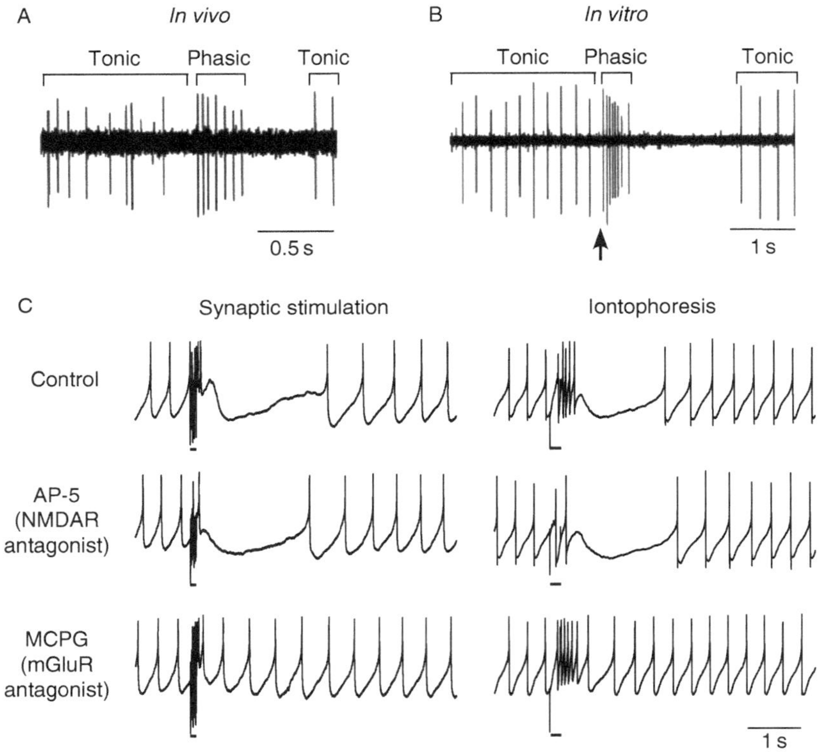

FIG. 1. Tonic and phasic DA neuron firing. (A) *In vivo* single-unit recording of DA neuron firing in an anesthetized rat. A spontaneous burst of APs is shown in the background of low-frequency tonic firing. There is a transient pause of activity before tonic firing recommences. Adapted from Overton and Clark (1997). (B) Loose-patch extracellular recording of DA neuron firing in a rat midbrain slice. Iontophoretic application of aspartate was made at the time indicated by the arrow (50 ms). (C) Traces illustrating the effects of the NMDA receptor antagonist AP-5 and the mGluR antagonist MCPG on the phasic burst–pause type of activity evoked by repetitive stimulation of glutamatergic afferents (eight stimuli at 66 Hz; left traces) or aspartate iontophoresis (200 ms; right traces). Also note the tonic pacemaker-like firing before and after the burst–pause. Adapted from Morikawa *et al.* (2003).

electrical stimulation of glutamatergic afferents or by rapid and focal application of glutamate or aspartate onto the recorded DA neuron (Fig. 1B and C) (Blythe *et al.*, 2009; Deister *et al.*, 2009; Morikawa *et al.*, 2003).

In awake, behaving animals (primates and rodents), unexpected presentation of primary rewards or reward-predicting stimuli triggers a switch from tonic single-spike activity to transient bursts (~200 ms) in a large population of DA neurons (55–80%) in both the VTA and substantia nigra pars compacta (SNc) (Pan *et al.*, 2005; Schultz, 1998). These DA neuron bursts are thought to generate phasic DA signals in projection areas. In line with this idea, intra-VTA infusion of the *N*-methyl-D-aspartate (NMDA) receptor antagonist AP-5, which would suppress DA neuron burst firing (see below), has been shown to inhibit rapid DA transients (~2–3 s) in the NAc detected with voltammetry (Sombers *et al.*, 2009). These phasic DA signals will promote the induction of long-term synaptic plasticity in DA projection areas (Arbuthnott and Wickens, 1996; Mahon *et al.*, 2004; Wickens *et al.*, 2003; Wolf *et al.*, 2004), thereby providing the "teaching signal" for animals to learn the valence of environmental stimuli and behaviors associated with rewards (Montague *et al.*, 2004; Schultz, 1998). In addition, phasic DA transients may also act as an incentive salience signal to temporarily direct cognitive and behavioral resources toward biologically significant environmental stimuli that elicit DA neuron burst firing, thereby attracting attention and triggering approach behavior toward those stimuli (Berridge, 2007; Nicola, 2007; Niv, 2007). This effect of phasic DA is most likely achieved through its immediate short-term actions on synaptic transmission and neuronal excitability in DA projection areas, resulting in an enhanced signal-to-noise ratio in the firing activity among neurons (Horvitz, 2002; Nicola *et al.*, 2004).

The tonic, asynchronous single-spike activity of DA neurons gives rise to global DA tone in the projection areas (Venton *et al.*, 2003). It has been suggested that there are "silent" DA neurons *in vivo* and that a proportion of active DA neurons, termed the "population activity," regulate tonic DA levels measured with microdialysis (Floresco *et al.*, 2003; but see Dai and Tepper, 1998). Interestingly, intra-VTA infusion of the NMDA receptor antagonist AP-5, which inhibits phasic DA signals in the NAc (Sombers *et al.*, 2009), can also reduce tonic DA levels in the NAc (Karreman *et al.*, 1996; Kretschmer, 1999) and attenuate the slow rise (~1 min) in NAc DA levels following intravenous injection of cocaine (Sombers *et al.*, 2009). These observations raise the possibility that tonic DA may, at least partially, reflect integration of phasic DA transients over time (Niv *et al.*, 2007; Stuber *et al.*, 2005). Although the precise role of tonic DA signals remains to be determined, they may act to motivate and energize the performance of already-learned goal-directed and habitual behaviors via short-term modulatory effects on synaptic transmission and neuronal excitability (Cagniard *et al.*, 2006; Goto and Grace, 2005; Niv, 2007; Salamone *et al.*, 2007; Yin *et al.*, 2009).

Intriguingly, animals that have learned to self-administer cocaine regulate their operant behavior (e.g., lever pressing) as if to maintain tonic DA levels in the NAc within a certain range (Pettit and Justice, 1989; Stuber *et al.*, 2005; Weiss *et al.*, 1992a; Wise *et al.*, 1995). Hence, decline in DA levels below a "threshold" will trigger operant responding, while animals will stop responding when DA levels are elevated beyond a certain upper limit. This is also supported by the well-known bell-shaped dose–response function of drug self-administration. Here, animals will compensate for an increase in unitary dose, where each operant response would produce a larger increase in DA levels, by a decrease in the rate of responding if the unitary doses are on the descending limb of the bell-shaped dose–response function (Gonzales *et al.*, 2004; Pettit and Justice, 1991; Piazza *et al.*, 2000). Furthermore, certain doses of DA receptor antagonists can cause an increase in operant responding in a manner analogous to a reduction in unitary dose (Bergman *et al.*, 1990; Caine and Koob, 1994; Gonzales *et al.*, 2004; Maldonado *et al.*, 1993). Therefore, there seems to be a mechanism to titrate DA levels, or more specifically the degree of DA receptor activation, in the NAc. This idea apparently cannot be accounted for by the behavior-facilitating function of DA described in the previous paragraphs. Although one might argue that animals are attempting to maintain an optimal "hedonic state" encoded by tonic DA levels (Wise, 2008), a further mechanistic explanation, including the actual neural substrate being sensed in maintaining tonic DA levels, seems warranted (Ahmed and Koob, 2005; Goto *et al.*, 2007; Tsibulsky and Norman, 1999). In this regard, it is well documented that tonic DA levels correlate with the self-report of "high" or "euphoria" in humans (Volkow *et al.*, 2004), although whether tonic DA actually mediates the subjective experience of pleasure ("hedonia hypothesis of DA") remains a matter of debate (Berridge, 2007; Wise, 2008).

B. Ionic Mechanisms Underlying Spontaneous Pacemaker Activity of DA Neurons

Numerous studies have investigated the roles of various ionic conductances in the generation of intrinsic pacemaker activity of DA neurons. It should be noted that the vast majority of these studies has focused on SNc DA neurons in the past, and it has been generally assumed that DA neurons possess rather homogeneous properties in both the SNc and the VTA. However, it has become increasingly evident in recent years that certain populations of VTA DA neurons, depending on their location and projection targets, exhibit electrophysiological and pharmacological profiles that do not conform to the conventional description of DA neurons (Cameron *et al.*, 1997; Ford *et al.*, 2006; Lammel *et al.*, 2008; Margolis *et al.*, 2006a, 2008b; Neuhoff *et al.*, 2002; Wolfart *et al.*, 2001). The three major criteria used by *in vitro* electrophysiologists to identify DA neurons are (1) low-frequency (<5 Hz) pacemaker firing with long-duration APs (~2–3 ms); (2) the presence of hyperpolarization-activated slow inward current (I_h), which mediates the characteristic depolarizing sag in response to membrane hyperpolarization; and (3) hyperpolarization/outward current produced by activation of D2-type DA autoreceptors (Grace and Onn, 1989; Johnson and North, 1992b; Lacey, 1993). Although most DA neurons in the lateral part of the VTA, likely projecting to the core and lateral shell regions of the NAc and to the basolateral amygdala (BLA), appear to display these properties (Ford *et al.*, 2006; Ikemoto, 2007; Labouebe *et al.*, 2007; Lammel *et al.*, 2008; Riegel and Williams, 2008; Wanat *et al.*, 2008; but see Margolis *et al.*, 2006b, 2008b), the readers of this chapter should be cautioned that the description of ionic mechanisms of DA neuron firing in this section would not be entirely valid for all VTA DA neurons, especially those in the medial VTA that projects to the prefrontal cortex (PFC) (Lammel *et al.*, 2008). In the following subsections, we will describe four major ion channels that have been shown to play critical roles in generating/controlling intrinsic DA neuron pacemaker activity.

1. *L-Type Voltage-Gated Ca^{2+} Channel*

Studies in the late 1980s have demonstrated that the spontaneous pacemaker firing of DA neurons can be completely suppressed by cadmium or cobalt, which inhibit all voltage-gated Ca^{2+} channels (Fujimura and Matsuda, 1989; Grace and Onn, 1989; Harris *et al.*, 1989). This is in sharp contrast to the Ca^{2+}-independent pacemaking mechanisms in other central nervous system (CNS) neurons (Bevan and Wilson, 1999; Llinas and Alonso, 1992; Raman and Bean, 1999). It has been further shown that dihydropyridine compounds that selectively inhibit L-type Ca^{2+} channels can block DA neuron pacemaker activity (Fig. 2) (Chan *et al.*, 2007; Mercuri *et al.*, 1994; Nedergaard *et al.*, 1993; Puopolo *et al.*, 2007). L-type Ca^{2+} channels expressed in DA neurons, encoded by the $Ca_V1.3$ gene, are unique in that they can be activated at relatively hyperpolarized membrane potentials (~-60 mV) and exhibit little inactivation (Chan *et al.*, 2007; Durante *et al.*, 2004), making them ideally suited to drive the subthreshold membrane potential oscillation underlying the pacemaker activity of these neurons. A modeling study has demonstrated that depolarization caused by L-type Ca^{2+} conductance combined with hyperpolarization mediated by Ca^{2+}-sensitive K^+ conductance is sufficient to produce intrinsic pacemaker activity (Wilson and Callaway, 2000). However, recent evidence calls this model into question and suggests that certain Ca^{2+}-independent mechanisms, for example, persistent Na^+ currents active at subthreshold membrane potentials, may also be

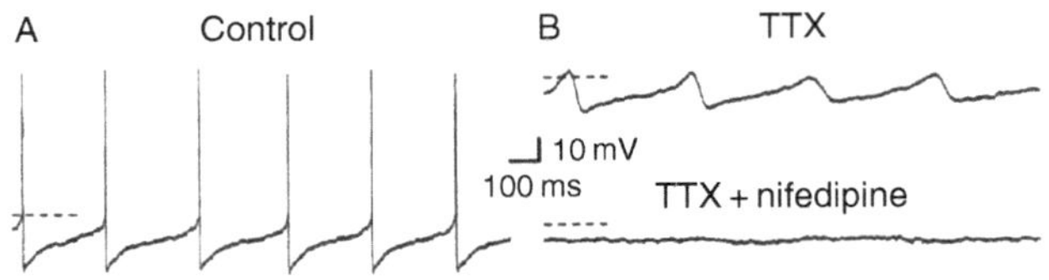

FIG. 2. Spontaneous activity of DA neurons. (A) Intracellular recording of spontaneous DA neuron firing in a brain slice. (B) The Na^+ channel blocker tetrodotoxin (TTX) abolished APs and revealed the spontaneous membrane potential oscillation, which was completely eliminated by the L-type Ca^{2+} channel antagonist nifedipine. Dotted lines indicate the AP threshold (−40 mV). Adapted from Nedergaard *et al.* (1993).

able to sustain DA neuron pacemaking (Chan *et al.*, 2007; Guzman *et al.*, 2009; Puopolo *et al.*, 2007; Putzier *et al.*, 2009).

2. *Small-Conductance Ca^{2+}-Sensitive K^+ Channel*

The bee venom peptide toxin apamin, which selectively blocks small-conductance Ca^{2+}-sensitive K^+ (SK) channels, reduces the large afterhyperpolarization (AHP), which follows each AP and dominates the first part of the interspike interval, in DA neurons (Ji *et al.*, 2009; Shepard and Bunney, 1991; Wolfart *et al.*, 2001). Here, Ca^{2+} entry during the subthreshold depolarization preceding each AP, together with Ca^{2+} influx triggered by the large, brief depolarization of the AP itself, elicits activation of SK-mediated outward K^+ current causing large hyperpolarization, that is, the AHP. There appear to be multiple sources of Ca^{2+} responsible for SK channel activation, including L-type Ca^{2+} channels (Chan *et al.*, 2007; Wilson and Callaway, 2000), T-type Ca^{2+} channels (Wolfart and Roeper, 2002), P/Q-type Ca^{2+} channels (Puopolo *et al.*, 2007), and Ca^{2+}-induced Ca^{2+} release from intracellular Ca^{2+} stores (Cui *et al.*, 2007; Wolfart and Roeper, 2002). It has been reported that SK channels do not play a major role in controlling the firing activity of VTA DA neurons, which have relatively low levels of SK channel expression compared to SNc neurons (Wolfart *et al.*, 2001).

3. *Hyperpolarization-Activated Cation (I_h) Channel*

I_h was first described as a current driving the pacemaker activity of cells in the sinoatrial node of the heart (Brown and Difrancesco, 1980; Noma and Irisawa, 1976). Since then, numerous studies have investigated the expression and function of I_h in neurons, where I_h contributes to the autonomous firing (Chan *et al.*, 2004; Luthi and McCormick, 1998; Maccaferri and McBain, 1996) and is also involved in setting the resting membrane potential and resting membrane conductance, thereby controlling the postsynaptic response to synaptic inputs (George *et al.*, 2009; Magee, 1998; Poolos *et al.*, 2002). Four genes encoding I_h channels, termed HCN (for hyperpolarization-activated cyclic nucleotide-gated) 1–4, have been identified, each having distinct biophysical properties and expression profiles (Kaupp and Seifert, 2001; Robinson and Siegelbaum, 2003). I_h channels have a mixed Na^+/K^+ conductance and are activated by hyperpolarization beyond –50 to –60 mV, producing slowly developing inward currents. As described earlier in this section, the presence of I_h has been widely used as an electrophysiological marker to identify DA neurons in the VTA/SNc. The activation of I_h takes place over hundreds of milliseconds in DA neurons, which express HCN2–4 but lack fast-activating HCN1 (Franz *et al.*, 2000; Notomi and Shigemoto, 2004; Santoro *et al.*, 2000). In spontaneously firing DA neurons, the AHP following each AP activates I_h, which accelerates depolarization back to the AP threshold during the interspike interval. Accordingly, inhibiting I_h with ZD7288, a relatively specific I_h blocker (Chevaleyre and Castillo, 2002; Harris and Constanti, 1995), prolongs AHPs and reduces the frequency of spontaneous DA neuron firing in both the VTA and SNc (Beckstead and Phillips, 2009; Okamoto *et al.*, 2006; Seutin *et al.*, 2001). It should be noted that the role of I_h in controlling DA neuron pacemaker activity may vary depending on the age and species of animals used, as well as on the location and neurochemical characteristics of DA neurons recorded (Chan *et al.*, 2007; McDaid *et al.*, 2008; Neuhoff *et al.*, 2002).

4. *A-Type K^+ (I_A) Channel*

A number of studies have characterized the voltage-gated transient K^+ current, termed I_A, in DA neurons (Khaliq and Bean, 2008; Koyama and Appel, 2006a; Liss *et al.*, 2001; Liu *et al.*, 1994; Silva *et al.*, 1990). I_A is a rapidly inactivating, transient K^+ current that is activated by depolarization (above –50 to –60 mV) from hyperpolarized membrane potentials where inactivation is removed. Activation is extremely rapid (<10 ms), whereas inactivation and the recovery from inactivation occur in the order of tens of milliseconds. I_A generally acts to exert negative control over the frequency of relatively slow, repetitive firing (1–100 Hz) (Connor and Stevens, 1971; Rudy, 1988). Indeed, blocking I_A with 4-aminopyridine (4-AP) has been shown to increase the firing frequency of DA neurons (Hahn *et al.*, 2003; Koyama and Appel, 2006a; Liss *et al.*, 2001; Yang *et al.*, 2001). It is thought that the large AHP removes the inactivation of I_A, which becomes inactivated by the end of each AP. Subsequently, I_A will be transiently activated during the depolarizing phase of AHP. The resulting outward (hyperpolarizing) current will slow down the return of the membrane potential to the AP threshold, thereby prolonging the interspike interval. Therefore, I_A and I_h, which are both active in the subthreshold voltage range during the interspike interval, play opposing roles in controlling tonic DA neuron firing.

C. The Influence of Neurotransmitter Inputs on DA Neuron Activity

The firing pattern of DA neurons *in vivo* significantly deviates from the uniform pacemaker activity observed in brain slices. In particular, DA neurons exhibit phasic bursts, that is, transient increases in the firing frequency (~10–30 Hz) for ~200 ms, in response to unexpected presentation of primary rewards or reward-predicting stimuli. These reward-predicting stimuli that elicit DA neuron bursts can be of a variety of sensory modalities and characteristics, which illustrates the wide divergence of neurotransmitter inputs that are integrated to regulate the activity of DA neurons (Geisler and Zahm, 2005). Here, we will summarize the sources and potential roles of various neurotransmitter inputs into the VTA.

1. *Glutamate*

Feedback input from the PFC provides the sole major cortical projection to the VTA (Geisler *et al.*, 2007; Sesack and Pickel, 1992). Interestingly, it has been shown that excitatory PFC afferents preferentially target DA neurons that project back to the PFC and GABAergic neurons that project to the NAc, demonstrating considerable target specificity of PFC input into the VTA (Carr and Sesack, 2000).

Previous studies have identified a limited number of subcortical structures providing glutamatergic afferents into the VTA. These include the pedunculopontine (PPTg) and laterodorsal tegmental (LDTg) nuclei (Charara *et al.*, 1996; Clements *et al.*, 1991; Lavoie and Parent, 1994a), the bed nucleus of the stria terminals (BNST) (Georges and Aston-Jones, 2002), and the superior colliculus (SC) (Comoli *et al.*, 2003; Dommett *et al.*, 2005). However, a recent thorough analysis using retrograde tract tracing combined with *in situ* hybridization for vesicular glutamate transporters has revealed a large number of previously unknown subcortical sources of glutamatergic inputs into the VTA, such as the lateral hypothalamic and preoptic areas, periaqueductal gray, and the dorsal and median raphe (Geisler *et al.*, 2007). Furthermore, glutamatergic synapses from these subcortical structures outnumber those from the PFC (Geisler *et al.*, 2007; Omelchenko and Sesack, 2007). These diverse sources of glutamatergic afferents may enable DA neurons to respond to a wide range of environmental stimuli. For example, electrophysiological evidence implicates glutamatergic inputs primarily from the SC and PPTg in mediating DA neuron burst responses to visual and auditory cues, respectively (Comoli *et al.*, 2003; Pan and Hyland, 2005).

Glutamatergic inputs activate α-amino-3-hydroxy-5-methyl-4-isoxazolepropionic acid (AMPA)- and NMDA-type ionotropic glutamate receptors in DA neurons. Accordingly, iontophoretic application of both AMPA and NMDA receptor agonists can stimulate DA neuron firing (Christoffersen and Meltzer, 1995; Zhang *et al.*, 1997). A number of *in vivo* electrophysiological studies in anesthetized animals have reported that DA neuron bursts occurring spontaneously or those triggered by electrical stimulation of the PFC, PPTg, or subthalamic nucleus are suppressed by local application of NMDA receptor antagonists but not by AMPA receptor antagonists (Chergui *et al.*, 1994, 1993; Christoffersen and Meltzer, 1995; Overton and Clark, 1992; Tong *et al.*, 1996). In line with this, it has been shown recently that genetic inactivation of NMDA receptors selectively in DA neurons impairs spontaneous bursts as well as those evoked by PPN stimulation (Zweifel *et al.*, 2009). Furthermore, DA neuron bursts triggered by electrical stimulation of glutamatergic afferents or iontophoretic application of glutamate/aspartate in brain slices are also blocked by NMDA receptor antagonists (Deister *et al.*, 2009; Morikawa *et al.*, 2003). These converging lines of evidence indicate the critical role of NMDA receptors in the generation of DA neuron bursts. However, it should be noted that AMPA receptor antagonists have also been shown to attenuate DA neuron burst firing evoked by stimulation of glutamatergic inputs both *in vivo* and *in vitro* (Blythe *et al.*, 2007; Di Loreto *et al.*, 1992; Georges and Aston-Jones, 2002). Therefore, the exact contribution of these two types of ionotropic glutamate receptors to DA neuron bursting under different conditions, and maybe in different populations of DA neurons, remains to be fully determined.

Glutamate also activates metabotropic glutamate receptors (mGluRs), mainly type 1 mGluR ($mGluR_1$), in DA neurons (Hubert *et al.*, 2001; Testa *et al.*, 1994). Tonic activation of mGluRs produces membrane depolarization that is mediated, at least partly, by opening of a nonselective cationic conductance (Guatteo *et al.*, 1999; Tozzi *et al.*, 2003). In contrast, rapid activation of mGluRs leads to membrane hyperpolarization via opening of SK channels following Ca^{2+} release from intracellular stores (Fiorillo and Williams, 1998; Morikawa *et al.*, 2003). This mGluR-induced hyperpolarization mediates the transient pause of activity following NMDA-induced bursting observed in brain slices (Fig. 1B and C), and is desensitized by tonic mGluR activation through a mechanism likely involving Ca^{2+} store depletion. It has recently been reported that tonic mGluR activation causes amplification of AP-evoked Ca^{2+} signals, thereby driving activity-dependent synaptic plasticity of NMDA receptor–mediated transmission (Cui *et al.*, 2007; Harnett *et al.*, 2009).

2. *GABA*

The powerful influence of GABAergic inhibition on DA neuron activity is illustrated by the observation that over 70% of synapses onto SNc DA neurons are GABAergic (Bolam and Smith, 1990; Tepper and Lee, 2007), although the proportion of GABAergic synapses seems to be lower in the VTA (Smith *et al.*, 1996). There are major GABAergic feedback projections from the NAc and the ventral pallidum into the VTA (Conrad and Pfaff, 1976; Kalivas *et al.*, 1993; Walaas and Fonnum, 1980; Zahm, 1989). Interestingly, the feedback loops between the VTA/SNc and NAc/striatum form a cascading spiral of projections (Haber *et al.*, 2000, 1990; Heimer *et al.*, 1991). Here, projections from the NAc shell, that is, the medial part of the NAc, to the VTA influence DA neurons that in turn project to the more lateral core of the NAc, which provides projections to the SNc that projects to the dorsal striatum. This anatomical arrangement has formed a conceptual framework to account for the transition from a goal-directed action to a stimulus-response habit during the progression of addiction, in which more lateral part of the VTA/SNc complex, together with more dorsal part of the striatum, are gradually recruited (Belin and Everitt, 2008; Everitt and Robbins, 2005).

DA neurons also receive GABAergic inputs from local GABA neurons within the VTA (Johnson and North, 1992b; Omelchenko and Sesack, 2009; Phillipson, 1979a). Furthermore, a region located at the caudal tail of the VTA, termed rostromedial tegmental nucleus (RMTg), has been recently shown to be an important source of GABAergic input to the VTA and SNc (Jhou *et al.*, 2009a, b; Kaufling *et al.*, 2009).

Both $GABA_A$ and $GABA_B$ receptors mediate the inhibitory action of GABA on DA neurons (Brazhnik *et al.*, 2008; Johnson and North, 1992b). $GABA_B$ receptor–mediated inhibition is achieved through activation of G protein–gated inwardly rectifying K^+ (GIRK) channels (Labouebe *et al.*, 2007; Lacey *et al.*, 1988). Furthermore, activation of presynaptic $GABA_B$ receptors inhibits glutamate and GABA release onto DA neurons (Hausser and Yung, 1994; Manzoni and Williams, 1999). GABAergic inhibition via both $GABA_A$ and $GABA_B$ receptors can suppress burst firing of DA neurons (Erhardt *et al.*, 2002; Paladini *et al.*, 1999), and recent evidence indicates that a disinhibitory mechanism, that is, transient removal of GABAergic inhibition, may contribute to the generation of phasic DA neuron bursts in behaving animals (Jhou *et al.*, 2009a; Matsumoto and Hikosaka, 2007; Tepper and Lee, 2007).

3. *Glycine*

The VTA receives sparse glycinergic input (Rampon *et al.*, 1996). Exogenous application of glycine has been shown to activate Cl^--mediated currents in dissociated VTA neurons, indicating the presence of functional ionotropic glycine receptors (Ye, 2000; Ye *et al.*, 1998). Furthermore, glycine-induced presynaptic inhibition of GABA release has been reported in VTA DA neurons, an effect that may lead to disinhibition of the firing activity (Ye *et al.*, 2004).

4. *Dopamine*

Dendrodendritic synapses, as well as a small number of axodendritic synapses, formed between DA neurons have been observed with electron microscopy in the VTA (Bayer and Pickel, 1990). Vesicular monoamine transporter 2 (VMAT2)-positive vesicles, which would underlie Ca^{2+}-dependent vesicular DA release from the somoatodendritic area (Chen and Rice, 2001; Kita *et al.*, 2009; Rice *et al.*, 1997), have also been detected in DA neuron dendrites (Nirenberg *et al.*, 1996). Released DA exerts feedback inhibition of DA neuron activity via activation of D2 autoreceptors causing GIRK-mediated hyperpolarization (Groves *et al.*, 1975; Lacey *et al.*, 1987). Indeed, D2 receptor–mediated inhibitory postsynaptic currents (IPSCs) elicited by electrical stimulation of dendrodendritic synapses have been recently reported in DA neurons of the SNc and VTA (Beckstead *et al.*, 2007, 2004). In addition, DA inhibits glutamate release and facilitates GABA release onto DA neurons via activation of presynaptic D2 and D1 receptors, respectively (Cameron and Williams, 1993; Koga and Momiyama, 2000). Postsynaptic D5 receptor-mediated slow potentiation of NMDA EPSCs has also been reported recently, suggesting that DA may promote burst firing of DA neurons (Schilstrom *et al.*, 2006).

5. *Norepinephrine*

The locus coeruleus and other noradrenergic nuclei in the medulla give rise to norepinephrine (NE)-containing terminals in the VTA (Mejias-Aponte *et al.*, 2009; Phillipson, 1979b). Overall, NE exerts excitatory control over DA neuron activity via activation of $\alpha 1$ adrenergic receptors causing membrane depolarization or suppressing the mGluR-induced pause (Grenhoff *et al.*, 1995; Paladini *et al.*, 2001). However, transient activation of $\alpha 1$ adrenergic receptors produces SK-mediated membrane hyperpolarization (Paladini and Williams, 2004). NE can also bind to D2 DA receptors and cause membrane hyperpolarization in DA neurons (Grenhoff *et al.*, 1995; Guiard *et al.*, 2008).

6. *Serotonin*

Serotoninergic neurons originating in the midbrain dorsal and median raphe nuclei provide dense projections to the VTA (Azmitia and Segal, 1978; Herve *et al.*, 1987; Parent *et al.*, 1981; Phillipson, 1979b). Different subtypes of serotonin (5-HT) receptors are present in the VTA, such as 5-HT_{1B}, 5-HT_{2A}, and 5-HT_{2C} receptors (Bubar and Cunningham, 2007; Doherty and Pickel, 2000; Pazos and

Palacios, 1985; Waeber *et al.*, 1989a). It has been shown that 5-HT application produces membrane depolarization and suppresses the mGluR-mediated pause in DA neurons, most likely via activation of 5-HT_{2A} receptors (Paolucci *et al.*, 2003; Pessia *et al.*, 1994). Furthermore, activation of presynaptic 5-HT_{1B} receptors on GABAergic terminals leads to a reduction in GABA release, an effect that may also contribute to the excitatory action of 5-HT on DA neuron activity (Johnson *et al.*, 1992; Morikawa *et al.*, 2000). In contrast, activation of 5-HT_{2C} receptors on local GABAergic neurons is thought to mediate inhibition of DA neuron output (Di Matteo *et al.*, 2001; Pessia *et al.*, 1994). 5-HT-induced depression of glutamate release onto DA neurons has also been reported (Jones and Kauer, 1999).

Numerous studies have implicated 5-HT_3 receptors, the only ionotropic receptor among 5-HT receptor subtypes, in the VTA in stimulation of the functional output of the mesolimbic DA system (reviewed by Engleman *et al.*, 2008). However, previous radioligand binding studies have failed to detect 5-HT_3 receptor expression in the VTA (Gehlert *et al.*, 1991; Waeber *et al.*, 1989b). Interestingly, 5-HT_3 receptors appear to be preferentially localized on nerve terminals (Miquel *et al.*, 2002). The exact mechanism underlying 5-HT_3 receptor–mediated stimulation of DA neurons (e.g., presynaptic vs. postsynaptic) remains to be determined.

7. *Acetylcholine*

Cholinergic projections to the VTA mainly derive from the PPTg and LDTg nuclei (Mena-Segovia *et al.*, 2008; Oakman *et al.*, 1995; Omelchenko and Sesack, 2005; Semba and Fibiger, 1992). It has been shown that glutamate or GABA is co-released with acetylcholine (ACh) from these projections (Clements *et al.*, 1991; Jia *et al.*, 2003; Lavoie and Parent, 1994b; Omelchenko and Sesack, 2006), although whether there is co-release from the same terminals appears to remain controversial (Mena-Segovia *et al.*, 2008). Cholinergic terminals in the VTA have been shown to preferentially innervate DA neurons that project to the NAc (Omelchenko and Sesack, 2006). DA neurons generally exhibit excitatory responses to ACh via activation of nicotinic and muscarinic ACh receptors.

Different subtypes of nicotinic ACh receptors (nAChRs), including α3–7 and β2–4, are expressed in DA neurons, although β2-containing nAChRs mediate the majority of postsynaptic nicotinic responses (Azam *et al.*, 2002; Champtiaux *et al.*, 2003; Wooltorton *et al.*, 2003). Activation of β2-containing nAChRs by the levels of nicotine experienced in smokers transiently excites DA neurons followed by desensitization in seconds to minutes (Pidoplichko *et al.*, 1997). Recent evidence highlights the important roles of presynaptic nAChRs in regulating DA neuron activity, where persistent activation of presynaptic α7-containing nAChRs enhances glutamatergic inputs onto DA neurons, while desensitization of presynaptic β2-containing nAChRs on GABAergic terminals may lead to disinhibition (Mansvelder and McGehee, 2002).

Activation of postsynaptic M5 muscarinic ACh receptors causes excitation of DA neurons (Gronier and Rasmussen, 1998; Lacey *et al.*, 1990; Yamada *et al.*, 2003; Yeomans *et al.*, 2001), although rapid activation of these receptors can evoke SK-mediated membrane hyperpolarization via intracellular Ca^{2+} release (Fiorillo and Williams, 2000). Presynaptic inhibition of glutamate and GABA release by activation of mAChRs has also been reported (Grillner *et al.*, 2000, 1999).

8. *Opioids*

β-Endorphin-containing fibers originating from the arcuate nucleus of the hypothalamus constitute the major opioid input to the VTA (Mansour *et al.*, 1988). Enkephalin-containing terminals have also been detected in the VTA (Garzon and Pickel, 2002). Endorphins and enkephalins bind to both μ- and δ-opioid receptors. It is generally believed that μ-opioid receptor–mediated inhibition of local GABAergic neurons, which leads to excitation of DA neurons via a disinhibitory mechanism, is the predominant mode of action of these endogenous opioids, as well as exogenous opiate drugs, within the VTA (Johnson and North, 1992a,b). In line with this, both systemic and intra-VTA administrations of μ-opioid receptor agonists stimulate VTA DA neuron firing and NAc DA release (Di Chiara and Imperato, 1988; Latimer *et al.*, 1987; Leone *et al.*, 1991; Matthews and German, 1984; Spanagel *et al.*, 1992).

The VTA also contains terminals that release the κ-opioid receptor–selective opioid, dynorphin (Fallon *et al.*, 1985; Pickel *et al.*, 1993). These terminals originate from the NAc, the central nucleus of the amygdala, and the lateral hypothalamus (Fallon *et al.*, 1985; Meredith, 1999). Recent studies demonstrate that activation of κ-opioid receptors can directly inhibit a subset of VTA DA neurons (Margolis *et al.*, 2003, 2006a), an effect that may contribute to the aversive action of κ-opioid receptor agonists in the VTA (Bals-Kubik *et al.*, 1993).

A number of studies have demonstrated that activation of all three opioid receptors can produce presynaptic inhibition of glutamate and/or GABA release onto VTA DA neurons (Ford *et al.*, 2006; Manzoni and Williams, 1999; Margolis *et al.*, 2008a, 2005; Shoji *et al.*, 1999). These presynaptic opioid receptors may also play a significant role in regulating DA neuron activity.

9. *Other Neuropeptides*

Recent evidence highlights the importance of inputs containing corticotropin-releasing factor (CRF) and orexins in driving drug-seeking behavior and preference for drug-associated stimuli (Aston-Jones *et al.*, 2010; Narita *et al.*, 2006; Wise and Morales, 2010).

The BNST, the central nucleus of the amygdala, and the paraventricular nucleus of the hypothalamus are sources of CRF input to the VTA (Rodaros *et al.*, 2007), where CRF is co-released with glutamate or GABA (Tagliaferro and Morales, 2008). CRF also appears to be produced in certain DA neurons in the VTA (Korotkova *et al.*, 2006). Recent studies demonstrate that CRF regulates glutamatergic and GABAergic transmission onto DA neurons and their intrinsic pacemaker activity via multiple signaling cascades and receptor subtypes (Beckstead *et al.*, 2009; Korotkova *et al.*, 2006; Riegel and Williams, 2008; Ungless *et al.*, 2003; Wanat *et al.*, 2008).

Orexin input to the VTA arises from the lateral hypothalamus (Balcita-Pedicino and Sesack, 2007; Fadel and Deutch, 2002; Mileykovskiy *et al.*, 2005), which may co-release glutamate with orexins (Rosin *et al.*, 2003). The two types of orexins (orexin A and orexin B) can both directly activate VTA DA neurons (Korotkova *et al.*, 2003). Furthermore, orexin A has been shown to produce prolonged potentiation of NMDA receptor–mediated excitatory transmission onto DA neurons (Borgland *et al.*, 2006). These effects may underlie the increased release of DA in projection areas caused by intra-VTA orexin injection (Narita *et al.*, 2006; Vittoz and Berridge, 2006).

10. *Cannabinoids*

Recent evidence indicates that endocannabinoids released from DA neurons themselves inhibit both glutamate and GABA release onto DA neurons via activation of presynaptic CB1 receptors (Melis *et al.*, 2004a,b; Riegel and Lupica, 2004). Release of endocannabinoids appears to be triggered by a rise in cytosolic Ca^{2+} levels due to increased firing and bursting activity or activation of mGluRs causing Ca^{2+} mobilization from intracellular stores.

IV. Acute Ethanol Action on DA Neuron Activity

The original studies in the 1980s found that i.v. administration of ethanol (0.5–2 g/kg) dose-dependently increases the firing frequency of VTA and SNc DA neurons recorded *in vivo* using awake (unanesthetized), but pharmacologically paralyzed, rats (Gessa *et al.*, 1985; Mereu *et al.*, 1984). It was subsequently shown that ethanol (10–200 mM) increases the spontaneous firing frequency of VTA DA neurons in a concentration-dependent manner using an *in vitro* brain slice preparation (Brodie *et al.*, 1990). This stimulatory effect of ethanol can be observed when the influence of synaptic inputs are blocked by low Ca^{2+}/high Mg^{2+} solution (Brodie *et al.*, 1990) or by a cocktail of ionotropic neurotransmitter receptor antagonists (Okamoto *et al.*, 2006; but also see Xiao *et al.*, 2007), and even when DA neurons are dissociated into individual cells (Brodie *et al.*, 1999b). These observations demonstrate that ethanol can influence the intrinsic mechanisms underlying the spontaneous pacemaker activity of these neurons. However, several lines of evidence indicate that the effects of ethanol on neurotransmitter inputs, either via its action on synapses at DA neurons or that on the activity of the neurons providing those inputs, should also play a significant role in ethanol stimulation of DA neuron firing observed *in vivo*. First, although not quantified, Mereu *et al.* (1984) reported that ethanol promoted the occurrence of bursting in SNc DA neurons *in vivo* (Figs. 3A and 4). This has been confirmed in a recent study demonstrating that ethanol increases the percentage of spikes occurring in bursts in VTA DA neurons (Foddai *et al.*, 2004). Furthermore, it has been shown that i.v. administration of ethanol in awake rats leads to an increase in the frequency of phasic DA transients in the NAc detected with voltammetry (Cheer *et al.*, 2007; Robinson *et al.*, 2009), supporting the idea that ethanol promotes the occurrence of DA neuron bursts.

Second, the magnitude of the ethanol effect appears to be significantly larger *in vivo* than *in vitro*. For example, i.v. injection of 1 g/kg ethanol, which would produce 60–70 mM peak ethanol concentrations in the blood and brain (Robinson *et al.*, 2002, 2000), increases DA neuron firing frequency by ~60–70% *in vivo* (Gessa *et al.*, 1985; Mereu *et al.*, 1984), whereas similar concentrations of ethanol cause only modest increases (~10–40%) in pacemaker frequency recorded *in vitro* (Brodie and Appel, 2000; Brodie *et al.*, 1999b, 1990; Okamoto *et al.*, 2006). Finally, it has been shown that VTA DA neurons are significantly more sensitive to ethanol than SNc DA neurons, especially at low ethanol doses (Gessa *et al.*, 1985). However, DA neurons in these two

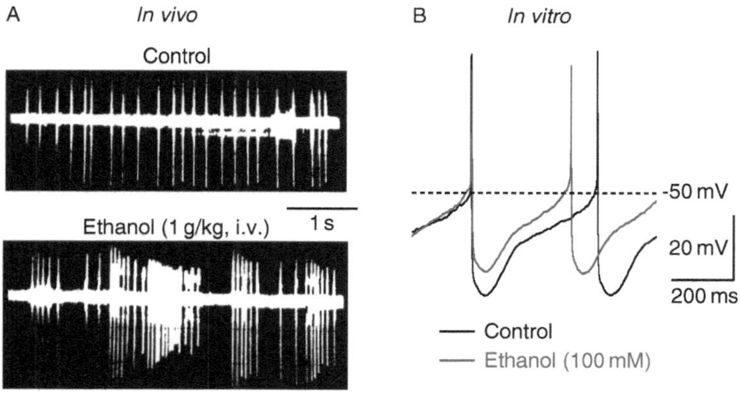

FIG. 3. Ethanol stimulates DA neuron firing. (A) Example traces of single-unit recording of DA neuron firing before (top) and 5 min after i.v. injection of ethanol (bottom) in an awake rat. Adapted from Mereu *et al.* (1984). (B) Representative traces of DA neuron firing recorded with a perforated-patch configuration in a brain slice. Note the effect of ethanol on the membrane potential trajectory during the interspike interval. Adapted from Okamoto *et al.* (2006).

areas appear to be equally affected by ethanol in brain slices (Okamoto *et al.*, 2006), although full concentration dependence was not determined in this study. Taken together, it is most likely that ethanol stimulates DA neuron firing through its action on both the intrinsic pacemaker mechanisms of these neurons and the neurotransmitter inputs they receive.

A. Acute Ethanol Action on Intrinsic Pacemaker Activity of DA Neurons

How does ethanol stimulate the pacemaker activity of DA neurons? The first clue to this question comes from the effect of ethanol on the membrane potential trajectory in spontaneously firing DA neurons, where ethanol reduces the size and duration of the AHP and accelerates the depolarizing ramp toward AP threshold during the interspike interval (Fig. 3B) (Brodie and Appel, 1998; Brodie *et al.*, 1999a; Okamoto *et al.*, 2006). As described in the previous section, there are a number of ion channel species (L-type Ca^{2+} channel, SK channel, I_h, I_A, and others) that become activated in the subthreshold voltage range and shape the membrane potential trajectory during the interspike interval. Therefore, we will next discuss the potential involvement of these ion channels in ethanol stimulation of DA neuron firing.

1. *L-Type Voltage-Gated Ca^{2+} Channel*

It has been shown that acute ethanol exposure inhibits L-type Ca^{2+} channels via modulation of voltage gating in several neuronal preparations (Walter and Messing, 1999; Wang *et al.*, 1994); however, the effect of ethanol on L-type Ca^{2+} currents has not been tested in DA neurons. Reduction in L-type Ca^{2+} channel conductance would (1) slow down the depolarizing ramp preceding APs and (2) decrease the size of SK-dependent AHPs due to a reduction in Ca^{2+} influx required for SK channel activation. Recent evidence indicates that DA neurons may be able to maintain their pacemaker activity through a mechanism dependent on I_h and persistent Na^+ conductance active during the interspike interval even when L-type Ca^{2+} channels are completely blocked by dihydropyridine antagonists (Chan *et al.*, 2007; Guzman *et al.*, 2009). Although L-type Ca^{2+} channels are unlikely to be the major target mediating ethanol excitation of DA neurons, analyzing the effect of ethanol on pacemaker activity under L-type Ca^{2+} channel blockade might help elucidate the roles of other ionic conductances in ethanol stimulation of DA neuron firing.

2. *Small-Conductance Ca^{2+}-Sensitive K^+ Channel*

Ethanol (≥ 10 mM) has been shown to inhibit SK channel (SK2 subtype)–mediated currents in a heterologous expression system (Dreixler *et al.*, 2000).

Reduction in SK conductance may well lead to an enhancement of DA neuron pacemaker firing by suppressing AHPs. However, blockade of SK channels with apamin failed to reduce the magnitude of ethanol excitation of DA neuron firing, ruling out a major role of SK channels in the ethanol effect (Brodie *et al.*, 1999a). It is possible that SK3 channel, which is the major SK channel subtype expressed in DA neurons (Wolfart *et al.*, 2001), might be less sensitive to ethanol compared to the SK2 channel described above.

3. *Hyperpolarization-Activated Cation (I_h) Channel*

A recent study has demonstrated that acute ethanol exposure (50–100 mM) enhances I_h in DA neurons of the VTA and SNc (Okamoto *et al.*, 2006). Consistent with this, ethanol increases the size of the I_h-dependent depolarizing sag response to membrane hyperpolarization (Brodie and Appel, 1998). Ethanol enhances I_h partly via augmenting cyclic adenosine monophosphate (cAMP)-dependent facilitation of voltage gating (Okamoto *et al.*, 2006). There appears to be species difference in the contribution of I_h to the effect of ethanol. Thus, ZD7288, an I_h blocker, significantly suppresses ethanol stimulation of DA neuron firing in mice but not in rats (Appel *et al.*, 2003; McDaid *et al.*, 2008; Okamoto *et al.*, 2006). It has been reported lately that SNc DA neurons from a line of mice bred for high sensitivity to the locomotor-stimulating effect of ethanol possess large I_h density and high sensitivity to ethanol stimulation of firing, suggesting the potential role of I_h in locomotor activation caused by ethanol (Beckstead and Phillips, 2009). It is interesting to note that a similar ethanol-induced enhancement of I_h has been implicated in ethanol stimulation of spontaneous pacemaker firing in hippocampal interneurons (Yan *et al.*, 2009).

4. *A-Type K^+ (I_A) Channel*

The effect of ethanol on I_A has been tested in non-mammalian neurons or in a heterologous expression system using high ethanol concentrations ($\geq$200 mM) (Alekseev *et al.*, 1997; Anantharam *et al.*, 1992; Treistman and Wilson, 1987a,b). In these investigations, ethanol produced a reduction in peak I_A amplitude and/or deceleration of inactivation without much change in the voltage dependence of activation, although the ethanol effect appears to vary depending on the cell type examined. A recent study reported that low concentrations ($\leq$1 μM) of acetaldehyde, the first metabolite of ethanol, increased I_A partly via facilitating the voltage dependence of activation in VTA DA neurons (Melis *et al.*, 2007). The effect of ethanol itself on I_A was not examined in this study. However, treatment of the slice with the catalase inhibitor 3-aminotriazole completely prevented ethanol stimulation of DA neuron pacemaker activity, suggesting that local conversion of ethanol to acetaldehyde by catalase might be a necessary step for ethanol actions on DA neurons. The involvement of acetaldehyde in the pharmacological and behavioral actions of ethanol has been extensively reviewed recently (Melis *et al.*, 2009).

5. *Other Ion Channels*

It has been shown that ethanol inhibits a type of delayed rectifier K^+ current, termed M-current (I_M), in VTA DA neurons (Koyama *et al.*, 2007). I_M is a non-inactivating K^+ current that is active at subthreshold membrane potentials and has been implicated in dampening repetitive firing and general neuronal excitability (Brown and Passmore, 2009). Pharmacological blockade of I_M has a marginal effect on the pacemaker firing of DA neurons recorded in brain slices, although I_M appears to have a significant role in controlling the firing of dissociated DA neurons, where dendrites are mostly severed from the soma (Koyama and Appel, 2006b; Koyama *et al.*, 2007). Accordingly, pharmacological blockade of I_M has little effect on the magnitude of ethanol stimulation of DA neuron firing in brain slices, ruling out I_M as a major ethanol target mediating stimulation of firing (Koyama *et al.*, 2007). It has also been reported that quinidine reduces the size of the early component of the AHP and suppresses the ethanol excitation of DA neurons (Appel *et al.*, 2003); however, the exact ion channel species affected by quinidine remains to be determined.

Another type of ethanol-sensitive ion channel that should be mentioned here is the GIRK channel, which is an important mediator of inhibitory neurotransmitter action that works through $G_{i/o}$ subtype of G proteins (Dascal, 1997). In DA neurons, GIRK-mediated hyperpolarization is responsible for the inhibition produced by activation of certain G protein–coupled receptors, such as D2 DA receptors and $GABA_B$ receptors (Beckstead *et al.*, 2004; Labouebe *et al.*, 2007; Lacey *et al.*, 1987). Ethanol has been shown to directly bind to GIRK channels and activate them without the involvement of G proteins (Aryal *et al.*, 2009; Kobayashi *et al.*, 1999; Lewohl *et al.*, 1999). In line with this, it has been reported recently that barium, which blocks GIRK channels, suppresses ethanol-induced inhibition of DA neuron firing uncovered after pharmacological blockade of I_h (McDaid *et al.*, 2008). This observation suggests that ethanol-induced activation of GIRK channels may mediate an inhibitory component of ethanol action on DA neuron pacemaker activity.

It is increasingly clear that ethanol acts on multiple ion channel species to influence the intrinsic pacemaker firing of DA neurons. Furthermore, the ionic mechanisms of ethanol action may well vary depending on the location and projection targets of individual DA neurons. Therefore, some of the discrepancies among different studies may be partly due to sampling of different populations of DA neurons, in addition to the difference in the age and species/strains of animals used. It should also be noted that the plane of brain slices (e.g., coronal vs. horizontal), in which different dendritic branches would be severed from the soma, might affect the ionic mechanisms of DA neuron firing and ethanol action.

B. Acute Ethanol Action on Neurotransmitter Inputs onto DA Neurons

As described earlier in this chapter, ethanol action on neurotransmitter inputs clearly plays a pivotal role in regulating VTA DA neuron activity *in vivo*. As throughout the mammalian CNS, glutamate and GABA form the major excitatory and inhibitory synaptic inputs onto VTA DA neurons. Indeed, many of the pharmacological effects of ethanol are mediated by its actions on these neurotransmitter systems in the brain (Lobo and Harris, 2008; Lovinger, 1997; Vengeliene *et al.*, 2008; Weight *et al.*, 1992). Although previous mechanistic investigations on ethanol stimulation of DA neuron activity have mostly, and curiously, focused on the effects of ethanol on the intrinsic properties of these neurons, a number of recent studies have examined the interactions of ethanol with neurotransmitter inputs in the VTA.

1. *Glutamate*

It has been reported recently that acute ethanol exposure can indirectly enhance glutamatergic transmission onto VTA DA neurons, likely via promoting somatodendritic release of DA (Deng *et al.*, 2009; Xiao *et al.*, 2009). Direct ethanol action on postsynaptic AMPA receptors was not detected in these studies. In the proposed scenario, released DA would subsequently activate presynaptic D_1 DA receptors on glutamatergic terminals, resulting in enhanced glutamate release. However, it should be noted that small *inhibition* of glutamate release onto VTA DA neurons via presynaptic D_1-like receptors has been reported in another study (Schilstrom *et al.*, 2006). Although it is well established that ethanol inhibits NMDA receptor–mediated currents in a variety of neurons/preparations (Lovinger *et al.*, 1989; Maldve *et al.*, 2002; Ronald *et al.*, 2001; Valenzuela *et al.*, 1998), the acute effect of ethanol on NMDA receptors in DA neurons is currently unknown.

Regarding the mGluR-mediated transmission, a preliminary study has shown that ethanol suppresses the mGluR- and IP_3-induced Ca^{2+} release from intracellular stores, thereby inhibiting the SK-mediated hyperpolarization resulting from rise in Ca^{2+} (Bernier *et al.*, 2007). This ethanol effect may promote the occurrence of DA neuron burst firing by suppressing the pause in firing (Fig. 4).

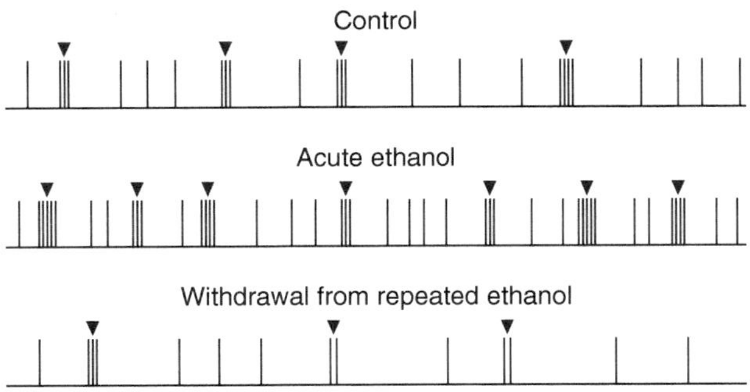

FIG. 4. Schematic diagram illustrating the effects of acute and repeated ethanol exposure on DA neuron firing. Vertical lines indicate individual spikes. Phasic bursts are marked by arrowheads.

2. *GABA*

An early *in vivo* recording study demonstrated that systemic administration of ethanol suppresses the firing of neurons in the substantia nigra pars reticulata, which provide GABAergic afferents to SNc DA neurons (Mereu and Gessa, 1985). This study raised the possibility that ethanol might stimulate DA neuron firing via a disinhibitory mechanism. In recent years, a series of work performed by Steffensen and colleagues have investigated the effect of ethanol on GABAergic neurons in the VTA recorded *in vivo* (Gallegos *et al.*, 1999; Steffensen *et al.*, 2000, 2009; Stobbs *et al.*, 2004). In their initial study, systemic administration of ethanol (0.2–2.0 g/kg, i.p.) dose-dependently inhibited VTA GABA neuron firing recorded in freely behaving rats (Gallegos *et al.*, 1999). This inhibitory effect can also be observed with local application of ethanol directly into the VTA, and likely involves ethanol inhibition of NMDA receptor–mediated excitation of these GABA neurons (Steffensen *et al.*, 2000; Stobbs *et al.*, 2004). Additionally, systemic administration of a DA D2 receptor antagonist has been shown to block ethanol inhibition of VTA GABA neuron firing (Ludlow *et al.*, 2009), although the location of D2 receptors involved is not clear. Interestingly, i.v. administration of low-dose ethanol (0.01–0.1 g/kg) stimulates VTA GABA neuron firing (Steffensen *et al.*, 2009), demonstrating dual effects of ethanol on the activity of these neurons depending on dosage/concentration. However, the contribution of ethanol effects on GABA neurons to the regulation of DA neuron firing *in vivo* has not been directly examined in these studies. It should also be noted that no change in VTA GABA levels, monitored with *in vivo* microdialysis, was detected after acute ethanol administration (1–2 g/kg, i.p.) in alcohol-preferring and non-preferring rat lines (Kemppainen *et al.*, 2010).

In line with these *in vivo* recording studies, Xiao *et al.* have recently reported ethanol concentration (20–80 mM)–dependent inhibition of GABA neuron firing, together with enhancement of DA neuron firing, recorded from the VTA in brain slices (Xiao *et al.*, 2007). In this study, ethanol stimulation of DA neuron firing was significantly reduced by application of $GABA_A$ receptor antagonists, consistent with the major role of a disinhibitory mechanism in ethanol excitation of DA neurons. Interestingly, the nonselective opioid receptor antagonist naloxone largely abolished the ethanol inhibition of GABA neuron firing. Furthermore, the stimulatory effect of ethanol on DA neuron firing was attenuated by naloxone and was also substantially occluded by the μ-opioid receptor agonist DAMGO. It is well known that activation of μ-opioid receptors on VTA GABA neurons causes membrane hyperpolarization and suppression of firing (Johnson and North, 1992a). Therefore, ethanol may disinhibit VTA DA neurons via promoting μ-opioid receptor–mediated suppression of GABAergic inputs onto DA neurons.

Direct recordings of GABAergic synaptic transmission onto VTA DA neurons have revealed some contradictory results. Theile *et al.* have shown that ethanol increases the frequency of $GABA_A$ miniature IPSCs (mIPSCs) recorded in the presence of tetrodotoxin (TTX) in VTA DA neurons, demonstrating that ethanol directly facilitates GABA release from presynaptic terminals (Theile *et al.*, 2008, 2009). No changes in mIPSC decay kinetics or baseline holding currents were observed following ethanol application, suggesting that DA neurons do not express slowly inactivating, tonically active $GABA_A$ receptors that are highly sensitive to ethanol (Wallner *et al.*, 2003; but also see Borghese and Harris, 2007). Theile *et al.* have further shown that ethanol-induced presynaptic facilitation of GABA release involves 5-HT_{2C} receptor activation causing release of Ca^{2+} from intracellular stores (Theile *et al.*, 2009). Similar enhancement of GABAergic transmission onto VTA DA neurons by acute ethanol exposure has been noted previously as an unpublished observation (Melis *et al.*, 2002), and has also been reported in substantia nigra reticulata neurons (Criswell *et al.*, 2008). In contrast, Xiao and Ye have reported that ethanol "inhibits" evoked and spontaneous $GABA_A$ IPSCs, most likely via reduction in VTA GABA neuron excitability, as evidenced by the lack of ethanol effect on AP-independent mIPSCs (Xiao and Ye, 2008). This observation is consistent with the previous finding by this group described above (Xiao *et al.*, 2007). However, the same group further detected presynaptic "enhancement" of $GABA_A$ IPSCs by ethanol in the presence of the μ-opioid receptor agonist, DAMGO, which would silence VTA GABA neuron activity (Xiao and Ye, 2008). Therefore, different levels of endogenous opioids present in the slice might, at least partially, account for the discrepancy in the observed ethanol effects on GABAergic transmission. In this regard, it is of note that Theile *et al.* used horizontal midbrain slices, while Xiao *et al.* used coronal slices.

Recently, more complex interactions involving purinergic type 2 (P2) receptors have been reported in VTA DA neurons (Xiao *et al.*, 2008). Here, ethanol appears to attenuate $P2X_1$- and $P2X_3$-mediated increase in GABAergic transmission, while enhancing P2Y-mediated inhibition of GABAergic transmission. Both of these ethanol effects would lead to inhibition of GABAergic inputs onto DA neurons.

Concerning $GABA_B$ receptor-mediated transmission, it has been shown that ethanol potentiates $GABA_B$ receptor-induced inhibition of VTA DA neurons (Federici *et al.*, 2009), most likely via enhancement of GIRK channels, as reported previously in expression systems and cerebellar granule cells (Kobayashi *et al.*, 1999; Lewohl *et al.*, 1999). Interestingly, $GABA_B$ receptor–induced inhibition of VTA GABA neuron firing recorded *in vivo* can also be augmented by acute ethanol administration (Steffensen *et al.*, 2000).

3. *Glycine*

Acute effects of ethanol on glycine-induced currents have been reported in immature VTA neurons freshly dissociated from neonatal rats (PND 5–14) (Ye *et al.*, 2001a,b). These studies have shown that ethanol either potentiates or inhibits glycine-induced currents in 35 and 45%, respectively, of VTA neurons tested. Both of these effects appear to involve changes in glycine receptor affinity through a protein kinase C–dependent mechanism (Jiang and Ye, 2003; Tao and Ye, 2002). The contribution of these actions on glycine-induced currents to ethanol stimulation of DA neuron firing has not been addressed in these studies.

4. *Dopamine*

Ethanol, either administered systemically or applied locally, has been shown to increase DA levels in the VTA (Kohl *et al.*, 1998; Yan *et al.*, 1996). Yan *et al.* have further reported that the ability of local ethanol application to increase DA levels in the VTA is not blocked by co-infusion of Ca^{2+}-free medium or TTX, suggesting that a simple mechanism involving ethanol excitation of DA neuron firing and subsequent Ca^{2+}-dependent somatodendritic DA release may not entirely account for the effect of ethanol on VTA DA release. The involvement of other neurotransmitter systems in ethanol-induced DA release in the VTA will be described below.

5. *Serotonin*

As serotonergic antidepressants may be therapeutic for alcoholism (Johnson, 2008; Uzbay, 2008), there is a strong impetus to investigate the involvement of serotoninergic systems in the pharmacological and behavioral actions of ethanol.

Indeed, local microinfusion of ethanol directly into the VTA induces release of 5-HT along with DA (Yan *et al.*, 1996). Thus, it would be important to understand how 5-HT interacts with ethanol to regulate VTA DA neuron activity. Brodie *et al.* have reported that, whereas 5-HT itself has no consistent effect on VTA DA neuron firing recorded in midbrain slice preparations, 5-HT, as well as the selective 5-HT_2 agonists 1-(2,5-dimethoxy-4-iodophenyl)-2-aminopropane (DOI) and α-methyl-5-HT, markedly potentiate ethanol enhancement of DA neuron firing (Brodie *et al.*, 1995). Furthermore, a moderate concentration of the 5-HT reuptake inhibitor clomipramine has also been shown to significantly enhance ethanol- and 5-HT-induced increases in VTA DA neuron firing (Trifunovic and Brodie, 1996). In line with the role of 5-HT_2 receptors in ethanol stimulation of DA neuron activity, a recent study has demonstrated that self-administration of ethanol directly into the VTA is attenuated by coinfusion of a 5-HT_{2A} antagonist (Ding *et al.*, 2009b). Furthermore, stimulation of 5-HT_{2C} receptors on GABAergic terminals have been shown to play a role in ethanol-induced facilitation of GABA release onto VTA DA neurons (Theile *et al.*, 2009).

Other types of 5-HT receptors have also been implicated in ethanol actions in the VTA. It has been shown that local intra-VTA administration of a 5-HT_{1B} antagonist, but not antagonists of 5HT_{1A} or 5-HT_{1D} receptors, significantly diminishes DA release in the VTA and NAc produced by systemic administration of ethanol (1–2 g/kg, i.p.) (Yan *et al.*, 2005). Furthermore, intra-VTA administration of a 5-HT_{1B} agonist also prolongs ethanol-induced DA release in the NAc. Ethanol-induced 5-HT release in the VTA might activate 5-HT_{1B} receptors located on GABAergic terminals, thus causing disinhibition of DA neurons via presynaptic inhibition of GABA release (Johnson *et al.*, 1992; Morikawa *et al.*, 2000).

The 5-HT_3 subtype is unique among 5-HT receptors in that it forms a ligand-gated cation channel as opposed to a G protein–coupled receptor. McBride and colleagues have demonstrated that intra-VTA infusion of the selective 5-HT_3 receptor agonist chlorophenylbiguanide (CPBG) via reverse microdialysis produces Ca^{2+}-dependent somatodendritic DA release in the VTA in a concentration-dependent manner (Campbell *et al.*, 1996). In this study, a maximal concentration of CPBG (100 μM) increased VTA DA concentrations by ~400% above baseline levels, and this effect was completely blocked by a 5-HT_3 antagonist (ICS 205-930). Meanwhile, ethanol (2 g/kg, i.p.) stimulated DA release in the VTA by only ~50% over baseline levels, and, most importantly, this stimulatory action of ethanol on somatodendritic DA release was completely blocked by the 5-HT_3 antagonist. These findings indicate that the potentiating effect of ethanol on 5-HT_3 receptor function within the VTA may be critically involved in the activation of mesolimbic DA circuits (Lovinger, 1991). In support of this idea, it has been shown that both the acquisition and maintenance of intra-VTA self-administration of ethanol can be completely prevented by co-administration of a 5-HT_3 antagonist (Rodd *et al.*, 2005c; Rodd-Henricks *et al.*, 2003).

6. *Acetylcholine*

Clearly, there exists a unique interaction between the reinforcing actions of ethanol and nicotine (Dani and Harris, 2005); thus, it is logical to investigate how co-administration of nicotine and ethanol may interact to modulate VTA DA release. Söderpalm and colleagues first demonstrated that systemic blockade of nAChRs completely prevented the ability of systemic ethanol to increase DA concentration in the NAc (Blomqvist *et al.*, 1993). It has been further shown that intra-VTA, but not intra-NAc, infusion of nAChR antagonists abolishes the NAc DA release produced by systemic ethanol administration (Blomqvist *et al.*, 1997; Tizabi *et al.*, 2002) and even that elicited by intra-NAc administration of ethanol (Ericson *et al.*, 2003). The latter study raises the possibility that ethanol action in the NAc might stimulate cholinergic inputs into the VTA, thereby activating DA neuron activity. The potent effects of nAChR antagonists within the VTA, together with those of 5-HT_3 antagonists described previously, provide evidence that modulation of ligand-gated ion channels operated by ACh or 5-HT can play a permissive role in ethanol excitation of VTA DA neurons.

Tizabi *et al.* (2002) have extended these works and demonstrated that direct intra-VTA administration of a low dose of nicotine alone induces NAc DA release and enhances the ability of a low systemic dose of ethanol (0.5 g/kg, i.p.) to increase NAc DA levels in an additive manner. Essentially similar additive effects of ethanol and nicotine on VTA DA neuron firing have also been reported in brain slices (Clark and Little, 2004), which may contribute to the positive association between alcohol drinking and cigarette smoking in humans (Dani and Harris, 2005).

7. *Opioids*

As the general opioid antagonist, naloxone, is FDA approved for alleviating alcohol craving, it is important to understand opioid regulation of the mesolimbic DA output. Behaviorally relevant doses of ethanol have been shown to increase β-endorphin levels in the VTA, which appears to be particularly sensitive to ethanol compared to other brain areas in terms of β-endorphin release (Jarjour *et al.*, 2009; Rasmussen *et al.*, 1998). Furthermore, a significant reduction (~40%) in the binding of [^{3}H]DAMGO, a μ-opioid receptor–selective radioligand, was observed in the VTA within 30 min of *in vivo* administration of ethanol (2.5 g/kg, i.g.), likely due to displacement by released β-endorphin and/or rapid internalization of μ-opioid receptors (Mendez *et al.*, 2001). As discussed previously, ethanol may stimulate VTA DA output via μ-opioid receptor–mediated inhibition of local GABA neuron firing (Xiao *et al.*, 2007). Indeed, systemic injection of the nonselective opioid receptor antagonist naltrexone can prevent ethanol-induced DA release in the NAc associated with ethanol self-administration

(Gonzales and Weiss, 1998). Furthermore, motor-activating actions of intra-VTA administration of ethanol and DAMGO were completely blocked by prior treatment with naltrexone or the μ-selective antagonist β-fulnaltrexamine (Sanchez-Catalan *et al.*, 2009).

In addition to these μ-opioid mechanisms, it has been recently reported that intra-VTA injection of a δ-selective "agonist" decreases ethanol consumption, potentially via presynaptic inhibition of GABA release onto VTA DA neurons (Margolis *et al.*, 2008a). Furthermore, a number of studies have demonstrated that stimulation of κ-opioid receptors suppresses ethanol consumption and preference (Lindholm *et al.*, 2001; Logrip *et al.*, 2009; Nestby *et al.*, 1999); however, the exact site(s) of action have not been addressed in these studies.

8. *Cannabinoids*

A single-unit recording study in anesthetized rats has assessed the involvement of cannabinoids in ethanol stimulation of VTA DA neuron firing (Perra *et al.*, 2005). In this study, NAc-projecting VTA DA neurons were indentified by antidromic stimulation of the NAc; likewise, NAc neurons receiving inputs from the BLA were identified by their response to orthodromic stimulation of the BLA. Systemic administration of a relatively low dose of ethanol (0.5 g/kg, i.v.) enhanced VTA DA neuron firing and inhibited BLA-stimulated NAc neuron firing. Prior administration of a CB1 antagonist completely blocked the actions of ethanol in both brain areas. Furthermore, treatment with the CB1 antagonist 4 min after ethanol administration also completely reversed the ethanol enhancement of DA neuron firing back to pre-ethanol baseline levels. This post-treatment paradigm may be the only example in the literature showing reversal of ethanol stimulation of DA neuron firing by pharmacological manipulation of any neurotransmitter system. However, it remains to be determined how ethanol stimulates the endocannabinoid system in the VTA and other brain areas.

V. Plastic Changes in DA Neuron Activity after Previous Ethanol Exposure

A number of studies have demonstrated that there is a marked decrease in tonic DA levels in the NAc in animals withdrawn from repeated/chronic ethanol exposure (Diana *et al.*, 1993; Rossetti *et al.*, 1992; Weiss *et al.*, 1996). This hypodopaminergic state in the NAc after withdrawal from repeated/chronic exposure appears to be a common feature that is also observed among other drugs of abuse, such as opiates (Acquas *et al.*, 1991; Crippens and Robinson, 1994; Rossetti *et al.*, 1992) and psychostimulants (Imperato *et al.*, 1992, 1996;

Robertson *et al.*, 1991; Rossetti *et al.*, 1992; but see Crippens and Robinson, 1994; Kalivas and Duffy, 1993; Weiss *et al.*, 1992b). Furthermore, human imaging studies have found reduced DA release in the striatum of alcoholics (Martinez *et al.*, 2005; Volkow *et al.*, 2007) as well as in the striatum of addicts to other drugs of abuse (Volkow *et al.*, 2004). It has been hypothesized that animals may compensate for the DA deficit by increasing the amount of drug intake (Melis *et al.*, 2005; Weiss *et al.*, 1992a), which seems to be congruent with the idea that there may be a mechanism to titrate DA levels in the NAc (see Section III.A.). Indeed, ethanol-withdrawn rats self-administer ethanol until NAc DA concentrations are restored to the pre-withdrawal baseline levels (Weiss *et al.*, 1996), indicating that the hypodopaminergic state may be an important component of ethanol dependence. However, it should be noted that other brain systems, such as the one involving CRF, have also been implicated in compulsive drug intake driven by the "negative affective state" during withdrawal (Koob, 2009).

In vivo recording studies have found reduced firing activity of VTA DA neurons after withdrawal from repeated/chronic ethanol exposure, which may account for the decrease in NAc DA levels described above (Diana *et al.*, 2003; Melis *et al.*, 2005). A series of work by Diana and colleagues have demonstrated that both tonic and phasic firing of DA neurons is dramatically decreased in rats withdrawn from 6-day ethanol administration (2–5 g/kg, intragastric, four times/day) (Diana *et al.*, 1993, 1995). Importantly, not only the overall firing frequency but also the burst frequency, that is, the number of bursts per second, and the number of APs per burst were diminished in ethanol-withdrawn rats (Fig. 4), indicating that the influence of neurotransmitter inputs on DA neuron activity are altered after repeated ethanol exposure. This decrease in VTA DA neuron activity can be observed up to 3 days after the final ethanol administration (Diana *et al.*, 1996). In contrast, Shen *et al.* have reported a decrease in VTA DA neuron population activity (i.e., the proportion of spontaneously active neurons) without any change in the firing frequency or firing pattern of remaining spontaneously active DA neurons in ethanol-withdrawn rats (Shen, 2003; Shen and Chiodo, 1993; Shen *et al.*, 2007). This can be observed 1 day to 6 weeks after withdrawal from 10-day to 8-week ethanol administration (4–5 g/kg, intragastric, twice/day). The reduction in DA neuron population activity has been ascribed to depolarization inactivation (i.e., too much depolarization causing Na^+ channel inactivation), as evidenced by reversal with apomorphine or amphetamine which hyperpolarize DA neurons via D2 receptor activation. The reason for the discrepancy between these two groups remains to be resolved.

It has been shown that the threshold for brain stimulation reward, assessed with intracranial self-stimulation (ICSS), is elevated after withdrawal from prolonged ethanol vapor exposure (≥ 17 days) (Schulteis *et al.*, 1995). Similar increases in "reward threshold," that is, decreases in "reward sensitivity," can be observed during withdrawal from all major drugs of abuse (Koob, 2009), and may reflect reduction in the functional output of the mesolimbic DA system.

In contrast to the studies described above, increases in basal DA levels in the NAc after withdrawal from repeated/chronic ethanol exposure have been reported in a line of rats selectively bred for high alcohol preference, termed alcohol-preferring (P) rats (Engleman *et al.*, 2003, 2000; Smith and Weiss, 1999; Thielen *et al.*, 2004). In addition, it has been shown that these rats become sensitized to the reinforcing effect of ethanol within the VTA after repeated ethanol exposure (Ding *et al.*, 2009a; Rodd *et al.*, 2005a,b). These studies in P rats illustrate genetic differences in ethanol-induced neuroadaptations of the mesolimbic DA system.

A. Plastic Changes in Intrinsic Properties of DA Neurons

No change in baseline pacemaker firing frequency of VTA DA neurons has been found in *in vitro* recording studies using VTA slices prepared from rats or mice that received repeated *in vivo* administration of ethanol (2–3.5 g/kg, i.p., once or twice daily, 5–21 days) (Brodie, 2002; Hopf *et al.*, 2007; Okamoto *et al.*, 2006). However, Bailey *et al.* reported a significant decrease in baseline firing frequency of VTA DA neurons recorded in brain slices 1–6 days after cessation of chronic ethanol consumption (>3 weeks) (Bailey *et al.*, 1998, 2001). The amount of ethanol intake was ~25–30 g/kg/day in these studies, suggesting that prolonged exposure to large amounts of ethanol may lead to suppression of DA neuron pacemaking. The mechanism of this decrease in intrinsic DA neuron activity is yet to be investigated.

Downregulation of I_h density without alteration in voltage dependence of I_h activation has been detected in VTA and SNc DA neurons 1 day after repeated *in vivo* ethanol injection (2 g/kg, i.p., once/day, 5 days) (Okamoto *et al.*, 2006). Reduction in I_h has also been observed in VTA DA neurons after 7 days of withdrawal from repeated ethanol administration (2 g/kg, i.p., twice/day, 5 days) (Hopf *et al.*, 2007). As stated above, these studies failed to detect alterations in baseline pacemaker frequency. However, the contribution of I_h to pacemaker activity appeared to be diminished in ethanol-treated animals, as evidenced by a smaller inhibition of firing frequency by pharmacological blockade of I_h (Okamoto *et al.*, 2006). These observations suggest that changes in other ionic conductances may compensate for I_h reduction to maintain the pacemaker activity of DA neurons (Guzman *et al.*, 2009). In line with this idea, Hopf *et al.* (2007) detected a reduction in SK-mediated conductance activated by step depolarization in ethanol-treated animals, a change that may counter the effect of I_h downregulation.

The study by Okamoto *et al.* (2006) has found that the magnitude of ethanol stimulation of DA neuron firing is reduced 1 day after 5-day ethanol exposure. This tolerance to ethanol excitation of DA neurons may be a consequence of I_h downregulation, because I_h mediates, at least partially, the excitatory effect of ethanol on pacemaker activity (see Section IV.A.). In contrast, an increase in ethanol stimulation of VTA DA neuron firing has been reported after prolonged ethanol exposure ($\geq$ 21 days) (Brodie, 2002). Therefore, the function/expression of ethanol-sensitive ion channels in DA neurons may undergo differential changes with different ethanol treatment protocols, resulting in tolerance or sensitization to the stimulatory effect of ethanol. In line with this, both tolerance and sensitization to ethanol stimulation of NAc DA release have been reported after repeated ethanol exposure (Ding *et al.*, 2009a; Zapata *et al.*, 2006), although alterations in neurotransmitter inputs to DA neurons may also play a significant role in these studies conducted *in vivo*.

B. Plastic Changes in Neurotransmitter Inputs onto DA Neurons

1. *Glutamate*

No change in NMDA receptor function, assessed with bath perfusion of NMDA onto the slice, has been observed in VTA DA neurons from C57BL/6 mice after a single ethanol exposure (2 g/kg, i.p.) (Wanat *et al.*, 2009), repeated ethanol exposure (2 g/kg, i.p., twice daily for 5 days) (Hopf *et al.*, 2007), or even after more extended ethanol exposure (3.5 g/kg, i.p., twice daily for >21 days) (Brodie, 2002). Nevertheless, NMDA-induced burst firing appears to be enhanced due to downregulation of SK channels (Hopf *et al.*, 2007). Furthermore, prolonged volitional ethanol intake for 35–50 days has been shown to increase AMPA-mediated synaptic transmission onto VTA DA neurons in rats (Stuber *et al.*, 2008), as is well documented after single or repeated exposure to other drugs of abuse (Jones and Bonci, 2005; Kauer and Malenka, 2007). In contrast, a reduction in both AMPA receptor and NMDA receptor functions has been reported after a single ethanol exposure in DBA mice, demonstrating strain specificity of ethanol-induced neuroadaptations (Wanat *et al.*, 2009).

A biochemical study has analyzed the expression levels of the obligatory NMDA receptor subunit, NR1, and AMPA receptor subunits GluR1 and GluR2 in VTA slice homogenates from rats following up to 12 weeks of 5% ethanol liquid diet feeding (Ortiz *et al.*, 1995). Both NR1 and GluR1, but not GluR2, levels were increased by nearly 30% following 12 weeks of ethanol diet, whereas no changes in any of these subunits were detected after a week of ethanol consumption. GluR1 upregulation is consistent with the increase in AMPA-mediated synaptic transmission after prolonged ethanol drinking (Stuber *et al.*, 2008). The functional significance of the increased NR1 levels (e.g., synaptic vs. extrasynaptic) remains to be determined. It should also be cautioned that biochemical assays using tissue homogenates do not discriminate different cell types present in the brain areas being analyzed.

2. *GABA*

Brodie was the first to directly investigate the potential for neuroadaptive changes in major neurotransmitter systems in the VTA after previous ethanol exposure (Brodie, 2002). In this study, reduced sensitivity to GABA-induced inhibition of VTA DA neuron firing, probably mediated by both $GABA_A$ and $GABA_B$ receptors, was observed in midbrain slices prepared from mice chronically treated with high-dose ethanol (3.5 g/kg, i.p., 21 days). It is not clear if this reduced sensitivity to GABA contributed to the sensitization to ethanol excitation of DA neurons observed in the same study. Conversely, Bonci's group has demonstrated a persistent increase in GABA release onto VTA DA neurons recorded in brain slices 24 h after a single exposure to ethanol (2 g/kg, i.p.) in C57BL/6 and DBA/2 mice (Melis *et al.*, 2002; Wanat *et al.*, 2009). This enhanced GABA release, likely resulting from upregulation of the cAMP–protein kinase A (PKA) signaling cascade at GABAergic terminals, can be observed even a week after the single ethanol exposure, suggesting that it may have enduring behavioral consequences.

A potential increase in GABAergic inhibition of DA neurons has also been suggested in an *in vivo* recording study of VTA GABA neurons following chronic ethanol exposure. Utilizing either i.p. injection (2 g/kg, twice daily) or intermittent vapor exposure, VTA GABA neurons in freely moving rats displayed an increase in basal firing rate, together with tolerance to the acute inhibitory effect of ethanol, following 14 days of ethanol exposure (Gallegos *et al.*, 1999). However, recordings of VTA GABA neurons in anesthetized rats fed with a liquid diet containing 5% ethanol for 2–3 weeks revealed no differences in baseline firing rate and in the acute inhibitory effect of ethanol (Ludlow *et al.*, 2009). Therefore, neuroadaptive responses of VTA GABA neurons to chronic ethanol appear to vary with different treatment and recording protocols. The contribution of these alterations in GABAergic transmission to the change in DA neuron activity, for example, reduced tonic and phasic firing (Diana *et al.*, 1993, 1995) (Fig. 4), has yet to be addressed.

A reduction in the levels of the $GABA_A$ receptor $\alpha 1$ subunit has been documented in the VTA; however, the duration of ethanol exposure required to detect this reduction appears to be significantly longer (12 weeks) than that required to observe a similar change in other brain areas (2–4 weeks) (Charlton *et al.*, 1997; Ortiz *et al.*, 1995; Papadeas *et al.*, 2001). It should be noted here that the $GABA_A$ receptor $\alpha 1$ subunit is selectively expressed in GABA neurons, but not in DA neurons, in the VTA (Okada *et al.*, 2004; Tan *et al.*, 2010). Therefore, examination of changes in other $GABA_A$ receptor subunit isoforms (e.g., α2, α3, α4) may be warranted in the future.

3. *Dopamine (DA)*

Prolonged (12-week), but not short-term (1- or 6-week), ethanol (5%) consumption has been shown to increase levels of tyrosine hydroxylase, the rate-limiting enzyme for DA synthesis, by ~70% in the VTA but not in the SNc (Ortiz *et al.*, 1995). Furthermore, an increase in tyrosine hydroxylase (TH) mRNA levels has also been reported after chronic ethanol consumption (5%, 4 weeks) in the VTA with no change in the SNc (Lee *et al.*, 2005). This increase in TH mRNA was completely blocked by concurrent daily administration of the opioid receptor antagonist naltrexone. Consistent with the change in TH levels, increased levels of DA and the DA metabolite DOPAC have been detected in the VTA, but not in the SNc, following consumption of ethanol (6.6%)-containing lipid diet for 1 month (Pellegrino and Druse, 1992). In contrast, a decrease in VTA DA levels has been documented following a very similar ethanol liquid diet administration (6.6%, 6 weeks) in relatively old rats (5 months) (Woods and Druse, 1996). Thus, the effect of chronic ethanol on DA production/turnover in the VTA may depend on the age of animals.

4. *Opioids*

It has been shown that κ-opioid receptor mRNA is decreased (to ~15% of baseline levels) in rats following repeated ethanol exposure (2 g/kg, i.p., twice daily, 14 days) (Rosin *et al.*, 1999). This may reflect compensatory downregulation of the receptor in response to increased levels of dynorphin, the endogenous κ-opioid, in the mesolimbic system observed after withdrawal from repeated ethanol exposure (Lindholm *et al.*, 2000; Przewlocka *et al.*, 1997). The increase in dynorphin levels might contribute to the aversive emotional state of ethanol withdrawal (Bals-Kubik *et al.*, 1993; Koob, 2009; Trevisan *et al.*, 1998; Wee and Koob, 2010).

VI. Concluding Remarks

We started this chapter by asking: "Why do we drink and why do we keep drinking?" Clearly, regulation of the mesolimbic DA output plays a critical role in alcohol drinking in nondependent individuals and also in driving compulsive alcohol consumption in dependent individuals (alcoholics). Based on this postulate, a number of potentially important "ethanol targets" have been identified over the past couple of decades, both in VTA DA neurons themselves and in the synaptic inputs impinging on them. These include ion channels, neurotransmitter receptors, intracellular signaling molecules, and machinery controlling neurotransmitter/neuromodulator release, as reviewed in this chapter. One of the challenges is to determine which of these ethanol targets in the VTA contribute to specific ethanol actions in behaving animals? Alternatively, could the effects of ethanol on neurons projecting to DA neurons play more important roles than ethanol targets within the VTA? Recent advances in manipulating specific neurons in the brain, such as optogenetic control of neuronal activity (Zhang *et al.*, 2007) or cell type–specific knockin strategies (Skvorak *et al.*, 2006), might give us fresh approaches to these questions over the next decade. Finally, because alcoholism, like addiction to other drugs of abuse, has a strong learning component (Hyman *et al.*, 2006), it is important to fully understand the influence of ethanol (acute and chronic) on the plasticity mechanisms involved in learning of the behaviors and environments associated with ethanol intake, both in the VTA and DA projection areas.

Acknowledgments

We thank Dr. R. Adron Harris and Brian Bernier for critical reading of the manuscript. This work was supported by NIH grants AA015521 to H.M. and AA16651, AA015167, and AA14874 to R.A.M.

References

Acquas, E., Carboni, E., and Di Chiara, G. (1991). Profound depression of mesolimbic dopamine release after morphine withdrawal in dependent rats. *Eur. J. Pharmacol.* **193**, 133–134.

Ahlenius, S., Carlsson, A., Engel, J., Svensson, T., and Sodersten, P. (1973). Antagonism by alpha methyltyrosine of the ethanol-induced stimulation and euphoria in man. *Clin. Pharmacol. Ther.* **14**, 586–591.

Ahmed, S. H., and Koob, G. F. (2005). Transition to drug addiction: A negative reinforcement model based on an allostatic decrease in reward function. *Psychopharmacology (Berl.)* **180**, 473–490.

Alekseev, S. I., Alekseev, A. S., and Ziskin, M. C. (1997). Effects of alcohols on A-type K^+ currents in *Lymnaea* neurons. *J. Pharmacol. Exp. Ther.* **281**, 84–92.

Anantharam, V., Bayley, H., Wilson, A., and Treistman, S. N. (1992). Differential effects of ethanol on electrical properties of various potassium channels expressed in oocytes. *Mol. Pharmacol.* **42**, 499–505.

Appel, S. B., Liu, Z., McElvain, M. A., and Brodie, M. S. (2003). Ethanol excitation of dopaminergic ventral tegmental area neurons is blocked by quinidine. *J. Pharmacol. Exp. Ther.* **306**, 437–446.

Arbuthnott, G. W., and Wickens, J. R. (1996). Dopamine cells are neurones too! *Trends. Neurosci.* **19**, 279–280.

Aryal, P., Dvir, H., Choe, S., and Slesinger, P. A. (2009). A discrete alcohol pocket involved in GIRK channel activation. *Nat. Neurosci.* **12**, 988–995.

Aston-Jones, G., Smith, R. J., Sartor, G. C., Moorman, D. E., Massi, L., Tahsili-Fahadan, P., and Richardson, K. A. (2010). Lateral hypothalamic orexin/hypocretin neurons: A role in reward-seeking and addiction. *Brain Res.* **1314**, 74–90.

Azam, L., Winzer-Serhan, U. H., Chen, Y., and Leslie, F. M. (2002). Expression of neuronal nicotinic acetylcholine receptor subunit mRNAs within midbrain dopamine neurons. *J. Comp. Neurol.* **444**, 260–274.

Azmitia, E. C., and Segal, M. (1978). An autoradiographic analysis of the differential ascending projections of the dorsal and median raphe nuclei in the rat. *J. Comp. Neurol.* **179**, 641–667.

Bailey, C. P., Manley, S. J., Watson, W. P., Wonnacott, S., Molleman, A., and Little, H. J. (1998). Chronic ethanol administration alters activity in ventral tegmental area neurons after cessation of withdrawal hyperexcitability. *Brain Res.* **803**, 144–152.

Bailey, C. P., O'Callaghan, M. J., Croft, A. P., Manley, S. J., and Little, H. J. (2001). Alterations in mesolimbic dopamine function during the abstinence period following chronic ethanol consumption. *Neuropharmacology* **41**, 989–999.

Balcita-Pedicino, J. J., and Sesack, S. R. (2007). Orexin axons in the rat ventral tegmental area synapse infrequently onto dopamine and gamma-aminobutyric acid neurons. *J. Comp. Neurol.* **503**, 668–684.

Bals-Kubik, R., Ableitner, A., Herz, A., and Shippenberg, T. S. (1993). Neuroanatomical sites mediating the motivational effects of opioids as mapped by the conditioned place preference paradigm in rats. *J. Pharmacol. Exp. Ther.* **264**, 489–495.

Bayer, V. E., and Pickel, V. M. (1990). Ultrastructural localization of tyrosine hydroxylase in the rat ventral tegmental area: Relationship between immunolabeling density and neuronal associations. *J. Neurosci.* **10**, 2996–3013.

Bechtholt, A. J., and Cunningham, C. L. (2005). Ethanol-induced conditioned place preference is expressed through a ventral tegmental area dependent mechanism. *Behav. Neurosci.* **119**, 213–223.

Beckstead, M. J., Ford, C. P., Phillips, P. E., and Williams, J. T. (2007). Presynaptic regulation of dendrodendritic dopamine transmission. *Eur. J. Neurosci.* **26**, 1479–1488.

Beckstead, M. J., Gantz, S. C., Ford, C. P., Stenzel-Poore, M. P., Phillips, P. E., Mark, G. P., and Williams, J. T. (2009). CRF enhancement of GIRK channel-mediated transmission in dopamine neurons. *Neuropsychopharmacology* **34**, 1926–1935.

Beckstead, M. J., Grandy, D. K., Wickman, K., and Williams, J. T. (2004). Vesicular dopamine release elicits an inhibitory postsynaptic current in midbrain dopamine neurons. *Neuron* **42**, 939–946.

Beckstead, M. J., and Phillips, T. J. (2009). Mice selectively bred for high- or low-alcohol-induced locomotion exhibit differences in dopamine neuron function. *J. Pharmacol. Exp. Ther.* **329**, 342–349.

Belin, D., and Everitt, B. J. (2008). Cocaine seeking habits depend upon dopamine-dependent serial connectivity linking the ventral with the dorsal striatum. *Neuron* **57**, 432–441.

Bergman, J., Kamien, J. B., and Spealman, R. D. (1990). Antagonism of cocaine self-administration by selective dopamine D(1) and D(2) antagonists. *Behav. Pharmacol.* **1**, 355–363.

Bernier, B., Cui, G. H., Harnett, M., and Morikawa, H. (2007). Ethanol action on metabotropic glutamate receptor-mediated pause in dopamine neurons. *Soc. Neurosci. Abstr.* **909**, 16.

Berridge, K. C. (2007). The debate over dopamine's role in reward: The case for incentive salience. *Psychopharmacology (Berl.)* **191**, 391–431.

Bevan, M. D., and Wilson, C. J. (1999). Mechanisms underlying spontaneous oscillation and rhythmic firing in rat subthalamic neurons. *J. Neurosci.* **19**, 7617–7628.

Blomqvist, O., Engel, J. A., Nissbrandt, H., and Soderpalm, B. (1993). The mesolimbic dopamine-activating properties of ethanol are antagonized by mecamylamine. *Eur. J. Pharmacol.* **249**, 207–213.

Blomqvist, O., Ericson, M., Engel, J. A., and Soderpalm, B. (1997). Accumbal dopamine overflow after ethanol: Localization of the antagonizing effect of mecamylamine. *Eur. J. Pharmacol.* **334**, 149–156.

Blythe, S. N., Atherton, J. F., and Bevan, M. D. (2007). Synaptic activation of dendritic AMPA and NMDA receptors generates transient high-frequency firing in substantia nigra dopamine neurons *in vitro*. *J. Neurophysiol.* **97**, 2837–2850.

Blythe, S. N., Wokosin, D., Atherton, J. F., and Bevan, M. D. (2009). Cellular mechanisms underlying burst firing in substantia nigra dopamine neurons. *J. Neurosci.* **29**, 15531–15541.

Bolam, J. P., and Smith, Y. (1990). The GABA and substance P input to dopaminergic neurones in the substantia nigra of the rat. *Brain Res.* **529**, 57–78.

Borghese, C. M., and Harris, R. A. (2007). Studies of ethanol actions on recombinant delta-containing gamma-aminobutyric acid type A receptors yield contradictory results. *Alcohol* **41**, 155–162.

Borgland, S. L., Taha, S. A., Sarti, F., Fields, H. L., and Bonci, A. (2006). Orexin A in the VTA is critical for the induction of synaptic plasticity and behavioral sensitization to cocaine. *Neuron* **49**, 589–601.

Brazhnik, E., Shah, F., and Tepper, J. M. (2008). GABAergic afferents activate both GABAA and GABAB receptors in mouse substantia nigra dopaminergic neurons *in vivo*. *J. Neurosci.* **28**, 10386–10398.

Brodie, M. S. (2002). Increased ethanol excitation of dopaminergic neurons of the ventral tegmental area after chronic ethanol treatment. *Alcohol. Clin. Exp. Res.* **26**, 1024–1030.

Brodie, M. S., and Appel, S. B. (1998). The effects of ethanol on dopaminergic neurons of the ventral tegmental area studied with intracellular recording in brain slices. *Alcohol. Clin. Exp. Res.* **22**, 236–244.

Brodie, M. S., and Appel, S. B. (2000). Dopaminergic neurons in the ventral tegmental area of C57BL/6J and DBA/2J mice differ in sensitivity to ethanol excitation. *Alcohol. Clin. Exp. Res.* **24**, 1120–1124.

Brodie, M. S., McElvain, M. A., Bunney, E. B., and Appel, S. B. (1999a). Pharmacological reduction of small conductance calcium-activated potassium current (SK) potentiates the excitatory effect of ethanol on ventral tegmental area dopamine neurons. *J. Pharmacol. Exp. Ther.* **290**, 325–333.

Brodie, M. S., Pesold, C., and Appel, S. B. (1999b). Ethanol directly excites dopaminergic ventral tegmental area reward neurons. *Alcohol. Clin. Exp. Res.* **23**, 1848–1852.

Brodie, M. S., Shefner, S. A., and Dunwiddie, T. V. (1990). Ethanol increases the firing rate of dopamine neurons of the rat ventral tegmental area *in vitro*. *Brain Res.* **508**, 65–69.

Brodie, M. S., Trifunovic, R. D., and Shefner, S. A. (1995). Serotonin potentiates ethanol-induced excitation of ventral tegmental area neurons in brain slices from three different rat strains. *J. Pharmacol. Exp. Ther.* **273**, 1139–1146.

Brown, D. A., and Passmore, G. M. (2009). Neural KCNQ (Kv7) channels. *Br. J. Pharmacol.* **156**, 1185–1195.

Brown, H., and Difrancesco, D. (1980). Voltage-clamp investigations of membrane currents underlying pace-maker activity in rabbit sino-atrial node. *J. Physiol.* **308**, 331–351.

Bubar, M. J., and Cunningham, K. A. (2007). Distribution of serotonin 5-HT2C receptors in the ventral tegmental area. *Neuroscience* **146**, 286–297.

Budygin, E. A., Phillips, P. E., Wightman, R. M., and Jones, S. R. (2001). Terminal effects of ethanol on dopamine dynamics in rat nucleus accumbens: An *in vitro* voltammetric study. *Synapse* **42**, 77–79.

Cagniard, B., Beeler, J. A., Britt, J. P., McGehee, D. S., Marinelli, M., and Zhuang, X. (2006). Dopamine scales performance in the absence of new learning. *Neuron* **51**, 541–547.

Caine, S. B., and Koob, G. F. (1994). Effects of dopamine D-1 and D-2 antagonists on cocaine self-administration under different schedules of reinforcement in the rat. *J. Pharmacol. Exp. Ther.* **270**, 209–218.

Cameron, D. L., Wessendorf, M. W., and Williams, J. T. (1997). A subset of ventral tegmental area neurons is inhibited by dopamine, 5-hydroxytryptamine and opioids. *Neuroscience* **77**, 155–166.

Cameron, D. L., and Williams, J. T. (1993). Dopamine D1 receptors facilitate transmitter release. *Nature* **366**, 344–347.

Campbell, A. D., Kohl, R. R., and McBride, W. J. (1996). Serotonin-3 receptor and ethanol-stimulated somatodendritic dopamine release. *Alcohol* **13**, 569–574.

Carr, D. B., and Sesack, S. R. (2000). Projections from the rat prefrontal cortex to the ventral tegmental area: Target specificity in the synaptic associations with mesoaccumbens and mesocortical neurons. *J. Neurosci.* **20**, 3864–3873.

Champtiaux, N., Gotti, C., Cordero-Erausquin, M., David, D. J., Przybylski, C., Lena, C., Clementi, F., Moretti, M., Rossi, F. M., Le Novere, N., McIntosh, J. M., Gardier, A. M., and Changeux, J. P. (2003). Subunit composition of functional nicotinic receptors in dopaminergic neurons investigated with knock-out mice. *J. Neurosci.* **23**, 7820–7829.

Chan, C. S., Guzman, J. N., Ilijic, E., Mercer, J. N., Rick, C., Tkatch, T., Meredith, G. E., and Surmeier, D. J. (2007). "Rejuvenation" protects neurons in mouse models of Parkinson's disease. *Nature* **447**, 1081–1086.

Chan, C. S., Shigemoto, R., Mercer, J. N., and Surmeier, D. J. (2004). HCN2 and HCN1 channels govern the regularity of autonomous pacemaking and synaptic resetting in globus pallidus neurons. *J. Neurosci.* **24**, 9921–9932.

Charara, A., Smith, Y., and Parent, A. (1996). Glutamatergic inputs from the pedunculopontine nucleus to midbrain dopaminergic neurons in primates: Phaseolus vulgaris-leucoagglutinin anterograde labeling combined with postembedding glutamate and GABA immunohistochemistry. *J. Comp. Neurol.* **364**, 254–266.

Charlton, M. E., Sweetnam, P. M., Fitzgerald, L. W., Terwilliger, R. Z., Nestler, E. J., and Duman, R. S. (1997). Chronic ethanol administration regulates the expression of GABAA receptor alpha 1 and alpha 5 subunits in the ventral tegmental area and hippocampus. *J. Neurochem.* **68**, 121–127.

Cheer, J. F., Wassum, K. M., Sombers, L. A., Heien, M. L., Ariansen, J. L., Aragona, B. J., Phillips, P. E., and Wightman, R. M. (2007). Phasic dopamine release evoked by abused substances requires cannabinoid receptor activation. *J. Neurosci.* **27**, 791–795.

Chen, B. T., and Rice, M. E. (2001). Novel Ca^{2+} dependence and time course of somatodendritic dopamine release: Substantia nigra versus striatum. *J. Neurosci.* **21**, 7841–7847.

Chergui, K., Akaoka, H., Charlety, P. J., Saunier, C. F., Buda, M., and Chouvet, G. (1994). Subthalamic nucleus modulates burst firing of nigral dopamine neurones via NMDA receptors. *Neuroreport* **5**, 1185–1188.

Chergui, K., Charlety, P. J., Akaoka, H., Saunier, C. F., Brunet, J. L., Buda, M., Svensson, T. H., and Chouvet, G. (1993). Tonic activation of NMDA receptors causes spontaneous burst discharge of rat midbrain dopamine neurons *in vivo*. *Eur. J. Neurosci.* **5**, 137–144.

Chevaleyre, V., and Castillo, P. E. (2002). Assessing the role of Ih channels in synaptic transmission and mossy fiber LTP. *Proc. Natl. Acad. Sci. USA* **99**, 9538–9543.

Christoffersen, C. L., and Meltzer, L. T. (1995). Evidence for *N*-methyl-D-aspartate and AMPA subtypes of the glutamate receptor on substantia nigra dopamine neurons: Possible preferential role for *N*-methyl-D-aspartate receptors. *Neuroscience* **67**, 373–381.

Clark, A., and Little, H. J. (2004). Interactions between low concentrations of ethanol and nicotine on firing rate of ventral tegmental dopamine neurones. *Drug Alcohol Depend.* **75**, 199–206.

Clements, J. R., Toth, D. D., Highfield, D. A., and Grant, S. J. (1991). Glutamate-like immunoreactivity is present within cholinergic neurons of the laterodorsal tegmental and pedunculopontine nuclei. *Adv. Exp. Med. Biol.* **295**, 127–142.

Comoli, E., Coizet, V., Boyes, J., Bolam, J. P., Canteras, N. S., Quirk, R. H., Overton, P. G., and Redgrave, P. (2003). A direct projection from superior colliculus to substantia nigra for detecting salient visual events. *Nat. Neurosci.* **6**, 974–980.

Connor, J. A., and Stevens, C. F. (1971). Prediction of repetitive firing behaviour from voltage clamp data on an isolated neurone soma. *J. Physiol.* **213**, 31–53.

Conrad, L. C., and Pfaff, D. W. (1976). Autoradiographic tracing of nucleus accumbens efferents in the rat. *Brain Res.* **113**, 589–596.

Crippens, D., and Robinson, T. E. (1994). Withdrawal from morphine or amphetamine: Different effects on dopamine in the ventral-medial striatum studied with microdialysis. *Brain Res.* **650**, 56–62.

Criswell, H. E., Ming, Z., Kelm, M. K., and Breese, G. R. (2008). Brain regional differences in the effect of ethanol on GABA release from presynaptic terminals. *J. Pharmacol. Exp. Ther.* **326**, 596–603.

Cui, G., Bernier, B. E., Harnett, M. T., and Morikawa, H. (2007). Differential regulation of action potential- and metabotropic glutamate receptor-induced Ca^{2+} signals by inositol 1,4,5-trisphosphate in dopaminergic neurons. *J. Neurosci.* **27**, 4776–4785.

Dai, M., and Tepper, J. M. (1998). Do silent dopaminergic neurons exist in rat substantia nigra *in vivo*? *Neuroscience* **85**, 1089–1099.

Dani, J. A., and Harris, R. A. (2005). Nicotine addiction and comorbidity with alcohol abuse and mental illness. *Nat. Neurosci.* **8**, 1465–1470.

Dascal, N. (1997). Signalling via the G protein-activated K^+ channels. *Cell Signal* **9**, 551–573.

Deister, C. A., Teagarden, M. A., Wilson, C. J., and Paladini, C. A. (2009). An intrinsic neuronal oscillator underlies dopaminergic neuron bursting. *J. Neurosci.* **29**, 15888–15897.

Deng, C., Li, K. Y., Zhou, C., and Ye, J. H. (2009). Ethanol enhances glutamate transmission by retrograde dopamine signaling in a postsynaptic neuron/synaptic bouton preparation from the ventral tegmental area. *Neuropsychopharmacology* **34**, 1233–1244.

Di Chiara, G., and Imperato, A. (1988). Drugs abused by humans preferentially increase synaptic dopamine concentrations in the mesolimbic system of freely moving rats. *Proc. Natl. Acad. Sci. USA* **85**, 5274–5278.

Di Loreto, S., Florio, T., and Scarnati, E. (1992). Evidence that non-NMDA receptors are involved in the excitatory pathway from the pedunculopontine region to nigrostriatal dopaminergic neurons. *Exp. Brain Res.* **89**, 79–86.

Di Matteo, V., De Blasi, A., Di Giulio, C., and Esposito, E. (2001). Role of 5-HT(2C) receptors in the control of central dopamine function. *Trends Pharmacol. Sci.* **22**, 229–232.

Diana, M., Brodie, M., Muntoni, A., Puddu, M. C., Pillolla, G., Steffensen, S., Spiga, S., and Little, H. J. (2003). Enduring effects of chronic ethanol in the CNS: Basis for alcoholism. *Alcohol Clin. Exp. Res.* **27**, 354–361.

Diana, M., Pistis, M., Carboni, S., Gessa, G. L., and Rossetti, Z. L. (1993). Profound decrement of mesolimbic dopaminergic neuronal activity during ethanol withdrawal syndrome in rats: Electrophysiological and biochemical evidence. *Proc. Natl. Acad. Sci. USA* **90**, 7966–7969.

Diana, M., Pistis, M., Muntoni, A., and Gessa, G. (1996). Mesolimbic dopaminergic reduction outlasts ethanol withdrawal syndrome: Evidence of protracted abstinence. *Neuroscience* **71**, 411–415.

Diana, M., Pistis, M., Muntoni, A. L., and Gessa, G. L. (1995). Ethanol withdrawal does not induce a reduction in the number of spontaneously active dopaminergic neurons in the mesolimbic system. *Brain Res.* **682**, 29–34.

Ding, Z. M., Rodd, Z. A., Engleman, E. A., and McBride, W. J. (2009a). Sensitization of ventral tegmental area dopamine neurons to the stimulating effects of ethanol. *Alcohol. Clin. Exp. Res.* **33**, 1571–1581.

Ding, Z. M., Toalston, J. E., Oster, S. M., McBride, W. J., and Rodd, Z. A. (2009b). Involvement of local serotonin-2A but not serotonin-1B receptors in the reinforcing effects of ethanol within the posterior ventral tegmental area of female Wistar rats. *Psychopharmacology (Berl.)* **204**, 381–390.

Doherty, M. D., and Pickel, V. M. (2000). Ultrastructural localization of the serotonin 2A receptor in dopaminergic neurons in the ventral tegmental area. *Brain Res.* **864**, 176–185.

Dommett, E., Coizet, V., Blaha, C. D., Martindale, J., Lefebvre, V., Walton, N., Mayhew, J. E., Overton, P. G., and Redgrave, P. (2005). How visual stimuli activate dopaminergic neurons at short latency. *Science* **307**, 1476–1479.

Dreixler, J. C., Jenkins, A., Cao, Y. J., Roizen, J. D., and Houamed, K. M. (2000). Patch-clamp analysis of anesthetic interactions with recombinant SK2 subtype neuronal calcium-activated potassium channels. *Anesth. Analg.* **90**, 727–732.

Durante, P., Cardenas, C. G., Whittaker, J. A., Kitai, S. T., and Scroggs, R. S. (2004). Low-threshold L-type calcium channels in rat dopamine neurons. *J. Neurophysiol.* **91**, 1450–1454.

Dyr, W., McBride, W. J., Lumeng, L., Li, T. K., and Murphy, J. M. (1993). Effects of D1 and D2 dopamine receptor agents on ethanol consumption in the high-alcohol-drinking (HAD) line of rats. *Alcohol* **10**, 207–212.

Engleman, E. A., McBride, W. J., Li, T. K., Lumeng, L., and Murphy, J. M. (2003). Ethanol drinking experience attenuates (–)sulpiride-induced increases in extracellular dopamine levels in the nucleus accumbens of alcohol-preferring (P) rats. *Alcohol Clin. Exp. Res.* **27**, 424–431.

Engleman, E. A., McBride, W. J., Wilber, A. A., Shaikh, S. R., Eha, R. D., Lumeng, L., Li, T. K., and Murphy, J. M. (2000). Reverse microdialysis of a dopamine uptake inhibitor in the nucleus accumbens of alcohol-preferring rats: Effects on dialysate dopamine levels and ethanol intake. *Alcohol. Clin. Exp. Res.* **24**, 795–801.

Engleman, E. A., Rodd, Z. A., Bell, R. L., and Murphy, J. M. (2008). The role of 5-HT3 receptors in drug abuse and as a target for pharmacotherapy. *CNS Neurol. Disord. Drug Targets* **7**, 454–467.

Erhardt, S., Mathe, J. M., Chergui, K., Engberg, G., and Svensson, T. H. (2002). GABA(B) receptor-mediated modulation of the firing pattern of ventral tegmental area dopamine neurons *in vivo*. *Naunyn Schmiedebergs Arch. Pharmacol.* **365**, 173–180.

Ericson, M., Molander, A., Lof, E., Engel, J. A., and Soderpalm, B. (2003). Ethanol elevates accumbal dopamine levels via indirect activation of ventral tegmental nicotinic acetylcholine receptors. *Eur. J. Pharmacol.* **467**, 85–93.

Everitt, B. J., and Robbins, T. W. (2005). Neural systems of reinforcement for drug addiction: From actions to habits to compulsion. *Nat. Neurosci.* **8**, 1481–1489.

Fadel, J., and Deutch, A. Y. (2002). Anatomical substrates of orexin-dopamine interactions: Lateral hypothalamic projections to the ventral tegmental area. *Neuroscience* **111**, 379–387.

Fallon, J. H., Leslie, F. M., and Cone, R. I. (1985). Dynorphin-containing pathways in the substantia nigra and ventral tegmentum: A double labeling study using combined immunofluorescence and retrograde tracing. *Neuropeptides* **5**, 457–460.

Federici, M., Nistico, R., Giustizieri, M., Bernardi, G., and Mercuri, N. B. (2009). Ethanol enhances GABAB-mediated inhibitory postsynaptic transmission on rat midbrain dopaminergic neurons by facilitating GIRK currents. *Eur. J. Neurosci.* **29**, 1369–1377.

Fiorillo, C. D., and Williams, J. T. (1998). Glutamate mediates an inhibitory postsynaptic potential in dopamine neurons. *Nature* **394**, 78–82.

Fiorillo, C. D., and Williams, J. T. (2000). Cholinergic inhibition of ventral midbrain dopamine neurons. *J. Neurosci.* **20**, 7855–7860.

Floresco, S. B., West, A. R., Ash, B., Moore, H., and Grace, A. A. (2003). Afferent modulation of dopamine neuron firing differentially regulates tonic and phasic dopamine transmission. *Nat. Neurosci.* **6**, 968–973.

Foddai, M., Dosia, G., Spiga, S., and Diana, M. (2004). Acetaldehyde increases dopaminergic neuronal activity in the VTA. *Neuropsychopharmacology* **29**, 530–536.

Ford, C. P., Mark, G. P., and Williams, J. T. (2006). Properties and opioid inhibition of mesolimbic dopamine neurons vary according to target location. *J. Neurosci.* **26**, 2788–2797.

Franz, O., Liss, B., Neu, A., and Roeper, J. (2000). Single-cell mRNA expression of HCN1 correlates with a fast gating phenotype of hyperpolarization-activated cyclic nucleotide-gated ion channels (Ih) in central neurons. *Eur. J. Neurosci.* **12**, 2685–2693.

Fujimura, K., and Matsuda, Y. (1989). Autogenous oscillatory potentials in neurons of the guinea pig substantia nigra pars compacta *in vitro*. *Neurosci. Lett.* **104**, 53–57.

Gallegos, R. A., Lee, R. S., Criado, J. R., Henriksen, S. J., and Steffensen, S. C. (1999). Adaptive responses of gamma-aminobutyric acid neurons in the ventral tegmental area to chronic ethanol. *J. Pharmacol. Exp. Ther.* **291**, 1045–1053.

Garzon, M., and Pickel, V. M. (2002). Ultrastructural localization of enkephalin and mu-opioid receptors in the rat ventral tegmental area. *Neuroscience* **114**, 461–474.

Gatto, G. J., McBride, W. J., Murphy, J. M., Lumeng, L., and Li, T. K. (1994). Ethanol self-infusion into the ventral tegmental area by alcohol-preferring rats. *Alcohol* **11**, 557–564.

Gehlert, D. R., Gackenheimer, S. L., Wong, D. T., and Robertson, D. W. (1991). Localization of 5-HT3 receptors in the rat brain using [3H]LY278584. *Brain Res.* **553**, 149–154.

Geisler, S., Derst, C., Veh, R. W., and Zahm, D. S. (2007). Glutamatergic afferents of the ventral tegmental area in the rat. *J. Neurosci.* **27**, 5730–5743.

Geisler, S., and Zahm, D. S. (2005). Afferents of the ventral tegmental area in the rat-anatomical substratum for integrative functions. *J. Comp. Neurol.* **490**, 270–294.

George, M. S., Abbott, L. F., and Siegelbaum, S. A. (2009). HCN hyperpolarization-activated cation channels inhibit EPSPs by interactions with M-type K(+) channels. *Nat. Neurosci.* **12**, 577–584.

Georges, F., and Aston-Jones, G. (2002). Activation of ventral tegmental area cells by the bed nucleus of the stria terminalis: A novel excitatory amino acid input to midbrain dopamine neurons. *J. Neurosci.* **22**, 5173–5187.

Gessa, G. L., Muntoni, F., Collu, M., Vargiu, L., and Mereu, G. (1985). Low doses of ethanol activate dopaminergic neurons in the ventral tegmental area. *Brain Res.* **348**, 201–203.

Gonzales, R. A., Job, M. O., and Doyon, W. M. (2004). The role of mesolimbic dopamine in the development and maintenance of ethanol reinforcement. *Pharmacol. Ther.* **103**, 121–146.

Gonzales, R. A., and Weiss, F. (1998). Suppression of ethanol-reinforced behavior by naltrexone is associated with attenuation of the ethanol-induced increase in dialysate dopamine levels in the nucleus accumbens. *J. Neurosci.* **18**, 10663–10671.

Goto, Y., and Grace, A. A. (2005). Dopaminergic modulation of limbic and cortical drive of nucleus accumbens in goal-directed behavior. *Nat. Neurosci.* **8**, 805–812.

Goto, Y., Otani, S., and Grace, A. A. (2007). The Yin and Yang of dopamine release: A new perspective. *Neuropharmacology* **53**, 583–587.

Grace, A. A., and Bunney, B. S. (1984a). The control of firing pattern in nigral dopamine neurons: Burst firing. *J. Neurosci.* **4**, 2877–2890.

Grace, A. A., and Bunney, B. S. (1984b). The control of firing pattern in nigral dopamine neurons: Single spike firing. *J. Neurosci.* **4**, 2866–2876.

Grace, A. A., and Onn, S. P. (1989). Morphology and electrophysiological properties of immunocytochemically identified rat dopamine neurons recorded *in vitro*. *J. Neurosci.* **9**, 3463–3481.

Grenhoff, J., North, R. A., and Johnson, S. W. (1995). Alpha 1-adrenergic effects on dopamine neurons recorded intracellularly in the rat midbrain slice. *Eur. J. Neurosci.* **7**, 1707–1713.

Grillner, P., Berretta, N., Bernardi, G., Svensson, T. H., and Mercuri, N. B. (2000). Muscarinic receptors depress GABAergic synaptic transmission in rat midbrain dopamine neurons. *Neuroscience* **96**, 299–307.

Grillner, P., Bonci, A., Svensson, T. H., Bernardi, G., and Mercuri, N. B. (1999). Presynaptic muscarinic (M3) receptors reduce excitatory transmission in dopamine neurons of the rat mesencephalon. *Neuroscience* **91**, 557–565.

Gronier, B., and Rasmussen, K. (1998). Activation of midbrain presumed dopaminergic neurones by muscarinic cholinergic receptors: An *in vivo* electrophysiological study in the rat. *Br. J. Pharmacol.* **124**, 455–464.

Groves, P. M., Wilson, C. J., Young, S. J., and Rebec, G. V. (1975). Self-inhibition by dopaminergic neurons. *Science* **190**, 522–528.

Guatteo, E., Mercuri, N. B., Bernardi, G., and Knopfel, T. (1999). Group I metabotropic glutamate receptors mediate an inward current in rat substantia nigra dopamine neurons that is independent from calcium mobilization. *J. Neurophysiol.* **82**, 1974–1981.

Guiard, B. P., El Mansari, M., and Blier, P. (2008). Cross-talk between dopaminergic and noradrenergic systems in the rat ventral tegmental area, locus ceruleus, and dorsal hippocampus. *Mol. Pharmacol.* **74**, 1463–1475.

Guzman, J. N., Sanchez-Padilla, J., Chan, C. S., and Surmeier, D. J. (2009). Robust pacemaking in substantia nigra dopaminergic neurons. *J. Neurosci.* **29**, 11011–11019.

Haber, S. N., Fudge, J. L., and McFarland, N. R. (2000). Striatonigrostriatal pathways in primates form an ascending spiral from the shell to the dorsolateral striatum. *J. Neurosci.* **20**, 2369–2382.

Haber, S. N., Lynd, E., Klein, C., and Groenewegen, H. J. (1990). Topographic organization of the ventral striatal efferent projections in the rhesus monkey: An anterograde tracing study. *J. Comp. Neurol.* **293**, 282–298.

Hahn, J., Tse, T. E., and Levitan, E. S. (2003). Long-term K^+ channel-mediated dampening of dopamine neuron excitability by the antipsychotic drug haloperidol. *J. Neurosci.* **23**, 10859–10866.

Hainsworth, A. H., Roper, J., Kapoor, R., and Ashcroft, F. M. (1991). Identification and electrophysiology of isolated pars compacta neurons from guinea-pig substantia nigra. *Neuroscience* **43**, 81–93.

Harnett, M. T., Bernier, B. E., Ahn, K. C., and Morikawa, H. (2009). Burst-timing-dependent plasticity of NMDA receptor-mediated transmission in midbrain dopamine neurons. *Neuron* **62**, 826–838.

Harris, N. C., and Constanti, A. (1995). Mechanism of block by ZD 7288 of the hyperpolarization-activated inward rectifying current in guinea pig substantia nigra neurons *in vitro*. *J. Neurophysiol.* **74**, 2366–2378.

Harris, N. C., Webb, C., and Greenfield, S. A. (1989). A possible pacemaker mechanism in pars compacta neurons of the guinea-pig substantia nigra revealed by various ion channel blocking agents. *Neuroscience* **31**, 355–362.

Hausser, M. A., and Yung, W. H. (1994). Inhibitory synaptic potentials in guinea-pig substantia nigra dopamine neurones *in vitro*. *J. Physiol.* **479**(Pt 3), 401–422.

Heimer, L., Zahm, D. S., Churchill, L., Kalivas, P. W., and Wohltmann, C. (1991). Specificity in the projection patterns of accumbal core and shell in the rat. *Neuroscience* **41**, 89–125.

Herve, D., Pickel, V. M., Joh, T. H., and Beaudet, A. (1987). Serotonin axon terminals in the ventral tegmental area of the rat: Fine structure and synaptic input to dopaminergic neurons. *Brain Res.* **435**, 71–83.

Hodge, C. W., Haraguchi, M., Erickson, H., and Samson, H. H. (1993). Ventral tegmental microinjections of quinpirole decrease ethanol and sucrose-reinforced responding. *Alcohol. Clin. Exp. Res.* **17**, 370–375.

Hodge, C. W., Samson, H. H., and Chappelle, A. M. (1997). Alcohol self-administration: Further examination of the role of dopamine receptors in the nucleus accumbens. *Alcohol. Clin. Exp. Res.* **21**, 1083–1091.

Hopf, F. W., Martin, M., Chen, B. T., Bowers, M. S., Mohamedi, M. M., and Bonci, A. (2007). Withdrawal from intermittent ethanol exposure increases probability of burst firing in VTA neurons *in vitro*. *J. Neurophysiol.* **98**, 2297–2310.

Horvitz, J. C. (2002). Dopamine gating of glutamatergic sensorimotor and incentive motivational input signals to the striatum. *Behav. Brain Res.* **137**, 65–74.

Hubert, G. W., Paquet, M., and Smith, Y. (2001). Differential subcellular localization of mGluR1a and mGluR5 in the rat and monkey substantia nigra. *J. Neurosci.* **21**, 1838–1847.

Hyland, B. I., Reynolds, J. N., Hay, J., Perk, C. G., and Miller, R. (2002). Firing modes of midbrain dopamine cells in the freely moving rat. *Neuroscience* **114**, 475–492.

Hyman, S. E., Malenka, R. C., and Nestler, E. J. (2006). Neural mechanisms of addiction: The role of reward-related learning and memory. *Annu. Rev. Neurosci.* **29**, 565–598.

Ikemoto, S. (2007). Dopamine reward circuitry: Two projection systems from the ventral midbrain to the nucleus accumbens–olfactory tubercle complex. *Brain Res. Rev.* **56**, 27–78.

Ikemoto, S., McBride, W. J., Murphy, J. M., Lumeng, L., and Li, T. K. (1997). 6-OHDA-lesions of the nucleus accumbens disrupt the acquisition but not the maintenance of ethanol consumption in the alcohol-preferring P line of rats. *Alcohol. Clin. Exp. Res.* **21**, 1042–1046.

Imperato, A., and Di Chiara, G. (1986). Preferential stimulation of dopamine release in the nucleus accumbens of freely moving rats by ethanol. *J. Pharmacol. Exp. Ther.* **239**, 219–228.

Imperato, A., Mele, A., Scrocco, M. G., and Puglisi-Allegra, S. (1992). Chronic cocaine alters limbic extracellular dopamine. Neurochemical basis for addiction. *Eur. J. Pharmacol.* **212**, 299–300.

Imperato, A., Obinu, M. C., Carta, G., Mascia, M. S., Casu, M. A., and Gessa, G. L. (1996). Reduction of dopamine release and synthesis by repeated amphetamine treatment: Role in behavioral sensitization. *Eur. J. Pharmacol.* **317**, 231–237.

Jarjour, S., Bai, L., and Gianoulakis, C. (2009). Effect of acute ethanol administration on the release of opioid peptides from the midbrain including the ventral tegmental area. *Alcohol. Clin. Exp. Res.* **33**, 1033–1043.

Jhou, T. C., Fields, H. L., Baxter, M. G., Saper, C. B., and Holland, P. C. (2009a). The rostromedial tegmental nucleus (RMTg), a GABAergic afferent to midbrain dopamine neurons, encodes aversive stimuli and inhibits motor responses. *Neuron* **61**, 786–800.

Jhou, T. C., Geisler, S., Marinelli, M., Degarmo, B. A., and Zahm, D. S. (2009b). The mesopontine rostromedial tegmental nucleus: A structure targeted by the lateral habenula that projects to the ventral tegmental area of Tsai and substantia nigra compacta. *J. Comp. Neurol.* **513**, 566–596.

Ji, H., Hougaard, C., Herrik, K. F., Strobaek, D., Christophersen, P., and Shepard, P. D. (2009). Tuning the excitability of midbrain dopamine neurons by modulating the Ca^{2+} sensitivity of SK channels. *Eur. J. Neurosci.* **29**, 1883–1895.

Jia, H. G., Yamuy, J., Sampogna, S., Morales, F. R., and Chase, M. H. (2003). Colocalization of gamma-aminobutyric acid and acetylcholine in neurons in the laterodorsal and pedunculopontine tegmental nuclei in the cat: A light and electron microscopic study. *Brain Res.* **992**, 205–219.

Jiang, Z. L., and Ye, J. H. (2003). Protein kinase C epsilon is involved in ethanol potentiation of glycine-gated Cl(−) current in rat neurons of ventral tegmental area. *Neuropharmacology* **44**, 493–502.

Johnson, B. A. (2008). Update on neuropharmacological treatments for alcoholism: Scientific basis and clinical findings. *Biochem. Pharmacol.* **75**, 34–56.

Johnson, S. W., Mercuri, N. B., and North, R. A. (1992). 5-hydroxytryptamine1B receptors block the GABAB synaptic potential in rat dopamine neurons. *J. Neurosci.* **12**, 2000–2006.

Johnson, S. W., and North, R. A. (1992a). Opioids excite dopamine neurons by hyperpolarization of local interneurons. *J. Neurosci.* **12**, 483–488.

Johnson, S. W., and North, R. A. (1992b). Two types of neurone in the rat ventral tegmental area and their synaptic inputs. *J. Physiol.* **450**, 455–468.

Jones, S., and Bonci, A. (2005). Synaptic plasticity and drug addiction. *Curr. Opin. Pharmacol.* **5**, 20–25.

Jones, S., and Kauer, J. A. (1999). Amphetamine depresses excitatory synaptic transmission via serotonin receptors in the ventral tegmental area. *J. Neurosci.* **19**, 9780–9787.

Kalivas, P. W., Churchill, L., and Klitenick, M. A. (1993). GABA and enkephalin projection from the nucleus accumbens and ventral pallidum to the ventral tegmental area. *Neuroscience* **57**, 1047–1060.

Kalivas, P. W., and Duffy, P. (1993). Time course of extracellular dopamine and behavioral sensitization to cocaine: I. Dopamine axon terminals. *J. Neurosci.* **13**, 266–275.

Karreman, M., Westerink, B. H., and Moghaddam, B. (1996). Excitatory amino acid receptors in the ventral tegmental area regulate dopamine release in the ventral striatum. *J. Neurochem.* **67**, 601–607.

Kauer, J. A., and Malenka, R. C. (2007). Synaptic plasticity and addiction. *Nat. Rev. Neurosci.* **8**, 844–858.

Kaufling, J., Veinante, P., Pawlowski, S. A., Freund-Mercier, M. J., and Barrot, M. (2009). Afferents to the GABAergic tail of the ventral tegmental area in the rat. *J. Comp. Neurol.* **513**, 597–621.

Kaupp, U. B., and Seifert, R. (2001). Molecular diversity of pacemaker ion channels. *Annu. Rev. Physiol.* **63**, 235–257.

Kemppainen, H., Raivio, N., Nurmi, H., and Kiianmaa, K. (2010). GABA and glutamate overflow in the VTA and ventral pallidum of alcohol-preferring AA and alcohol-avoiding ANA rats after ethanol. *Alcohol Alcohol.* **45**, 111–118.

Khaliq, Z. M., and Bean, B. P. (2008). Dynamic, nonlinear feedback regulation of slow pacemaking by A-type potassium current in ventral tegmental area neurons. *J. Neurosci.* **28**, 10905–10917.

Kita, J. M., Kile, B. M., Parker, L. E., and Wightman, R. M. (2009). *in vitro* measurement of somatodendritic release of dopamine in the ventral tegmental area. *Synapse* **63**, 951–960.

Kobayashi, T., Ikeda, K., Kojima, H., Niki, H., Yano, R., Yoshioka, T., and Kumanishi, T. (1999). Ethanol opens G-protein-activated inwardly rectifying K^+ channels. *Nat. Neurosci.* **2**, 1091–1097.

Koga, E., and Momiyama, T. (2000). Presynaptic dopamine D2-like receptors inhibit excitatory transmission onto rat ventral tegmental dopaminergic neurones. *J. Physiol.* **523**(Pt 1), 163–173.

Kohl, R. R., Katner, J. S., Chernet, E., and McBride, W. J. (1998). Ethanol and negative feedback regulation of mesolimbic dopamine release in rats. *Psychopharmacology (Berl.)* **139**, 79–85.

Koob, G. F. (2009). Neurobiological substrates for the dark side of compulsivity in addiction. *Neuropharmacology* **56**(Suppl. 1), 18–31.

Korotkova, T. M., Brown, R. E., Sergeeva, O. A., Ponomarenko, A. A., and Haas, H. L. (2006). Effects of arousal- and feeding-related neuropeptides on dopaminergic and GABAergic neurons in the ventral tegmental area of the rat. *Eur. J. Neurosci.* **23**, 2677–2685.

Korotkova, T. M., Sergeeva, O. A., Eriksson, K. S., Haas, H. L., and Brown, R. E. (2003). Excitation of ventral tegmental area dopaminergic and nondopaminergic neurons by orexins/hypocretins. *J. Neurosci.* **23**, 7–11.

Koyama, S., and Appel, S. B. (2006a). A-type K^+ current of dopamine and GABA neurons in the ventral tegmental area. *J. Neurophysiol.* **96**, 544–554.

Koyama, S., and Appel, S. B. (2006b). Characterization of M-current in ventral tegmental area dopamine neurons. *J. Neurophysiol.* **96**, 535–543.

Koyama, S., Brodie, M. S., and Appel, S. B. (2007). Ethanol inhibition of m-current and ethanol-induced direct excitation of ventral tegmental area dopamine neurons. *J. Neurophysiol.* **97**, 1977–1985.

Kretschmer, B. D. (1999). Modulation of the mesolimbic dopamine system by glutamate: Role of NMDA receptors. *J. Neurochem.* **73**, 839–848.

Labouebe, G., Lomazzi, M., Cruz, H. G., Creton, C., Lujan, R., Li, M., Yanagawa, Y., Obata, K., Watanabe, M., Wickman, K., Boyer, S. B., Slesinger, P. A., and Luscher, C. (2007). RGS2 modulates coupling between GABAB receptors and GIRK channels in dopamine neurons of the ventral tegmental area. *Nat. Neurosci.* **10**, 1559–1568.

Lacey, M. G. (1993). Neurotransmitter receptors and ionic conductances regulating the activity of neurones in substantia nigra pars compacta and ventral tegmental area. *Prog. Brain Res.* **99**, 251–276.

Lacey, M. G., Calabresi, P., and North, R. A. (1990). Muscarine depolarizes rat substantia nigra zona compacta and ventral tegmental neurons *in vitro* through M1-like receptors. *J. Pharmacol. Exp. Ther.* **253**, 395–400.

Lacey, M. G., Mercuri, N. B., and North, R. A. (1987). Dopamine acts on D2 receptors to increase potassium conductance in neurones of the rat substantia nigra zona compacta. *J. Physiol.* **392**, 397–416.

Lacey, M. G., Mercuri, N. B., and North, R. A. (1988). On the potassium conductance increase activated by GABAB and dopamine D2 receptors in rat substantia nigra neurones. *J. Physiol.* **401**, 437–453.

Lammel, S., Hetzel, A., Hackel, O., Jones, I., Liss, B., and Roeper, J. (2008). Unique properties of mesoprefrontal neurons within a dual mesocorticolimbic dopamine system. *Neuron* **57**, 760–773.

Latimer, L. G., Duffy, P., and Kalivas, P. W. (1987). Mu opioid receptor involvement in enkephalin activation of dopamine neurons in the ventral tegmental area. *J. Pharmacol. Exp. Ther.* **241**, 328–337.

Lavoie, B., and Parent, A. (1994a). Pedunculopontine nucleus in the squirrel monkey: Cholinergic and glutamatergic projections to the substantia nigra. *J. Comp. Neurol.* **344**, 232–241.

Lavoie, B., and Parent, A. (1994b). Pedunculopontine nucleus in the squirrel monkey: Distribution of cholinergic and monoaminergic neurons in the mesopontine tegmentum with evidence for the presence of glutamate in cholinergic neurons. *J. Comp. Neurol.* **344**, 190–209.

Lee, Y. K., Park, S. W., Kim, Y. K., Kim, D. J., Jeong, J., Myrick, H., and Kim, Y. H. (2005). Effects of naltrexone on the ethanol-induced changes in the rat central dopaminergic system. *Alcohol Alcohol.* **40**, 297–301.

Leone, P., Pocock, D., and Wise, R. A. (1991). Morphine-dopamine interaction: Ventral tegmental morphine increases nucleus accumbens dopamine release. *Pharmacol. Biochem. Behav.* **39**, 469–472.

Lewohl, J. M., Wilson, W. R., Mayfield, R. D., Brozowski, S. J., Morrisett, R. A., and Harris, R. A. (1999). G-protein-coupled inwardly rectifying potassium channels are targets of alcohol action. *Nat. Neurosci.* **2**, 1084–1090.

Lindholm, S., Ploj, K., Franck, J., and Nylander, I. (2000). Repeated ethanol administration induces short- and long-term changes in enkephalin and dynorphin tissue concentrations in rat brain. *Alcohol* **22**, 165–171.

Lindholm, S., Werme, M., Brene, S., and Franck, J. (2001). The selective kappa-opioid receptor agonist U50,488H attenuates voluntary ethanol intake in the rat. *Behav. Brain Res.* **120**, 137–146.

Liss, B., Franz, O., Sewing, S., Bruns, R., Neuhoff, H., and Roeper, J. (2001). Tuning pacemaker frequency of individual dopaminergic neurons by Kv4.3L and KChip3.1 transcription. *EMBO J.* **20**, 5715–5724.

Liu, L., Shen, R. Y., Kapatos, G., and Chiodo, L. A. (1994). Dopamine neuron membrane physiology: Characterization of the transient outward current (IA) and demonstration of a common signal transduction pathway for IA and IK. *Synapse* **17**, 230–240.

Llinas, R. R., and Alonso, A. (1992). Electrophysiology of the mammillary complex *in vitro*: I. Tuberomammillary and lateral mammillary neurons. *J. Neurophysiol.* **68**, 1307–1320.

Lobo, I. A., and Harris, R. A. (2008). GABA(A) receptors and alcohol. *Pharmacol. Biochem. Behav.* **90**, 90–94.

Logrip, M. L., Janak, P. H., and Ron, D. (2009). Blockade of ethanol reward by the kappa opioid receptor agonist U50,488H. *Alcohol* **43**, 359–365.

Lovinger, D. M. (1991). Ethanol potentiation of 5-HT3 receptor-mediated ion current in NCB-20 neuroblastoma cells. *Neurosci. Lett.* **122**, 57–60

Lovinger, D. M. (1997). Alcohols and neurotransmitter gated ion channels: Past, present and future. *Naunyn Schmiedebergs Arch. Pharmacol.* **356**, 267–282.

Lovinger, D. M., White, G., and Weight, F. F. (1989). Ethanol inhibits NMDA-activated ion current in hippocampal neurons. *Science* **243**, 1721–1724.

Ludlow, K. H., Bradley, K. D., Allison, D. W., Taylor, S. R., Yorgason, J. T., Hansen, D. M., Walton, C. H., Sudweeks, S. N., and Steffensen, S. C. (2009). Acute and chronic ethanol modulate dopamine D2-subtype receptor responses in ventral tegmental area GABA neurons. *Alcohol. Clin. Exp. Res.* **33**, 804–811.

Luthi, A., and McCormick, D. A. (1998). Periodicity of thalamic synchronized oscillations: The role of Ca^{2+}-mediated upregulation of Ih. *Neuron* **20**, 553–563.

Maccaferri, G., and McBain, C. J. (1996). The hyperpolarization-activated current (Ih) and its contribution to pacemaker activity in rat CA1 hippocampal stratum oriens-alveus interneurones. *J. Physiol.* **497**(Pt 1), 119–130.

Magee, J. C. (1998). Dendritic hyperpolarization-activated currents modify the integrative properties of hippocampal CA1 pyramidal neurons. *J. Neurosci.* **18**, 7613–7624.

Mahon, S., Deniau, J. M., and Charpier, S. (2004). Corticostriatal plasticity: Life after the depression. *Trends Neurosci.* **27**, 460–467.

Maldonado, R., Robledo, P., Chover, A. J., Caine, S. B., and Koob, G. F. (1993). D1 dopamine receptors in the nucleus accumbens modulate cocaine self-administration in the rat. *Pharmacol. Biochem. Behav.* **45**, 239–242.

Maldve, R. E., Zhang, T. A., Ferrani-Kile, K., Schreiber, S. S., Lippmann, M. J., Snyder, G. L., Fienberg, A. A., Leslie, S. W., Gonzales, R. A., and Morrisett, R. A. (2002). DARPP-32 and regulation of the ethanol sensitivity of NMDA receptors in the nucleus accumbens. *Nat. Neurosci.* **5**, 641–648.

Mansour, A., Khachaturian, H., Lewis, M. E., Akil, H., and Watson, S. J. (1988). Anatomy of CNS opioid receptors. *Trends Neurosci.* **11**, 308–314.

Mansvelder, H. D., and McGehee, D. S. (2002). Cellular and synaptic mechanisms of nicotine addiction. *J. Neurobiol.* **53**, 606–617.

Manzoni, O. J., and Williams, J. T. (1999). Presynaptic regulation of glutamate release in the ventral tegmental area during morphine withdrawal. *J. Neurosci.* **19**, 6629–6636.

Margolis, E. B., Fields, H. L., Hjelmstad, G. O., and Mitchell, J. M. (2008a). Delta-opioid receptor expression in the ventral tegmental area protects against elevated alcohol consumption. *J. Neurosci.* **28**, 12672–12681.

Margolis, E. B., Hjelmstad, G. O., Bonci, A., and Fields, H. L. (2003). Kappa-opioid agonists directly inhibit midbrain dopaminergic neurons. *J. Neurosci.* **23**, 9981–9986.

Margolis, E. B., Hjelmstad, G. O., Bonci, A., and Fields, H. L. (2005). Both kappa and mu opioid agonists inhibit glutamatergic input to ventral tegmental area neurons. *J. Neurophysiol.* **93**, 3086–3093.

Margolis, E. B., Lock, H., Chefer, V. I., Shippenberg, T. S., Hjelmstad, G. O., and Fields, H. L. (2006a). Kappa opioids selectively control dopaminergic neurons projecting to the prefrontal cortex. *Proc. Natl. Acad. Sci. USA* **103**, 2938–2942.

Margolis, E. B., Lock, H., Hjelmstad, G. O., and Fields, H. L. (2006b). The ventral tegmental area revisited: Is there an electrophysiological marker for dopaminergic neurons? *J. Physiol.* **577**, 907–924.

Margolis, E. B., Mitchell, J. M., Ishikawa, J., Hjelmstad, G. O., and Fields, H. L. (2008b). Midbrain dopamine neurons: Projection target determines action potential duration and dopamine D(2) receptor inhibition. *J. Neurosci.* **28**, 8908–13.

Martinez, D., Gil, R., Slifstein, M., Hwang, D. R., Huang, Y., Perez, A., Kegeles, L., Talbot, P., Evans, S., Krystal, J., Laruelle, M., and Abi-Dargham, A. (2005). Alcohol dependence is associated with blunted dopamine transmission in the ventral striatum. *Biol. Psychiatry* **58**, 779–786.

Matsumoto, M., and Hikosaka, O. (2007). Lateral habenula as a source of negative reward signals in dopamine neurons. *Nature* **447**, 1111–1115.

Matthews, R. T., and German, D. C. (1984). Electrophysiological evidence for excitation of rat ventral tegmental area dopamine neurons by morphine. *Neuroscience* **11**, 617–625.

McDaid, J., McElvain, M. A., and Brodie, M. S. (2008). Ethanol effects on dopaminergic ventral tegmental area neurons during block of Ih: Involvement of barium-sensitive potassium currents. *J. Neurophysiol.* **100**, 1202–1210.

Mejias-Aponte, C. A., Drouin, C., and Aston-Jones, G. (2009). Adrenergic and noradrenergic innervation of the midbrain ventral tegmental area and retrorubral field: Prominent inputs from medullary homeostatic centers. *J. Neurosci.* **29**, 3613–3626.

Melis, M., Camarini, R., Ungless, M. A., and Bonci, A. (2002). Long-lasting potentiation of GABAergic synapses in dopamine neurons after a single *in vivo* ethanol exposure. *J. Neurosci.* **22**, 2074–2082.

Melis, M., Diana, M., Enrico, P., Marinelli, M., and Brodie, M. S. (2009). Ethanol and acetaldehyde action on central dopamine systems: Mechanisms, modulation, and relationship to stress. *Alcohol* **43**, 531–539.

Melis, M., Enrico, P., Peana, A. T., and Diana, M. (2007). Acetaldehyde mediates alcohol activation of the mesolimbic dopamine system. *Eur. J. Neurosci.* **26**, 2824–2833.

Melis, M., Perra, S., Muntoni, A. L., Pillolla, G., Lutz, B., Marsicano, G., Di Marzo, V., Gessa, G. L., and Pistis, M. (2004a). Prefrontal cortex stimulation induces 2-arachidonoyl-glycerol-mediated suppression of excitation in dopamine neurons. *J. Neurosci.* **24**, 10707–10715.

Melis, M., Pistis, M., Perra, S., Muntoni, A. L., Pillolla, G., and Gessa, G. L. (2004b). Endocannabinoids mediate presynaptic inhibition of glutamatergic transmission in rat ventral tegmental area dopamine neurons through activation of CB1 receptors. *J. Neurosci.* **24**, 53–62.

Melis, M., Spiga, S., and Diana, M. (2005). The dopamine hypothesis of drug addiction: Hypodopaminergic state. *Int. Rev. Neurobiol.* **63**, 101–154.

Mena-Segovia, J., Winn, P., and Bolam, J. P. (2008). Cholinergic modulation of midbrain dopaminergic systems. *Brain Res. Rev.* **58**, 265–271.

Mendez, M., Leriche, M., and Calva, J. C. (2001). Acute ethanol administration differentially modulates mu opioid receptors in the rat meso-accumbens and mesocortical pathways. *Brain Res. Mol. Brain Res.* **94**, 148–156.

Mercuri, N. B., Bonci, A., Calabresi, P., Stratta, F., Stefani, A., and Bernardi, G. (1994). Effects of dihydropyridine calcium antagonists on rat midbrain dopaminergic neurones. *Br. J. Pharmacol.* **113**, 831–838.

Meredith, G. E. (1999). The synaptic framework for chemical signaling in nucleus accumbens. *Ann. NY Acad. Sci.* **877**, 140–156.

Mereu, G., Fadda, F., and Gessa, G. L. (1984). Ethanol stimulates the firing rate of nigral dopaminergic neurons in unanesthetized rats. *Brain Res.* **292**, 63–69.

Mereu, G., and Gessa, G. L. (1985). Low doses of ethanol inhibit the firing of neurons in the substantia nigra, pars reticulata: A GABAergic effect? *Brain Res.* **360**, 325–330.

Mileykovskiy, B. Y., Kiyashchenko, L. I., and Siegel, J. M. (2005). Behavioral correlates of activity in identified hypocretin/orexin neurons. *Neuron* **46**, 787–798.

Miquel, M. C., Emerit, M. B., Nosjean, A., Simon, A., Rumajogee, P., Brisorgueil, M. J., Doucet, E., Hamon, M., and Verge, D. (2002). Differential subcellular localization of the 5-HT3-As receptor subunit in the rat central nervous system. *Eur. J. Neurosci.* **15**, 449–457.

Montague, P. R., Hyman, S. E., and Cohen, J. D. (2004). Computational roles for dopamine in behavioural control. *Nature* **431**, 760–767.

Morikawa, H., Khodakhah, K., and Williams, J. T. (2003). Two intracellular pathways mediate metabotropic glutamate receptor-induced Ca^{2+} mobilization in dopamine neurons. *J. Neurosci.* **23**, 149–157.

Morikawa, H., Manzoni, O. J., Crabbe, J. C., and Williams, J. T. (2000). Regulation of central synaptic transmission by 5-HT(1B) auto- and heteroreceptors. *Mol. Pharmacol.* **58**, 1271–1278.

Myers, R. D., and Robinson, D. E. (1999). Mmu and D2 receptor antisense oligonucleotides injected in nucleus accumbens suppress high alcohol intake in genetic drinking HEP rats. *Alcohol* **18**, 225–233.

Narita, M., Nagumo, Y., Hashimoto, S., Khotib, J., Miyatake, M., Sakurai, T., Yanagisawa, M., Nakamachi, T., Shioda, S., and Suzuki, T. (2006). Direct involvement of orexinergic systems in the activation of the mesolimbic dopamine pathway and related behaviors induced by morphine. *J. Neurosci.* **26**, 398–405.

Nedergaard, S., Flatman, J. A., and Engberg, I. (1993). Nifedipine- and omega-conotoxin-sensitive Ca^{2+} conductances in guinea-pig substantia nigra pars compacta neurones. *J. Physiol.* **466**, 727–747.

Nestby, P., Schoffelmeer, A. N., Homberg, J. R., Wardeh, G., De Vries, T. J., Mulder, A. H., and Vanderschuren, L. J. (1999). Bremazocine reduces unrestricted free-choice ethanol self-administration in rats without affecting sucrose preference. *Psychopharmacology (Berl.)* **142**, 309–317.

Neuhoff, H., Neu, A., Liss, B., and Roeper, J. (2002). I(h) channels contribute to the different functional properties of identified dopaminergic subpopulations in the midbrain. *J. Neurosci* .**22**, 1290–1302.

Nicola, S. M. (2007). The nucleus accumbens as part of a basal ganglia action selection circuit. *Psychopharmacology (Berl.)* **191**, 521–550.

Nicola, S. M., Woodward Hopf, F., and Hjelmstad, G. O. (2004). Contrast enhancement: A physiological effect of striatal dopamine? *Cell Tissue Res.* **318**, 93–106.

Nirenberg, M. J., Chan, J., Liu, Y., Edwards, R. H., and Pickel, V. M. (1996). Ultrastructural localization of the vesicular monoamine transporter-2 in midbrain dopaminergic neurons: Potential sites for somatodendritic storage and release of dopamine. *J. Neurosci.* **16**, 4135–4145.

Niv, Y. (2007). Cost, benefit, tonic, phasic: What do response rates tell us about dopamine and motivation? *Ann. N Y Acad. Sci.* **1104**, 357–376.

Niv, Y., Daw, N. D., Joel, D., and Dayan, P. (2007). Tonic dopamine: Opportunity costs and the control of response vigor. *Psychopharmacology (Berl.)* **191**, 507–520.

Noma, A., and Irisawa, H. (1976). Membrane currents in the rabbit sinoatrial node cell as studied by the double microelectrode method. *Pflugers Arch.* **364**, 45–52.

Notomi, T., and Shigemoto, R. (2004). Immunohistochemical localization of Ih channel subunits, HCN1-4, in the rat brain. *J. Comp. Neurol.* **471**, 241–276.

Nowak, K. L., McBride, W. J., Lumeng, L., Li, T. K., and Murphy, J. M. (2000). Involvement of dopamine D2 autoreceptors in the ventral tegmental area on alcohol and saccharin intake of the alcohol-preferring P rat. *Alcohol. Clin. Exp. Res.* **24**, 476–483.

Oakman, S. A., Faris, P. L., Kerr, P. E., Cozzari, C., and Hartman, B. K. (1995). Distribution of pontomesencephalic cholinergic neurons projecting to substantia nigra differs significantly from those projecting to ventral tegmental area. *J. Neurosci.* **15**, 5859–5869.

Okada, H., Matsushita, N., and Kobayashi, K. (2004). Identification of GABAA receptor subunit variants in midbrain dopaminergic neurons. *J. Neurochem.* **89**, 7–14.

Okamoto, T., Harnett, M. T., and Morikawa, H. (2006). Hyperpolarization-activated cation current (Ih) is an ethanol target in midbrain dopamine neurons of mice. *J. Neurophysiol.* **95**, 619–626.

Omelchenko, N., and Sesack, S. R. (2005). Laterodorsal tegmental projections to identified cell populations in the rat ventral tegmental area. *J. Comp. Neurol.* **483**, 217–235.

Omelchenko, N., and Sesack, S. R. (2006). Cholinergic axons in the rat ventral tegmental area synapse preferentially onto mesoaccumbens dopamine neurons. *J. Comp. Neurol.* **494**, 863–875.

Omelchenko, N., and Sesack, S. R. (2007). Glutamate synaptic inputs to ventral tegmental area neurons in the rat derive primarily from subcortical sources. *Neuroscience* **146**, 1259–1274.

Omelchenko, N., and Sesack, S. R. (2009). Ultrastructural analysis of local collaterals of rat ventral tegmental area neurons: GABA phenotype and synapses onto dopamine and GABA cells. *Synapse* **63**, 895–906.

Ortiz, J., Fitzgerald, L. W., Charlton, M., Lane, S., Trevisan, L., Guitart, X., Shoemaker, W., Duman, R. S., and Nestler, E. J. (1995). Biochemical actions of chronic ethanol exposure in the mesolimbic dopamine system. *Synapse* **21**, 289–298.

Overton, P., and Clark, D. (1992). Iontophoretically administered drugs acting at the *N*-methyl-D-aspartate receptor modulate burst firing in A9 dopamine neurons in the rat. *Synapse* **10**, 131–140.

Overton, P. G., and Clark, D. (1997). Burst firing in midbrain dopaminergic neurons. *Brain Res. Brain Res. Rev.* **25**, 312–334.

Paladini, C. A., Fiorillo, C. D., Morikawa, H., and Williams, J. T. (2001). Amphetamine selectively blocks inhibitory glutamate transmission in dopamine neurons. *Nat. Neurosci.* **4**, 275–281.

Paladini, C. A., Iribe, Y., and Tepper, J. M. (1999). GABAA receptor stimulation blocks NMDA-induced bursting of dopaminergic neurons *in vitro* by decreasing input resistance. *Brain Res.* **832**, 145–151.

Paladini, C. A., and Williams, J. T. (2004). Noradrenergic inhibition of midbrain dopamine neurons. *J. Neurosci.* **24**, 4568–4575.

Pan, W. X., and Hyland, B. I. (2005). Pedunculopontine tegmental nucleus controls conditioned responses of midbrain dopamine neurons in behaving rats. *J. Neurosci.* **25**, 4725–4732.

Pan, W. X., Schmidt, R., Wickens, J. R., and Hyland, B. I. (2005). Dopamine cells respond to predicted events during classical conditioning: Evidence for eligibility traces in the reward-learning network. *J. Neurosci.* **25**, 6235–6242.

Paolucci, E., Berretta, N., Tozzi, A., Bernardi, G., and Mercuri, N. B. (2003). Depression of mGluR-mediated IPSCs by 5-HT in dopamine neurons of the rat substantia nigra pars compacta. *Eur. J. Neurosci.* **18**, 2743–2750.

Papadeas, S., Grobin, A. C., and Morrow, A. L. (2001). Chronic ethanol consumption differentially alters GABA(A) receptor alpha1 and alpha4 subunit peptide expression and GABA(A) receptor-mediated 36 Cl(–) uptake in mesocorticolimbic regions of rat brain. *Alcohol. Clin. Exp. Res.* **25**, 1270–1275.

Parent, A., Descarries, L., and Beaudet, A. (1981). Organization of ascending serotonin systems in the adult rat brain. A radioautographic study after intraventricular administration of [3*H*]5-hydroxytryptamine. *Neuroscience* **6**, 115–138.

Pazos, A., and Palacios, J. M. (1985). Quantitative autoradiographic mapping of serotonin receptors in the rat brain: I. Serotonin-1 receptors. *Brain Res.* **346**, 205–230.

Pellegrino, S. M., and Druse, M. J. (1992). The effects of chronic ethanol consumption on the mesolimbic and nigrostriatal dopamine systems. *Alcohol. Clin. Exp. Res.* **16**, 275–280.

Perra, S., Pillolla, G., Melis, M., Muntoni, A. L., Gessa, G. L., and Pistis, M. (2005). Involvement of the endogenous cannabinoid system in the effects of alcohol in the mesolimbic reward circuit: Electrophysiological evidence *in vivo*. *Psychopharmacology (Berl.)* **183**, 368–377.

Pessia, M., Jiang, Z. G., North, R. A., and Johnson, S. W. (1994). Actions of 5-hydroxytryptamine on ventral tegmental area neurons of the rat *in vitro*. *Brain Res.* **654**, 324–330.

Pettit, H. O., and Justice, J. B., Jr. (1989). Dopamine in the nucleus accumbens during cocaine self-administration as studied by *in vivo* microdialysis. *Pharmacol. Biochem. Behav.* **34**, 899–904.

Pettit, H. O., and Justice, J. B., Jr. (1991). Effect of dose on cocaine self-administration behavior and dopamine levels in the nucleus accumbens. *Brain Res.* **539**, 94–102.

Pfeffer, A. O., and Samson, H. H. (1986). Effect of pimozide on home cage ethanol drinking in the rat: Dependence on drinking session length. *Drug Alcohol Depend.* **17**, 47–55.

Pfeffer, A. O., and Samson, H. H. (1988). Haloperidol and apomorphine effects on ethanol reinforcement in free feeding rats. *Pharmacol. Biochem. Behav.* **29**, 343–350.

Phillipson, O. T. (1979a). A Golgi study of the ventral tegmental area of Tsai and interfascicular nucleus in the rat. *J. Comp. Neurol.* **187**, 99–115.

Phillipson, O. T. (1979b). Afferent projections to the ventral tegmental area of Tsai and interfascicular nucleus: A horseradish peroxidase study in the rat. *J. Comp. Neurol.* **187**, 117–143.

Piazza, P. V., Deroche-Gamonent, V., Rouge-Pont, F., and Le Moal, M. (2000). Vertical shifts in self-administration dose-response functions predict a drug-vulnerable phenotype predisposed to addiction. *J. Neurosci.* **20**, 4226–4232.

Pickel, V. M., Chan, J., and Sesack, S. R. (1993). Cellular substrates for interactions between dynorphin terminals and dopamine dendrites in rat ventral tegmental area and substantia nigra. *Brain Res.* **602**, 275–289.

Pidoplichko, V. I., DeBiasi, M., Williams, J. T., and Dani, J. A. (1997). Nicotine activates and desensitizes midbrain dopamine neurons. *Nature* **390**, 401–404.

Poolos, N. P., Migliore, M., and Johnston, D. (2002). Pharmacological upregulation of h-channels reduces the excitability of pyramidal neuron dendrites. *Nat. Neurosci.* **5**, 767–774.

Przewlocka, B., Turchan, J., Lason, W., and Przewlocki, R. (1997). Ethanol withdrawal enhances the prodynorphin system activity in the rat nucleus accumbens. *Neurosci. Lett.* **238**, 13–16.
Puopolo, M., Raviola, E., and Bean, B. P. (2007). Roles of subthreshold calcium current and sodium current in spontaneous firing of mouse midbrain dopamine neurons. *J. Neurosci.* **27**, 645–656.
Putzier, I., Kullmann, P. H., Horn, J. P., and Levitan, E. S. (2009). Cav1.3 channel voltage dependence, not Ca^{2+} selectivity, drives pacemaker activity and amplifies bursts in nigral dopamine neurons. *J. Neurosci.* **29**, 15414–15419.
Raman, I. M., and Bean, B. P. (1999). Ionic currents underlying spontaneous action potentials in isolated cerebellar Purkinje neurons. *J. Neurosci.* **19**, 1663–1674.
Rampon, C., Luppi, P. H., Fort, P., Peyron, C., and Jouvet, M. (1996). Distribution of glycine-immunoreactive cell bodies and fibers in the rat brain. *Neuroscience* **75**, 737–755.
Rasmussen, D. D., Bryant, C. A., Boldt, B. M., Colasurdo, E. A., Levin, N., and Wilkinson, C. W. (1998). Acute alcohol effects on opiomelanocortinergic regulation. *Alcohol. Clin. Exp. Res.* **22**, 789–801.
Rassnick, S., Pulvirenti, L., and Koob, G. F. (1992). Oral ethanol self-administration in rats is reduced by the administration of dopamine and glutamate receptor antagonists into the nucleus accumbens. *Psychopharmacology (Berl.)* **109**, 92–98.
Rassnick, S., Stinus, L., and Koob, G. F. (1993). The effects of 6-hydroxydopamine lesions of the nucleus accumbens and the mesolimbic dopamine system on oral self-administration of ethanol in the rat. *Brain Res.* **623**, 16–24.
Rice, M. E., Cragg, S. J., and Greenfield, S. A. (1997). Characteristics of electrically evoked somatodendritic dopamine release in substantia nigra and ventral tegmental area *in vitro*. *J. Neurophysiol.* **77**, 853–862.
Riegel, A. C., and Lupica, C. R. (2004). Independent presynaptic and postsynaptic mechanisms regulate endocannabinoid signaling at multiple synapses in the ventral tegmental area. *J. Neurosci.* **24**, 11070–11078.
Riegel, A. C., and Williams, J. T. (2008). CRF facilitates calcium release from intracellular stores in midbrain dopamine neurons. *Neuron* **57**, 559–570.
Robertson, M. W., Leslie, C. A., and Bennett, J. P. Jr. (1991). Apparent synaptic dopamine deficiency induced by withdrawal from chronic cocaine treatment. *Brain Res.* **538**, 337–339.
Robinson, D. L., Brunner, L. J., and Gonzales, R. A. (2002). Effect of gender and estrous cycle on the pharmacokinetics of ethanol in the rat brain. *Alcohol. Clin. Exp. Res.* **26**, 165–172.
Robinson, D. L., Howard, E. C., McConnell, S., Gonzales, R. A., and Wightman, R. M. (2009). Disparity between tonic and phasic ethanol-induced dopamine increases in the nucleus accumbens of rats. *Alcohol. Clin. Exp. Res.* **33**, 1187–1196.
Robinson, D. L., Lara, J. A., Brunner, L. J., and Gonzales, R. A. (2000). Quantification of ethanol concentrations in the extracellular fluid of the rat brain: *In vivo* calibration of microdialysis probes. *J. Neurochem.* **75**, 1685–1693.
Robinson, R. B., and Siegelbaum, S. A. (2003). Hyperpolarization-activated cation currents: From molecules to physiological function. *Annu. Rev. Physiol.* **65**, 453–480.
Rodaros, D., Caruana, D. A., Amir, S., and Stewart, J. (2007). Corticotropin-releasing factor projections from limbic forebrain and paraventricular nucleus of the hypothalamus to the region of the ventral tegmental area. *Neuroscience* **150**, 8–13.
Rodd, Z. A., Bell, R. L., McQueen, V. K., Davids, M. R., Hsu, C. C., Murphy, J. M., Li, T. K., Lumeng, L., and McBride, W. J. (2005a). Chronic ethanol drinking by alcohol-preferring rats increases the sensitivity of the posterior ventral tegmental area to the reinforcing effects of ethanol. *Alcohol. Clin. Exp. Res.* **29**, 358–366.
Rodd, Z. A., Bell, R. L., McQueen, V. K., Davids, M. R., Hsu, C. C., Murphy, J. M., Li, T. K., Lumeng, L., and McBride, W. J. (2005b). Prolonged increase in the sensitivity of the posterior ventral tegmental area to the reinforcing effects of ethanol following repeated exposure to cycles of ethanol access and deprivation. *J. Pharmacol. Exp. Ther.* **315**, 648–657.

Rodd, Z. A., Bell, R. L., Zhang, Y., Murphy, J. M., Goldstein, A., Zaffaroni, A., Li, T. K., and McBride, W. J. (2005c). Regional heterogeneity for the intracranial self-administration of ethanol and acetaldehyde within the ventral tegmental area of alcohol-preferring (P) rats: Involvement of dopamine and serotonin. *Neuropsychopharmacology* **30**, 330–338.

Rodd, Z. A., Melendez, R. I., Bell, R. L., Kuc, K. A., Zhang, Y., Murphy, J. M., and McBride, W. J. (2004). Intracranial self-administration of ethanol within the ventral tegmental area of male Wistar rats: Evidence for involvement of dopamine neurons. *J. Neurosci.* **24**, 1050–1057.

Rodd-Henricks, Z. A., McKinzie, D. L., Melendez, R. I., Berry, N., Murphy, J. M., and McBride, W. J. (2003). Effects of serotonin-3 receptor antagonists on the intracranial self-administration of ethanol within the ventral tegmental area of Wistar rats. *Psychopharmacology (Berl.)* **165**, 252–259.

Ron, D., and Jurd, R. (2005). The "ups and downs" of signaling cascades in addiction. *Sci. STKE* **2005**, re14.

Ronald, K. M., Mirshahi, T., and Woodward, J. J. (2001). Ethanol inhibition of *N*-methyl-D-aspartate receptors is reduced by site-directed mutagenesis of a transmembrane domain phenylalanine residue. *J. Biol. Chem.* **276**, 44729–44735.

Rosin, A., Lindholm, S., Franck, J., and Georgieva, J. (1999). Downregulation of kappa opioid receptor mRNA levels by chronic ethanol and repetitive cocaine in rat ventral tegmentum and nucleus accumbens. *Neurosci. Lett.* **275**, 1–4.

Rosin, D. L., Weston, M. C., Sevigny, C. P., Stornetta, R. L., and Guyenet, P. G. (2003). Hypothalamic orexin (hypocretin) neurons express vesicular glutamate transporters VGLUT1 or VGLUT2. *J. Comp. Neurol.* **465**, 593–603.

Rossetti, Z. L., Melis, F., Carboni, S., and Gessa, G. L. (1992). Dramatic depletion of mesolimbic extracellular dopamine after withdrawal from morphine, alcohol or cocaine: A common neurochemical substrate for drug dependence. *Ann. NY Acad. Sci.* **654**, 513–516.

Rudy, B. (1988). Diversity and ubiquity of K channels. *Neuroscience* **25**, 729–749.

Salamone, J. D., Correa, M., Farrar, A., and Mingote, S. M. (2007). Effort-related functions of nucleus accumbens dopamine and associated forebrain circuits. *Psychopharmacology (Berl.)* **191**, 461–482.

Samson, H. H., Hodge, C. W., Tolliver, G. A., and Haraguchi, M. (1993). Effect of dopamine agonists and antagonists on ethanol-reinforced behavior: The involvement of the nucleus accumbens. *Brain Res. Bull.* **30**, 133–141.

Sanchez-Catalan, M. J., Hipolito, L., Zornoza, T., Polache, A., and Granero, L. (2009). Motor stimulant effects of ethanol and acetaldehyde injected into the posterior ventral tegmental area of rats: Role of opioid receptors. *Psychopharmacology* (*Berl.*) **204**, 641–653.

Santoro, B., Chen, S., Luthi, A., Pavlidis, P., Shumyatsky, G. P., Tibbs, G. R., and Siegelbaum, S. A. (2000). Molecular and functional heterogeneity of hyperpolarization-activated pacemaker channels in the mouse CNS. *J. Neurosci.* **20**, 5264–5275.

Schilstrom, B., Yaka, R., Argilli, E., Suvarna, N., Schumann, J., Chen, B. T., Carman, M., Singh, V., Mailliard, W. S., Ron, D., and Bonci, A. (2006). Cocaine enhances NMDA receptor-mediated currents in ventral tegmental area cells via dopamine D5 receptor-dependent redistribution of NMDA receptors. *J. Neurosci.* **26**, 8549–8558.

Schulteis, G., Markou, A., Cole, M., and Koob, G. F. (1995). Decreased brain reward produced by ethanol withdrawal. *Proc. Natl. Acad. Sci. USA* **92**, 5880–5884.

Schultz, W. (1998). Predictive reward signal of dopamine neurons. *J. Neurophysiol.* **80**, 1–27.

Semba, K., and Fibiger, H. C. (1992). Afferent connections of the laterodorsal and the pedunculopontine tegmental nuclei in the rat: A retro- and antero-grade transport and immunohistochemical study. *J. Comp. Neurol.* **323**, 387–410.

Sesack, S. R., and Pickel, V. M. (1992). Prefrontal cortical efferents in the rat synapse on unlabeled neuronal targets of catecholamine terminals in the nucleus accumbens septi and on dopamine neurons in the ventral tegmental area. *J. Comp. Neurol.* **320**, 145–160.

Seutin, V., Massotte, L., Renette, M. F., and Dresse, A. (2001). Evidence for a modulatory role of Ih on the firing of a subgroup of midbrain dopamine neurons. *Neuroreport* **12**, 255–258.

Shen, R. Y. (2003). Ethanol withdrawal reduces the number of spontaneously active ventral tegmental area dopamine neurons in conscious animals. *J. Pharmacol. Exp. Ther.* **307**, 566–572.

Shen, R. Y., and Chiodo, L. A. (1993). Acute withdrawal after repeated ethanol treatment reduces the number of spontaneously active dopaminergic neurons in the ventral tegmental area. *Brain Res.* **622**, 289–293.

Shen, R. Y., Choong, K. C., and Thompson, A. C. (2007). Long-term reduction in ventral tegmental area dopamine neuron population activity following repeated stimulant or ethanol treatment. *Biol. Psychiatry* **61**, 93–100.

Shepard, P. D., and Bunney, B. S. (1991). Repetitive firing properties of putative dopamine-containing neurons *in vitro*: Regulation by an apamin-sensitive $Ca^{(2+)}$-activated K^+ conductance. *Exp. Brain Res.* **86**, 141–150.

Shoji, Y., Delfs, J., and Williams, J. T. (1999). Presynaptic inhibition of GABA(B)-mediated synaptic potentials in the ventral tegmental area during morphine withdrawal. *J. Neurosci.* **19**, 2347–2355.

Silva, N. L., Pechura, C. M., and Barker, J. L. (1990). Postnatal rat nigrostriatal dopaminergic neurons exhibit five types of potassium conductances. *J. Neurophysiol.* **64**, 262–272.

Skvorak, K., Vissel, B., and Homanics, G. E. (2006). Production of conditional point mutant knockin mice. *Genesis* **44**, 345–353.

Smith, A. D., and Weiss, F. (1999). Ethanol exposure differentially alters central monoamine neurotransmission in alcohol-preferring versus -nonpreferring rats. *J. Pharmacol. Exp. Ther.* **288**, 1223–1228.

Smith, Y., Charara, A., and Parent, A. (1996). Synaptic innervation of midbrain dopaminergic neurons by glutamate-enriched terminals in the squirrel monkey. *J. Comp. Neurol.* **364**, 231–253.

Sombers, L. A., Beyene, M., Carelli, R. M., and Wightman, R. M. (2009). Synaptic overflow of dopamine in the nucleus accumbens arises from neuronal activity in the ventral tegmental area. *J. Neurosci.* **29**, 1735–1742.

Spanagel, R., Herz, A., and Shippenberg, T. S. (1992). Opposing tonically active endogenous opioid systems modulate the mesolimbic dopaminergic pathway. *Proc. Natl. Acad. Sci. USA* **89**, 2046–2050.

Steffensen, S. C., Nie, Z., Criado, J. R., and Siggins, G. R. (2000). Ethanol inhibition of *N*-methyl-D-aspartate responses involves presynaptic gamma-aminobutyric acid(B) receptors. *J. Pharmacol. Exp. Ther.* **294**, 637–647.

Steffensen, S. C., Walton, C. H., Hansen, D. M., Yorgason, J. T., Gallegos, R. A., and Criado, J. R. (2009). Contingent and non-contingent effects of low-dose ethanol on GABA neuron activity in the ventral tegmental area. *Pharmacol. Biochem. Behav.* **92**, 68–75.

Stobbs, S. H., Ohran, A. J., Lassen, M. B., Allison, D. W., Brown, J. E., and Steffensen, S. C. (2004). Ethanol suppression of ventral tegmental area GABA neuron electrical transmission involves *N*-methyl-D-aspartate receptors. *J. Pharmacol. Exp. Ther.* **311**, 282–289.

Stuber, G. D., Hopf, F. W., Hahn, J., Cho, S. L., Guillory, A., and Bonci, A. (2008). Voluntary ethanol intake enhances excitatory synaptic strength in the ventral tegmental area. *Alcohol. Clin. Exp. Res.* **32**, 1714–1720.

Stuber, G. D., Wightman, R. M., and Carelli, R. M. (2005). Extinction of cocaine self-administration reveals functionally and temporally distinct dopaminergic signals in the nucleus accumbens. *Neuron* **46**, 661–669.

Tagliaferro, P., and Morales, M. (2008). Synapses between corticotropin-releasing factor-containing axon terminals and dopaminergic neurons in the ventral tegmental area are predominantly glutamatergic. *J. Comp. Neurol.* **506**, 616–626.

Tan, K. R., Brown, M., Labouebe, G., Yvon, C., Creton, C., Fritschy, J. M., Rudolph, U., and Luscher, C. (2010). Neural bases for addictive properties of benzodiazepines. *Nature* **463**, 769–774.

Tao, L., and Ye, J. H. (2002). Protein kinase C modulation of ethanol inhibition of glycine-activated current in dissociated neurons of rat ventral tegmental area. *J. Pharmacol. Exp. Ther.* **300**, 967–975.

Tepper, J. M., and Lee, C. R. (2007). GABAergic control of substantia nigra dopaminergic neurons. *Prog. Brain Res.* **160**, 189–208.

Testa, C. M., Standaert, D. G., Young, A. B., and Penney, J. B. Jr. (1994). Metabotropic glutamate receptor mRNA expression in the basal ganglia of the rat. *J. Neurosci.* **14**, 3005–3018.

Theile, J. W., Morikawa, H., Gonzales, R. A., and Morrisett, R. A. (2008). Ethanol enhances GABAergic transmission onto dopamine neurons in the ventral tegmental area of the rat. *Alcohol. Clin. Exp. Res.* **32**, 1040–1048.

Theile, J. W., Morikawa, H., Gonzales, R. A., and Morrisett, R. A. (2009). Role of 5-hydroxytryptamine2C receptors in Ca^{2+}-dependent ethanol potentiation of GABA release onto ventral tegmental area dopamine neurons. *J. Pharmacol. Exp. Ther.* **329**, 625–633.

Thielen, R. J., Engleman, E. A., Rodd, Z. A., Murphy, J. M., Lumeng, L., Li, T. K., and McBride, W. J. (2004). Ethanol drinking and deprivation alter dopaminergic and serotonergic function in the nucleus accumbens of alcohol-preferring rats. *J. Pharmacol. Exp. Ther.* **309**, 216–225.

Tizabi, Y., Copeland, R. L., Jr., Louis, V. A., and Taylor, R. E. (2002). Effects of combined systemic alcohol and central nicotine administration into ventral tegmental area on dopamine release in the nucleus accumbens. *Alcohol. Clin. Exp. Res.* **26**, 394–399.

Tong, Z. Y., Overton, P. G., and Clark, D. (1996). Antagonism of NMDA receptors but not AMPA/kainate receptors blocks bursting in dopaminergic neurons induced by electrical stimulation of the prefrontal cortex. *J. Neural. Transm.* **103**, 889–904.

Tozzi, A., Bengtson, C. P., Longone, P., Carignani, C., Fusco, F. R., Bernardi, G., and Mercuri, N. B. (2003). Involvement of transient receptor potential-like channels in responses to mGluR-I activation in midbrain dopamine neurons. *Eur. J. Neurosci.* **18**, 2133–2145.

Treistman, S. N., and Wilson, A. (1987a). Alkanol effects on early potassium currents in *Aplysia* neurons depend on chain length. *Proc. Natl. Acad. Sci. USA* **84**, 9299–9303.

Treistman, S. N., and Wilson, A. (1987b). Effects of ethanol on early potassium currents in *Aplysia*: Cell specificity and influence of channel state. *J. Neurosci.* **7**, 3207–3214.

Trevisan, L. A., Boutros, N., Petrakis, I. L., and Krystal, J. H. (1998). Complications of alcohol withdrawal: Pathophysiological insights. *Alcohol Health Res. World* **22**, 61–66.

Trifunovic, R. D., and Brodie, M. S. (1996). The effects of clomipramine on the excitatory action of ethanol on dopaminergic neurons of the ventral tegmental area *in vitro*. *J. Pharmacol. Exp. Ther.* **276**, 34–40.

Tsibulsky, V. L., and Norman, A. B. (1999). Satiety threshold: A quantitative model of maintained cocaine self-administration. *Brain Res.* **839**, 85–93.

Ungless, M. A., Singh, V., Crowder, T. L., Yaka, R., Ron, D., and Bonci, A. (2003). Corticotropin-releasing factor requires CRF binding protein to potentiate NMDA receptors via CRF receptor 2 in dopamine neurons. *Neuron* **39**, 401–407.

Uzbay, I. T. (2008). Serotonergic anti-depressants and ethanol withdrawal syndrome: A review. *Alcohol Alcohol.* **43**, 15–24.

Valenzuela, C. F., Bhave, S., Hoffman, P., and Harris, R. A. (1998). Acute effects of ethanol on pharmacologically isolated kainate receptors in cerebellar granule neurons: Comparison with NMDA and AMPA receptors. *J. Neurochem.* **71**, 1777–1780.

Vengeliene, V., Bilbao, A., Molander, A., and Spanagel, R. (2008). Neuropharmacology of alcohol addiction. *Br. J. Pharmacol.* **154**, 299–315.

Venton, B. J., Zhang, H., Garris, P. A., Phillips, P. E., Sulzer, D., and Wightman, R. M. (2003). Real-time decoding of dopamine concentration changes in the caudate-putamen during tonic and phasic firing. *J. Neurochem.* **87**, 1284–1295.

Vittoz, N. M., and Berridge, C. W. (2006). Hypocretin/orexin selectively increases dopamine efflux within the prefrontal cortex: Involvement of the ventral tegmental area. *Neuropsychopharmacology* **31**, 384–395.

Volkow, N. D., Fowler, J. S., Wang, G. J., and Swanson, J. M. (2004). Dopamine in drug abuse and addiction: Results from imaging studies and treatment implications. *Mol. Psychiatry* **9**, 557–569.

Volkow, N. D., Wang, G. J., Telang, F., Fowler, J. S., Logan, J., Jayne, M., Ma, Y., Pradhan, K., and Wong, C. (2007). Profound decreases in dopamine release in striatum in detoxified alcoholics: Possible orbitofrontal involvement. *J. Neurosci.* **27**, 12700–12706.

Waeber, C., Dietl, M. M., Hoyer, D., and Palacios, J. M. (1989a). 5.HT1 receptors in the vertebrate brain. Regional distribution examined by autoradiography. *Naunyn Schmiedebergs Arch. Pharmacol.* **340**, 486–494.

Waeber, C., Hoyer, D., and Palacios, J. M. (1989b). 5-hydroxytryptamine3 receptors in the human brain: Autoradiographic visualization using [3*H*]ICS 205-930. *Neuroscience* **31**, 393–400.

Walaas, I., and Fonnum, F. (1980). Biochemical evidence for gamma-aminobutyrate containing fibres from the nucleus accumbens to the substantia nigra and ventral tegmental area in the rat. *Neuroscience* **5**, 63–72.

Wallner, M., Hanchar, H. J., and Olsen, R. W. (2003). Ethanol enhances alpha 4 beta 3 delta and alpha 6 beta 3 delta gamma-aminobutyric acid type A receptors at low concentrations known to affect humans. *Proc. Natl. Acad. Sci. USA* **100**, 15218–15223.

Walter, H. J., and Messing, R. O. (1999). Regulation of neuronal voltage-gated calcium channels by ethanol. *Neurochem. Int.* **35**, 95–101.

Wanat, M. J., Hopf, F. W., Stuber, G. D., Phillips, P. E., and Bonci, A. (2008). Corticotropin-releasing factor increases mouse ventral tegmental area dopamine neuron firing through a protein kinase C-dependent enhancement of Ih. *J. Physiol.* **586**, 2157–2170.

Wanat, M. J., Sparta, D. R., Hopf, F. W., Bowers, M. S., Melis, M., and Bonci, A. (2009). Strain specific synaptic modifications on ventral tegmental area dopamine neurons after ethanol exposure. *Biol. Psychiatry* **65**, 646–653.

Wang, X., Wang, G., Lemos, J. R., and Treistman, S. N. (1994). Ethanol directly modulates gating of a dihydropyridine-sensitive Ca^{2+} channel in neurohypophysial terminals. *J. Neurosci.* **14**, 5453–5460.

Wee, S., and Koob, G.F. (2010). The role of the dynorphin-kappa opioid system in the reinforcing effects of drugs of abuse. *Psychopharmacology (Berl.)* **210**, 121–135.

Weight, F. F., Aguayo, L. G., White, G., Lovinger, D. M., and Peoples, R. W. (1992). GABA- and glutamate-gated ion channels as molecular sites of alcohol and anesthetic action. *Adv. Biochem. Psychopharmacol.* **47**, 335–347.

Weiss, F., Hurd, Y. L., Ungerstedt, U., Markou, A., Plotsky, P. M., and Koob, G. F. (1992a). Neurochemical correlates of cocaine and ethanol self-administration. *Ann. NY Acad. Sci.* **654**, 220–241.

Weiss, F., Lorang, M. T., Bloom, F. E., and Koob, G. F. (1993). Oral alcohol self-administration stimulates dopamine release in the rat nucleus accumbens: Genetic and motivational determinants. *J. Pharmacol. Exp. Ther.* **267**, 250–258.

Weiss, F., Parsons, L. H., Schulteis, G., Hyytia, P., Lorang, M. T., Bloom, F. E., and Koob, G. F. (1996). Ethanol self-administration restores withdrawal-associated deficiencies in accumbal dopamine and 5-hydroxytryptamine release in dependent rats. *J. Neurosci.* **16**, 3474–3485.

Weiss, F., Paulus, M. P., Lorang, M. T., and Koob, G. F. (1992b). Increases in extracellular dopamine in the nucleus accumbens by cocaine are inversely related to basal levels: Effects of acute and repeated administration. *J. Neurosci.* **12**, 4372–4380.

Wickens, J. R., Reynolds, J. N., and Hyland, B. I. (2003). Neural mechanisms of reward-related motor learning. *Curr. Opin. Neurobiol.* **13**, 685–690.

Wilson, C. J., and Callaway, J. C. (2000). Coupled oscillator model of the dopaminergic neuron of the substantia nigra. *J. Neurophysiol.* **83**, 3084–3100.

Wise, R. A. (2008). Dopamine and reward: The anhedonia hypothesis 30 years on. *Neurotox. Res.* **14**, 169–183.

Wise, R. A., and Morales, M. (2010). A ventral tegmental CRF–glutamate–dopamine interaction in addiction. *Brain Res.* **1314**, 38–43.

Wise, R. A., Newton, P., Leeb, K., Burnette, B., Pocock, D., and Justice, J. B., Jr. (1995). Fluctuations in nucleus accumbens dopamine concentration during intravenous cocaine self-administration in rats. *Psychopharmacology (Berl.)* **120**, 10–20.

Wolf, M. E., Sun, X., Mangiavacchi, S., and Chao, S. Z. (2004). Psychomotor stimulants and neuronal plasticity. *Neuropharmacology* **47**(Suppl. 1), 61–79.

Wolfart, J., Neuhoff, H., Franz, O., and Roeper, J. (2001). Differential expression of the small-conductance, calcium-activated potassium channel SK3 is critical for pacemaker control in dopaminergic midbrain neurons. *J. Neurosci.* **21**, 3443–3456.

Wolfart, J., and Roeper, J. (2002). Selective coupling of T-type calcium channels to SK potassium channels prevents intrinsic bursting in dopaminergic midbrain neurons. *J. Neurosci.* **22**, 3404–3413.

Woods, J. M., and Druse, M. J. (1996). Effects of chronic ethanol consumption and aging on dopamine, serotonin, and metabolites. *J. Neurochem.* **66**, 2168–2178.

Wooltorton, J. R., Pidoplichko, V. I., Broide, R. S., and Dani, J. A. (2003). Differential desensitization and distribution of nicotinic acetylcholine receptor subtypes in midbrain dopamine areas. *J. Neurosci.* **23**, 3176–3185.

Xiao, C., Shao, X. M., Olive, M. F., Griffin, W. C., 3rd., Li, K. Y., Krnjevic, K., Zhou, C., and Ye, J. H. (2009). Ethanol facilitates glutamatergic transmission to dopamine neurons in the ventral tegmental area. *Neuropsychopharmacology* **34**, 307–318.

Xiao, C., and Ye, J. H. (2008). Ethanol dually modulates GABAergic synaptic transmission onto dopaminergic neurons in ventral tegmental area: Role of mu-opioid receptors. *Neuroscience* **153**, 240–248.

Xiao, C., Zhang, J., Krnjevic, K., and Ye, J. H. (2007). Effects of ethanol on midbrain neurons: Role of opioid receptors. *Alcohol. Clin. Exp. Res.* **31**, 1106–1113.

Xiao, C., Zhou, C., Li, K., Davies, D. L., and Ye, J. H. (2008). Purinergic type 2 receptors at GABAergic synapses on ventral tegmental area dopamine neurons are targets for ethanol action. *J. Pharmacol. Exp. Ther.* **327**, 196–205.

Yamada, M., Basile, A. S., Fedorova, I., Zhang, W., Duttaroy, A., Cui, Y., Lamping, K. G., Faraci, F. M., Deng, C. X., and Wess, J. (2003). Novel insights into M5 muscarinic acetylcholine receptor function by the use of gene targeting technology. *Life Sci.* **74**, 345–353.

Yan, H., Li, Q., Fleming, R., Madison, R. D., Wilson, W. A., and Swartzwelder, H. S. (2009). Developmental sensitivity of hippocampal interneurons to ethanol: Involvement of the hyperpolarization-activated current, Ih. *J. Neurophysiol.* **101**, 67-83.

Yan, Q. S., Reith, M. E., Jobe, P. C., and Dailey, J. W. (1996). Focal ethanol elevates extracellular dopamine and serotonin concentrations in the rat ventral tegmental area. *Eur. J. Pharmacol.* **301**, 49–57.

Yan, Q. S., Zheng, S. Z., Feng, M. J., and Yan, S. E. (2005). Involvement of 5-HT1B receptors within the ventral tegmental area in ethanol-induced increases in mesolimbic dopaminergic transmission. *Brain Res.* **1060**, 126–137.

Yang, F., Feng, L., Zheng, F., Johnson, S. W., Du, J., Shen, L., Wu, C. P., and Lu, B. (2001). GDNF acutely modulates excitability and A-type K(+) channels in midbrain dopaminergic neurons. *Nat. Neurosci.* **4**, 1071–1078.

Ye, J. (2000). Physiology and pharmacology of native glycine receptors in developing rat ventral tegmental area neurons. *Brain Res.* **862**, 74–82.

Ye, J. H., Ren, J., Liu, P. L., and McArdle, J. J. (1998). Glycine-activated chloride currents of neurons freshly isolated from the ventral tegmental area of rats. *Brain Res.* **796**, 53–62.

Ye, J. H., Tao, L., Ren, J., Schaefer, R., Krnjevic, K., Liu, P. L., Schiller, D. A., and McArdle, J. J. (2001a). Ethanol potentiation of glycine-induced responses in dissociated neurons of rat ventral tegmental area. *J. Pharmacol. Exp. Ther.* **296**, 77–83.

Ye, J. H., Tao, L., Zhu, L., Krnjevic, K., and McArdle, J. J. (2001b). Ethanol inhibition of glycine-activated responses in neurons of ventral tegmental area of neonatal rats. *J. Neurophysiol.* **86**, 2426–2434.

Ye, J. H., Wang, F., Krnjevic, K., Wang, W., Xiong, Z. G., and Zhang, J. (2004). Presynaptic glycine receptors on GABAergic terminals facilitate discharge of dopaminergic neurons in ventral tegmental area. *J. Neurosci.* **24**, 8961–8974.

Yeomans, J., Forster, G., and Blaha, C. (2001). M5 muscarinic receptors are needed for slow activation of dopamine neurons and for rewarding brain stimulation. *Life Sci.* **68**, 2449–2456.

Yim, H. J., and Gonzales, R. A. (2000). Ethanol-induced increases in dopamine extracellular concentration in rat nucleus accumbens are accounted for by increased release and not uptake inhibition. *Alcohol* **22**, 107–115.

Yin, H. H., Mulcare, S. P., Hilario, M. R., Clouse, E., Holloway, T., Davis, M. I., Hansson, A. C., Lovinger, D. M., and Costa, R. M. (2009). Dynamic reorganization of striatal circuits during the acquisition and consolidation of a skill. *Nat. Neurosci.* **12**, 333–341.

Zahm, D. S. (1989). The ventral striatopallidal parts of the basal ganglia in the rat: II. Compartmentation of ventral pallidal efferents. *Neuroscience* **30**, 33–50.

Zapata, A., Gonzales, R. A., and Shippenberg, T. S. (2006). Repeated ethanol intoxication induces behavioral sensitization in the absence of a sensitized accumbens dopamine response in C57BL/6J and DBA/2J mice. *Neuropsychopharmacology* **31**, 396–405.

Zhang, F., Aravanis, A. M., Adamantidis, A., de Lecea, L., and Deisseroth, K. (2007). Circuit-breakers: Optical technologies for probing neural signals and systems. *Nat. Rev. Neurosci.* **8**, 577–581.

Zhang, T. A., Maldve, R. E., and Morrisett, R. A. (2006). Coincident signaling in mesolimbic structures underlying alcohol reinforcement. *Biochem. Pharmacol.* **72**, 919–927.

Zhang, X. F., Hu, X. T., White, F. J., and Wolf, M. E. (1997). Increased responsiveness of ventral tegmental area dopamine neurons to glutamate after repeated administration of cocaine or amphetamine is transient and selectively involves AMPA receptors. *J. Pharmacol. Exp. Ther.* **281**, 699–706.

Zweifel, L. S., Parker, J. G., Lobb, C. J., Rainwater, A., Wall, V. Z., Fadok, J. P., Darvas, M., Kim, M. J., Mizumori, S. J., Paladini, C. A., Phillips, P. E., and Palmiter, R. D. (2009). Disruption of NMDAR-dependent burst firing by dopamine neurons provides selective assessment of phasic dopamine-dependent behavior. *Proc. Natl. Acad. Sci. USA* **106**, 7281–7288.

ALCOHOL AND THE PREFRONTAL CORTEX

Kenneth Abernathy, L. Judson Chandler, and John J. Woodward

Department of Neurosciences and Center for Drug and Alcohol Programs, Medical University of South Carolina, Charleston, SC 29425, USA

The prefrontal cortex occupies the anterior portion of the frontal lobes and is thought to be one of the most complex anatomical and functional structures of the mammalian brain. Its major role is to integrate and interpret inputs from cortical and sub-cortical structures and use this information to develop purposeful responses that reflect both present and future circumstances. This includes both action-oriented sequences involved in obtaining rewards and inhibition of behaviors that pose undue risk or harm to the individual. Given the central role in initiating and regulating these often complex cognitive and behavioral responses, it is no surprise that alcohol has profound effects on the function of the prefrontal cortex. In this chapter, we review the basic anatomy and physiology of the prefrontal cortex and discuss what is known about the actions of alcohol on the function of this brain region. This includes a review of both the human and animal literature including information on the electrophysiological and behavioral effects that follow acute and chronic exposure to alcohol. The chapter concludes with a discussion of unanswered questions and areas needing further investigation.

INTERNATIONAL REVIEW OF
NEUROBIOLOGY, VOL. 91
DOI: 10.1016/S0074-7742(10)91009-X

0074-7742/10 $35.00

I. Introduction

It is now widely accepted that alcohol and other addictive drugs act within the mesolimbic dopamine (DA) system of the brain. This system originates in the ventral tegmental area (VTA) and projects to various limbic structures, including the nucleus accumbens (NAc), amygdala, and hippocampus. Most notably, it is thought that the positive reinforcing effects of drugs of abuse relate to enhanced DA neurotransmission particularly within the NAc. Therefore, virtually all models of the addiction neurocircuitry feature the mesolimbic DA system as central to the addictive process. However, evidence gained over the past decade or more suggests that drug-induced changes in the prefrontal cortex (PFC) also critically regulate drug and alcohol addiction (Everitt and Robbins, 2005; Kalivas and Volkow, 2005; Kalivas, 2009). This evidence comes from diverse studies that include human and animal behavioral work, brain imaging, electrophysiology, and molecular and cellular observations. Whereas a comprehensive review of the role of the PFC in addiction is beyond the scope of this chapter, our aim is to provide a review of the literature regarding the effects of acute and chronic alcohol exposure on PFC structure and function. Consistent with the accumulating evidence implicating the PFC in addiction to opiates, cocaine, and other psychostimulants, it is our contention that current models of the neurocircuitry of alcohol addiction should prominently include ethanol-induced alterations in PFC structure and function. We further suggest that long-lasting alterations in the executive function of the PFC and its associated networks may play an equal or even greater role in alcohol addiction and relapse to drinking than changes within the mesolimbic DA system.

II. Anatomy of the Prefrontal Cortex

The cerebral cortex of humans and nonhuman primates is separated into a frontal and posterior region by the Rolandic fissure. The primary motor cortex (area 4) is a relatively narrow cortical area that lies immediately anterior to this fissure, and immediately anterior to the primary motor cortex is the premotor cortex (area 6). The PFC makes up the remaining anterior pole of the frontal cortex (Fig. 1, shaded area) and is divided into three interconnected subregions known as the lateral PFC, the medial/cingulate PFC, and the orbital PFC (commonly referred to as the orbitofrontal cortex (OFC)). Although overlapping but not identical, the term ventromedial PFC is sometimes encountered in the literature instead of OFC.

The PFC was originally termed the "frontal granular cortex" based upon the presence of a cortical granular layer IV and a location rostral to the agranular

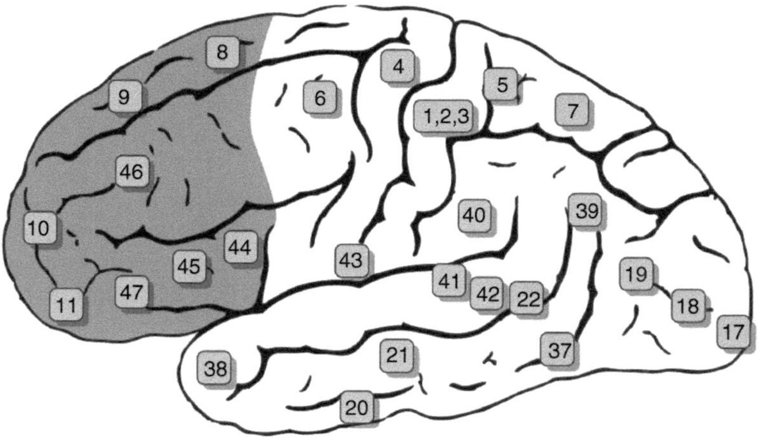

FIG. 1. Brodmann areas of the human brain. Schematic shows the Brodmann areas of the human brain with those of the prefrontal cortex shaded in orange. From Grays Anatomy.

premotor cortex. However, subsequent comparisons across species of other cortical areas with known functional homology soon revealed that defining cortical regions based upon a granular versus agranular cytoarchitecture was invalid, and current definitions of the PFC utilize functional homology and anatomical connectivity across species. With regard to connectivity, the presence of dense reciprocal projections between the mediodorsal nucleus of the thalamus along with certain cortico-cortical connections is now considered the major defining structural feature of the PFC that is applicable to all mammalian species (Fuster, 2008).

While it must be kept in mind that there are generally no neat boundaries separating the subregions of the PFC, the human PFC can be grossly localized to specific Brodmann areas of the frontal cortex. The lateral PFC is composed principally of parts or all of Brodmann areas 8, 9, 10, and 46, and the medial PFC is composed principally of parts of areas 8, 9, 10, and 12, with areas 24, 25, and 32 forming the anterior cingulate (ACC). The orbitofrontal cortex (OFC) is the ventral portion of the frontal cortex on the dorsal surface of the orbit and is comprised of areas 11, 12, 13, and 14.

The PFC is extensively interconnected not only between cortical layers but also between subregions, and this connectivity helps to define distinct neural networks that exert behavioral control. The portion of the lateral PFC dorsal to the fundus of the principal sulcus receives afferent projections mainly from the medial PFC and belongs to the dorsolateral PFC network, whereas that portion that lies ventral to the principal sulcus receives afferent projections mostly from the OFC and belongs to the ventrolateral (sometimes referred to as orbitoventral) PFC network (Barbas and Pandya, 1989; Ongur and Price, 2000). In addition to reciprocal connections with the thalamus, the lateral PFC is connected either directly or indirectly with virtually all areas of the neocortex and hippocampus and has dense efferent projections to the dorsal caudate nucleus (e.g., dorsal striatum). There are distinct medial and orbital PFC networks that are also characterized by cortico-cortical connections and connections with brain regions outside the PFC. The medial PFC is connected with the thalamus and sends efferents to the hypothalamus and periaqueductal gray, and belongs to a network that plays a major role in autonomic and somatic responses to emotional stimuli. The ACC receives similar reciprocal projection as the medial PFC but is thought to belong to a network involved in attentional processing and conflict monitoring. The OFC is densely connected with the basal ganglia, amygdala, and other prefrontal areas, and belongs to a network that mediates motivational and affective aspects of behavior. The medial and OFC networks are sometimes grouped together as the orbitomedial PFC (OMPFC). The medial PFC and OFC are phylogenetically older regions, whereas the lateral PFC is a phylogenetically newer structure and provide substrates for executive cognitive functions. The role of the PFC in behavioral control is discussed in greater detail in the following section.

The extent to which rats have a PFC has been the subject of debate (Uylings *et al.*, 2003; Seamans *et al.*, 2008). The main issue of controversy is whether rats have a prefrontal area that is comparable with the dorsolateral PFC of primates. In their review of the extant human, nonhuman primate, and rodent literature, Uylings *et al.* (2003) concluded that, similar to primates, rodents do in fact have a region of the frontal cortex that can be defined both anatomically and functionally as PFC. They provide a strong argument that rats have a functionally divided prefrontal cortex that includes not only features of the medial and orbital areas in primates, but also some features of the primate dorsolateral PFC. In a more recent review, Seamans *et al.* (2008) took this a step further by suggesting that the rat medial PFC combines elements of the primate dorsolateral PFC and ACC at a rudimentary level that in primates may have formed the building blocks required for abstract rule encoding during evolutionary expansion of the PFC dorsally.

As in primates, the rodent PFC is located in the anterior pole of the frontal cortex and is loosely defined as the ACC, the medial PFC, and the OFC, as well as portions of the agranular insular cortex. The rodent medial PFC is subdivided into a dorsally located prelimbic mPFC (PrL-mPFC) and ventrally located infralimbic mPFC (IfL-mPFC) subregion. The OFC is also subdivided into the medial, ventral, lateral, and dorsolateral orbitofrontal regions. The connectivity of the rodent PFC has common features with that of primates, including a dorsal-to-ventral gradient of projection patterns. For example, the PrL-mPFC sends dense projections to the dorsal striatum that transitions to the ventral stratum (nucleus accumbens) as one moves ventrally through the IfL-mPFC to the OFC.

III. The Executive Function of the Prefrontal Cortex

The perception–action cycle, originally described by Arbib in 1985, is a behavioral construct of circular information flow that describes the interaction of an organism with its environment allowing it to carry out the orderly sequencing of goal-directed actions (Arbib, 1985). As pointed out by Fuster (2008), an important element of this construct is that it includes internal feedback from effectors to sensors to provide representations of current actions to sensory structures to modulate further input. In anatomical terms, there are two cognitive networks of this cycle (Fuster, 2008). One is the perceptual network of the posterior cortex and the other is the executive network of the frontal cortex. Although these networks are structurally distinct, they are not functionally independent and there is a great deal of information exchange between them. Furthermore, the cortical components of these networks do not act independent of subcortical structures. At the apex of the hierarchical organization of the perception–action cycle is the PFC.

Before proceeding further with a discussion of executive function, we should note there are different models of cognitive networks of the PFC. Prominent are variations of modular concepts where cognitive representations are essentially ascribed to different neural connections and anatomical locations (e.g., localization of working memory to the dorsolateral PFC) (Norman and Shallice, 1986). In contrast is the model of large-scale cortical networks first suggested by Bressler and expanded upon by Fuster (Bressler, 1995; Fuster, 2008). In this model, while cognitive networks may have foci, to one degree or another, in specific subregions of the PFC, they are widely distributed and bind together neuronal assemblies from widespread regions of the cortex during information processing. Importantly, a network can subserve different cognitive needs (e.g., memories), and different neurons or groups of neurons can participate in, or belong to, different networks depending upon the particular ongoing cognitive demands. The formation of these networks is presumably dependent upon attractor dynamics of cell assemblies that encode specific actions or memories.

The primary function of the PFC can be summarized as the temporal organization of behavior. Executive function involves attention, planning, and decision making, and is how the PFC organizes behavior in a sequential manner. This occurs in large part through the dynamic interaction of two parallel networks of the PFC—an "executive" network with a primary location in the dorsolateral PFC (or medial PFC in rodents) and a "limbic" network primarily contained in the OFC (Fig. 2). These networks mediate higher-order behaviors such as purposeful goal-directed actions, language, and reasoning. These networks are required to perform a number of functions for appropriate control of goal-directed behavior (outlined by Moghaddam and Homayoun, 2008). In summary, they must (1) detect situations that demand mediation,

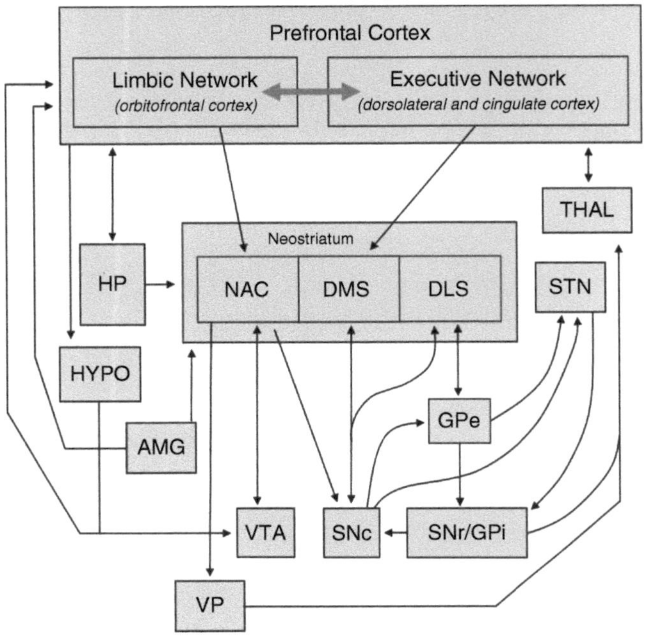

FIG. 2. Network organization of the PFC. Schematic shows the important connections of the limbic (orbitofrontal) and executive (dorsolateral and cingulate) divisions of the PFC. Abbreviations: HP (hippocampus), NAC (nucleus accumbens), DMS (dorsomedial striatum), DLS (dorsolateral striatum), THAL (thalamus), STN (sub-thalamic nuclei), HYPO (hypothalamus), AMG (amygdala), GPe (globus pallidus-external), VTA (ventral tegmental area), SNc (substantia nigra-compacta), SNr/GPi (substantia nigra-reticulata/globus pallidus-internal), VP (ventral pallidum).

(2) direct selective attention to stimuli relevant to this situation, (3) suppress distractions caused by irrelevant stimuli, (4) bring on line relevant past memories, (5) plan behavioral sequence based upon these memories and the present relevant stimuli, and (6) encode the preparatory set that leads to motor execution of the appropriate behavior. As noted above, the executive network is thought to be primarily localized within the dorsolateral PFC, and as such this structure is considered to be the ultimate regulator of goal-directed actions. However, the ACC also critically contributes to many aspects of executive functioning by mediating attention to action and conflict monitoring. Also as noted previously, the OFC is extensively connected with sensory cortical areas and limbic structures, and it is thought that this network plays a role in integrating and determining the salience of information about environmental contingencies. It plays a particularly important role in inhibitory control of inappropriate behaviors by relaying processed information to the executive control network of

the dorsolateral PFC. Thus, the executive and limbic networks function in concert to orchestrate behaviors particularly related to context and expected outcomes (e.g., expectation of a goal-directed action).

Any discussion of the executive function of the PFC would be remiss without at least mentioning working memory, one of the major executive cognitive functions of the PFC. Working memory is a form of sustained attention for the processing of *prospective* action. Thus, working memory involves the maintenance and manipulation of task-relevant information in the service of planning, problem solving, and predicting forthcoming events (Unterrainer and Owen, 2006; D'Esposito, 2007). Working memory is therefore a form of "active" memory involving sustained attention that is focused on an internal representation and can be distinguished from short-term memory. Although short-term memory often functions in the service of working memory (e.g., bringing a short-term memory online for the purpose of planning and predicting forthcoming events), it is classically viewed as temporary memory storage prior to its storage in long-term memory. Closely related to working memory is attentional "set", which is essentially attention to motor activity and is used to plan a sequence of forthcoming actions. In a sense, working memory is the representation of the near past whereas set is a representation of the near future (Fuster, 2008), and together they are critical for the temporal organization of behavior.

Glutamatergic neurotransmission is the primary regulator of activity-dependent synaptic modifications that underlie experience-dependent brain plasticity (Chandler, 2003). As will be discussed in the next section, both acute and chronic alcohol exposure significantly impact glutamatergic neurotransmission in the PFC. In addition, results from several studies suggest that plastic processes of associative learning overlap and interact with those that underlie the development of addictive behaviors (Berke and Hyman, 2000). Addiction is clearly a complex disease that involves motivational and higher-order cognitive processes that initiate and control goal-directed behaviors, and accumulating evidence implicates altered glutamatergic neurotransmission mediated by projections to and from the prefrontal cortex in the neuroplasticity of addiction (Jentsch and Taylor, 1999; Kalivas, 2009). The PFC is highly integrated into the addiction neurocircuitry. Addictive behaviors, such as those associated with alcohol abuse and alcoholism, include loss of control over consumption and relapse to drinking. The PFC normally exerts "top-down" (e.g., information derived from prior experience) inhibitory control over internal and external sensory-driven compulsive behaviors. Increasing evidence suggests that continued drug exposure leads to attenuation of the ability of the PFC to monitor and inhibit these behaviors, with eventual loss of inhibitory control over drinking. Volkow and colleagues have conceptualized this loss of inhibitory control by the PFC as a syndrome of "impaired response inhibition and salience attribution" (Goldstein and Volkow, 2002; Volkow *et al.*, 2003). This syndrome is envisioned to encompass an integrated cluster of addictive behaviors that depend upon interaction with the PFC. An important feature of the PFC is that it is functionally and structurally adaptable, and a major component of cortical cognitive processing is that it is highly influenced by "knowledge" of past experiences. Thus, reward information such as that provided by the VTA dopaminergic system has a pervasive and lasting influence on the activity of the PFC.

IV. Electrophysiological Properties of the PFC

A. Firing Properties of Cortical Pyramidal and Interneurons

1. *Pyramidal Neurons*

In the neocortex, glutamatergic pyramidal neurons represent approximately 80% of the neuronal population and have been best studied in sensory cortices, especially the visual cortex. Neocortical pyramidal neurons are arranged in layers (I–VI in most parts of cortex, rodent PFC lacks a well-defined layer IV) and make connections with both glutamatergic and γ-amino-butyric acid (GABA) neurons within the cortex and those in subcortical areas. Although less well characterized than interneurons (see below), cortical pyramidal neurons also show diversity with respect to size, dendritic morphology, and firing properties. In general, cortical pyramidal neurons are grouped into two general classes, regular spiking (RS) neurons that show little or slow adaptation during repetitive firing and intrinsically bursting (IB) neurons that display more complex periods of firing. In the prefrontal cortex of both rodents and primates, most RS neurons (RS1 subtype) respond to current injection by smooth increases in action potential (AP) frequency with little change in AP threshold or amplitude (Zhang, 2004; Chang and Luebke, 2007). RS2 neurons show decreasing action-potential height and an increase in AP firing threshold whereas fast-accommodating (FA) RS neurons show similar changes in AP characteristics along with a robust accommodation in spiking. Intrinsically bursting (IB) neurons make up only a small percentage of neocortical pyramidal neurons and generate three to six spikes during threshold amounts of stimulation.

In rodents, these patterns of activity develop during the first two postnatal weeks and are not complete until adulthood (Zhang, 2004). With respect to morphology, two general classes of pyramidal neurons have been proposed, with Type I neurons having thick apical dendrites that terminate in tufts in superficial cortical layers. These neurons project axonally to primarily subcortical structures. Type II neurons have less apical branching and project intracortically as well as to the striatum. Generally, Type II neurons are regular spiking whereas Type I neurons show bursting patterns of action potentials (Molnar and Cheung, 2006). All cortical pyramidal neurons show complex dendritic branching patterns and asymmetrical synapses characterized by the dendritic spine, the specialized structure that receives excitatory input. In general, PFC pyramidal neurons show more branching and greater density of spines compared with those in visual cortex and other sensory cortices (Elston, 2003).

2. *Interneurons*

Although GABAergic neurons within neocortex represent only 20% of the neuronal population, there is a great deal of diversity with respect to their anatomical, biochemical, and functional properties. The Petilla Interneuron Nomenclature Group (named in honor of Ramon Y Cajal's hometown) has recently suggested a set of common terms to classify the various subtypes of interneurons (Ascoli *et al.*, 2008). Although beyond the scope of the present review, the conclusions of this group are summarized below. GABAergic neurons within the neocortex (interneurons) usually have short axons that bifurcate within a cortical column but can project across columns. Despite this general scheme, there are long-axon GABA neurons that project to distal subcortical structures although less is known about these types of neurons. Cortical GABAergic interneurons show a wide variety of axonal and dendritic branching patterns that have historically been grouped into several categories, including basket cells, chandelier cells, bipolar cells, double bouquet cells, bitufted cells, and neurogliaform cells (Markram *et al.*, 2004). These different cell types specialize in their location of axon innervation of target neurons, including soma and proximal dendrites (basket cells), more distal dendrites (double bouquet, bipolar, bitufted, neurogliaform, Martinotti cell), and axon initial segments (chandelier). These distinct patterns of target neuron innervation allow for precise control of both input (dendritic processing and integration) and output (somatic and initial segment) functions of the postsynaptic neuron. Interneurons are also classified according to their expression of different calcium binding proteins (parvalbumin, calbindin, and calretinin) and neuropeptides (neuropeptide Y, vasoactive intestinal peptide, somatostatin, and cholecystokinin). In terms of electrophysiological responses, neocortical interneurons display a wide range of spiking patterns in response to steady-state current injection. These include non-accommodating, accommodating, stuttering, irregular spiking, and bursting. Within the prefrontal cortex, most studies have recognized four different classes of interneurons based on spiking patterns during current injection (Kawaguchi, 1995; Gorelova *et al.*, 2002). Fast-spiking (FS) neurons are the most numerous and display very rapid action potentials ($\sim$0.5 ms half-width) each followed by a deep after-hyperpolarization. FS neuron firing is non-accommodating and rarely do these neurons show bursting type of activity. Late spiking (LS) interneurons show a slowly developing depolarization during current injection that eventually triggers spikes. LS cells also are non-accommodating and have spike widths approximately twice that of FS neurons. Low threshold spike (LTS) neurons generate action potentials upon current injection only when cells are held at fairly hyperpolarized potentials ($< -70\,\text{mV}$). Action potentials also arise at the end of a hyperpolarizing pulse. Regular spiking nonpyramidal (RSNP) neurons show a slow adaptation to firing and have prominent hyperpolarization-induced inward currents (IH). When combined with morphological and biochemical analysis, FS neurons include some basket cells, PV positive neurons that are also positive for GABA, and some chandelier cells (Kawaguchi, 1995; Markram *et al.*, 2004). LS neurons are neurogliaform cells while LTS neurons have axons terminating on Layer I neurons. RSNP neurons are not easily classified into any of these groups and likely are comprised of multiple and diverse members.

B. Network Activity

1. *Persistent Activity-In Vitro Models of PFC Function*

Recordings from prefrontal cortex in awake behaving monkeys revealed sustained periods of firing during delay periods in visual-guided tasks designed to assess working memory demands (Williams and Goldman-Rakic, 1995; Ichihara-Takeda and Funahashi, 2007). These persistent patterns of firing are maintained in the absence of visual cues and were interpreted as a cellular correlate of the PFC's ability to hold and manipulate information in anticipation of a future action. Work from Steriade and others also described a slow oscillation (0.5–2 Hz) in prefrontal cortex during periods of quiet wakefulness or sleep (Steriade *et al.*, 1993). The slow oscillation consisted of periods of hyperpolarized down-states interrupted by brief transitions into depolarized up-states during which action potentials often arose. One hallmark of this activity is the relatively constant amount of depolarization produced during an up-state ($\sim$10–15 mV) that persists for several seconds. During active waking periods or REM sleep, neurons reside mostly at up-state membrane potentials consistent with enhanced sensory input and synaptic drive. In acute slices of prefrontal cortex, neurons usually do not display spontaneous oscillations of membrane potentials (termed bistability) or action potential firing although these patterns can be elicited under certain conditions. For example, acute slices from the ferret visual or prefrontal cortex showed spontaneous oscillations in membrane potential when the ionic composition of the bathing solution was modified to mimic that of brain cerebrospinal fluid (Sanchez-Vives and McCormick, 2000). In rodents, slices of prefrontal cortex that are maintained in organotypic cell culture develop spontaneous oscillations in membrane potential during the first 2 weeks that are maintained for the life of the culture (Seamans *et al.*, 2003; Tu *et al.*, 2007). Similar patterns of activity can be induced in acute rodent slices by supplementing the bathing solution with optimal concentrations of NMDA and dopamine (Tseng and O'Donnell, 2004). Up-states in organotypic PFC cultures can also be evoked by applying brief trains of electrical pulses through a stimulating electrode and are accompanied by robust action potential firing. Paired recordings of neurons in PFC cultures show that the initiation and termination of up-states are highly synchronized between pyramidal neurons and fast-spiking interneurons although action potential firing is typically asynchronous (Seamans *et al.*, 2003; Woodward and Pava, 2009). Pharmacological analysis of PFC up-states in PFC cultures reveals several components including an initial EPSP:IPSP sequence driven by AMPA and $GABA_A$ receptors followed a prolonged inward current composed of both AMPA- and NMDA-mediated events (Seamans *et al.*, 2003).

V. Neuromodulation of PFC Neuron Activity

A. Dopamine

Dopamine modulation of PFC function has long been known to be important for optimizing cognitive performance, especially in tasks requiring working memory. Disruption of this important modulatory input is associated with various neurological and psychiatric illnesses. PFC dopamine arises from terminals provided by dopaminergic neurons whose cell bodies reside in the ventral tegmental area (VTA). These neurons also provide dopaminergic tone to other limbic structures including nucleus accumbens and recent evidence suggests that there are discrete subpopulations of VTA dopamine neurons that display both anatomical and functional specializations depending on their pattern of innervation (Lammel *et al.*, 2008). There is a rich literature regarding the effects of dopamine on PFC-associated behaviors and neuronal function and the reader is referred to several excellent reviews of this material (Tzschentke, 2001; Seamans and Yang, 2004). A brief summary of the key findings as they relate to PFC electrophysiology is presented here. In the prefrontal cortex, dopamine D1 receptors are expressed to a higher degree than other subtypes and are located primarily on dendritic spines and shafts of target neurons. D2 dopamine receptors are expressed both pre- and postsynaptically, and receptors on axon terminals reduce the release of both glutamate and GABA. In general, most studies have shown that dopamine transiently reduces neuronal excitability followed by a long-lasting enhancement in firing activity. These effects arise from both modulation of intrinsic mediators of excitability (voltage-sensitive Na+ and K+ channels) and changes in glutamatergic and GABAergic synaptic transmission. D1-mediated responses are associated primarily with a long-lasting increase in excitability, with NMDA currents being particularly sensitive to low concentrations of dopamine that potentiate NMDA EPSCs. At higher doses, dopamine and certain D1 agonists may actually inhibit NMDA function via direct channel blockade suggesting caution in interpreting results when high concentrations of these compounds are used (Cui *et al.*, 2006). Dopamine also enhances the firing activity of fast-spiking but not other types of interneurons in the prefrontal cortex (Gorelova *et al.*, 2002). This effect is blocked by D1 antagonists and involves the cAMP:PKA signaling pathway (Trantham-Davidson *et al.*, 2004). Higher concentrations of dopamine lead to suppression of interneuron excitability via a D2-receptor-mediated pathway that leads to enhanced phosphatase activity and reduced release of GABA (Trantham-Davidson *et al.*, 2004, 2008). In a recent study that utilized the slice culture model described above, Kroener *et al.* (2009) investigated the impact of dopamine on persistent activity of deep-layer pyramidal neurons. These cultures contained slices of PFC and VTA, and dopaminergic neurons established a strong innervation of the PFC slice consistent with that observed *in vivo.* Adding 1 uM or more exogenous dopamine reduced up-state duration and spiking activity whereas dopamine enhanced up-state duration and firing in cultures lacking endogenous DA tone. Action potentials induced by direct current injection during down-state periods were enhanced by dopamine even at concentrations as low as 100 nM. Dopamine also enhanced the ability of stimulus-evoked EPSPs to induce firing when these events were generated during an up-state. This effect was concentration dependent as up-state duration and spiking were unaffected by 10 nM DA but were inhibited by 10 μM DA. These data are consistent with dopamine's role in tuning PFC function during working memory tasks and support the idea that dopamine's actions are optimal over a narrow concentration range (Goldman-Rakic, 1999).

B. Norepinephrine

Norepinephrine (NE) is also an important catecholamine modulator of PFC function and NE neurons that innervate the PFC originate in the locus ceruleus. NE content is generally lower in the PFC than dopamine and NE fibers appear to target more superficial layers as compared to dopamine. NE's effects on PFC neuronal function are mediated primarily by $\alpha1$, $\alpha2a$, and $\beta1$ adrenergic receptors. Its affinity for these subtypes is greatest for $\alpha2a$ receptors and this subtype is thought to underlie most of NE's modulation of PFC function during normal, nonstressed waking conditions. The effects of methylphenidate, an indirect NE agonist, on the intrinsic excitability of deep-layer PFC pyramidal neurons have recently been studied using whole-cell patch clamp electrophysiology (Andrews and Lavin, 2006). Methylphenidate enhanced current-evoked spike activity and this effect was blocked by α2a antagonists but not those selective for α1 or β1 receptors. Interestingly, the effect of methylphenidate on firing rate was abolished when blockers of synaptic transmission were added to the bath suggesting that NE may suppress the activity of fast-spiking interneurons that normally regulate pyramidal neuron excitability. In addition to these findings, the effect of activating adrenergic receptors on NMDA EPSCs recorded from deep-layer pyramidal neurons in the rodent PFC has also been investigated. Norepinephrine itself reduced stimulus-evoked NMDA EPSCs and this effect was mimicked by both α1 and α2 receptors agonists (Liu *et al.*, 2006). This study also demonstrated that adrenergic receptor suppression of NMDA responses involved different intracellular signaling pathways with α1 effects mediated via the PLC-IP3-Ca2+ pathway and α2 effects occurring through inhibition of the cAMP-PKA signal transduction process. The α1 calcium-mediated inhibition of PFC activity may also arise via activation of calcium-sensitive potassium channels (SK) that contribute to the regulation of the after-hyperpolarization. Hyperpolarization-activated cyclic-nucleotide gated channels (HCN) may reduce activity through shunting inhibition when levels of cAMP are high. Modulating these channels pharmacologically improves working memory performance in rodents consistent with the idea that optimal levels of NE signaling are required for proper PFC function (Arnsten, 2009).

C. Serotonin

Serotonin (5-hydroxytryptamine, 5-HT) neurons originate in the raphe nuclei of the pons and cerebellum and project widely to subcortical and cortical structures, including both medial and orbital aspects of the prefrontal cortex. The 5-HT2A receptor subtype is thought to mediate many of the effects of psychedelic hallucinogens such as lysergic acid diethylamide, and dysregulation of 5-HT signaling in PFC is thought to contribute to the pathology of schizophrenia and mood disorders. Although 5-HT activates a large number of receptors including an ion channel subtype (5-HT3), 5-HT1A and 5-HT2A receptors are thought to be the predominant forms expressed by PFC neurons (Kia *et al.*, 1996; Willins *et al.*, 1997; Santana *et al.*, 2004). Activation of 5-HT2A receptors enhances the frequency of both inhibitory and excitatory postsynaptic currents recorded in pyramidal neurons, with less effect noted for interneurons (Zhou and Hablitz, 1999). The 5-HT2A-dependent increase in spontaneous events was not observed in the presence of TTX suggesting that these receptors do not alter presynaptic release of glutamate. Whereas 5-HT or 5-HT2A agonists generate only small, subthreshold depolarizations in most PFC pyramidal neurons tested, approximately a third of large pyramidal neurons in layer V/VI were strongly depolarized by these treatments (Beique *et al.*, 2007). This is consistent with data from *in situ* hybridization studies showing that approximately 50% of glutamatergic neurons in prelimbic PFC express 5-HT2A mRNA with 26% of layer VIa showing strong expression (Santana *et al.*, 2004). These findings suggest that the 5-HT2A-dependent increase in spontaneous events recorded in PFC neurons is due to direct excitation of a small number of excitatory neurons that spreads throughout the PFC due to the strong recurrent connections among deep-layer pyramidal neurons.

In contrast to these results, activation of 5-HT1A receptors inhibits neuronal excitation through enhancement of potassium currents and inhibition of calcium channels (Araneda and Andrade, 1991). 5-HT receptors also appear to directly influence GABA and NMDA-mediated conductances with 5-HT2A/C receptors inhibiting GABA-induced currents via a PKC-dependent pathway (Liu *et al.*, 2006). NMDA receptor currents are inhibited by 5-HT1A receptors whereas 5-HT2A activation enhances NMDA function. These effects appear to result from changes in NMDA receptor trafficking with inhibition and activation, respectively, of ERK phosphorylation of the MAP kinase pathway (Yuen *et al.*, 2008). Although PFC neurons express both 5-HT1A and 5-HT2A receptors, these subtypes have a differential affinity for serotonin suggesting that activation of raphe serotonin neurons may initiate opposing changes in NMDA function in the PFC that depends on the extent and rate of firing. 5-HT also influences glutamatergic plasticity of PFC pyramidal neurons. Activation of 5-HT2A/C receptors during tetanic stimulation of deep-layer pyramidal neurons led to long-term depression of AMPA-mediated synaptic EPSCs (Zhong *et al.*, 2008). This effect was NMDA independent but was blocked by antagonists of group I metabotropic glutamate receptors as well as inhibitors of P38 MAP kinase.

D. Endocannabinoids

Endocannabinoids (ECs) are neuromodulators derived from membrane phospholipids during periods of neuronal activity. Synthesis occurs in the postsynaptic neuron and is calcium dependent while EC action takes place on the presynaptic terminal, thus linking neuronal activity to the production of a retrograde messenger. ECs suppress the presynaptic release of GABA and glutamate through a process termed depolarization-induced suppression of inhibition or excitation (DSI, DSE) that requires activation of cannabinoid-type I receptors (CB1). ECs are rapidly degraded by enzymatic pathways involving fatty acid amide hydrolase (FAAH) or monoacylglycerol lipase (MGL) that preferentially inactivate anandamide and 2-acylglycerol (2-AG), the two best characterized endocannabinoids in the brain. EC modulation of synaptic transmission has been studied in most brain regions examined, including the cerebral cortex where both DSE and DSI have been described. However, most of these studies are confined to somatosensory cortex and show that in this region, EC modulation of signaling is layer specific with DSE and DSI occurring in layer II/III neurons while only DSE is observed in layer V pyramidal neurons (Fortin and Levine, 2007). In one study of layer V PFC neurons from juvenile rats, bath application of the CB1 agonist WIN55212-1 significantly decreased stimulus-evoked EPSCs and this effect was blocked by the CB1 antagonist SR141716A (Auclair *et al.*, 2000). The decrease in excitatory transmission was mediated by a reduction in presynaptic release as WIN reduced the frequency but not the amplitude of evoked asynchronous EPSCs. WIN also shifted the plasticity of layer V PFC neurons toward long-term depression, consistent with its effect on presynaptic glutamate release. Endogenous cannabinoids also mediate PFC neuronal plasticity as stimulation of layer II/III afferents induces a long-lasting LTD in layer V pyramidal neurons from mouse PFC (Lafourcade *et al.*, 2007). This depression was prevented by AM251, a CB1 antagonist, and was again mediated by a reduction in the frequency but not the amplitude of spontaneous EPSCs. The mechanism underlying the induction of EC synthesis was also investigated and was found to require mGluR5 receptors and activation of intracellular phospholipase C. Taken together, the results of these studies suggest that the endocannabinoid system is an important regulator of PFC neuronal activity and support results from behavioral studies showing impairments in working memory following use of cannabis (Riedel and Davies, 2005).

VI. Effects of Alcohol on PFC Neuron Function—*In Vitro* Studies

Although the effects of biogenic amines and endogenous cannabinoids on PFC neuron excitability have been examined using *in vitro* electrophysiological techniques, there is much less known about the actions of alcohol on these neurons. Recent studies from the authors' laboratories have investigated the effects of acute ethanol exposure on persistent activity patterns in PFC neurons in organotypic cell culture and on synaptic transmission in slices from rat prefrontal cortex. Tu *et al.* (2007) used triple-slice organotypic cultures that contained prefrontal cortex, hippocampus, and ventral tegmental area. As described earlier, pyramidal neurons in the prefrontal cortex portion of these cultures develop spontaneous patterns of persistent activity within 2 weeks of culture, and up-states can also be evoked by direct electrical stimulation of these cultures (Fig. 3). Ethanol, applied via the bath solution, had no effect on firing of PFC pyramidal neurons induced by direct current injection but reduced the

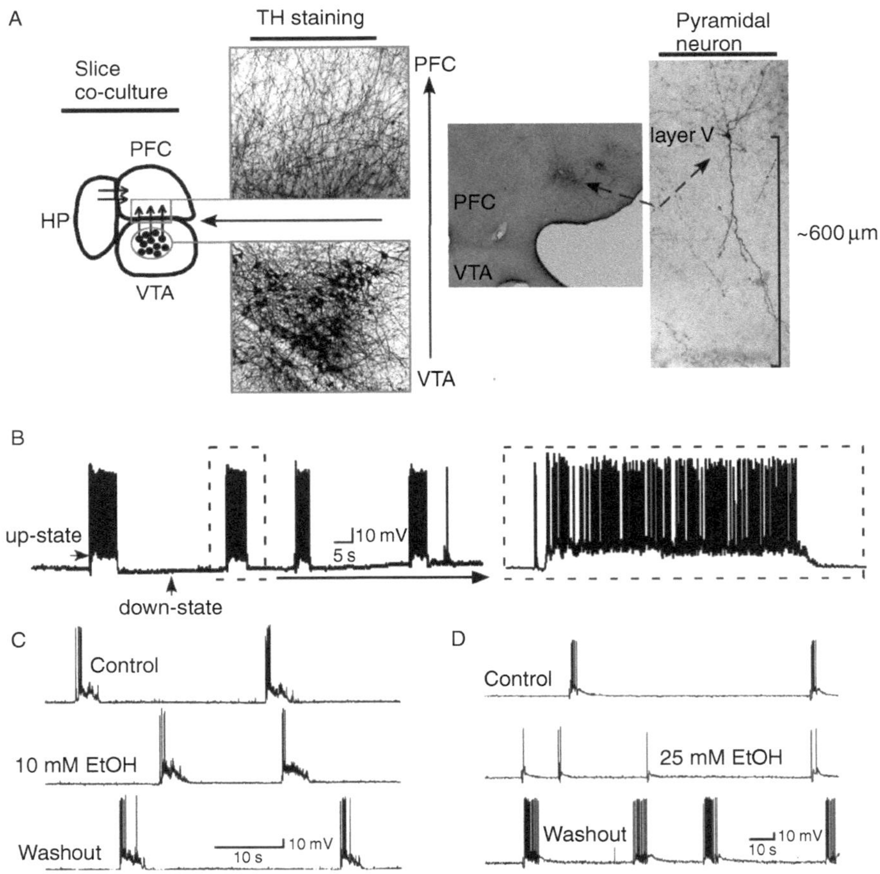

FIG. 3. Up-states in PFC slice cultures are inhibited by ethanol. Panel A shows arrangement of prefrontal cortex (PFC), hippocampus (HP), and ventral tegmental area (VTA) in brain slice co-culture. Picture shows dopamine neurons stained for tyrosine hydroxylase (TH). Panel B shows example of up-states recorded from a deep-layer PFC pyramidal neuron. Ethanol reversibly inhibits up-state activity at a concentration of 25 mM (panel D) but not 10 mM (panel C). Taken from Tu *et al.* (2007).

duration, amplitude, and spike activity of spontaneous up-states. These effects occurred at the beginning at 17 mM (~0.08% blood ethanol concentration) and persistent activity of PFC neurons was almost totally suppressed at 50 mM ethanol. Up-state duration was affected at lower concentrations of ethanol (25 mM) than was up-state amplitude, although both measures were reduced as ethanol concentrations were increased. Following washout of the ethanol solution, up-state duration and spiking were often enhanced as compared to pre-ethanol control values and this rebound in activity persisted for 10–15 min post washout.

Ethanol also inhibited up-states evoked by brief electrical stimulation of the VTA designed to mimic the *in vivo* bursting pattern of DA neurons upon presentation of rewarding stimuli (Schultz, 1986; Bayer *et al.*, 2007). As compared to spontaneous up-states, there was less inhibition of VTA-stimulated persistent activity and this difference was eliminated when the D1 receptor antagonist SCH23390 was included in the bath solution. These results suggest that the phasic release of dopamine induced by DA neuron bursting can partially offset the inhibitory actions of ethanol on prefrontal neuron function. The mechanism underlying this effect is not known but may involve potentiation of NMDA receptors that are an important component of up-states. In a follow-up study, we examined the effects of acute ethanol on pharmacologically isolated synaptic currents in deep-layer pyramidal neurons of rat prefrontal cortex (Weitlauf and Woodward, 2008). Ethanol, at concentrations up to 88 mM (~0.4% blood ethanol concentration), had no significant effect on evoked or spontaneous AMPA-mediated EPSCs (Fig. 4). Inhibitory currents mediated by the synaptic release of GABA were also largely insensitive to high concentrations of ethanol and no effects were observed in the paired pulse ratio of evoked GABA responses or in the frequency or amplitude of spontaneous $GABA_A$ IPSCs. These studies also showed no evidence for a tonic GABA-mediated current in deep-layer PFC pyramidal neurons or any effect of ethanol on the holding current that would have been expected if ethanol-sensitive extrasynaptic $GABA_A$ receptors were present. In contrast to these effects, ethanol inhibited stimulus-evoked NMDA-mediated EPSCs with significant effects occurring at concentrations of 22 mM (~0.1% blood ethanol concentration) and above. Unlike that found in the slice cultures, there was no evidence for any rebound in activity following ethanol washout for any of the synaptic currents examined. These results suggest that NMDA receptors are a particularly important target for the acute effects of ethanol on PFC pyramidal neuron function and that ethanol's inhibition of persistent activity in slice cultures is a result of this effect. This is consistent with the ability of a low concentration of the NMDA antagonist AP5 to reduce up-state duration, amplitude, and spiking in PFC cultures (Tu *et al.*, 2007).

In a recent study, we recorded from both pyramidal and fast-spiking GABA interneurons in PFC cultures and examined how ethanol affects up-state activity

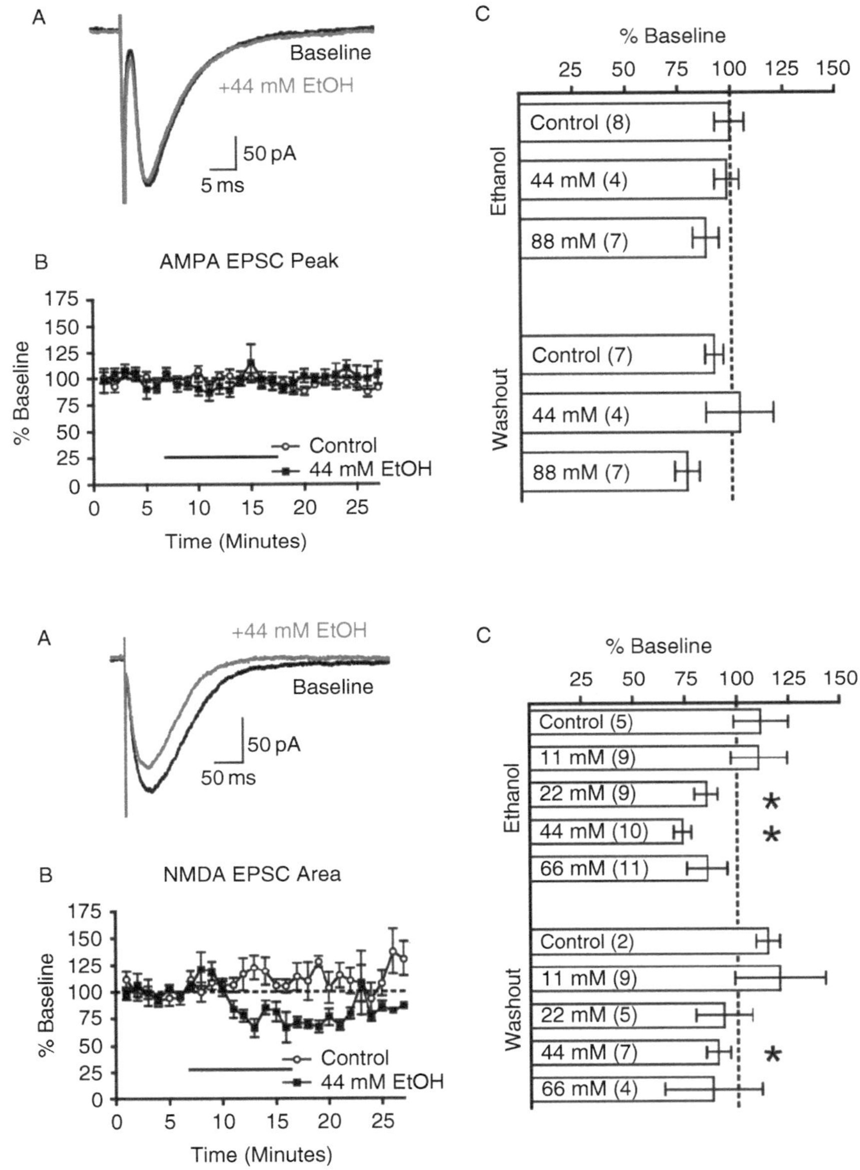

FIG. 4. Acute ethanol inhibits NMDA but not AMPA-mediated EPSCs in PFC pyramidal neurons. Left panels (A–C) show lack of significant effect of ethanol on AMPA-mediated EPSCs from deep-layer PFC pyramidal neurons. Right panels (A–C) shows that ethanol inhibits NMDA-mediated EPSCs from deep-layer PFC pyramidal neurons. Taken from Weitlauf and Woodward, 2008.

in these neurons (Woodward and Pava, 2009). During dual patch-clamp recordings, neurons entered and exited the up-state simultaneously regardless of the type of neuron pairs recorded (pyramidal:pyramidal; interneuron:interneuron; pyramidal:interneuron). These findings are consistent with the idea of highly interconnected recurrent networks of excitatory and inhibitory neurons in the PFC. In FS GABAergic inteneurons, fast EPSPs could be observed during up-states whereas the up-state depolarization in pyramidal neurons was smooth and relatively constant as previously reported (McCormick *et al.*, 2003; Seamans *et al.*, 2003). However, individual EPSPs in pyramidal neurons were revealed by intracellular perfusion of the NMDA channel blocker MK801 supporting the suggestion from Seamans *et al.* (2003) that the slower kinetics of NMDA receptors bridge the fast transients gated by AMPA receptors. Intracellular MK801 had little effect on the up-state dynamics of FS neurons consistent with the reduced expression of NMDA receptors expression in this subpopulation of PFC interneurons (Wang and Gao, 2009). Despite these differences in functional NMDA receptors, bath application of ethanol reduced up-state duration by approximately the same amount in both pyramidal and interneurons. This finding again underscores the highly interconnected nature of prefrontal neurons and how they contribute to network activity.

VII. Effects of Alcohol on PFC Neuron Function—*In Vivo* Animal Studies

Acute administration of ethanol to experimental animals is associated with changes in various neurotransmitter systems in the PFC. Several studies have examined these effects using outbred strains of rats. For example, μ-opioid receptors were upregulated in frontal cortex 2 h after intragastric administration of 2.5 g/kg ethanol (Mendez *et al.*, 2001). In Wistar rats, ethanol (1 g/kg) increased the extracellular levels of 5-HT as measured by *in vivo* microdialysis (Langen *et al.*, 2002). One day of exposure to an ethanol-containing liquid diet decreased CB1 receptor levels in rat PFC (Rubio *et al.*, 2009). Measures of the immediate-early gene *Fos* have been measured in rat PFC neurons following acute administration of ethanol to gauge its effects on activity. Ethanol increased *Fos* staining with apparently equal increases in both GABAergic and glutamatergic neurons (Leriche *et al.*, 2008). In contrast to these results, spontaneous spike activity of prefrontal neurons in Sprague-Dawley rats as measured by *in vivo* multi-array electrodes (Fig. 5) was decreased in a dose-dependent manner by ethanol (Tu *et al.*, 2007). The differences between the results of the *Fos* and spike activity studies may be due to the use of different sampling times or the use of an anesthetized animal in the recording studies.

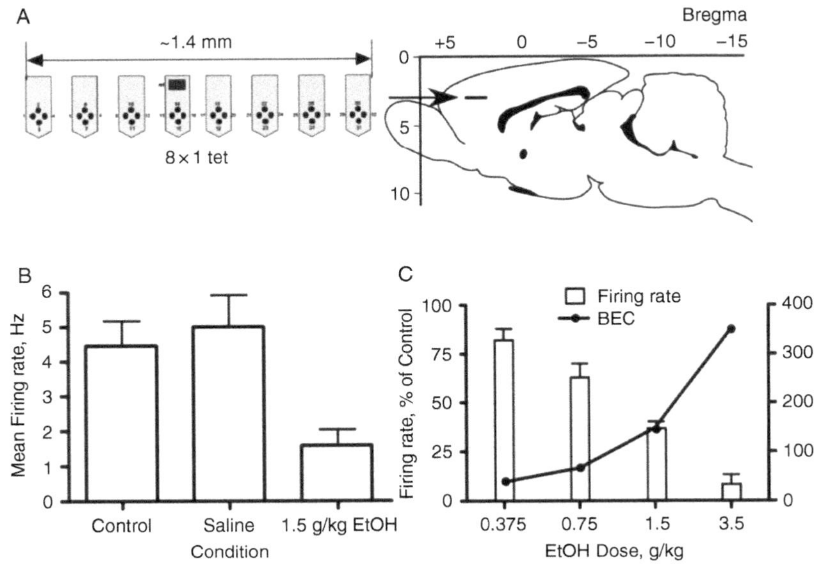

FIG. 5. Ethanol inhibits spike firing of PFC neurons *in vivo*. Panel A shows the arrangement of multi-electrode array and location in the brain. Panel B shows effects of saline or ethanol on spike firing of PFC neurons. Panel C demonstrates that ethanol inhibits PFC firing in a dose-dependent manner and is correlated with blood ethanol concentration (BEC). Taken from Tu *et al.* (2007).

In addition to these studies with outbred rats, a variety of rodent models have been developed through selective breeding to generate animals with a high propensity for ethanol self-administration and dependence. These animals may better mimic the drinking patterns of human alcoholics and provide an opportunity to investigate how genetic, structural, and functional differences in individuals may predispose them to alcohol dependence. For example, the alcohol-preferring P rat consumes high amounts of alcohol naturally unlike outbred strains that require time-consuming behavioral training sessions to overcome their initial resistance to alcohol consumption. P rats show a higher level of glucose utilization in PFC than non-alcohol-preferring rats in a drug-naive state (Smith *et al.*, 2001). In response to an acute EtOH challenge, P rats show a divergent response with low doses increasing PFC glucose utilization, and higher doses producing a decrease. Alcohol non-preferring rats, on the other hand, showed no change to the acute ethanol challenge (Strother *et al.*, 2005), indicating that enhanced sensitivity to ethanol in PFC may play a role in the development of alcohol dependence.

With respect to mice, the DBA/2J (DBA) strain shows relatively low levels of voluntary ethanol consumption as compared to C57BL/6J (C57) mice. Microarray analysis has demonstrated very different patterns of gene activation in PFC of these two strains in response to acute alcohol exposure. DBA mice showed a greater change in the number of genes that were upregulated in response to EtOH than their C57 counterparts (Kerns *et al.*, 2005). Among those transcripts upregulated were a number of genes that encode for glucocorticoid-binding proteins and cortisol levels are known to increase after acute alcohol consumption (King *et al.*, 2006). This suggests that the divergent response in PFC gene expression between the two strains following ethanol challenge may indicate a differential response to circulating glucocorticoids. Other genes that were upregulated in the microarray study included those related to synaptic plasticity and myelination. Interestingly, there was grouping of upregulated genes around specific chromosomal regions suggesting some common element that underlies ethanol-responsive genes. Chronic EtOH has been shown to alter the acetylation of histones (Pascual *et al.*, 2009), so that chromatin rearrangement following ethanol exposure may play a role in these changes in gene expression. Alcohol-preferring AA rats show increased phosphorylation and subsequent inhibition of GSK-3β, involved in metabolism and possibly neurotransmission, in response to increased phosphorylation of AKT, a regulator of GSK-3β (Neznanova *et al.*, 2009).

Other studies suggest differences in neurotransmission in the PFC between different rat strains. The Lewis rat strain self-administers alcohol at greater levels than the Fisher rat. The Lewis rats demonstrated lower basal levels of glutamate in the PFC than did the Fisher strain (Selim and Bradberry, 1996). In addition, alcohol-naive P rats have a lower extracellular concentration of DA in medial PFC than do non-EtOH preferring Wistar rats (Engleman *et al.*, 2006). In the same study, however, acute EtOH administration did not alter DA levels in the PFC of either group. By contrast, the Sardinian alcohol-preferring rat (sP) showed greater basal levels of DA in the mPFC than did the Sardinian alcohol-nonpreferring (sNP) strain (Leggio *et al.*, 2003). Although this does not rule out a role for PFC DA basal levels in alcohol seeking or the development of alcohol dependence, it is clear that factors other than basal DA levels are involved. For example, alcohol-preferring AA rats showed higher levels of mu-opioid binding in PFC under drug-naive conditions than did alcohol-avoiding ANA rats, as well as higher levels of proenkephalin mRNA (Marinelli *et al.*, 2000). Differences have also been observed in PFC 5-HT receptor levels in PFC between strains and sP rats demonstrated lower levels of basal 5-HT2A-binding sites than sNP animals (Ciccocioppo *et al.*, 1997). Given the powerful effects of these neurotransmitters on PFC function, it is likely that differences in expression or functional activity of these neuromodulators is important in shaping the PFC's response to ethanol.

VIII. Effects of Alcohol on PFC Neuron Function—Human Studies

A. Acute Ethanol

Acute ethanol administration in humans has been shown to cause deficits in executive activities that are thought to require the PFC. For example, performance in a spatial recognition task and a planning task was decreased in social drinkers while inebriated (Weissenborn and Duka, 2003) and acute EtOH has been shown to cause poorer decision making using a gambling task to assess PFC function (George *et al.*, 2005). Working memory (WM) tasks are also commonly used to test executive function given the role that the PFC has been shown to play in mediating this activity. However, the effects of acute EtOH on working memory performance in humans are mixed. One study observed alcohol-induced WM deficits in a backward digit span task only among those who demonstrated a high baseline WM performance when sober (Finn *et al.*, 1999). The study that demonstrated deficits in a spatial recognition task (Weissenborn and Duka, 2003) failed to show a decrease in spatial WM performance in alcohol-exposed subjects, whereas impairments in a memory scanning task following acute ethanol challenge have been reported (Grattan-Miscio and Vogel-Sprott, 2005). To try and reconcile these conflicting data, Saults and colleagues examined the effects of ethanol on different types of working memory tasks that varied in two dimensions of substrates that needed to be retained for completion of the task. These authors showed that consumption of alcohol impaired memory for auditory and visual sequences but not that for simultaneous presentations of auditory or visual stimuli (Saults *et al.*, 2007). They concluded that ethanol impaired certain strategies for holding information online rather than causing a general decrease in the size of the working memory buffer.

A number of studies have sought to examine the effects of EtOH on PFC function in humans using brain imaging techniques. Acute administration of alcohol increases blood flow in the right prefrontal cortex of healthy patients (Tiihonen *et al.*, 1994) and this effect appears dose dependent. PFC blood flow was significantly increased after doses of 0.7 g/kg or 1 g/kg EtOH (Volkow *et al.*, 1988; Sano *et al.*, 1993) whereas flow was reduced when 1.5 g/kg EtOH was administered. Although cerebral blood flow is a useful proxy for measuring ongoing neuronal activity in awake humans, it can be problematic when studying alcohol, as alcohol acts as a direct vasodilator. An alternative approach for studying changes in brain activity during ethanol exposure is to measure changes in glucose metabolism. Whole brain glucose utilization has been shown to decrease during experiments in which the subject has self-reported feelings of "highness" (Wang *et al.*, 2000) and mood alteration (de Wit *et al.*, 1990). This decrease is observed in cortical areas, including PFC, with little change seen in basal ganglia and corpus callosum (Volkow *et al.*, 1990). While these studies were performed with moderate doses of EtOH, low doses have also been shown to reduce glucose metabolism, although behavioral deficits were not reported at this low dose (Volkow *et al.*, 2006). Transcranial magnetic stimulation (TMS) has been combined with brain imaging to probe changes in PFC activity, and healthy social drinkers show decreased TMS-induced prefrontal activity after 0.8 g/kg EtOH (Kahkonen *et al.*, 2003). These results are congruent with studies that have shown decreases in prefrontal glucose metabolism in the PFC after acute EtOH administration (de Wit *et al.*, 1990; Volkow *et al.*, 1990). The deleterious effects of acute EtOH on PFC-dependent behaviors may also result from the breakdown of functional specificity in different brain regions. Indeed, recent work has demonstrated a decrease in functional heterogeneity in right prefrontal cortex following ethanol administration (Volkow *et al.*, 2008). A decrease in asymmetry has also been observed while inebriated patients participated in a verbal fluency task (Wendt and Risberg, 2001). In this case, the decrease in lateral activation of the left dorsolateral PFC, a region associated with language performance, correlated with decreased performance on the task. Another study examined the effects of acute EtOH on memory encoding that ordinarily depends upon activation of the right PFC (Soderlund *et al.*, 2007). Again, ethanol administration decreased performance on the task, and this was associated with reduced bilateral PFC activation.

B. Chronic Alcohol

Chronic alcohol use in humans has been linked to deficits in executive function that depend on the PFC. One study showed that a significant number of substance-dependent individuals, including alcohol-dependent subjects, showed deficits in a gambling task that were similar to patients with lesions of the ventromedial PFC (Bechara *et al.*, 2001). Long-term use of alcohol has been shown to be more detrimental than cocaine in attention and executive functioning tasks (Goldstein *et al.*, 2004). Even in alcoholics who do not demonstrate deficits in certain executive tasks, differential patterns of activity in PFC are observed that suggest changes in the way that the brain performs the tasks (Pfefferbaum *et al.*, 2001). Similarly, adolescents who engage in binge drinking performed adequately on a visual working memory task, yet there was hypoactivation of the right anterior PFC when compared with controls, suggesting a reorganization or compensatory mechanism in problem solving (Crego *et al.*, 2010).

These alterations in PFC activity may go unnoticed under low-level cognitive demands but may underlie deficits associated with higher-order cognitive function. Reduced glucose metabolism in alcohol-dependent subjects has been identified in mediofrontal and left dorsolateral PFC (Dao-Castellana *et al.*, 1998). Decreases in medial frontal cortex glucose metabolism have been correlated with poor performance on the Wisconson Card Sorting Task in alcoholic subjects (Adams *et al.*, 1993). It seems, though, that frontal glucose metabolism increases after a period of abstinence, suggesting that some behavioral consequences of chronic alcohol use may be reversible (Volkow *et al.*, 1994). Indeed, abstinent subjects in one study demonstrated an increase in cognitive and executive functioning that was correlated with an increase in frontal glucose metabolism (Johnson-Greene *et al.*, 1997).

Changes in the structural morphology and integrity of the PFC have also been observed that may underlie the cognitive deficits associated with chronic alcohol exposure. Alcohol-dependent subjects show reduced gray matter in the dorsolateral PFC (Jernigan *et al.*, 1991). Chronic alcohol use is associated with reduced white matter volume throughout the cortex (de la Monte, 1988). Frontal lobes seem to be especially susceptible to volume loss following long-term chronic alcohol exposure (Pfefferbaum *et al.*, 1997). Reductions in levels of *N*-acetylaspartate, an abundant CNS metabolite, have also been observed in prefrontal white matter of detoxified alcoholic subjects (Schweinsburg *et al.*, 2001), perhaps indicating an atrophy of white matter during withdrawal. Additionally, the integrity of white matter in the right orbitofrontal cortex is significantly impaired in chronic alcoholics (Pfefferbaum and Sullivan, 2005; Harris *et al.*, 2008). A study using stereology to monitor cell number demonstrated no significant cell loss through the neocortex, even though overall cortical volume was decreased, suggesting atrophy or dendritic retraction rather than frank cell death (Jensen and Pakkenberg, 1993). Significant reduction of soma size in frontal cingulate cortex of alcoholics, without a significant change in cell number, has also been observed (Kril *et al.*, 1989). Changes in glial cell count following chronic ethanol exposure have also been reported and indicate decreases in both density and size of glia in dorsolateral PFC (Miguel-Hidalgo *et al.*, 2002) and orbitofrontal cortex (Miguel-Hidalgo *et al.*, 2006). Whether the loss of volume in PFC in alcohol-dependent subjects is reversible is still undetermined. In one study, subjects who were abstinent for 6–9 months after entering a rehab program showed decreased ventricle size, but there was no increase in frontal lobe volume when compared to their entry into the rehab program (Wobrock *et al.*, 2009).

Genomic and proteomic studies in humans have demonstrated alterations in gene expression and protein levels in alcoholics as opposed to nonalcoholic controls. In one study, genes for mitochondrial proteins were downregulated in the PFC of alcoholics and this was paralleled by an increase in repair proteins that are activated due to oxidative damage, possibly in response to mitochondrial dysfunction (Flatscher-Bader *et al.*, 2005). There were also region-specific alterations in transcription factors and genes that encode DNA-binding proteins, protein trafficking, and apoptotic factors. Reverse transcription polymerase chain reaction (RT-PCR) analysis has demonstrated a downregulation in *PCP4/PEP19*, a gene that encodes for a protein that may have anti-apoptotic properties (Iwamoto *et al.*, 2004). This study also showed alterations in expression of *NEFH*, which encodes a neurofilament protein, suggesting that changes in the expression of this protein could lead to altered organization of the cytoskeleton. Similar to animal studies, PFC tissue collected from chronic alcoholics shows changes in the expression of a number of genes involved in myelination (Mayfield *et al.*, 2002). Levels of key energy-regulating and metabolism proteins are also downregulated in PFC white matter of alcoholics as opposed to nonalcoholic controls (Alexander-Kaufman *et al.*, 2006). Other proteins show enhanced expression, including midkine, a repair protein, and EAAT1, a glutamate transporter (Flatscher-Bader and Wilce, 2008). Changes in midkine may indicate a response to oxidative damage after chronic alcohol use, whereas the change in transporter levels may be a compensatory mechanism in response to elevated levels of extracellular glutamate during withdrawal. Proteins related to the thiamine pathway that are involved in carbohydrate metabolism were downregulated in the PFC of non-Wernicke-Korsokoff alcoholics (Matsumoto, 2009). The transcription factor NF-kappaB and its p50 subunit, which inhibit gene transcription, were also downregulated in postmortem prefrontal tissue of alcoholics as compared to controls, suggesting increased expression of genes related to synaptic plasticity that may play a role in the development and maintenance of alcohol dependence (Okvist *et al.*, 2007). Taken together, these postmortem studies indicate a number of differences in the PFC of alcoholics that could lead to compromised function of the PFC and increased cognitive deficits.

IX. Summary

It is clear from the above discussion that exposure to alcohol, either acutely or chronically, has significant effects on the functional and structural status of the prefrontal cortex. Given the key role that this brain region plays in the integration, manipulation, and evaluation of incoming sensory and cognitive information, it is not surprising that alcoholics display deficits in executive control, decision making, and risk management. Despite these observations, however, there are still many unanswered questions that need to be addressed regarding the role of the PFC in the etiology and treatment of alcoholism. At a basic cellular level, there is still a paucity of data with respect to the acute effects of ethanol on subtypes of excitatory and inhibitory neurons within the diverse anatomical structures that make up the prefrontal cortex. Functional mapping studies using both electrophysiological and multineuron imaging techniques are needed to assess the relative ethanol sensitivity of neurons within prefrontal cortical microciruits and across the various subdivisions of this area (e.g., anterior cingulate, prelimbic, infralimbic, and orbitofrontal). There is also little known at the network level about how alcohol disrupts normal patterns of activity that are thought to be critical for processes such as working memory and decision making and whether the chronically alcohol-exposed brain can recover from deficits observed in these behaviors. The continuing development of awake animal array recording techniques coupled with sophisticated multidimensional analytical approaches promises to provide information that can address this key question. In terms of human PFC function and involvement in alcohol-related behaviors, results from imaging studies have provided a great deal of information regarding the effects of chronic alcohol exposure on brain structure and processing. However, these approaches are limited by the rather coarse level of anatomical, cellular, and temporal resolution of functional imaging. Advances in small animal imaging techniques offer at least a chance to address some of these limitations and multi-investigator teams with expertise in both animal and human imaging need to be established to help bridge the gap between clinical and preclinical studies. Finally, it is clear that advances in cognitive neuroscience, neuropsychology, and psychiatry need to be coordinated with basic science findings in order to spur development of new approaches to treat alcohol-induced alterations in prefrontal cortical function. Recent findings in the fear and post-traumatic stress disorder literature suggest that neural mechanisms of plasticity involved in learning can be used to overcome or reset maladaptive behaviors associated with prior emotional, physical, or chemical alterations in brain function. Applying these concepts and ideas to the treatment of alcoholics and those with alcohol-use disorders may augment the few pharmacological approaches that are currently available for treating these alcohol-related problems.

References

Adams, K. M., Gilman, S., Koeppe, R. A., Kluin, K. J., Brunberg, J. A., Dede, D., Berent, S., and Kroll, P. D. (1993). Neuropsychological deficits are correlated with frontal hypometabolism in positron emission tomography studies of older alcoholic patients. *Alcohol. Clin. Exp. Res.* **17**, 205–210.

Alexander-Kaufman, K., James, G., Sheedy, D., Harper, C., and Matsumoto, I. (2006). Differential protein expression in the prefrontal white matter of human alcoholics: A proteomics study. *Mol. Psychiatry* **11**, 56–65.

Andrews, G. D., and Lavin, A. (2006). Methylphenidate increases cortical excitability via activation of alpha-2 noradrenergic receptors. *Neuropsychopharmacology* **31**, 594–601.

Araneda, R., and Andrade, R. (1991). 5-Hydroxytryptamine2 and 5-hydroxytryptamine 1A receptors mediate opposing responses on membrane excitability in rat association cortex. *Neuroscience* **40**, 399–412.

Arbib, M. A. (1985). Schemas for the temporal organization of behaviour. *Hum. Neurobiol.* **4**, 63–72.

Arnsten, A. F. (2009). Stress signalling pathways that impair prefrontal cortex structure and function. *Nat. Rev. Neurosci.* **10**, 410–422.

Ascoli, G. A., Alonso-Nanclares, L., Anderson, S. A., Barrionuevo, G., Benavides-Piccione, R., Burkhalter, A., Buzsaki, G., Cauli, B., Defelipe, J., Fairen, A., Feldmeyer, D., Fishell, G., Fregnac, Y., Freund, T. F., Gardner, D., Gardner, E. P., Goldberg, J. H., Helmstaedter, M., Hestrin, S., Karube, F., Kisvarday, Z. F., Lambolez, B., Lewis, D. A., Marin, O., Markram, H., Munoz, A., Packer, A., Petersen, C. C., Rockland, K. S., Rossier, J., Rudy, B., Somogyi, P., Staiger, J. F., Tamas, G., Thomson, A. M., Toledo-Rodriguez, M., Wang, Y., West, D. C., and Yuste, R. (2008). Petilla terminology: Nomenclature of features of GABAergic interneurons of the cerebral cortex. *Nat. Rev. Neurosci.* **9**, 557–568.

Auclair, N., Otani, S., Soubrie, P., and Crepel, F. (2000). Cannabinoids modulate synaptic strength and plasticity at glutamatergic synapses of rat prefrontal cortex pyramidal neurons. *J. Neurophysiol.* **83**, 3287–3293.

Barbas, H., and Pandya, D. N. (1989). Architecture and intrinsic connections of the prefrontal cortex in the rhesus monkey. *J. Comp. Neurol.* **286**, 353–375.

Bayer, H. M., Lau, B., and Glimcher, P. W. (2007). Statistics of midbrain dopamine neuron spike trains in the awake primate. *J. Neurophysiol.* **98**, 1428–1439.

Bechara, A., Dolan, S., Denburg, N., Hindes, A., Anderson, S. W., and Nathan, P. E. (2001). Decision-making deficits, linked to a dysfunctional ventromedial prefrontal cortex, revealed in alcohol and stimulant abusers. *Neuropsychologia* **39**, 376–389.

Beique, J. C., Imad, M., Mladenovic, L., Gingrich, J. A., and Andrade, R. (2007). Mechanism of the 5-hydroxytryptamine 2A receptor-mediated facilitation of synaptic activity in prefrontal cortex. *Proc. Natl. Acad. Sci. U. S. A.* **104**, 9870–9875.

Berke, J. D., and Hyman, S. E. (2000). Addiction, dopamine, and the molecular mechanisms of memory. *Neuron* **25**, 515–532.

Bressler, S. L. (1995). Large-scale cortical networks and cognition. *Brain Res. Brain Res. Rev.* **20**, 288–304.

Chandler, L. J. (2003). Ethanol and brain plasticity: Receptors and molecular networks of the postsynaptic density as targets of ethanol. *Pharmacol. Ther.* **99**, 311–326.

Chang, Y. M., and Luebke, J. I. (2007). Electrophysiological diversity of layer 5 pyramidal cells in the prefrontal cortex of the rhesus monkey: *In vitro* slice studies. *J. Neurophysiol.* **98**, 2622–2632.

Ciccocioppo, R., Ge, J., Barnes, N. M., and Cooper, S. J. (1997). Autoradiographic mapping of brain 5-HT2A binding sites in P and in AA alcohol-preferring rats. *Brain Res. Bull.* **44**, 33–37.

Crego, A., Rodriguez-Holguin, S., Parada, M., Mota, N., Corral, M., and Cadaveira, F. (2010). Reduced anterior prefrontal cortex activation in young binge drinkers during a visual working memory task. *Drug Alcohol Depend.* **109**, 45–46.

Cui, C., Xu, M., and Atzori, M. (2006). Voltage-dependent block of *N*-methyl-D-aspartate receptors by dopamine D1 receptor ligands. *Mol. Pharmacol.* **70**, 1761–1770.

D'Esposito, M. (2007). From cognitive to neural models of working memory. *Philos. Trans. R. Soc. Lond. B. Biol. Sci.* **362**, 761–772.

Dao-Castellana, M. H., Samson, Y., Legault, F., Martinot, J. L., Aubin, H. J., Crouzel, C., Feldman, L., Barrucand, D., Rancurel, G., Feline, A., and Syrota, A. (1998). Frontal dysfunction in neurologically normal chronic alcoholic subjects: Metabolic and neuropsychological findings. *Psychol. Med.* **28**, 1039–1048.

de la Monte, S. M. (1988). Disproportionate atrophy of cerebral white matter in chronic alcoholics. *Arch. Neurol.* **45**, 990–992.

de Wit, H., Metz, J., Wagner, N., and Cooper, M. (1990). Behavioral and subjective effects of ethanol: Relationship to cerebral metabolism using PET. *Alcohol. Clin. Exp. Res.* **14**, 482–489.

Elston, G. N. (2003). Cortex, cognition and the cell: New insights into the pyramidal neuron and prefrontal function. *Cereb. Cortex* **13**, 1124–1138.

Engleman, E. A., Ingraham, C. M., McBride, W. J., Lumeng, L., and Murphy, J. M. (2006). Extracellular dopamine levels are lower in the medial prefrontal cortex of alcohol-preferring rats compared to Wistar rats. *Alcohol* **38**, 5–12.

Everitt, B. J., and Robbins, T. W. (2005). Neural systems of reinforcement for drug addiction: From actions to habits to compulsion. *Nat. Neurosci.* **8**, 1481–1489.

Finn, P. R., Justus, A., Mazas, C., and Steinmetz, J. E. (1999). Working memory, executive processes and the effects of alcohol on Go/No-Go learning: Testing a model of behavioral regulation and impulsivity. *Psychopharmacology (Berl)* **146**, 465–472.

Flatscher-Bader, T., van der Brug, M., Hwang, J. W., Gochee, P. A., Matsumoto, I., Niwa, S., and Wilce, P. A. (2005). Alcohol-responsive genes in the frontal cortex and nucleus accumbens of human alcoholics. *J. Neurochem.* **93**, 359–370.

Flatscher-Bader, T., and Wilce, P. A. (2008). Impact of alcohol abuse on protein expression of midkine and excitatory amino acid transporter 1 in the human prefrontal cortex. *Alcohol. Clin. Exp. Res.* **32**, 1849–1858.

Fortin, D. A., and Levine, E. S. (2007). Differential effects of endocannabinoids on glutamatergic and GABAergic inputs to layer 5 pyramidal neurons. *Cereb. Cortex* **17**, 163–174.

Fuster, J. M. (2008). The Prefrontal Cortex. Academic Press, London.

George, S., Rogers, R. D., and Duka, T. (2005). The acute effect of alcohol on decision making in social drinkers. *Psychopharmacology (Berl)* **182**, 160–169.

Goldman-Rakic, P. S. (1999). The physiological approach: Functional architecture of working memory and disordered cognition in schizophrenia. *Biol. Psychiatry* **46**, 650–661.

Goldstein, R. Z., Leskovjan, A. C., Hoff, A. L., Hitzemann, R., Bashan, F., Khalsa, S. S., Wang, G. J., Fowler, J. S., and Volkow, N. D. (2004). Severity of neuropsychological impairment in cocaine and alcohol addiction: Association with metabolism in the prefrontal cortex. *Neuropsychologia* **42**, 1447–1458.

Goldstein, R. Z., and Volkow, N. D. (2002). Drug addiction and its underlying neurobiological basis: Neuroimaging evidence for the involvement of the frontal cortex. *Am. J. Psychiatry* **159**, 1642–1652.

Gorelova, N., Seamans, J. K., and Yang, C. R. (2002). Mechanisms of dopamine activation of fast-spiking interneurons that exert inhibition in rat prefrontal cortex. *J. Neurophysiol.* **88**, 3150–3166.

Grattan-Miscio, K. E., and Vogel-Sprott, M. (2005). Effects of alcohol and performance incentives on immediate working memory. *Psychopharmacology (Berl)* **181**, 188–196.

Harris, G. J., Jaffin, S. K., Hodge, S. M., Kennedy, D., Caviness, V. S., Marinkovic, K., Papadimitriou, G. M., Makris, N., and Oscar-Berman, M. (2008). Frontal white matter and cingulum diffusion tensor imaging deficits in alcoholism. *Alcohol. Clin. Exp. Res.* **32**, 1001–1013.

Ichihara-Takeda, S., and Funahashi, S. (2007). Activity of primate orbitofrontal and dorsolateral prefrontal neurons: Task-related activity during an oculomotor delayed-response task. *Exp. Brain Res.* **181**, 409–425.

Iwamoto, K., Bundo, M., Yamamoto, M., Ozawa, H., Saito, T., and Kato, T. (2004). Decreased expression of NEFH and PCP4/PEP19 in the prefrontal cortex of alcoholics. *Neurosci. Res.* **49**, 379–385.

Jensen, G. B., and Pakkenberg, B. (1993). Do alcoholics drink their neurons away? *Lancet* **342**, 1201–1204.

Jentsch, J. D., and Taylor, J. R. (1999). Impulsivity resulting from frontostriatal dysfunction in drug abuse: Implications for the control of behavior by reward-related stimuli. *Psychopharmacology (Berl)* **146**, 373–390.

Jernigan, T. L., Butters, N., DiTraglia, G., Schafer, K., Smith, T., Irwin, M., Grant, I., Schuckit, M., and Cermak, L. S. (1991). Reduced cerebral grey matter observed in alcoholics using magnetic resonance imaging. *Alcohol. Clin. Exp. Res.* **15**, 418–427.

Johnson-Greene, D., Adams, K. M., Gilman, S., Koeppe, R. A., Junck, L., Kluin, K. J., Martorello, S., and Heumann, M. (1997). Effects of abstinence and relapse upon neuropsychological function and cerebral glucose metabolism in severe chronic alcoholism. *J. Clin. Exp. Neuropsychol.* **19**, 378–385.

Kahkonen, S., Wilenius, J., Nikulin, V. V., Ollikainen, M., and Ilmoniemi, R. J. (2003). Alcohol reduces prefrontal cortical excitability in humans: A combined TMS and EEG study. *Neuropsychopharmacology* **28**, 747–754.

Kalivas, P. W. (2009). The glutamate homeostasis hypothesis of addiction. *Nat. Rev. Neurosci.* **10**, 561–572.

Kalivas, P. W., and Volkow, N. D. (2005). The neural basis of addiction: A pathology of motivation and choice. *Am. J. Psychiatry* **162**, 1403–1413.

Kawaguchi, Y. (1995). Physiological subgroups of nonpyramidal cells with specific morphological characteristics in layer II/III of rat frontal cortex. *J. Neurosci.* **15**, 2638–2655.

Kerns, R. T., Ravindranathan, A., Hassan, S., Cage, M. P., York, T., Sikela, J. M., Williams, R. W., and Miles, M. F. (2005). Ethanol-responsive brain region expression networks: Implications for behavioral responses to acute ethanol in DBA/2J versus C57BL/6J mice. *J. Neurosci.* **25**, 2255–2266.

Kia, H. K., Miquel, M. C., Brisorgueil, M. J., Daval, G., Riad, M., El Mestikawy, S., Hamon, M., and Verge, D. (1996). Immunocytochemical localization of serotonin1A receptors in the rat central nervous system. *J. Comp. Neurol.* **365**, 289–305.

King, A., Munisamy, G., de Wit, H., and Lin, S. (2006). Attenuated cortisol response to alcohol in heavy social drinkers. *Int. J. Psychophysiol.* **59**, 203–209.

Kril, J. J., Gundlach, A. L., Dodd, P. R., Johnston, G. A., and Harper, C. G. (1989). Cortical dihydropyridine binding sites are unaltered in human alcoholic brain. *Ann.Neurol.* **26**, 395–397.

Kroener, S., Chandler, L. J., Phillips, P. E., and Seamans, J. K. (2009). Dopamine modulates persistent synaptic activity and enhances the signal-to-noise ratio in the prefrontal cortex. *PLoS One* **4**, e6507.

Lafourcade, M., Elezgarai, I., Mato, S., Bakiri, Y., Grandes, P., and Manzoni, O. J. (2007). Molecular components and functions of the endocannabinoid system in mouse prefrontal cortex. *PLoS One* **2**, e709.

Lammel, S., Hetzel, A., Hackel, O., Jones, I., Liss, B., and Roeper, J. (2008). Unique properties of mesoprefrontal neurons within a dual mesocorticolimbic dopamine system. *Neuron* **57**, 760–773.

Langen, B., Dietze, S., and Fink, H. (2002). Acute effect of ethanol on anxiety and 5-HT in the prefrontal cortex of rats. *Alcohol* **27**, 135–141.

Leggio, B., Masi, F., Grappi, S., Nanni, G., Gambarana, C., Colombo, G., and de Montis, M. G. (2003). Sardinian alcohol-preferring and non-preferring rats show different reactivity to aversive stimuli and a similar response to a natural reward. *Brain Res.* **973**, 275–284.

Leriche, M., Mendez, M., Zimmer, L., and Berod, A. (2008). Acute ethanol induces Fos in GABAergic and non-GABAergic forebrain neurons: A double-labeling study in the medial prefrontal cortex and extended amygdala. *Neuroscience* **153**, 259–267.

Liu, W., Yuen, E. Y., Allen, P. B., Feng, J., Greengard, P., and Yan, Z. (2006). Adrenergic modulation of NMDA receptors in prefrontal cortex is differentially regulated by RGS proteins and spinophilin. *Proc. Natl. Acad. Sci. U. S. A.* **103**, 18338–18343.

Marinelli, P. W., Kiianmaa, K., and Gianoulakis, C. (2000). Opioid propeptide mRNA content and receptor density in the brains of AA and ANA rats. *Life Sci.* **66**, 1915–1927.

Markram, H., Toledo-Rodriguez, M., Wang, Y., Gupta, A., Silberberg, G., and Wu, C. (2004). Interneurons of the neocortical inhibitory system. *Nat Rev Neurosci* **5**, 793–807.

Matsumoto, I. (2009). Proteomics approach in the study of the pathophysiology of alcohol-related brain damage. *Alcohol Alcohol.* **44**, 171–176.

Mayfield, R. D., Lewohl, J. M., Dodd, P. R., Herlihy, A., Liu, J., and Harris, R. A. (2002). Patterns of gene expression are altered in the frontal and motor cortices of human alcoholics. *J. Neurochem.* **81**, 802–813.

McCormick, D. A., Shu, Y., Hasenstaub, A., Sanchez-Vives, M., Badoual, M., and Bal, T. (2003). Persistent cortical activity: Mechanisms of generation and effects on neuronal excitability. *Cereb. Cortex* **13**, 1219–1231.

Mendez, M., Leriche, M., and Calva, J. C. (2001). Acute ethanol administration differentially modulates mu opioid receptors in the rat meso-accumbens and mesocortical pathways. *Brain Res. Mol. Brain Res.* **94**, 148–156.

Miguel-Hidalgo, J. J., Overholser, J. C., Meltzer, H. Y., Stockmeier, C. A., and Rajkowska, G. (2006). Reduced glial and neuronal packing density in the orbitofrontal cortex in alcohol dependence and its relationship with suicide and duration of alcohol dependence. *Alcohol. Clin. Exp. Res.* **30**, 1845–1855.

Miguel-Hidalgo, J. J., Wei, J., Andrew, M., Overholser, J. C., Jurjus, G., Stockmeier, C. A., and Rajkowska, G. (2002). Glia pathology in the prefrontal cortex in alcohol dependence with and without depressive symptoms. *Biol. Psychiatry* **52**, 1121–1133.

Moghaddam, B., and Homayoun, H. (2008). Divergent plasticity of prefrontal cortex networks. *Neuropsychopharmacology* **33**, 42–55.

Molnar, Z., and Cheung, A. F. (2006). Towards the classification of subpopulations of layer V pyramidal projection neurons. *Neurosci. Res.* **55**, 105–115.

Neznanova, O., Bjork, K., Rimondini, R., Hansson, A. C., Hyytia, P., Heilig, M., and Sommer, W. H. (2009). Acute ethanol challenge inhibits glycogen synthase kinase-3beta in the rat prefrontal cortex. *Int J Neuropsychopharmacol* **12**, 275–280.

Norman, D. A., and Shallice, T. (1986). Attention to action. In "Consciousness and Self-Regulation," (R. J. Davidson, G. E. Schawartz, and D. Shapiro, Eds.), pp. 1–18. Plenum, New York.

Okvist, A., Johansson, S., Kuzmin, A., Bazov, I., Merino-Martinez, R., Ponomarev, I., Mayfield, R. D., Harris, R. A., Sheedy, D., Garrick, T., Harper, C., Hurd, Y. L., Terenius, L., Ekstrom, T. J., Bakalkin, G., and Yakovleva, T. (2007). Neuroadaptations in human chronic alcoholics: Dysregulation of the NF-kappaB system. *PLoS ONE* **2**, e930.

Ongur, D., and Price, J. L. (2000). The organization of networks within the orbital and medial prefrontal cortex of rats, monkeys and humans. *Cereb Cortex* **10**, 206–219.

Pascual, M., Boix, J., Felipo, V., and Guerri, C. (2009). Repeated alcohol administration during adolescence causes changes in the mesolimbic dopaminergic and glutamatergic systems and promotes alcohol intake in the adult rat. *J. Neurochem.* **108**, 920–931.

Pfefferbaum, A., Desmond, J. E., Galloway, C., Menon, V., Glover, G. H., and Sullivan, E. V. (2001). Reorganization of frontal systems used by alcoholics for spatial working memory: An fMRI study. *Neuroimage* **14**, 7–20.

Pfefferbaum, A., and Sullivan, E. V. (2005). Disruption of brain white matter microstructure by excessive intracellular and extracellular fluid in alcoholism: Evidence from diffusion tensor imaging. *Neuropsychopharmacology* **30**, 423–432.

Pfefferbaum, A., Sullivan, E. V., Mathalon, D. H., and Lim, K. O. (1997). Frontal lobe volume loss observed with magnetic resonance imaging in older chronic alcoholics. *Alcohol. Clin. Exp. Res.* **21**, 521–529.

Riedel, G., and Davies, S. N. (2005). Cannabinoid function in learning, memory and plasticity. Handb. Exp. Pharmacol. 445–477.

Rubio, M., de Miguel, R., Fernandez-Ruiz, J., Gutierrez-Lopez, D., Carai, M. A., and Ramos, J. A. (2009). Effects of a short-term exposure to alcohol in rats on FAAH enzyme and CB1 receptor in different brain areas. *Drug Alcohol Depend.* **99**, 354–358.

Sanchez-Vives, M. V., and McCormick, D. A. (2000). Cellular and network mechanisms of rhythmic recurrent activity in neocortex. *Nat. Neurosci.* **3**, 1027–1034.

Sano, M., Wendt, P. E., Wirsen, A., Stenberg, G., Risberg, J., and Ingvar, D. H. (1993). Acute effects of alcohol on regional cerebral blood flow in man. *J. Stud. Alcohol* **54**, 369–376.

Santana, N., Bortolozzi, A., Serrats, J., Mengod, G., and Artigas, F. (2004). Expression of serotonin1A and serotonin2A receptors in pyramidal and GABAergic neurons of the rat prefrontal cortex. *Cereb. Cortex* **14**, 1100–1109.

Saults, J. S., Cowan, N., Sher, K. J., and Moreno, M. V. (2007). Differential effects of alcohol on working memory: Distinguishing multiple processes. *Exp. Clin. Psychopharmacol.* **15**, 576–587.

Schultz, W. (1986). Responses of midbrain dopamine neurons to behavioral trigger stimuli in the monkey. *J. Neurophysiol.* **56**, 1439–1461.

Schweinsburg, B. C., Taylor, M. J., Alhassoon, O. M., Videen, J. S., Brown, G. G., Patterson, T. L., Berger, F., and Grant, I. (2001). Chemical pathology in brain white matter of recently detoxified alcoholics: A 1H magnetic resonance spectroscopy investigation of alcohol-associated frontal lobe injury. *Alcohol. Clin. Exp. Res.* **25**, 924–934.

Seamans, J. K., Lapish, C. C., and Durstewitz, D. (2008). Comparing the prefrontal cortex of rats and primates: Insights from electrophysiology. *Neurotox Res* **14**, 249–262.

Seamans, J. K., Nogueira, L., and Lavin, A. (2003). Synaptic basis of persistent activity in prefrontal cortex *in vivo* and in organotypic cultures. *Cereb. Cortex* **13**, 1242–1250.

Seamans, J. K., and Yang, C. R. (2004). The principal features and mechanisms of dopamine modulation in the prefrontal cortex. *Prog. Neurobiol.* **74**, 1–58.

Selim, M., and Bradberry, C. W. (1996). Effect of ethanol on extracellular 5-HT and glutamate in the nucleus accumbens and prefrontal cortex: Comparison between the Lewis and Fischer 344 rat strains. *Brain Res.* **716**, 157–164.

Smith, D. G., Learn, J. E., McBride, W. J., Lumeng, L., Li, T. K., and Murphy, J. M. (2001). Alcohol-naive alcohol-preferring (P) rats exhibit higher local cerebral glucose utilization than alcohol-nonpreferring (NP) and Wistar rats. *Alcohol. Clin. Exp. Res.* **25**, 1309–1316.

Soderlund, H., Grady, C. L., Easdon, C., and Tulving, E. (2007). Acute effects of alcohol on neural correlates of episodic memory encoding. *Neuroimage* **35**, 928–939.

Steriade, M., Nunez, A., and Amzica, F. (1993). A novel slow (< 1 Hz) oscillation of neocortical neurons *in vivo*: Depolarizing and hyperpolarizing components. *J. Neurosci.* **13**, 3252–3265.

Strother, W. N., McBride, W. J., Lumeng, L., and Li, T. K. (2005). Effects of acute administration of ethanol on cerebral glucose utilization in adult alcohol-preferring and alcohol-nonpreferring rats. *Alcohol* **35**, 119–128.

Tiihonen, J., Kuikka, J., Hakola, P., Paanila, J., Airaksinen, J., Eronen, M., and Hallikainen, T. (1994). Acute ethanol-induced changes in cerebral blood flow. *Am. J. Psychiatry* **151**, 1505–1508.

Trantham-Davidson, H., Kroner, S., and Seamans, J. K. (2008). Dopamine modulation of prefrontal cortex interneurons occurs independently of DARPP-32. *Cereb. Cortex* **18**, 951–958.

Trantham-Davidson, H., Neely, L. C., Lavin, A., and Seamans, J. K. (2004). Mechanisms underlying differential D1 versus D2 dopamine receptor regulation of inhibition in prefrontal cortex. *J. Neurosci.* **24**, 10652–10659.

Tseng, K. Y., and O'Donnell, P. (2004). Post-pubertal Emergence of Prefrontal Cortical Up States Induced by D1-NMDA Co-activation. *Cereb. Cortex.* **15**, 49–57.

Tu, Y., Kroener, S., Abernathy, K., Lapish, C., Seamans, J., Chandler, L. J., and Woodward, J. J. (2007). Ethanol inhibits persistent activity in prefrontal cortical neurons. *J. Neurosci.* **27**, 4765–4775.

Tzschentke, T. M. (2001). Pharmacology and behavioral pharmacology of the mesocortical dopamine system. *Prog. Neurobiol.* **63**, 241–320.

Unterrainer, J. M., and Owen, A. M. (2006). Planning and problem solving: From neuropsychology to functional neuroimaging. *J. Physiol. Paris* **99**, 308–317.

Uylings, H. B., Groenewegen, H. J., and Kolb, B. (2003). Do rats have a prefrontal cortex? *Behav. Brain Res.* **146**, 3–17.

Volkow, N. D., Fowler, J. S., and Wang, G. J. (2003). The addicted human brain: Insights from imaging studies. *J. Clin. Invest.* **111**, 1444–1451.

Volkow, N. D., Hitzemann, R., Wolf, A. P., Logan, J., Fowler, J. S., Christman, D., Dewey, S. L., Schlyer, D., Burr, G., Vitkun, S., and *et al.* (1990). Acute effects of ethanol on regional brain glucose metabolism and transport. *Psychiatry Res.* **35**, 39–48.

Volkow, N. D., Ma, Y., Zhu, W., Fowler, J. S., Li, J., Rao, M., Mueller, K., Pradhan, K., Wong, C., and Wang, G. J. (2008). Moderate doses of alcohol disrupt the functional organization of the human brain. *Psychiatry Res.* **162**, 205–213.

Volkow, N. D., Mullani, N., Gould, L., Adler, S. S., Guynn, R. W., Overall, J. E., and Dewey, S. (1988). Effects of acute alcohol intoxication on cerebral blood flow measured with PET. *Psychiatry Res.* **24**, 201–209.

Volkow, N. D., Wang, G. J., Franceschi, D., Fowler, J. S., Thanos, P. P., Maynard, L., Gatley, S. J., Wong, C., Veech, R. L., Kunos, G., and Kai Li, T. (2006). Low doses of alcohol substantially decrease glucose metabolism in the human brain. *Neuroimage* **29**, 295–301.

Volkow, N. D., Wang, G. J., Hitzemann, R., Fowler, J. S., Overall, J. E., Burr, G., and Wolf, A. P. (1994). Recovery of brain glucose metabolism in detoxified alcoholics. *Am. J. Psychiatry* **151**, 178–183.

Wang, G. J., Volkow, N. D., Franceschi, D., Fowler, J. S., Thanos, P. K., Scherbaum, N., Pappas, N., Wong, C. T., Hitzemann, R. J., and Felder, C. A. (2000). Regional brain metabolism during alcohol intoxication. *Alcohol. Clin. Exp. Res.* **24**, 822–829.

Wang, H. X., and Gao, W. J. (2009). Cell Type-Specific Development of NMDA Receptors in the Interneurons of Rat Prefrontal Cortex. *Neuropsychopharmacology.* **34**, 2028–2040.

Weissenborn, R., and Duka, T. (2003). Acute alcohol effects on cognitive function in social drinkers: Their relationship to drinking habits. *Psychopharmacology (Berl)* **165**, 306–312.

Weitlauf, C., and Woodward, J. J. (2008). Ethanol selectively attenuates NMDAR-mediated synaptic transmission in the prefrontal cortex. *Alcohol. Clin. Exp. Res.* **32**, 690–698.

Wendt, P. E., and Risberg, J. (2001). Ethanol reduces rCFB activation of left dorsolateral prefrontal cortex during a verbal fluency task. *Brain Lang.* **77**, 197–215.

Williams, G. V., and Goldman-Rakic, P. S. (1995). Modulation of memory fields by dopamine D1 receptors in prefrontal cortex. *Nature* **376**, 572–576.

Willins, D. L., Deutch, A. Y., and Roth, B. L. (1997). Serotonin 5-HT2A receptors are expressed on pyramidal cells and interneurons in the rat cortex. *Synapse* **27**, 79–82.

Wobrock, T., Falkai, P., Schneider-Axmann, T., Frommann, N., Wolwer, W., and Gaebel, W. (2009). Effects of abstinence on brain morphology in alcoholism: A MRI study. *Eur. Arch. Psychiatry Clin. Neurosci.* **259**, 143–150.

Woodward, J. J., and Pava, M. J. (2009). Effects of Ethanol on Persistent Activity and Up-States in Excitatory and Inhibitory Neurons in Prefrontal Cortex. *Alcohol. Clin. Exp. Res.* **33**, 2134–2140.

Yuen, E. Y., Jiang, Q., Chen, P., Feng, J., and Yan, Z. (2008). Activation of 5-HT2A/C receptors counteracts 5-HT1A regulation of n-methyl-D-aspartate receptor channels in pyramidal neurons of prefrontal cortex. *J. Biol. Chem.* **283**, 17194–17204.

Zhang, Z. W. (2004). Maturation of layer V pyramidal neurons in the rat prefrontal cortex: Intrinsic properties and synaptic function. *J. Neurophysiol.* **91**, 1171–1182.

Zhong, P., Liu, W., Gu, Z., and Yan, Z. (2008). Serotonin facilitates long-term depression induction in prefrontal cortex via p38 MAPK/Rab5-mediated enhancement of AMPA receptor internalization. *J. Physiol.* **586**, 4465–4479.

Zhou, F. M., and Hablitz, J. J. (1999). Activation of serotonin receptors modulates synaptic transmission in rat cerebral cortex. *J. Neurophysiol.* **82**, 2989–2999.

BK CHANNEL AND ALCOHOL, A COMPLICATED AFFAIR

Gilles Erwan Martin

Department of Psychiatry, The Brudnick Neuropsychiatric Research Institute, University of Massachusetts Medical School, Worcester, MA 01605, USA

Alcohol is a fast acting molecule that alters behavior within a few minutes of absorption. Its rapid behavioral impact suggests early action on ion channels. Of all voltage-gated potassium ion channels, BK channels, a subcategory of potassium channels characterized by their large unitary conductance, and by their capacity of being activated synergistically by membrane potential and intracellular free calcium, are unique due to their high sensitivity to alcohol. In this review, we discuss BK channels structure and function, and how they help us understand the various ways BK channel mediates alcohol's effects on neuronal function and on behavior in the striatum.

I. Introduction

Alcohol, one of the few legal drugs of abuse, is a fast-acting molecule known to alter behavior within a few minutes of absorption. While alcohol has many targets (e.g., DNA expression and second messenger pathways), its rapid behavioral impact suggests early action on ion channels. If the most common neurotransmitters (e.g., GABA, Glu, and Gly) and their respective receptors have long been identified as prime alcohol targets due to their sensitivity to the drug, voltage-gated potassium ion channels, however, have not shown similar

INTERNATIONAL REVIEW OF NEUROBIOLOGY, VOL. 91
DOI: 10.1016/S0074-7742(10)91010-6

sensitivity. Over the past decade, the focus on BK channels, a subcategory of potassium channels characterized by their uniquely large unitary conductance and by their capacity of being activated synergistically by membrane potential and intracellular free calcium, has provided clear evidence of their keen sensitivity to physiological concentrations (10–50 mM) of alcohol. These studies also document their role in alcohol-related behavior in invertebrates and mammals. Here, we discuss BK channel structure and function, and how they help us understand the various ways BK channel mediates alcohol's effects on neuronal activity and on behavior in the striatum.

II. BK Channel Properties and Its Regulation by Alcohol

A. BK Channel Characteristics

In neurons of the central and peripheral nervous system, as well as in smooth muscle cells, BK channels form as a result of the association of four pore-forming α subunits (Quirk and Reinhart, 2001; Shen et al., 1994). Each α subunit is a seven transmembrane domain (S0–S6) protein highly selective for potassium (it conducts K^+ 10- and 200-fold more efficiently than other cations like Rb^+ and Cs^+, respectively) that is nearly impermeant to Na^+. The BK channel derives its name from its uniquely large unitary conductance (i.e., Big K^+) of about 250 pS (Latorre *et al.*, 1982; Marty, 1981; Pallotta *et al.*, 1981), depending on recording conditions, conductance that stands in marked contrast to many other types of voltage-dependent ion channels. Such an unusually large conductance provides the BK channel with the ability to disproportionally impact neuronal excitability. BK channel activity increases with more positive potentials across the membrane and with increases in free intracellular calcium concentrations. The synergistic action of voltage and calcium, along with their large conductance, sets BK channels apart from all other ion channels. Like all voltage-gated potassium channels from the Kv family, the BK channel α subunit comprises a voltage sensor (segment S4) and a pore region (S5–S6) (Adelman *et al.*, 1992; Atkinson *et al.*, 1991). However, unique to the BKs are the RCK region (Fig. 1; red segment) and a 28-amino-acid sequence known as the calcium bowl (Schreiber and Salkoff, 1997) (Fig. 1; blue segment) identified as key to the channel calcium sensitivity. More recent studies have shown that other regions of the channel also contribute to this property (Fig. 1A; green triangles). Unlike most other members of the potassium channel family, the pore-forming α subunit of the BK channel comes from only one gene known as *Slo1* or *Slowpoke*, first identified in the 1990s (Adelman *et al.*, 1992; Atkinson *et al.*, 1991). However, several α subunit splice variants with distinct biophysical properties (Ha *et al.*,

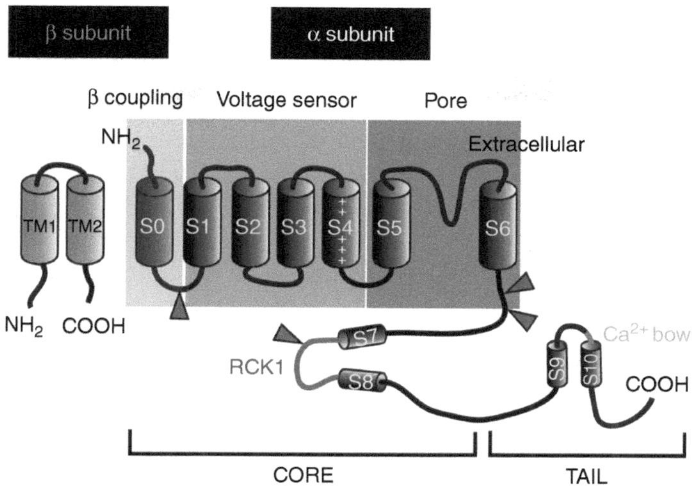

FIG. 1. Of all known ionic channels, BKs, also known as Slo and MaxiK (Gene ID, KCNMA1), display the largest unitary conductance (~250 pS) and are found in many members of the animal kingdom, including nematodes (*Caenorhabditis* (Crowder, 2004)), insects (*Drosophila* (Cowmeadow *et al.*, 2006; Lee *et al.*, 2008)), and mammals (Wanner *et al.*, 1999), a fact indicative of their importance in cell regulation. Structurally, the pore-forming α subunit consists of seven transmembrane spanning domains (S0–S6), the voltage-sensor (S1–S4), and the pore-forming loop (between S5 and S6). In addition, the channel displays four hydrophobic segments (S7–S10) within the long intracellular carboxyl (C) terminus where the two main calcium-sensitive domains (RCK and Ca^{2+} bowl) are found. The narrowest part of the pore, called the selectivity filter, determines the high flux rate of potassium in comparison to other ions. Tethering of β auxiliary subunits is ensured by the S0 domain. (Adapted from Treistman and Martin, 2009).

2000; Liu *et al.*, 2002; Pietrzykowski *et al.*, 2008; Tian *et al.*, 2001, 2004) and sensitivity to alcohol (Pietrzykowski *et al.*, 2008) have recently been identified, providing some degree of diversity for this subunit. Each α subunit forming BK channel homotetramers is associated with an auxiliary β subunit. Tethering of the β subunit is ensured by the α subunit S0 hydrophobic transmembrane domain (see Fig. 1). To date, four (β1–β4) subunits have been cloned. As set forth below, these auxiliary subunits are critical in contributing to BK channel sensitivity to alcohol. The BK channel α core subunit is ubiquitously expressed in the central and peripheral nervous system as well as in smooth muscle (Kamouchi *et al.*, 1997; McCobb *et al.*, 1995). In the brain, it is particularly abundant in the nucleus accumbens, hippocampus, cerebellum, and amygdala (Chang *et al.*, 1997), with its expression varying slightly between the types of brain studied (i.e., rats vs. mice) (Sausbier *et al.*, 2006). By contrast, BK β subunit expression in brain is much lower, which led early studies to conclude that brain BK channels were composed only of the core α subunit. Nevertheless, recent studies have shown that β4 expression, the most abundant brain β subunit, is found mostly in the forebrain, including the dorsal striatum and the nucleus accumbens, as well as the hippocampus and prefrontal cortex (Behrens *et al.*, 2000; Brenner *et al.*, 2000, 2005). Whereas β1 was initially thought to be exclusively expressed in peripheral smooth muscles (Behrens *et al.*, 2000; Zhang *et al.*, 2008), there is now strong evidence that it is also found in the brain (Martin *et al.*, 2004), although at very low levels compared to that of the β4 subunit. Finally, β2 and β3 subunits are nearly absent in the brain.

B. Regulation of Neuronal Activity by BK Channel

1. *Firing Pattern*

An effective way for a drug of abuse like alcohol to modify temporarily or permanently messages encoded by neurons is to act on the ultimate form of the neuronal encoding process, i.e., the action potential. BK channels, among other ion channels underlying action potentials, warrant particular attention as they are activated within a range of potentials (–60 to +40 mV) that correspond to the ascending phase of the action potential. The influx of calcium through voltage-gated calcium channels (VGCCs), activated by rapid depolarization induced by fast Na^+ currents, increases the open probability of these channels. BK channel activation in turn helps the cell repolarize and deactivates VGCCs, creating a powerful feedback mechanism controlling action potential duration and firing patterns (Brenner *et al.*, 2005; Edgerton and Reinhart, 2003; Faber and Sah, 2002, 2003). An elegant single-channel study carried out in cell-attached and inside-out patch-clamp recording modes in hippocampal neurons (Marrion and Tavalin, 1998) demonstrated that BK channels and voltage-gated N-type calcium channels co-localize on the same patch of membrane and that calcium influx through these calcium channels provides the source for BK activation, confirming a symbiotic relationship between these two types of ion channels (see also Loane *et al.*, 2007). In addition to modulating action potentials in the brain (Robitaille and Charlton, 1992; Shao *et al.*, 1999; Storm, 1990; Widmer *et al.*, 1998) and peripheral nervous system (Gruss *et al.*, 2001; Scholz *et al.*, 1998), BK channels also contribute to action potential after-hyperpolarization (AHP) (Adams *et al.*, 1982; Goldberg and Wilson, 2005; Lancaster and Nicoll, 1987; Shao *et al.*, 1999). Interestingly, the modulation of action potentials by BK channels does not appear to be limited to somatic action potentials. Evidence from immunohistochemical and electron microscopy studies suggests that BK channels are expressed in the membrane of axons and terminals (Kaufmann *et al.*, 2009; Knaus *et al.*, 1996; Misonou *et al.*, 2006; Sailer *et al.*, 2006) where they control action potential repolarization (Dopico *et al.*, 1996).

2. *Synaptic Integration*

Synaptic integration, which can be loosely defined as the neuronal computation of all synaptic inputs generated at any time points during synaptic transmission, depends on where (location) and when (synchronicity) synaptic events are generated on the dendritic arborization. A less appreciated component of this phenomenon is the kinetics of the synaptic event itself. Indeed, a widening of individual EPSPs increases the time window necessary for summation of synaptic events, whereas faster electrical events have the opposite effects. Until recently, it was commonly admitted that the blockade of the pore of the NMDA-type glutamate receptor by extracellular Mg^{2+} prevented this receptor from contributing significantly to synaptic transmission in conditions when EPSPs are evoked at low frequency. Evidence from studies on spike-timing-dependent plasticity (STDP), however, indicate that Mg^{2+}-mediated NMDA receptor blockade is easily overcome (Campanac and Debanne, 2008; Fuenzalida *et al.*, 2010; Markram *et al.*, 1997; Urakubo *et al.*, 2008). This synaptic plasticity paradigm requires the pairing of action potentials, generated in the soma and invading retrogradely the dendritic tree, and EPSPs. Thus, in structures like the prefrontal cortex and the hippocampus, when EPSPs follow APs with a delay of about 20 ms, a robust LTP is evoked (Dan and Poo, 2004). Taken together, these findings suggest that the initial depolarization provided by the activation of AMPA receptors (AMPARs) is sufficiently strong to relieve Mg^{2+} blockade and to recruit NMDA receptors. As indicated above, because of the synergistic action of membrane voltage and calcium, co-localization of BK channels and NMDA receptors in dendrites would provide BKs with sufficient leverage to shape EPSPs, and by extension STDP, as NMDA receptors are recruited during EPSPs depolarizing phase. The presence of BK channels in dendrites of neocortical pyramidal cells, magnocellular cells from the supraoptic nucleus, and in ventral striatum medium spiny neurons is supported by electrophysiological (Benhassine and Berger, 2005; Martin *et al.*, 2004; Wynne *et al.*, 2009) and anatomical (in spines, shaft, and dendrites) (Kaufmann *et al.*, 2009) evidence. To date, no data support the presence of BK channels in dendrites of the dorsal striatum medium spiny neurons. Nevertheless, the functional similarities between NAcc and dorsal striatum MSNs strongly suggest that BK channels likely share many characteristics, including their compartmental distribution.

In addition, electrophysiology experiments suggest that BK channels are co-localized with NMDA receptors (Backus and Trube, 1993; Isaacson and Murphy, 2001; Shah and Haylett, 2002). In a recent study combining whole-cell patch-clamp recording and calcium imaging, Benhassine and Berger (2008) showed BK—EPSPs co-localization in dendrites of neocortical neurons. The relatively low BK affinity for Ca^{2+} provides a likely explanation as to why these channels are located close to the calcium source (e.g., NMDARs, voltage-gated calcium channels—VGCCs), forming nanodomains, as opposed to small conductance calcium-activated potassium (SK) channels whose greater affinity position them much farther from calcium entry gate (Fakler and Adelman, 2008). By extension, potentiation of BK channels by EtOH would have the opposite effects on NMDA receptor calcium influx, the major permeant ion of these ionotropic receptors.

C. BK Channel and Alcohol Interactions

1. *Influence of Lipid Environment*

Before mutagenesis-based experiments demonstrated that the properties of membrane proteins were directly altered by alcohol, it was believed that alcohol's actions were mediated through its interactions with the lipid environment. Although this approach has been mostly replaced by an alcohol–protein interaction understanding, it would be unwise to discount the role of lipids as a mediator or modulator of alcohol effects. The role of lipids is supported by a number of studies using reconstituted lipid environments on membrane proteins in planar bilayer recording. In this preparation, proteins are "extracted" from biological membranes and incorporated into a highly simplified lipid film painted across an opening in a recording chamber. Channel structure and lipid composition are controlled using transfected cells for the host membrane. Single-channel activity can then be recorded, and it varies as a function of bilayer composition and presence of alcohol. Because ethanol activation of slo channels is replicated in simple, artificial bilayers made of two phospholipid species (Crowley *et al.*, 2003, 2005), it strongly suggests that cytoskeletal elements and regulatory proteins loosely bound to the channel are not necessary for drug action. Rather, ethanol modulation of BK channels only requires the slo subunit and a very simple lipid matrix. The critical role of slo is stressed by the differential ethanol responses found among several pituitary clonal cells (see above) and bslo versus mslohslo channels when expressed in a similar proteolipid environment. Thus, whereas a low EtOH concentration enhances mslo BK channel single-channel activity, it inhibits that of bslo channels (Dopico *et al.*, 1998; Dopico, 2003). Collectively, these studies are consistent with the idea that the slo subunit and its immediate lipid microenvironment represent the functional target of ethanol. This approach led to the conclusion that ethanol action does not require the continuous presence of cytosolic messengers, cell organelles, or cell metabolism.

2. *BK β Subunits Modulate Channel Sensitivity to Alcohol*

In addition to lipids, it has become increasingly apparent that BK channel auxiliary subunits play a critical role in modulating channel response to alcohol. Thus, in the nucleus accumbens, low alcohol concentrations enhance, in a dose-dependent fashion, the activity of single BK channels in membrane patches from medium spiny neuron cell body. By contrast, the same alcohol concentrations fail to alter open probability of dendritic BKs (Martin *et al.*, 2004). Examination of the dendritic and somatic biophysical properties revealed striking differences between the two channel populations, differences in line with a compartment-specific expression of β1 in dendrites and β4 in the cell body. This finding was confirmed using antibodies against these two subunits (Martin *et al.*, 2004). These results obtained in mammalian neurons, corroborated by data showing alcohol's effects on αβ1 and αβ4 BK channels expressed in HEK-293 cells (Feinberg-Zadek and Treistman, 2007), highlight BK's complex role in mediating alcohol effects on neuronal physiology.

3. *Phosphorylation*

The idea that protein kinases mediate alcohol effects is not new. Interactions between alcohol and kinases in neuronal and nonneuronal preparations have long been known (Aroor and Shukla, 2004; Chandler *et al.*, 1998; Gandhi and Ross, 1989; Newton and Messing, 2006). Genetically engineered mice were particularly useful tools to demonstrate the role of kinases (e.g., PKA and PKC) in a variety of EtOH-related behaviors. Thus, PKA and PKCγ KO mice show increased EtOH consumption in a two-bottle choice paradigm (Bowers and Wehner, 2001; Thiele *et al.*, 2000), whereas PKCϵ KO mice show the opposite effect (Olive *et al.*, 2000). Similarly, mice lacking the gene coding for Fyn kinase have reduced alcohol consumption (Boehm *et al.*, 2003; Stork *et al.*, 2002; Yaka *et al.*, 2003). These studies, and many others, clearly demonstrate the critical role played by kinases in controlling drinking behavior. Numerous studies showed that the BK channel is similarly regulated by a variety of kinases. Thus, at the single-channel level, the BK channel α subunit is phosphorylated by PKA (Chen *et al.*, 2005; Erxleben *et al.*, 2002; Gong *et al.*, 2002; Hall and Armstrong, 2000; Park *et al.*, 2007), PKC (Schubert *et al.*, 1999), and PKG (Fukao *et al.*, 1999; Fury *et al.*, 2002; Nara *et al.*, 2000; Swayze and Braun, 2001; White *et al.*, 1993), a list that is probably incomplete. Indeed, a careful analysis of BK α subunit putative phosphorylation sites (see Fig. 2; Preliminary data), using ScanSite (http://scansite.mit.edu) reveals that this core subunit intracellular c-terminus domain possesses two putative sites for Fyn/Src, in addition to sites for PKA (2), PKC (36), and CaMKII (4). In a recent detailed study of the role of CaMKII, this kinase not only controls the activity of bovine BK slo channel expressed in oocytes but also alcohol-modulatory actions through a Thr107 located on the intracellular loop between the transmembrane domains S0 and S1. Importantly, gradual phosphorylation of the Thr residue of the BK channel tetramer progressively increases the channel activity and switches alcohol effects from potentiation to inhibition (Liu *et al.*, 2006). This study clearly indicates that BK channel response to alcohol is not all-or-nothing but rather a gradual phenomenon controlled in part by BK phosphorylation level.

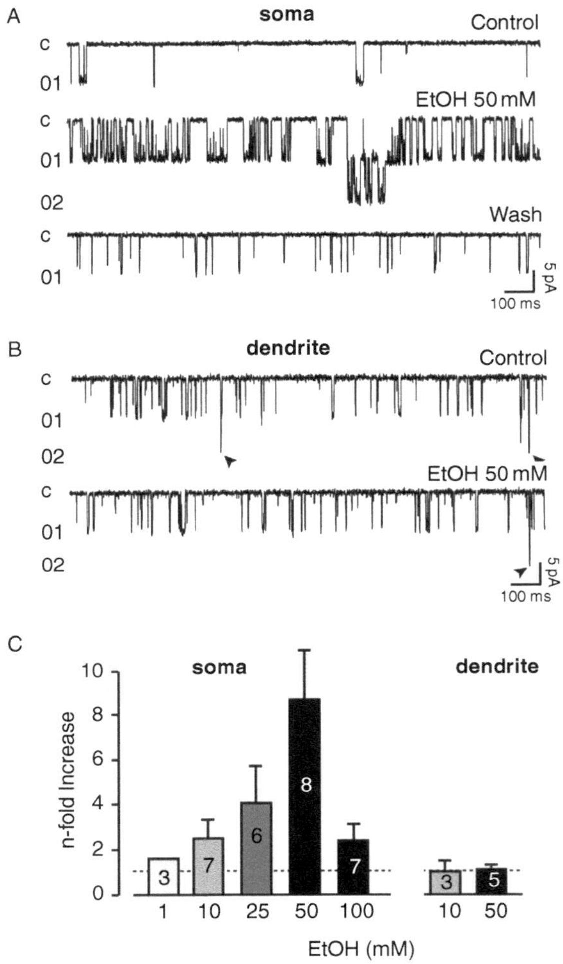

FIG. 2. Acute EtOH increases somatic but fails to alter dendritic BK channel activity. (A) Traces of somatic BK channels before, during (50 mM EtOH), and 2 min after (Wash) acute EtOH exposure (Vh = −30 mV; 5 μM free-Ca^{2+}) in inside-out patches. EtOH strongly potentiated the activity of somatic BK channels. (B) shows the lack of effect of 50 mM EtOH on dendritic BK channels recorded at Vh = −30 mV in the presence of 5 μM Ca^{2+} (wash not shown). The membrane patch was excised 25 μm from the soma. Downward deflections indicate inward currents; C and O represent the closed and open states of the channels, respectively. (C) Plot of averaged effects of various EtOH concentrations on somatic and dendritic BK channels. The dashed line indicates control level (one fold), and the numbers within bars represent the number of patches tested. (Adapted from Martin *et al.* 2004).

III. BK Channels in the Striatum

As indicated above, although the BK channel α subunit is ubiquitously expressed in both the central and peripheral nervous system, β subunit expression is much less abundant and is restricted to specific brain regions. In the adult rat brain, Knaus *et al.* (1996) and Chang *et al.*, (1997) showed that BK β subunit mRNA is found in the neo and olfactory cortices, hippocampus, cerebellum, striatum, thalamus, and amygdala, corroborating a number of electrophysiological studies (Bielefeldt *et al.*, 1992; Dopico *et al.*, 1996; Kang *et al.*, 1996). So far, β4 has been found to be the predominant form in the brain (Behrens *et al.*, 2000; Brenner *et al.*, 2000; Weiger *et al.*, 2000). The presence of the BK channel β1 subunit in the NAcc was unexpected, because it had been believed to be present exclusively in the smooth muscles of peripheral organs (Behrens *et al.*, 2000; Weiger *et al.*, 2000). However, it is unlikely that this finding is artifactual for several reasons: (1) positive identification was made both for mRNA, using reverse transcription - polymerase chain reaction (RT-PCR), and for protein, using an antibody; and (2) a highly selective antibody commonly used against the β1 subunit clearly showed that the staining originated from NAcc neurons. However, in order to fully appreciate the BK channel's role in the NAcc, it is important to note that within the same medium spiny neurons, BK channels β1 and β4 subunits loci of expression are segregated (Martin *et al.*, 2004). Thus, β4 appears to be mostly found in the soma whereas β1 subunits are confined to the dendritic arborization (Martin *et al.*, 2004). This finding suggests that BK channel phenotype responsible for shaping action potentials is different from that of BKs recruited by EPSPs at synapses in the dendritic arborization.

A. BK Channel Response to Acute Alcohol in the Nucleus Accumbens

Probing of BK channel sensitivity to alcohol in the NAcc was primarily carried out using single-channel inside-out patch-clamp recording. Instead of treating all BK channels as the same regardless of their subcompartmental origin, this technique provides the distinct advantage of isolating BK channels from either the soma or the dendrites, allowing for characterization of their biophysical properties. This work revealed that somatic and dendritic BK channels possess very different sensitivities to acute alcohol. Thus, in the soma, their activity was markedly enhanced, in a dose-dependent fashion, by alcohol (10–50 mM). By contrast, in the dendrites, they showed no sensitivity to the drug. A comparison of their biophysical properties (calcium sensitivity, kinetics, open time) with that of BK channels in heterologous expression systems supported the idea that dendritic and somatic BKs were mostly composed of αβ1 and αβ4 subunits, respectively.

Furthermore, the use of antibodies directed against β1 and β4 subunits confirmed their spatially segregated distribution in NAcc medium spiny neurons. Importantly, a recent study documenting β subunit expression pattern in dendrites and soma of magnocellular neurons (MCNs) from the supraoptic nucleus (SON) shows that BK channel sensitivity to alcohol and subunit composition in all three neuronal compartments (soma, dendrites, and terminals) appears to be unique to each neuronal population. Thus, in the SON, BK channel β subunit expression pattern presents some similarities and differences with the NAcc. As in MSNs, dendrites of SON neurons express β1. However, β1 subunit expression extends to the cell body where it is nearly absent in MSNs. Finally, early studies reported that low alcohol concentrations enhanced BK channel open probability in terminals of magnocellular (Dopico *et al.*, 1996), suggesting the presence of BK β4 subunits, a finding confirmed later using antibodies directed against this subunit (Wynne *et al.*, 2009). The comparison between MSNs and magnocellular neurons highlights the idea that alcohol modulation of neuronal function is not indiscriminate and that the neuronal compartment affected, and the various physiological functions it subserves (see above), is under the control of BK channel β subunit expression pattern.

B. BK Channels and Alcohol-Related Behavior

Among alcohol-related behaviors, tolerance, described as the gradual loss of drug effectiveness over time, is probably the most immediate and noticeable consequence of alcohol absorption. This phenomenon is particularly important because its magnitude in naive subjects can predict the likelihood to develop alcohol abuse (Chrostek and Szmitkowski, 1998; Erwin *et al.*, 1980; Fillmore *et al.*, 2005; Heath *et al.*, 1999; Schuckit, 1994). The neurobiology of tolerance is thought to involve several types of adaptation, ranging from alteration in membrane lipid composition (Yuan *et al.*, 2008) to neuroadaptative changes in target proteins (Crews *et al.*, 1996; Woodward *et al.*, 2006). Work on *Caenorhabditis elegans* showed that ethanol causes behavioral intoxication at the same doses that causes intoxication in humans (Davies *et al.*, 2003). Interestingly, a loss-of-function mutation of the gene (*slo1*) coding for the BK α subunit (the only BK channel subunit expressed in *C. elegans*) altered behavior as a strong tolerance to alcohol develops rapidly, suggesting that BK channels may be an important ion channel regulating the expression of behavioral alcohol tolerance. A similar conclusion was reached using *Drosophila melanogaster*, another invertebrate (Cowmeadow *et al.*, 2005, 2006).

A combined single-channel patch-clamp recording and behavioral study on mice further tested the role of BK channels in tolerance at the molecular, cellular, and behavioral levels. This work focused on the β4 subunit because it is highly expressed in striatum where, when co-assembled with the BK α, it forms functional ethanol-sensitive channels in the soma of MSNs. In striatal MSNs from β4 knockout (KO) mice, alcohol increases BK channel open probability, a response that disappeared 2–3 min after the beginning of drug exposure, indicative of a strong tolerance. By contrast, in wild-type (wt) mice, tolerance was not observed when alcohol-mediated enhancement of BK channel activity persisted for up to 9 min. A similar tolerance was reported at the cellular level by monitoring MSNs firing pattern. Thus, in wt mice, ethanol markedly depressed spike patterning, an effect that persisted for the duration of ethanol exposure (Fig. 3A). However, in KO mice, the initial reduction of spike frequency was followed by a return to control levels. Importantly, the stark difference in physiology at the single-channel and whole-cell levels was also observed at the behavioral level when comparing locomotor activity of wt and KO mice in response to a daily 2 g/kg EtOH challenge. Following EtOH injection on day 1, WT mice showed a marked (70%) reduction of their activity levels 5 min after injection. Activity remained depressed when monitored over the following 10- and 15-min time blocks (Fig. 3B). On day 4, the response of WT mice to EtOH 5 and 10 min after injection were comparable with that of day 1 and only showed tolerance at the 15-min time point (Fig. 3B). However, in sharp contrast to WT mice, ambulatory activity in KO mice displayed rapid tolerance to the locomotor suppressing effects of EtOH. Thus, although EtOH significantly depressed locomotor activity 5 min after injection compared with control, when tested at 10 and at 15 min after EtOH injection on day 1, ambulatory activity had returned to control levels. However, by the fourth day of ethanol challenge, nearly complete tolerance to ethanol-induced hypolocomotion was observed (Fig. 3C). Because rapid tolerance is predictive of increased alcohol consumption, voluntary ethanol intake in WT and β4 KO mice was compared using a restricted access EtOH self-administration paradigm termed "drinking in the dark" (Rhodes *et al.*, 2005, 2007). Remarkably, β4 KO mice consumed significantly greater levels of EtOH compared with consumption in WT mice during each of the first three nights of the assay.

From this work it was concluded that BK channel β4 expression in the soma of striatal MSNs is instrumental in preventing the development of tolerance to acute alcohol exposure.

In light of these results, and of BK channel β4 subunit expression in other brain regions (e.g., prefrontal cortex and hippocampus) connected to the mesolimbic pathway, it is unlikely that BK channel's role in mediating alcohol's action is confined to the nucleus accumbens. In recent work, Brenner *et al.* (2005) reported that BK β4 subunits serve as key regulators of hippocampal dentate granule cells' intrinsic firing properties. Thus, in β4 KO mice, AHPs and spiking frequency were markedly accelerated and increased, respectively. Considering the sensitivity of this subset of hippocampal neurons to early alcohol exposure (Livy, 2003; Miki *et al.*, 2000), we can envision a situation where dysfunctional αβ4 interactions, possibly

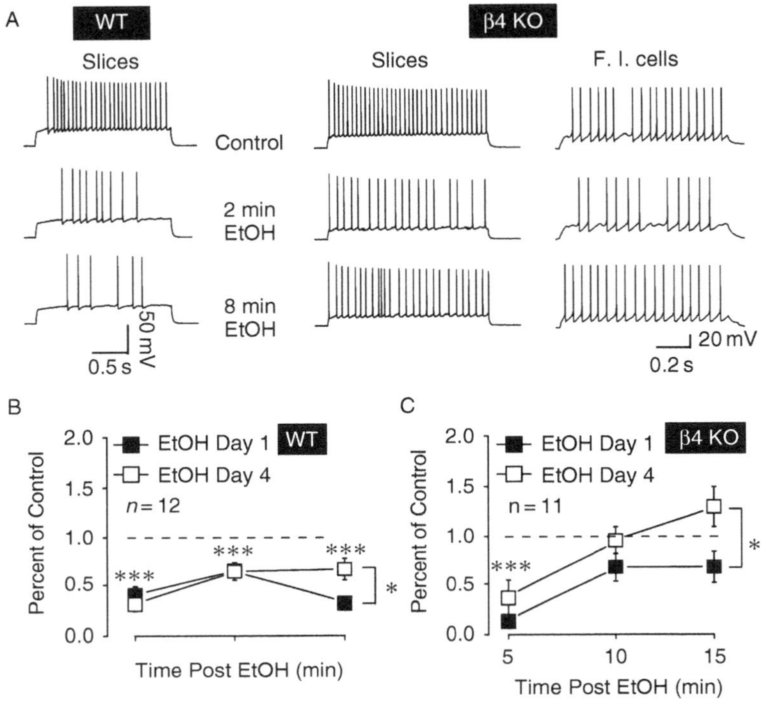

FIG. 3. EtOH-mediated decrease of MSN excitability exhibits tolerance in β4 KO, but not WT mice. (A) Representative action potential trains evoked in a slice preparation (Slices) by a single 100 pA current step in WT (Left) and KO (Middle) mice before (control) and after 50 mM EtOH exposure (2 or 8 min). Two minutes after EtOH, the number of APs is smaller in both WT and KO mice. While KO mice MSN excitability partially recovers 8 min after EtOH exposure (Bottom Trace; Right), WT neuronal excitability remains depressed (Left, Bottom Trace). Results obtained in slices were reproduced on freshly isolated neurons from KO mice (Right; β4 KO/F.I. cells). (B) Ambulatory activity of WT mice 5, 10, and 15 min after a 2 g/kg i.p. EtOH injection on days 0, 1, and 4. Each data point represents 5 min summed activity. Asterisks above symbols indicate significant locomotor activity on both days 1 and 4 of EtOH injection at each time point (5, 10, and 15 min) compared to saline injection baseline (represented by the dashed line). Asterisk next to bracket indicates difference between days 1 and 4 at 15 min time point only. (C) Effects of the same EtOH dose at the same time points in KO mice. (Adapted from Martin *et al.*, 2008).

caused by single nucleotide polymorphisms (SNPs), may contribute to the vulnerability of the hippocampus to alcohol, leading to severe behavioral consequences directly related not only to the tasks of learning and memory, but also more broadly to other cognitive and behavior disorders. In another recent study, Gollasch *et al.* (1996) reported the existence of four SNPs on the BK channel β1 subunit, which are mostly expressed in smooth muscles (e.g., heart). Interestingly, one, in particular, on exon 4b was able to induce noticeably higher heart rate variability. Although similar SNPs have not yet been identified on β4 subunits, these findings illustrate the potential functional consequences the single point mutations may have on BK channels, and more widely on behavior.

IV. Perspectives

The past decade has witnessed an explosion of information about BK channels, their expression, sensitivity to alcohol, and their potential role in shaping neuronal activity in a number of brain regions, including the striatum. Yet, important questions remain to be addressed in order to gain a better understanding of the role of BK in alcohol addiction. Given the prominent role of the BK channel β4 subunit in alcohol tolerance in mice, it would be interesting to determine whether this subunit plays a similar role in human populations of alcoholics. At a cellular level, there is strong evidence that somatic BK channels are instrumental in controlling action potential patterning, a key parameter likely to alter neuronal communication. However, despite their presence in dendrites, and their co-localization with NMDA receptors in dendritic spines, there is currently very little information regarding their role in synaptic plasticity and memory, and how their function in these neuronal compartments is influenced by chronic alcohol exposure. Finally, unlike other key ion channels (e.g., NMDA and GABA receptors), little is known about the role phosphorylation plays in alcohol effects on BK channel function.

Reference

Adams, P. R., Constanti, A., Brown, D. A., and Clark, R. B. (1982). Intracellular Ca2+ activates a fast voltage-sensitive K+ current in vertebrate sympathetic neurones. *Nature* **296**, 746–749.

Adelman, J. P., Shen, K. Z., Kavanaugh, M. P., Warren, R. A., Wu, Y. N., Lagrutta, A., Bond, C. T., and North, R. A. (1992). Calcium-activated potassium channels expressed from cloned complementary DNAs. *Neuron* **9**, 209–216.

Aroor, A. R., and Shukla, S. D. (2004). MAP kinase signaling in diverse effects of ethanol. *Life Sci.* **74**, 2339–2364.

Atkinson, N. S., Robertson, G. A., and Ganetzky, B. (1991). A component of calcium-activated potassium channels encoded by the *Drosophila* slo locus. *Science* **253**, 551–555.

Backus, K. H., and Trube, G. (1993). Single-channel activity in cultured cortical neurons of the rat in the presence of a toxic dose of glutamate. *Eur. J. Neurosci.* **5**, 174–185.

Behrens, R., Nolting, A., Reimann, F., Schwarz, M., Waldschutz, R., and Pongs, O. (2000). hKCNMB3 and hKCNMB4, cloning and characterization of two members of the large-conductance calcium-activated potassium channel beta subunit family. *FEBS Lett.* **474**, 99–106.

Benhassine, N., and Berger, T. (2008). Large-conductance calcium-dependent potassium channels prevent dendritic excitability in neocortical pyramidal neurons. *Pflugers Arch.* **457**, 1133–1145.

Benhassine, N., and Berger, T. (2005). Homogeneous distribution of large-conductance calcium-dependent potassium channels on soma and apical dendrite of rat neocortical layer 5 pyramidal neurons. *Eur. J. Neurosci.* **21**, 914–926.

Bielefeldt, K., Rotter, J. L., and Jackson, M. B. (1992). Three potassium channels in rat posterior pituitary nerve terminals. *J. Physiol.* **458**, 41–67.

Boehm, S. L. 2nd, Peden, L., Chang, R., Harris, R. A., and Blednov, Y. A. (2003). Deletion of the fyn-kinase gene alters behavioral sensitivity to ethanol. *Alcohol. Clin. Exp. Res.* **27**, 1033–1040.

Bowers, B. J., and Wehner, J. M. (2001). Ethanol consumption and behavioral impulsivity are increased in protein kinase Cgamma null mutant mice. *J. Neurosci.* **21**, RC180.

Brenner, R., Chen, Q. H., Vilaythong, A., Toney, G. M., Noebels, J. L., and Aldrich, R. W. (2005). BK channel beta4 subunit reduces dentate gyrus excitability and protects against temporal lobe seizures. *Nat. Neurosci.* **8**, 1752–1759.

Brenner, R., Jegla, T. J., Wickenden, A., Liu, Y., and Aldrich, R. W. (2000). Cloning and functional characterization of novel large conductance calcium-activated potassium channel beta subunits, hKCNMB3 and hKCNMB4. *J. Biol. Chem.* **275**, 6453–6461.

Campanac, E., and Debanne, D. (2008). Spike timing-dependent plasticity: A learning rule for dendritic integration in rat CA1 pyramidal neurons. *J. Physiol.* **586**, 779–793.

Chandler, L. J., Harris, R. A., and Crews, F. T. (1998). Ethanol tolerance and synaptic plasticity. *Trends Pharmacol. Sci.* **19**, 491–495.

Chang, C. P., Dworetzky, S. I., Wang, J., and Goldstein, M. E. (1997). Differential expression of the alpha and beta subunits of the large-conductance calcium-activated potassium channel: Implication for channel diversity. *Brain Res. Mol. Brain Res.* **45**, 33–40.

Chen, L., Tian, L., MacDonald, S. H., McClafferty, H., Hammond, M. S., Huibant, J. M., Ruth, P., Knaus, H. G., and Shipston, M. J. (2005). Functionally diverse complement of large conductance calcium- and voltage-activated potassium channel (BK) alpha-subunits generated from a single site of splicing. *J. Biol. Chem.* **280**, 33599–33609.

Chrostek, L., and Szmitkowski, M. (1998). Genetic evaluation of tolerance to alcohol. *Postepy Hig Med Dosw* **52**, 35–47.

Cowmeadow, R. B., Krishnan, H. R., Ghezzi, A., Al'Hasan, Y. M., Wang, Y. Z., and Atkinson, N. S. (2006). Ethanol tolerance caused by slowpoke induction in Drosophila. *Alcohol. Clin. Exp. Res.* **30**, 745–753.

Cowmeadow, R. B., Krishnan, H. R., and Atkinson, N. S. (2005). The slowpoke gene is necessary for rapid ethanol tolerance in Drosophila. *Alcohol. Clin. Exp. Res.* **29**, 1777–1786.

Crews, F. T., Morrow, A. L., Criswell, H., and Breese, G. (1996). Effects of ethanol on ion channels. *Int. Rev. Neurobiol.* **39**, 283–367.

Crowder, C. M. (2004). Ethanol targets: A BK channel cocktail in *C. elegans*. *Trends Neurosci.* **27**, 579–582.

Crowley, J. J., Treistman, S. N., and Dopico, A. M. (2003). Cholesterol antagonizes ethanol potentiation of human brain BKCa channels reconstituted into phospholipid bilayers. *Mol. Pharmacol.* **64**, 365–372.

Crowley, J. J., Treistman, S. N., and Dopico, A. M. (2005). Distinct structural features of phospholipids differentially determine ethanol sensitivity and basal function of BK channels. *Mol. Pharmacol.* **68**, 4–10.

Dan, Y., and Poo, M. M. (2004). Spike timing-dependent plasticity of neural circuits. *Neuron* **44**, 23–30.

Davies, A. G., Pierce-Shimomura, J. T., Kim, H., VanHoven, M. K., Thiele, T. R., Bonci, A., Bargmann, C. I., and McIntire, S. L. (2003). A central role of the BK potassium channel in behavioral responses to ethanol in *C. elegans*. *Cell* **115**, 655–666.

Dopico, A. M., Anantharam, V., and Treistman, S. N. (1998). Ethanol increases the activity of Ca (++)-dependent K+ (mslo) channels: Functional interaction with cytosolic Ca++. *J. Pharmacol. Exp. Ther.* **284**, 258–268.

Dopico, A. M., Lemos, J. R., and Treistman, S. N. (1996). Ethanol increases the activity of large conductance, Ca(2+)-activated K+ channels in isolated neurohypophysial terminals. *Mol. Pharmacol.* **49**, 40–48.

Dopico, A. M. (2003). Ethanol sensitivity of BK(Ca) channels from arterial smooth muscle does not require the presence of the beta 1-subunit. *Am. J. Physiol. Cell. Physiol.* **284**, C1468–1480.

Edgerton, J. R., and Reinhart, P. H. (2003). Distinct contributions of small and large conductance Ca2+-activated K+ channels to rat Purkinje neuron function. *J. Physiol.* **548**, 53–69.

Erwin, V. G., McClearn, G. E., and Kuse, A. R. (1980). Interrelationships of alcohol consumption, actions of alcohol, and biochemical traits. *Pharmacol. Biochem. Behav.* **13**(Suppl 1), 297–302.

Erxleben, C., Everhart, A. L., Romeo, C., Florance, H., Bauer, M. B., Alcorta, D. A., Rossie, S., Shipston, M. J., and Armstrong, D. L. (2002). Interacting effects of N-terminal variation and strex exon splicing on slo potassium channel regulation by calcium, phosphorylation, and oxidation. *J. Biol. Chem.* **277**, 27045–27052.

Faber, E. S., and Sah, P. (2002). Physiological role of calcium-activated potassium currents in the rat lateral amygdala. *J. Neurosci.* **22**, 1618–1628.

Faber, E. S., and Sah, P. (2003). Ca2+-activated K+ (BK) channel inactivation contributes to spike broadening during repetitive firing in the rat lateral amygdala. *J. Physiol.* **552**, 483–497.

Fakler, B., and Adelman, J. P. (2008). Control of K(Ca) channels by calcium nano/microdomains. *Neuron* **59**, 873–881.

Feinberg-Zadek, P. L., and Treistman, S. N. (2007). Beta-subunits are important modulators of the acute response to alcohol in human BK channels. *Alcohol. Clin. Exp. Res.* **31**, 737–744.

Fillmore, M. T., Marczinski, C. A., and Bowman, A. M. (2005). Acute tolerance to alcohol effects on inhibitory and activational mechanisms of behavioral control. *J. Stud. Alcohol* **66**, 663–672.

Fuenzalida, M., Fernandez de Sevilla, D., Couve, A., and Buno, W. (2010). Role of AMPA and NMDA receptors and back-propagating action potentials in spike timing-dependent plasticity. *J. Neurophysiol.* **103**, 47–54.

Fukao, M., Mason, H. S., Britton, F. C., Kenyon, J. L., Horowitz, B., and Keef, K. D. (1999). Cyclic GMP-dependent protein kinase activates cloned BKCa channels expressed in mammalian cells by direct phosphorylation at serine 1072. *J. Biol. Chem.* **274**, 10927–10935.

Fury, M., Marx, S. O., and Marks, A. R. (2002). Molecular BKology: The study of splicing and dicing. *Sci. STKE* **2002**, PE12.

Gandhi, C. R., and Ross, D. H. (1989). Influence of ethanol on calcium, inositol phospholipids and intracellular signalling mechanisms. *Experientia* **45**, 407–413.

Goldberg, J. A., and Wilson, C. J. (2005). Control of spontaneous firing patterns by the selective coupling of calcium currents to calcium-activated potassium currents in striatal cholinergic interneurons. *J. Neurosci.* **25**, 10230–10238.

Gollasch, M., Ried, C., Liebold, M., Haller, H., Hofmann, F., and Luft, F. C. (1996). High permeation of L-type Ca^{2+} channels at physiological $[Ca^{2+}]$: Homogeneity and dependence on the alpha 1-subunit. *Am. J. Physiol.* **271**, C842–C850.

Gong, L. W., Gao, T. M., Huang, H., Zhou, K. X., and Tong, Z. (2002). ATP modulation of large conductance Ca(2+)-activated K(+) channels via a functionally associated protein kinase A in CA1 pyramidal neurons from rat hippocampus. *Brain Res.* **951**, 130–134.

Gruss, M., Henrich, M., Konig, P., Hempelmann, G., Vogel, W., and Scholz, A. (2001). Ethanol reduces excitability in a subgroup of primary sensory neurons by activation of BK(Ca) channels. *Eur. J. Neurosci.* **14**, 1246–1256.

Ha, T. S., Jeong, S. Y., Cho, S. W., Jeon, H., Roh, G. S., Choi, W. S., and Park, C. S. (2000). Functional characteristics of two BKCa channel variants differentially expressed in rat brain tissues. *Eur. J. Biochem.* **267**, 910–918.

Hall, S. K., and Armstrong, D. L. (2000). Conditional and unconditional inhibition of calcium-activated potassium channels by reversible protein phosphorylation. *J. Biol. Chem.* **275**, 3749–3754.

Heath, A. C., Madden, P. A., Bucholz, K. K., Dinwiddie, S. H., Slutske, W. S., Bierut, L. J., Rohrbaugh, J. W., Statham, D. J., Dunne, M. P., Whitfield, J. B., and Martin, N. G. (1999). Genetic differences in alcohol sensitivity and the inheritance of alcoholism risk. *Psychol. Med.* **29**, 1069–1081.

Isaacson, J. S., and Murphy, G. J. (2001). Glutamate-mediated extrasynaptic inhibition: Direct coupling of NMDA receptors to Ca(2+)-activated K+ channels. *Neuron* **31**, 1027–1034.

Kamouchi, M., Trouet, D., De Greef, C., Droogmans, G., Eggermont, J., and Nilius, B. (1997). Functional effects of expression of hslo Ca2+ activated K+ channels in cultured macrovascular endothelial cells. *Cell Calcium* **22**, 497–506.

Kang, J., Huguenard, J. R., and Prince, D. A. (1996). Two types of BK channels in immature rat neocortical pyramidal neurons. *J. Neurophysiol.* **76**, 4194–4197.

Kaufmann, W. A., Ferraguti, F., Fukazawa, Y., Kasugai, Y., Shigemoto, R., Laake, P., Sexton, J. A., Ruth, P., Wietzorrek, G., Knaus, H. G., Storm, J. F., and Ottersen, O. P. (2009). Large-conductance calcium-activated potassium channels in Purkinje cell plasma membranes are clustered at sites of hypolemmal microdomains. *J. Comp. Neurol.* **515**, 215–230.

Knaus, H. G., Schwarzer, C., Koch, R. O., Eberhart, A., Kaczorowski, G. J., Glossmann, H., Wunder, F., Pongs, O., Garcia, M. L., and Sperk, G. (1996). Distribution of high-conductance Ca(2+)-activated K+ channels in rat brain: Targeting to axons and nerve terminals. *J. Neurosci.* **16**, 955–963.

Lancaster, B., and Nicoll, R. A. (1987). Properties of two calcium-activated hyperpolarizations in rat hippocampal neurones. *J. Physiol.* **389**, 187–203.

Latorre, R., Vergara, C., and Hidalgo, C. (1982). Reconstitution in planar lipid bilayers of a Ca2+-dependent K+ channel from transverse tubule membranes isolated from rabbit skeletal muscle. *Proc. Natl. Acad. Sci. U S A.* **79**, 805–809.

Lee, J., Ueda, A., and Wu, C. F. (2008). Pre- and post-synaptic mechanisms of synaptic strength homeostasis revealed by slowpoke and shaker K(+) channel mutations in Drosophila. *Neuroscience* **154**, 1283–1296.

Liu, J., Asuncion-Chin, M., Liu, P., and Dopico, A. M. (2006). CaM kinase II phosphorylation of slo Thr107 regulates activity and ethanol responses of BK channels. *Nat. Neurosci.* **9**, 41–49.

Liu, X., Chang, Y., Reinhart, P. H., Sontheimer, H., and Chang, Y. (2002). Cloning and characterization of glioma BK, a novel BK channel isoform highly expressed in human glioma cells. *J. Neurosci.* **22**, 1840–1849.

Livy, D. J., Miller, E. K., Maier, S. E., and West, J. R. (2003). Fetal alcohol exposure and temporal vulnerability: Effects of binge-like alcohol exposure on the developing rat hippocampus. *Neurotoxicol. Teratol.* **25**, 447–458.

Loane, D. J., Lima, P. A., and Marrion, N. V. (2007). Co-assembly of N-type Ca2+ and BK channels underlies functional coupling in rat brain. *J. Cell Sci.* **120**, 985–995.

Markram, H., Lubke, J., Frotscher, M., and Sakmann, B. (1997). Regulation of synaptic efficacy by coincidence of postsynaptic APs and EPSPs. *Science* **275**, 213–215.

Marrion, N. V., and Tavalin, S. J. (1998). Selective activation of Ca2+-activated K+ channels by co-localized Ca2+ channels in hippocampal neurons. *Nature* **395**, 900–905.

Martin, G., Puig, S., Pietrzykowski, A., Zadek, P., Emery, P., and Treistman, S. (2004). Somatic localization of a specific large-conductance calcium-activated potassium channel subtype controls compartmentalized ethanol sensitivity in the nucleus accumbens. *J. Neurosci.* **24**, 6563–6572.

Martin, G. E., and Treistman, S. N. (2009). BK channel: Mediator and model for alcohol tolerance. *Trends in Neurosci.* **32**, 629–637.

Martin, G. E., Hendrickson, L. M., Penta, K. L., Friesen, R. M., Pietrzykowski, A. Z., Tapper, A. R., and Treistman, S. N. (2008). Identification of a BK channel auxiliary protein controlling molecular and behavioral tolerance to alcohol. *P.N.A.S.* **105**, 17543–17548.

Marty, A. (1981). Ca-dependent K channels with large unitary conductance in chromaffin cell membranes. *Nature* **291**, 497–500.

McCobb, D. P., Fowler, N. L., Featherstone, T., Lingle, C. J., Saito, M., Krause, J. E., and Salkoff, L. (1995). A human calcium-activated potassium channel gene expressed in vascular smooth muscle. *Am. J. Physiol.* **269**, H767–777.

Miki, T., Harris, S. J., Wilce, P., Takeuchi, Y., Bedi, K. S. (2000). Neurons in the hilus region of the rat hippocampus are depleted in number by exposure to alcohol during early postnatal life. *Hippocampus* **10**, 284–295.

Misonou, H., Menegola, M., Buchwalder, L., Park, E. W., Meredith, A., Rhodes, K. J., Aldrich, R. W., and Trimmer, J. S. (2006). Immunolocalization of the Ca2+-activated K+ channel Slo1 in axons and nerve terminals of mammalian brain and cultured neurons. *J. Comp. Neurol.* **496**, 289–302.

Nara, M., Dhulipala, P. D., Ji, G. J., Kamasani, U. R., Wang, Y. X., Matalon, S., and Kotlikoff, M. I. (2000). Guanylyl cyclase stimulatory coupling to K(Ca) channels. *Am. J. Physiol. Cell Physiol.* **279**, C1938–1945.

Newton, P. M., and Messing, R. O. (2006). Intracellular signaling pathways that regulate behavioral responses to ethanol. *Pharmacol. Ther.* **109**, 227–237.

Olive, M. F., Mehmert, K. K., Messing, R. O., and Hodge, C. W. (2000). Reduced operant ethanol self-administration and *in vivo* mesolimbic dopamine responses to ethanol in PKCepsilon-deficient mice. *Eur. J. Neurosci.* **12**, 4131–4140.

Pallotta, B. S., Magleby, K. L., and Barrett, J. N. (1981). Single channel recordings of Ca2+-activated K+ currents in rat muscle cell culture. *Nature* **293**, 471–474.

Park, W. S., Son, Y. K., Kim, N., Youm, J. B., Warda, M., Ko, J. H., Ko, E. A., Kang, S. H., Kim, E., Earm, Y. E., and Han, J. (2007). Direct modulation of Ca(2+)-activated K(+) current by H-89 in rabbit coronary arterial smooth muscle cells. *Vascul. Pharmacol.* **46**, 105–113.

Pietrzykowski, A. Z., Friesen, R. M., Martin, G. E., Puig, S. I., Nowak, C. L., Wynne, P. M., Siegelmann, H. T., and Treistman, S. N. (2008). Posttranscriptional regulation of BK channel splice variant stability by miR-9 underlies neuroadaptation to alcohol. *Neuron* **59**, 274–287.

Quirk, J. C., and Reinhart, P. H. (2001). Identification of a novel tetramerization domain in large conductance K(ca) channels. *Neuron* **32**, 13–23.

Rhodes JS, et al. (2005) Evaluation of a simple model of ethanol drinking to intoxication in C57BL/6J mice. *Physiol. Behav.* **84**, 53–63.

Rhodes JS, et al. (2007) Mouse inbred strain differences in ethanol drinking to intoxication. Genes Brain Behav **6**, 1–18.

Robitaille, R., and Charlton, M. P. (1992). Presynaptic calcium signals and transmitter release are modulated by calcium-activated potassium channels. *J. Neurosci.* **12**, 297–305.

Sailer, C. A., Kaufmann, W. A., Kogler, M., Chen, L., Sausbier, U., Ottersen, O. P., Ruth, P., Shipston, M. J., and Knaus, H. G. (2006). Immunolocalization of BK channels in hippocampal pyramidal neurons. *Eur. J. Neurosci.* **24**, 442–454.

Sausbier, U., Sausbier, M., Sailer, C. A., Arntz, C., Knaus, H. G., Neuhuber, W., and Ruth, P. (2006). Ca2+-activated K+ channels of the BK-type in the mouse brain. *Histochem. Cell Biol.* **125**, 725–741.

Scholz, A., Gruss, M., and Vogel, W. (1998). Properties and functions of calcium-activated K+ channels in small neurones of rat dorsal root ganglion studied in a thin slice preparation. *J. Physiol.* **513**, 55–69.

Schreiber, M., and Salkoff, L. (1997). A novel calcium-sensing domain in the BK channel. *Biophys. J.* **73**, 1355–1363.

Schubert, R., Noack, T., and Serebryakov, V. N. (1999). Protein kinase C reduces the KCa current of rat tail artery smooth muscle cells. *Am. J. Physiol.* **276**, C648–658.

Schuckit, M. A. (1994) Low level of response to alcohol as a predictor of future alcoholism. *Am. J. Psychiatry* **151**, 184–189.

Shah, M. M., and Haylett, D. G. (2002). K+ currents generated by NMDA receptor activation in rat hippocampal pyramidal neurons. *J. Neurophysiol.* **87**, 2983–2989.

Shao, L. R., Halvorsrud, R., Borg-Graham, L., and Storm, J. F. (1999). The role of BK-type Ca2+-dependent K+ channels in spike broadening during repetitive firing in rat hippocampal pyramidal cells. *J. Physiol.* **521**(Pt 1), 135–146.

Shen, K. Z., Lagrutta, A., Davies, N. W., Standen, N. B., Adelman, J. P., and North, R. A. (1994). Tetraethylammonium block of Slowpoke calcium-activated potassium channels expressed in Xenopus oocytes: Evidence for tetrameric channel formation. *Pflugers Arch.* **426**, 440–445.

Stork, O., Kojima, N., Stork, S., Kume, N., and Obata, K. (2002). Resistance to alcohol withdrawal-induced behaviour in Fyn transgenic mice and its reversal by ifenprodil. *Brain Res. Mol. Brain Res.* **105**, 126–135.

Storm, J. F. (1990). Potassium currents in hippocampal pyramidal cells. *Prog. Brain Res.* **83**, 161–187.

Swayze, R. D., and Braun, A. P. (2001). A catalytically inactive mutant of type I cGMP-dependent protein kinase prevents enhancement of large conductance, calcium-sensitive K+ channels by sodium nitroprusside and cGMP. *J. Biol. Chem.* **276**, 19729–19737.

Thiele, T. E., Willis, B., Stadler, J., Reynolds, J. G., Bernstein, I. L., and McKnight, G. S. (2000). High ethanol consumption and low sensitivity to ethanol-induced sedation in protein kinase A-mutant mice. *J. Neurosci.* **20**, RC75.

Tian, L., Coghill, L. S., McClafferty, H., MacDonald, S. H., Antoni, F. A., Ruth, P., Knaus, H. G., and Shipston, M. J. (2004). Distinct stoichiometry of BKCa channel tetramer phosphorylation specifies channel activation and inhibition by cAMP-dependent protein kinase. *Proc. Natl. Acad. Sci. U S A.* **101**, 11897–11902.

Tian, L., Duncan, R. R., Hammond, M. S., Coghill, L. S., Wen, H., Rusinova, R., Clark, A. G., Levitan, I. B., and Shipston, M. J. (2001). Alternative splicing switches potassium channel sensitivity to protein phosphorylation. *J. Biol. Chem.* **276**, 7717–7720.

Urakubo, H., Honda, M., Froemke, R. C., and Kuroda, S. (2008). Requirement of an allosteric kinetics of NMDA receptors for spike timing-dependent plasticity. *J. Neurosci.* **28**, 3310–3323.

Wanner, S. G., Koch, R. O., Koschak, A., Trieb, M., Garcia, M. L., Kaczorowski, G. J., and Knaus, H. G. (1999). High-conductance calcium-activated potassium channels in rat brain: Pharmacology, distribution, and subunit composition. *Biochemistry* **38**, 5392–5400.

Weiger, T. M., Holmqvist, M. H., Levitan, I. B., Clark, F. T., Sprague, S., Huang, W. J., Ge, P., Wang, C., Lawson, D., Jurman, M. E., Glucksmann, M. A., Silos-Santiago, I., DiStefano, P. S., and Curtis, R. (2000). A novel nervous system beta subunit that downregulates human large conductance calcium-dependent potassium channels. *J. Neurosci.* **20**, 3563–3570.

White, R. E., Lee, A. B., Shcherbatko, A. D., Lincoln, T. M., Schonbrunn, A., and Armstrong, D. L. (1993). Potassium channel stimulation by natriuretic peptides through cGMP-dependent dephosphorylation. *Nature* **361**, 263–266.

Widmer, H., Lemos, J. R., and Treistman, S. N. (1998). Ethanol reduces the duration of single evoked spikes by a selective inhibition of voltage-gated calcium currents in acutely dissociated supraoptic neurons of the rat. *J. Neuroendocrinol.* **10**, 399–406.

Woodward, J. J., Ron, D., Winder, D., and Roberto, M. (2006). From blue states to up states: A regional view of NMDA-ethanol interactions. *Alcohol. Clin. Exp. Res.* **30**, 359–367.

Wynne, P. M., Puig, S. I., Martin, G. E., and Treistman, S. N. (2009). Compartmentalized beta subunit distribution determines characteristics and ethanol sensitivity of somatic, dendritic, and terminal large-conductance calcium-activated potassium channels in the rat central nervous system. *J. Pharmacol. Exp. Ther.* **329**, 978–986.

Yaka, R., Phamluong, K., and Ron, D. (2003). Scaffolding of Fyn kinase to the NMDA receptor determines brain region sensitivity to ethanol. *J. Neurosci.* **23**, 3623–3632.

Yuan, C., O'Connell, R. J., Wilson, A., Pietrzykowski, A. Z., and Treistman, S. N. (2008). Acute alcohol tolerance is intrinsic to the BKCa protein, but is modulated by the lipid environment. *J. Biol. Chem.* **283**, 5090–5098.

Zhang, X. Y., Wang, S., Yan, Z., Wan, Y., Wang, W., Cui, G. B., Du, P., Ma, K. J., Han, W., Zhang, Y. Q., and Wei, J. G. (2008). Molecular cloning, tissue distribution and bioinformatics analyses of the rabbit BK channel beta1 subunit gene. *Mol. Biol. Rep.* **35**, 649–655.

A REVIEW OF SYNAPTIC PLASTICITY AT PURKINJE NEURONS WITH A FOCUS ON ETHANOL-INDUCED CEREBELLAR DYSFUNCTION

C. Fernando Valenzuela, Britta Lindquist, and Paula A. Zamudio-Bulcock

Department of Neurosciences, School of Medicine, University of New Mexico Health Sciences Center, Albuquerque, NM 87131, USA

The cerebellum controls balance, posture, motor coordination, and cognition, and studies suggest that ethanol impairs these cerebellar functions. However, the mechanisms through which ethanol produces these effects are not fully understood. Here, we review evidence suggesting that ethanol acts, in part, by impairing synaptic plasticity mechanisms at cerebellar Purkinje neurons. We will primarily focus on recent experiments indicating that long-term depression at both parallel fiber– and climbing fiber–Purkinje cell synapses is inhibited by acute ethanol exposure. We will also discuss experimental evidence showing that chronic prenatal ethanol exposure converts long-term depression into long-term potentiation at parallel fiber–Purkinje cell synapses.

INTERNATIONAL REVIEW OF
NEUROBIOLOGY, VOL. 91
DOI: 10.1016/S0074-7742(10)91011-8

I. Introduction

The cerebellum (L. "little brain") is a specialized brain structure that is located posterior to the pons and medulla, below the occipital lobes. It is involved in motor coordination and planning, muscle tone maintenance, equilibrium control, and cognitive functions. A number of studies have demonstrated that normal cerebellar functioning is impaired by exposure to ethanol during both development and maturity. Long-lasting cerebellar degeneration is a common consequence of prenatal ethanol exposure and is, in part, responsible for the deficits in motor function and balance that characterize fetal alcohol spectrum disorders. Cerebellar dysfunction is the likely cause of many of the signs of acute ethanol intoxication, such as dysarthria, nystagmus, and ataxia. In chronic alcoholics, signs of cerebellar compromise are also frequently observed, particularly gait ataxia and tremor of the lower legs and trunk, which are indicative of alterations in the antero-superior region of the cerebellum. These individuals also frequently exhibit executive function alterations that are likely caused by damage of the connections between the cerebellar hemispheres and the frontal cortex. The mechanisms responsible for these acute and chronic effects of ethanol have been the subject of numerous studies, demonstrating that ethanol modulates multiple components of the cerebellar neuronal network via a myriad of mechanisms, including altered synaptic transmission and neuronal excitability, impaired energy metabolism, oxidative stress, and glial abnormalities (Jaatinen and Rintala, 2008). Recent evidence suggests that ethanol also acts by altering synaptic plasticity mechanisms that are essential for normal cerebellar functioning and we review here recent advancements in this area of research. We will begin by providing an overview of cerebellar structure, function, and development. We will then review the different types of synaptic plasticity mechanisms that have been identified in the cerebellar cortex and what is presently known about the effects of ethanol on these mechanisms.

II. Functional Cerebellar Neuroanatomy

A. Anatomy and Function

The cerebellum is phylogenetically ancient, being present, at least in rudimentary form, from fishes through higher vertebrates. As an essential contributor to associative motor learning, it has been the subject of fruitful research since the 1960s (Eccles *et al.*, 1967; Eccles, 1967; Marr, 1969). In recent years, the cerebellum has also been credited with roles in emotion and cognition, as well as non-motor forms of associative learning (Bellebaum and Daum, 2009; Gordon, 2007; Schmahmann, 2004; Steinlin, 2008; Strick *et al.*, 2009; but see Glickstein and Doron, 2008). Cerebellar damage due to developmental insults, trauma, toxins, compression, or disease is associated with significant behavioral and motor deficits (Walker, 1990).

The cerebellum is composed of two hemispheres joined in the midline by an elevated ridge of tissue called the vermis (L. "worm") (Fig. 1A). Its dorsal surface undulates with abundant small horizontally oriented ridges (folia) and is divided by prominent fissures into lobes: anterior, posterior, and flocculonodular. Viewed from the dorsum, the flocculonodular lobe is covered by visible portions of the cerebellum: the vermis and hemispheres. The ventral surface of the cerebellum contains white matter connections between the cerebellum and other parts of the brainstem, spinal cord, and cerebral cortex. These white matter connections ramify into superior, middle, and inferior peduncles. A cross section through the cerebellum reveals a trilaminar cortex, deep white matter, and deep cerebellar nuclei. The deep cerebellar nuclei are symmetrically paired structures named (from lateral to medial) the dentate, interposed, and fastigial nuclei (Ghez and Thach, 2000). Each deep cerebellar nucleus is associated with a portion of the overlying cortex. These anatomic features correlate with distinct functions carried out by each region. Sensory inputs and functional specialization in the cerebellum are organized somatotopically from medial to lateral. The head and

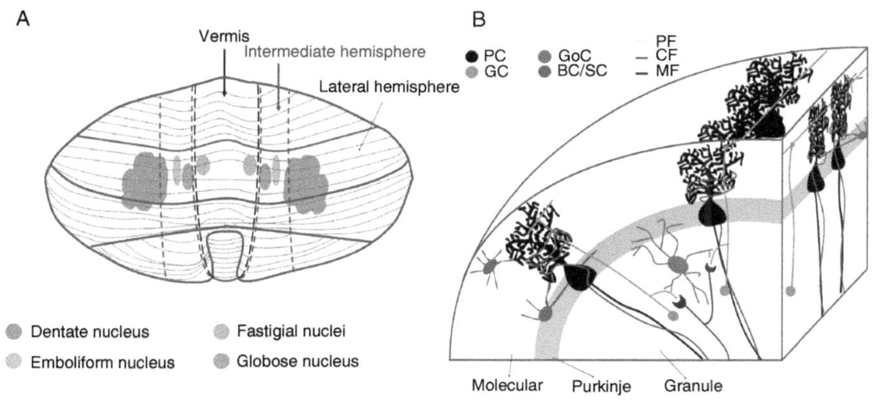

FIG. 1. Overview of cerebellar structure. (A) Schematic representation of the cerebellum viewed from the posterior side. The deep nuclei, which are located at the core of the cerebellum, are also shown. The ventral flocculonodular lobe and cerebellar peduncles are not shown. (B) Schematic representation of the cerebellar cortical layers and their cellular components. PC: Purkinje cell; GoC: Golgi cell, GC: granule cell, BC: basket cell (located more basally), SC: stellate cell (located more apically), MF: mossy fiber, CF: climbing fiber, PF: parallel fiber. The molecular, Purkinje, and granule cell layers are indicated at the bottom (color images available in online version).

proximal parts of the body are represented in the medial portions of the cerebellum, while the extremities are represented more laterally. Cerebellar cortical neurons in each lobe participate in a complete closed circuit, receiving peripheral information and projecting to a deep nucleus, which relays back to modify the same body or brain area that generated the original input. Functionally, the cerebellum can be divided into vestibulocerebellum, spinocerebellum, and cerebrocerebellum.

The vestibulocerebellum corresponds to the flocculonodular lobe and is responsible for balance and eye movements. It receives information about head position and movement from primary vestibular afferents and sends efferents both to the lateral vestibular nuclei and to the fastigial nucleus (Purves, 2001b). This lobe is unique in that it has direct reciprocal projections with the vestibular nucleus in the pons, in addition to its relays within a deep cerebellar nucleus (i.e., fastigial nucleus). Fibers from the vestibular nucleus descend in the vestibulospinal tract to excite lower motor neurons for extensor muscles in the legs and trunk which maintain an erect posture (Purves, 2001a). Dysfunction of the flocculonodular lobe—including that caused by acute ethanol intoxication—produces clinical signs related to impaired gaze (nystagmus and saccades) and balance (Walker, 1990). Dorsal to this lobe, in the midline, is the vermis (Fig. 1A), which controls gaze, posture, speech, and movements of the trunk and proximal limbs during locomotion. Its inputs include visual, auditory, and vestibular information, as well as sensory information from the head and proximal limbs. Its output, via the fastigial nucleus, projects to descending fibers in the brainstem. Vermal lesions cause clinical manifestations such as unstable posture, wide-based stance, and ataxic gait. Chronic alcohol consumption commonly produces degeneration in the vermis, particularly the anterior lobe (Walker, 1990). Lateral to the vermis, the intermediate hemispheres coordinate movements of the distal extremities and digits. Their input is somatosensory information from the limbs, and their output (via the interposed nucleus that is formed by the emboliform and globose nuclei; Fig. 1A) descends in the lateral corticospinal and rubrospinal tracts. Together, the vermis and intermediate hemispheres make up the spinocerebellum. The lateral hemispheres (Fig. 1A) receive input from the cerebral cortex, relay in the dentate nucleus, and give output to the motor, premotor, and prefrontal cortices. The lateral cerebellar hemispheres are known as the cerebrocerebellum. Dysfunction of the intermediate and lateral hemispheres—which is commonly caused by acute ethanol intoxication—can cause intention tremor, uncoordinated movement (so-called "decomposition of movement"), dysmetria, and difficulty with rapidly alternating movements (dysdiadochokinesis) (Walker, 1990).

Due to reciprocal connections with the brainstem, spinal cord, and cerebral cortex, the cerebellum is able to compare an intended action with its consequences, and thus to modify or fine-tune behaviors in progress. As further discussed below, the circuitry of the cerebellar cortex is well suited to comparing and correcting errors, and to optimizing future performances based on feedback from past attempts.

B. Circuitry and Histology

Three layers can be easily identified in the cerebellar cortex: molecular, Purkinje, and granule cell layers (Fig. 1B). Most superficial is the molecular layer, which contains GABAergic interneurons—basket and stellate cells—as well as axons and dendrites from deeper cell layers. The fibers of the molecular layer are very precisely oriented, as will be discussed below. Basket cells are located more basally and extend long axonal processes, while stellate cells are more apical and are characterized by short axons (Barmack and Yakhnitsa, 2008). Deep to the molecular layer is a monolayer of large GABAergic Purkinje cells, whose axons carry the sole output of the cerebellar cortex. Deep to the Purkinje cell layer is the granule cell layer, which is populated with abundant (10^{11}) small glutamatergic granule cells and occasional larger inhibitory GABAergic Golgi interneurons (D'Angelo and De Zeeuw, 2009). Intermediate between the Purkinje cell layer and the granule cell layer, GABAergic/glycinergic Lugaro cells have been described. In response to serotonin, Lugaro cells act in the granule cell layer to inhibit Golgi interneurons (Ito, 2006). The relationship between these cell types described above and the input and output fibers of the cerebellum can be described as an excitatory main loop regulated by an inhibitory side loop (Dean *et al.*, 2010).

The excitatory afferents to the cerebellar cortex consist of a mossy fiber–parallel fiber–Purkinje cell pathway and a climbing fiber–Purkinje cell pathway (Fig. 1B). Excitatory input from the periphery enters the cerebellum as mossy fibers, passing through the inferior (vestibular and spinal input) and middle (cerebral input) cerebellar peduncles. The mossy fibers terminate in the granule cell layer, forming excitatory glutamatergic synapses both on granule cells (up to 400–800) and Golgi cells (Dean *et al.*, 2010). If a granule cell reaches threshold, it fires an action potential that is transmitted up its ascending axon to the molecular layer, where the process bifurcates and travels for great distances in either direction along the long axis of the folium, as a parallel fiber (Fig. 1B). Along the way, the parallel fiber makes numerous glutamatergic synapses on 300 or more Purkinje cell dendrites, and on molecular layer interneurons and Golgi cell dendrites (Eccles, 1967; Ghez and Thach, 2000; Ito, 2006). Within the molecular layer, the elaborate dendritic tree of the Purkinje cell is oriented orthogonally to the parallel fiber beam, that is, in the parasagittal plane (Ghez and Thach, 2000). Each Purkinje cell receives 10^4–10^6 parallel fiber synapses (Ghez and Thach, 2000;

Ito, 2006; Jorntell and Hansel, 2006) distributed over its dendritic tree, most of which are silent under normal conditions (Dean *et al.*, 2010). Those parallel fiber–Purkinje cell synapses that are not silent are only weakly excitatory, capable of creating a subthreshold postsynaptic potential mediated by α-amino-3-hydroxy-5-methyl-4-isoxazolepropionic acid (AMPA) receptor activation. Spatiotemporal summation of parallel fiber inputs can generate a simple spike in the Purkinje cell (Ghez and Thach, 2000). This simple spike does not engage *N*-methyl-d-aspartate (NMDA) receptors and does not backpropagate. As a consequence, there is no significant sustained elevation of Ca^{2+} in the Purkinje cell downstream of parallel fiber excitation. Each Purkinje cell also receives glutamatergic input from a single climbing fiber that originates in the inferior olivary nucleus and provides a powerful AMPA receptor–mediated excitatory input that generates an all-or-none complex spike in the Purkinje cell (Dean *et al.*, 2010). Coincident firing of the climbing fiber and nearby parallel fibers leads to long-term depression (LTD) that, as discussed below, is an important target of ethanol. Purkinje cell activity produces GABAergic inhibition of the deep cerebellar (or vestibular) nuclei. Neurons in the deep cerebellar nuclei are tonically active and Purkinje cell–mediated inhibition or disinhibition can change their firing rate (Pugh and Raman, 2009). Their efferent projections exit the cerebellum via the superior and inferior peduncles.

The excitatory main loop described above is modulated by inhibition from local interneurons. As mossy fiber activity stimulates granule cells, it also increases the activity of Golgi cells, which in turn inhibit granule cell excitability. Such elicited, phasic inhibition accounts for an estimated 10% of inhibition in the granule cell layer; more significantly, GABA released from Golgi cells exerts tonic inhibition on the granule cells and this makes up 90% of the GABAergic inhibitory input to these neurons (Hamann *et al.*, 2002). Phasic inhibition is correlated with synaptic release of GABA and is mediated by low-affinity, desensitizing postsynaptic receptors in the granule cell. Tonic inhibition, on the other hand, is mediated by high-affinity, non-desensitizing extrasynaptic receptors that are engaged by ambient background levels of GABA in the extracellular space. The interactions of mossy fibers, Golgi cells, and granule cells are facilitated by their association in a glomerular structure. GABA levels are maintained locally around the fibers because diffusion is limited by a glial sheath (reviewed in Botta *et al.*, 2007a). Within the molecular layer, local activity by interneurons modifies Purkinje cell activity. Basket and stellate cells receiving excitatory input from parallel fibers—and climbing fibers via glutamate spillover (Szapiro and Barbour, 2007)—respond with an increase in their firing rate. Both of these interneurons then inhibit the activity of Purkinje cells, though in different ways. Stellate cells with their short axons inhibit Purkinje cells at the level of the distal dendrites, while the more basal basket cells make contact with the proximal dendrites and soma. Inhibitory GABAergic tone by stellate and basket cells in the molecular layer, and Golgi cells in the granule cell layer, powerfully modulates the output of the cerebellar cortex; importantly, acute ethanol exposure has been shown to increase GABA release from these interneurons in the cerebellar slice preparation (Botta *et al.*, 2007a; Carta *et al.*, 2004; Hanchar *et al.*, 2005; Hirono *et al.*, 2009; Kelm *et al.*, 2007; Mameli *et al.*, 2008).

The histological organization and functional relationships between cell types are preserved throughout the cerebellum. This circuitry provides the substrate for plasticity and motor learning. The essential components have been reduced to a microzone or microcircuit, useful in understanding the computations performed by this "neuronal machine." The microcircuit and its function have been reviewed elsewhere (Ito, 2006; Marr, 1969).

III. Overview of Cerebellar Development

As discussed below, fetal alcohol spectrum disorders are often associated with cerebellar damage, including alterations in granule cell migration and Purkinje cell survival (Green, 2004; Kumada *et al.*, 2007). Importantly, a recent study suggests that parallel fiber LTD is impaired in mice that were chronically exposed to ethanol during gestation (Servais *et al.*, 2007). Therefore, key steps of cerebellar development will be briefly reviewed in this section.

The formation of cerebellar circuits follows a complex series of steps that depend on neuronal intrinsic programs and extrinsic mechanisms. In rodents, migration and differentiation of cerebellar neuronal precursors occurs from embryonic day 9 to birth (Bosman and Konnerth, 2009). This period is approximately equivalent to the first and second trimesters of human pregnancy. During this period, the ventricular epithelium and the rhombic lip give rise to all cerebellar neuronal precursors. In the ventricular epithelium, Purkinje, basket, stellate, Golgi, and deep cerebellar nuclear cells are born, and soon after, they start migrating toward their final destinations. Purkinje cells, for instance, are born between embryonic days 13 and 17, and start migrating along radial glial fibers at embryonic day 14. In the rhombic lip, granule cells are born and form the external granule layer by embryonic day 17 (Swinny *et al.*, 2005). Although the cerebellar cortex starts to form during embryonic life in rodents, a large component of its development occurs postnatally, which is equivalent to the third trimester of pregnancy in humans. During the first 13 days of neonatal life in rodents, neurons proliferate, migrate to their final destination, and gradually grow and extend their axons and dendrites, leading to the formation of synapses (Bosman and Konnerth, 2009). However, the refining and maturation of synapses is only completed by end of the third postnatal week (Swinny *et al.*, 2005). Such a prolonged developmental program is also seen in the human cerebellar cortex, where, for instance, granule cell proliferation starts during the third gestational trimester and continues into the first year of life.

At birth, the rodent cerebellar cortex is a four-layered structure consisting of the external granule layer, the nascent molecular layer, the Purkinje cell plate, and the internal granule layer, located between the Purkinje cell plate and the prospective white matter (Fig. 2). The external granule layer contains proliferating granule cells and the nascent molecular layer mainly contains immature Purkinje cell dendrites. The Purkinje plate consists of Purkinje cell bodies arranged in a multilayered fashion, each innervated by multiple climbing fibers.

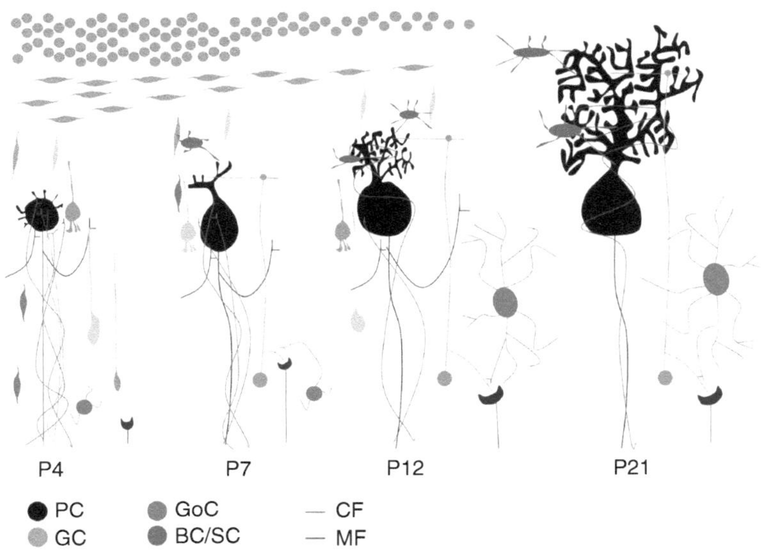

FIG. 2. Schematic representation of cerebellar cortical development. For definition of abbreviations, please see legend to Fig. 1. At P4, GCs are migrating from the external granule cell layer to the internal granule cell layer, where GoC and MF are already present. GCs migrate tangentially within the external granule cell layer and along the Bergmann glia in the nascent molecular layer, undifferentiated BC and SC are migrating toward the nascent molecular layer, and PCs are innervated by multiple CFs. Note the Purkinje cell axonal collaterals, which are thought to play a role in traveling waves of activity (see Section III). By P7, immature GCs as well as BCs/SCs continue to migrate, while their more mature counterparts start to innervate PCs. Additionally, GCs start to receive GoC and MF inputs. At this stage the CF elimination process has started. At P12, already differentiated BCs innervate the proximal dendrites of PCs whereas SCs innervate the distal ones, while immature GC migration and CF elimination are still taking place. By the end of the third postnatal week, P21, the external granule cell layer disappears and CF monoinnervation is achieved (color figure available in the online version).

The internal granule layer contains sparse granule cells, Golgi cells, and mossy fibers. Mossy fibers arrive at the inner granule cell layer around postnatal day 3; the Golgi cell is present at the time of birth and starts forming synapses onto granule cells after postnatal day 7 (Takayama and Inoue, 2006). Immunohistochemical studies have shown that by postnatal day 21 the external granule layer has disappeared and all granule neurons have migrated down to the inner granule cell layer (Sotelo, 2008).

Regarding Purkinje neurons, these become organized as a monolayer, grow an apical dendrite, and develop a flattened dendritic tree that extends laterally between postnatal days 5 and 13 (reviewed in Sotelo and Dusart, 2009) (Fig. 2). As early as P2, these cells grow provisional somatic synaptic protrusions, onto which the first synaptic glutamatergic connections will form (Eilers *et al.*, 2001; Kawa, 2003; Kurihara *et al.*, 1997). At P9, it was demonstrated that Purkinje neurons send axonal collaterals to neighboring Purkinje neurons and that these mediate $GABA_A$ receptor–dependent traveling waves of activity that may play a role in the maturation of cerebellar circuitry (Watt *et al.*, 2009). Around embryonic day 17, climbing fibers start to contact Purkinje cells. Initially, each Purkinje cell forms synapses with multiple climbing fibers; by the end of the third postnatal week, only the strongest of these fibers will remain after a process of synapse competition and elimination (Bosman and Konnerth, 2009; Hashimoto *et al.*, 2009). Synaptic plasticity is essential for the process of eliminating redundant climbing fibers, as discussed below; however, the effects of ethanol on climbing fiber elimination have yet to be characterized. Parallel fiber–Purkinje cell synapses are thought to develop during the third trimester equivalent—although the numbers of these synapses were not found to be affected by ethanol exposure during this developmental period, their numbers were shown to be increased by motor training in these ethanol-exposed rats when they reached adulthood and this may underlie the compensatory effects of training on motor performance that were observed in these animals (Klintsova *et al.*, 2000).

The establishment of GABAergic synapses onto Purkinje neurons also occurs early in rodent postnatal life and ends by P21; functional synapses have been detected as early as P2 (Kawa, 2002). During the period prior to the formation of GABAergic synapses in Purkinje neurons, $GABA_A$ receptors can depolarize the membrane potential and activate voltage-gated Ca^{2+} channels. These excitatory actions of $GABA_A$ receptors have been suggested to contribute to the activity-dependent maturation of cerebellar cortical circuitry components, including climbing fibers (Eilers *et al.*, 2001; Watt *et al.*, 2009). Importantly, ethanol has been shown to enhance the excitatory actions of $GABA_A$ receptors in the developing CA3 hippocampal region and it would be interesting to determine whether a similar effect is seen in the developing cerebellum (Galindo *et al.*, 2005).

IV. Synaptic Plasticity at Purkinje Cells

As we will review in Section V, acute ethanol exposure has been shown to inhibit LTD at parallel fiber–Purkinje cell and climbing fiber–Purkinje cell synapses, without affecting long-term potentiation (LTP) at parallel fiber–Purkinje cell synapses. In addition, parallel fiber LTD was converted into LTP by chronic prenatal exposure to ethanol. Therefore, we will review here the mechanisms that mediate these forms of synaptic plasticity.

A. Parallel Fiber-to-Purkinje Cell Synapses

1. *Long-Term Depression*

LTD of parallel fiber–Purkinje cell synapses was first described by Ito *et al.* (1982) and can be divided into three phases (Ogasawara *et al.*, 2008) (Fig. 3). The initial phase requires repetitive parallel fiber–Purkinje cell and climbing fiber–Purkinje cell synapse co-activation (for example at 1 Hz for 5 min). This procedure stimulates type 1 metabotropic glutamate receptors (mGluR1), leading to the activation of G_q and phospholipase C-β, which generates diacylglycerol and inositol 1,4,5-trisphosphate (IP_3) (Aiba *et al.*, 1994). IP_3 binds to its receptors, which are tethered to mGluR1 receptors via Homer proteins (Tu *et al.*, 1998), causing release of Ca^{2+} from intracellular stores that is likely complemented and facilitated by extracellular Ca^{2+} influx through voltage-gated Ca^{2+} channels, which are activated by the membrane potential depolarization triggered by climbing fiber–induced stimulation of AMPA receptors. Type 1 and/or 3 transient receptor potential cation channels (TRPC1/3) stimulated by mGluR1 may also contribute to the activation of voltage-gated Ca^{2+} channels and to the increase in $[Ca^{2+}]_i$, albeit to a lesser extent than the above-mentioned mechanisms (Hartmann *et al.*, 2008; Ohtsuki *et al.*, 2009). This results in a supralinear increase in $[Ca^{2+}]_i$ in Purkinje cell dendrites that disappears about 1 s after synaptic stimulation has been terminated and which, together with diacylglycerol, activates protein kinase C-α (PKCα) (Leitges *et al.*, 2004; Sarkisov and Wang, 2008; Tanaka and Augustine, 2008; Tanaka *et al.*, 2007; Wang *et al.*, 2000). PKCα phosphorylates the GluR2 subunit of AMPA receptors at serine 880, which disrupts the interaction of this subunit with glutamate receptor–interacting protein-1 and -2 (GRIP1/2), promoting GluR2 binding to protein interacting with C-kinase 1 (PICK1) and internalization via a clathrin-dependent pathway (Jorntell and Hansel, 2006; Takamiya *et al.*, 2008; Wang and Linden, 2000). Internalization of AMPA receptors is also mediated by tumor necrosis factor-α converting enzyme, a matrix metalloprotease that cleaves

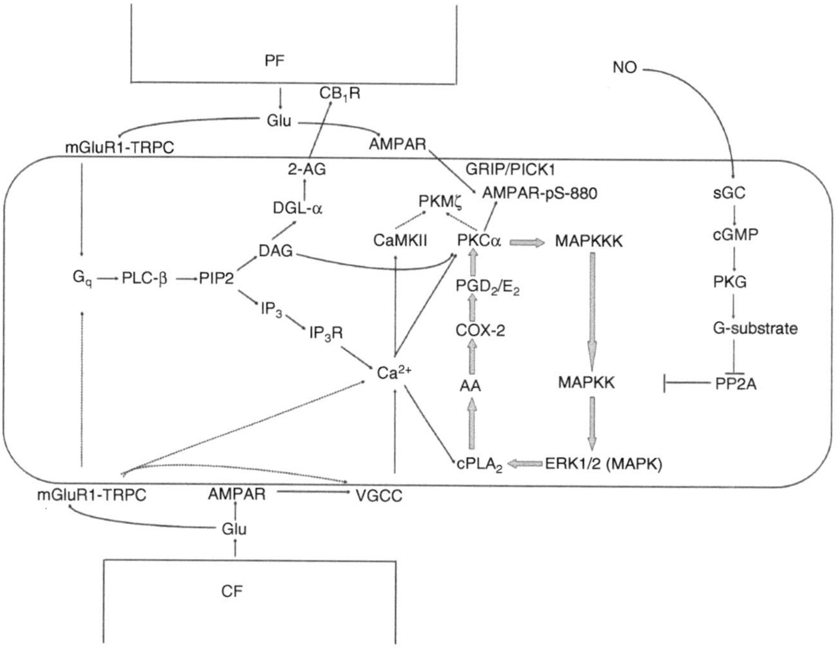

FIG. 3. Schematic representation of the signaling pathways involved in parallel fiber LTD. The sequence of events leading to parallel fiber LTD is described in detail in Section IV.A.1. AA: arachidonic acid, 2-AG: 2-arachidonoylglycerol, AMPAR: AMPA receptor, CaMKII: Ca^{2+}/calmodulin-dependent protein kinase II, CB_1R: cannabinoid receptor type 1, CF: climbing fiber, cGMP: cyclic guanosine monophosphate, COX: cyclooxygenase, cPLA2: Ca^{2+}-dependent phospholipase A2, DAG: diacylglycerol, DGL-α: diacylglycerol lipase-α, ERK: extracellular signal–activated kinase, Gq: type-q G protein, Glu: glutamate, GRIP: glutamate receptor–interacting protein, IP_3: inositol 1,4,5-trisphosphate, MAPK: mitogen-activated protein kinase, MAPKK: MAPK kinase, MAPKKK: MAPKK kinase, mGluR1: metabotropic glutamate receptor, NO: nitric oxide, PC: Purkinje cell, PF: parallel fiber, PG: prostaglandin, PICK1: protein interacting with C-kinase 1, PIP2: phosphatidylinositol 4,5-bisphosphate, PKCα, protein kinase C-alpha, PKG: cGMP-dependent protein kinase, PKMζ: catalytically active PKC-zeta, PLC: phospholipase C, PP2A: protein phosphatase 2A, sGC: soluble guanylate cyclase, TRPC: transient receptor potential cation channel, VGCC: voltage-gated Ca^{2+} channel.

neuronal pentraxin receptor and releases it from its N-terminal transmembrane domain (Cho *et al.*, 2008). The cleaved neuronal pentraxin receptor then associates with the extracellular portion of AMPA receptors, causing them to cluster, facilitating their internalization via the endosome pathway. It should be emphasized that the elevation in $[Ca^{2+}]_i$ activates not only PKC but also α-Ca^{2+}/calmodulin-dependent protein kinase II (α-CaMKII), which has also been shown to be essential for parallel fiber LTD induction (Hansel *et al.*, 2006).

The next phase of parallel fiber LTD is the intermediate phase, which lasts approximately 30 min and involves persistent endocytosis of AMPA receptors mediated by what is known as the mitogen-activated protein kinase (MAPK)–PKC positive feedback loop (Ogasawara *et al.*, 2008; Ogasawara and Kawato, 2009; Tanaka and Augustine, 2008) (Fig. 3). In this loop, PKC activates the MAPK kinase pathway and Ca^{2+}-dependent phospholipase A_2, leading to an increase in arachidonic acid levels (Ogasawara and Kawato, 2009; Tanaka and Augustine, 2008). Arachidonic acid activates cyclooxygenase type 2 (COX-2), increasing the synthesis of prostaglandins D_2 and E_2 (PGD2/E2) (Le *et al.*, 2010). Both arachidonic acid and PGD_2/E_2 feed back onto PKC to sustain the positive feedback loop engaged, contributing to maintenance of AMPA receptor phosphorylation and endocytosis. Nitric oxide released from stellate cells may also contribute to activation of the MAPK pathway via inhibition of protein phosphatase 2A (Fig. 3) (Feil *et al.*, 2003; Shin and Linden, 2005; reviewed in Ito, 2006; Ogasawara *et al.*, 2008). Endocannabinoids retrogradely released from Purkinje cells may also contribute to the intermediate phase of LTD by a mechanism that is not fully understood (Safo and Regehr, 2005).

The late phase of parallel fiber LTD lasts for up to several weeks and it requires the synthesis of new proteins (Linden, 1996) (Fig. 3). Although the identity of these proteins has yet to be determined, it has been postulated that the late phase of parallel fiber LTD could be mediated by increased expression of PKMζ, an autonomously active variant of the atypical PKCζ isoform. PKMζ can be directly and persistently synthesized from a PKMζ mRNA and this process is regulated by other PKC isoforms and CaMKII; importantly, PKMζ has been shown to play a key role in memory consolidation (Hernandez *et al.*, 2003; Ogasawara *et al.*, 2008; Sacktor, 2010). It is therefore important to determine whether PKMζ is involved in the late phase of parallel fiber LTD and if its expression shifts from Purkinje cells to cerebellar nuclear neurons during motor learning consolidation (Shutoh *et al.*, 2006).

2. *Long-Term Potentiation*

It is well established that parallel fiber–Purkinje cell synapses can undergo a form of presynaptic LTP (pre-LTP) after high-frequency stimulation of the parallel fibers (120 pulses at 4 Hz). Pre-LTP at parallel fibers is mediated by a persistent increase in glutamate release and depends on activation of protein kinase A (PKA) via Ca^{2+}-dependent adenylyl cyclase type I (Salin *et al.*, 1996; Shibuki and Okada, 1992; Storm *et al.*, 1998). Release of NO and PKA-dependent phosphorylation of the presynaptic active zone protein RIM1α are also required for this form of plasticity (Castillo *et al.*, 2002; Jacoby *et al.*, 2001; Kimura *et al.*, 1998; Lonart *et al.*, 2003; Qiu and Knopfel, 2007; but see Lev-Ram *et al.*, 2002). More recently, a study that made use of transgenic mice expressing a fluorescent Ca^{2+} sensor in parallel fibers demonstrated that pre-LTP at parallel fibers is associated with a PKA-dependent long-lasting potentiation of presynaptic Ca^{2+} transients (Qiu and Knopfel, 2007). Moreover, this study also found that pre-LTP at parallel fibers depends on the function of NMDA receptors located in molecular layer interneurons whose precise involvement in this process is not fully understood. It is noteworthy that pre-LTP at parallel fibers can be blocked by endocannabinoid release triggered by climbing fiber–evoked increases in $[Ca^{2+}]_i$, suggesting that pre-LTP at parallel fibers is disabled under conditions that favor parallel fiber LTD (i.e., co-stimulation of parallel fiber and climbing fiber; Section IV.A.1) (van Beugen *et al.*, 2006).

A postsynaptic form of parallel fiber LTP (post-LTP) can be induced by stimulation of the parallel fibers at 1 Hz for 5 min in the absence of climbing fiber co-stimulation and, unlike pre-LTP, it can reverse parallel fiber LTD while allowing potentiated synapses to undergo another cycle of parallel fiber LTD (Lev-Ram *et al.*, 2003, 2002). Moreover, parallel fiber LTD can depotentiate synapses that are in a state of post-LTP but not pre-LTP (Coesmans *et al.*, 2004; Lev-Ram *et al.*, 2003). Post-LTP at parallel fibers was shown to depend on NO, a $[Ca^{2+}]_i$ elevation of lower magnitude than that required for parallel fiber LTD, and on activation of protein phosphatases 1, 2A, and 2B (Fig. 4) (Belmeguenai and Hansel, 2005; Coesmans *et al.*, 2004; Lev-Ram *et al.*, 2003). Post-LTP at parallel fibers likely involves the insertion of AMPA receptors into the membrane

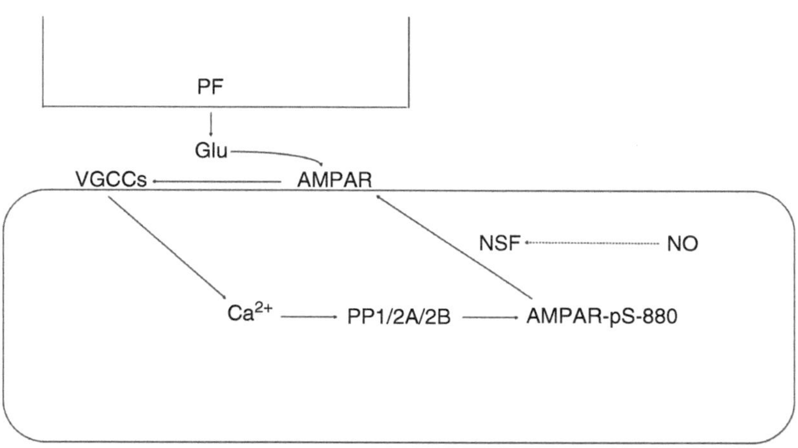

FIG. 4. Schematic representation of the signaling pathways involved in postsynaptic parallel fiber LTP. The sequence of events leading to parallel fiber LTP is described in detail in Section IV.A.2. AMPAR: AMPA receptor, Glu: glutamate, NO: nitric oxide, NSF: *N*-ethylmaleimide-sensitive factor, PC: Purkinje cell, PF: parallel fiber, PP1/2A/2B: protein phosphatase 1/2A/2B, VGCC: voltage-gated Ca^{2+} channel.

via a mechanism that depends on *N*-ethylmaleimide-sensitive factor, a process that could be triggered by NO-induced S-nitrosylation of this factor (compare with the potential role of NO in parallel fiber LTD illustrated in Fig. 3) (Huang *et al.*, 2005; Kakegawa and Yuzaki, 2005). It should be noted that in slices, post-LTP at parallel fibers does not depend on mGluR1 receptor activation (Belmeguenai *et al.*, 2008). However, a novel form of post-LTP at parallel fibers that depends on the activation of these receptors was recently demonstrated in mouse Purkinje cells *in vivo* (Wang *et al.*, 2009).

B. Climbing Fiber-to-Purkinje Cell Synapses

It was assumed for many years that the strong responses produced in Purkinje cell by climbing fiber activation were invariant, until Hansel and Linden (2000) demonstrated that this is not the case. These investigators showed that repetitive stimulation of climbing fibers (5 Hz for 30 s) produces LTD of climbing fiber responses in slices from juvenile rats. Similar to parallel fiber LTD, the mechanism of climbing fiber LTD was found to require an mGluR1-dependent increase in postsynaptic $[Ca^{2+}]_i$ and activation of PKC, with no change in glutamate release (Hansel and Linden, 2000; Shen *et al.*, 2002) (Fig. 5). In addition, climbing fiber LTD requires activation of PKA mediated by corticotropin-releasing factor (CRF), which is likely released from climbing fibers in an activity-dependent manner—it is noteworthy that CRF release from climbing fibers may also play a role in parallel fiber LTD (Schmolesky *et al.*, 2007; reviewed in Ohtsuki *et al.*, 2009). The amplitude of the secondary spikes in the complex spike was reduced during climbing fiber LTD and this was associated with a decrease in climbing fiber–triggered Ca^{2+} transients in both primary dendrites and spiny branchlets (Weber *et al.*, 2003). The amplitude of the afterhyperpolarization that follows the complex spike was also reduced during climbing fiber LTD (Schmolesky *et al.*, 2005). Importantly, it was found that climbing fiber LTD impaired the ability of the climbing fiber to promote induction of parallel fiber LTD—in fact, under conditions of climbing fiber LTD, climbing fiber plus parallel fiber co-stimulation actually produced parallel fiber LTP (Coesmans *et al.*, 2004).

Climbing fiber plasticity mechanisms may also play a central role during development, as demonstrated in two recent studies. As mentioned in Section III, Purkinje cells are innervated by multiple climbing fibers during development and weak climbing fibers are gradually eliminated in an activity-dependent manner, resulting in a mature pattern of Purkinje cell innervation, that is, by a single climbing fiber. In the first study, Bosman *et al.* (2008) found in slices from postnatal-day-6–8 rats that depolarization of a Purkinje cell combined with high-frequency stimulation (4 Hz for 5 min in voltage clamp or 100 s in current

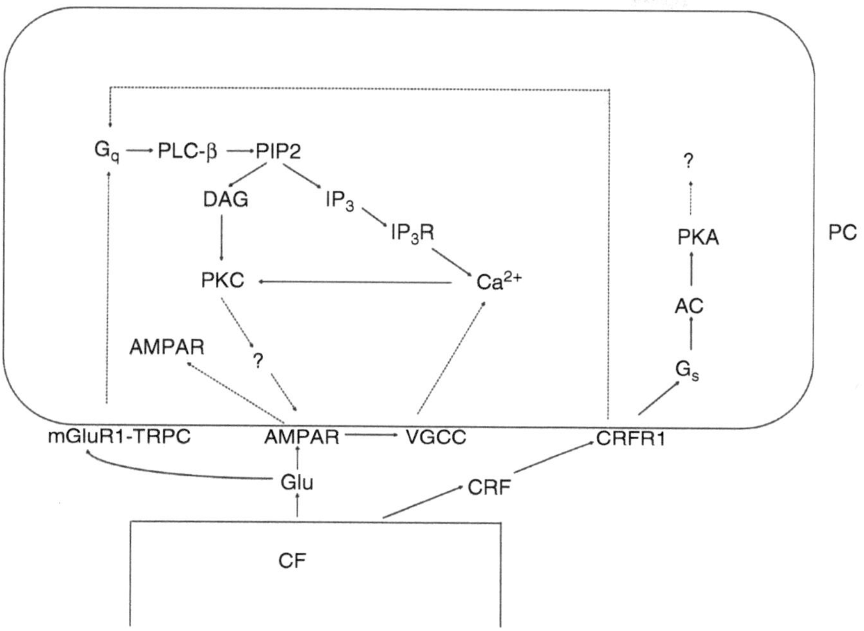

FIG. 5. Schematic representation of the signaling pathways involved in climbing fiber LTD. The sequence of events leading to climbing fiber LTD is described in detail in Section IV.2.B. AC, adenylate cyclase, AMPAR: AMPA receptor, CF: climbing fiber, CRF: corticotropin-releasing factor, CRFR: CRF receptor, DAG: diacylglycerol, Gq: type-q G protein, Gs: type-s G protein, Glu: glutamate, IP_3: inositol 1,4,5-trisphosphate, IP_3R: IP_3 receptor, PC: Purkinje cell, PIP2: phosphatidylinositol 4,5-bisphosphate, PKA: 3'-5'-cyclic adenosine monophosphate–dependent protein kinase, PKC: protein kinase C, PLC: phospholipase C, TRPC: transient receptor potential cation channel, VGCC: voltage-gated Ca^{2+} channel.

clamp) of its strongest climbing fiber input resulted in LTP of responses mediated by this input. This climbing fiber LTP was independent of NMDA receptor activation and required a rise in postsynaptic $[Ca^{2+}]_i$ that ultimately resulted in an increase in postsynaptic AMPA receptor conductance. Importantly, postsynaptic climbing fiber LTP was not observed at the weak climbing fiber input.

In the second study, Ohtsuki and Hirano (2008) used a similar approach to characterize climbing fiber plasticity in slices from developing mice. These investigators found that high-frequency stimulation (5 Hz for 20 s) induced LTP in strong climbing fibers and LTD in weak climbing fibers in slices from postnatal-day-5–9 mice. LTP and LTD were associated with a decrease and an increase in the paired-pulse ratio of the climbing fiber responses, respectively, suggesting that the mechanism responsible for these phenomena involves presynaptic changes in glutamate release probability. LTP, but not LTD, required a postsynaptic increase in $[Ca^{2+}]_i$ and only LTP spread to climbing fibers in the vicinity of the one that received high-frequency stimulation. Discrepancies in the results of these two studies may be related to differences in the experimental conditions, including the use of cerebellar slices from different species (rats vs. mice). Nevertheless, these studies suggest that LTP and LTD plasticity mechanisms may play a role in the activity-dependent stabilization and/or pruning of climbing fiber in maturing Purkinje cells, and it would be interesting to determine whether these mechanisms are targeted by ethanol.

C. Motor Learning and Plasticity Mechanisms

Plasticity-mediated changes in the strength of cerebellar synaptic connections are thought to play a central role in motor learning and this has been studied using eyeblink conditioning, a form of associative motor learning where conditioned behavior has been linked, in part, to parallel fiber LTD. As will be described in Section V, this behavioral paradigm has been shown to be affected by acute, chronic, and developmental ethanol exposure. Eyeblink conditioning requires the cortex of the anterior cerebellar lobe and the interposed nucleus (Garcia *et al.*, 1999; Kalmbach *et al.*; Lavond, 2002; Steinmetz *et al.*, 1992). In this paradigm, an animal is trained to associate a neutral (conditioned) stimulus, such as a tone or a light, with an aversive (unconditioned) stimulus, typically a puff of air delivered to the cornea. In the cerebellar training regimen, the tone anticipates and overlaps with the corneal stimulus. Without training, corneal stimulation causes a protective reflex, the eyeblink. After training, an animal blinks in response to the conditioned tone even without corneal stimulation.

The unconditioned corneal stimulus causes eyeblink by a monosynaptic reflex arc. Tactile sensation is carried by the ophthalmic branch of the trigeminal nerve to the trigeminal nucleus, and then onto the ipsilateral facial nucleus. There, the sensory neuron synapses with the lower motor neurons of cranial nerve VII that innervate the orbicularis oculi muscle to close the eyelid. In parallel, information about the corneal touch is relayed by interneurons in the trigeminal nucleus to the inferior olive that feeds back onto Purkinje cells via the climbing fibers. The conditioned auditory stimulus is carried by cochlear sensory fibers to the cochlear nucleus in the pons. There, it gives rise to mossy fibers which pass through the middle cerebellar peduncle to stimulate deep cerebellar nuclei, as well as terminate in the granule cell layer. The granule cells stimulate Purkinje cells via parallel fibers.

During training, information about the auditory stimulus and the corneal irritation is integrated at the Purkinje cell. Inferior olive climbing fibers encoding the unconditioned stimulus and mossy fiber–parallel fibers carrying information about the conditioned stimulus impinge on the same Purkinje cell, where plasticity changes are thought to play a role in changing its output during conditioning (Kalmbach *et al.*, 2010). Here, appropriate synaptic inputs can be strengthened by parallel fiber LTP and inappropriate synaptic inputs can be silenced by parallel fiber LTD, and, as a result, Purkinje cell output increases or decreases (Jorntell and Hansel, 2006). Moreover, extinction of eyeblink conditioning can be induced by repetitive application of the conditioning stimuli alone and it is thought that this can be mediated by parallel fiber LTP (Jorntell and Hansel, 2006).

Another site of integration occurs at the interposed nucleus, where plasticity mechanisms are also likely to play a role in eyeblink conditioning (Lavond, 2002). Here, glutamatergic mossy fibers from the cochlear nucleus (tone) and climbing fibers from the inferior olive (corneal irritation) converge with GABAergic Purkinje cell output. Axons originating in the interposed nucleus exit the cerebellum via the superior cerebellar peduncle and excite target neurons in the red nucleus in the midbrain. This vestigial motor nucleus provides an alternative motor pathway, the rubrospinal tract. Under naïve circumstances, the red nucleus stimulates the facial nucleus, but not enough to produce an eyeblink. After eyeblink conditioning, disinhibition of the interposed nucleus by the Purkinje cell allows transmission through the rubrospinal pathway. After training, the cerebellar cortex has decreased the threshold for the interposed nucleus to fire upon hearing a tone. With less inhibition by the Purkinje cells, the interposed nucleus is able to act on the rubrospinal pathway and effect the conditioned behavior.

V. Modulation of Cerebellar Synaptic Plasticity by Ethanol

A. Acute Effects

The first slice electrophysiological study of the acute effects of ethanol on cerebellar plasticity focused on climbing fiber LTD (Carta *et al.*, 2006). Initial experiments revealed that acute ethanol exposure reversibly decreases the area under the curve of the complex spike, mainly by reducing the late phase of this response. An ethanol concentration of 75 mM reduced the complex spike area by approximately 25%, and a smaller but statistically significant reduction of approximately 10% was also detected with a 10 mM concentration (the legal intoxication limit in the United States is 0.08 g/dl = 17.4 mM). It was hypothesized that mGluR1-TRPC-dependent currents would mediate the late phase of the complex spike and that ethanol could affect this phase via inhibition of mGluR1-dependent signaling. In support of this hypothesis, it was found that the mGluR1 antagonist LY367385 blocked the late phase of the complex spike and occluded the effect of ethanol on this response. Ethanol was also shown to inhibit mGluR1-dependent excitatory postsynaptic currents evoked by climbing fiber stimulation (Fig. 6A, B). This effect of ethanol was not an indirect consequence of a decrease in glutamate release probability, as ethanol did not affect paired-pulse plasticity of AMPA receptor–mediated excitatory postsynaptic currents evoked by climbing fiber stimulation. As shown in Fig. 5, climbing fiber LTD requires mGluR1 receptor activation and it was therefore hypothesized that ethanol would affect this form of synaptic plasticity. Indeed, it was found that a 50 mM ethanol concentration blocked climbing fiber LTD (Fig. 6C, D).

As mentioned above, parallel fiber LTD also depends on mGluR1 activation and this raised the possibility that this form of synaptic plasticity could be

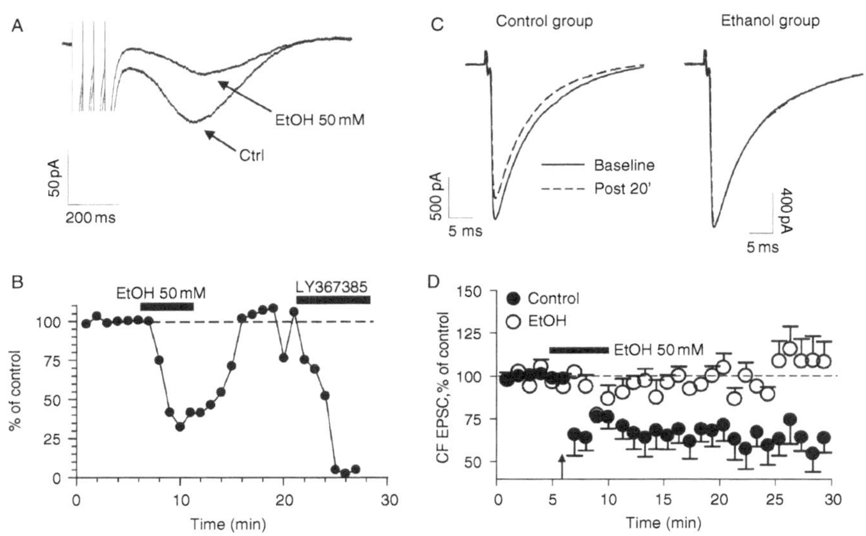

FIG. 6. Acute ethanol exposure inhibits both climbing fiber–evoked mGluR1-dependent currents and climbing fiber LTD in rat cerebellar slices. (A) 50 mM ethanol acutely inhibits mGluR1-dependent excitatory postsynaptic currents evoked by climbing fiber stimulation with a train of four pulses at a frequency of 20 Hz in the presence of 10 μM 2,3-dihydroxy-6-nitro-7-sulfamoyl-benzo[f]quinoxaline-2, 3-dione (NBQX) and 20 μM bicuculline. (B) Time course of the effect of ethanol corresponding to the same recording shown in (A). Also shown is the effect of LY367385 (50 μM), an mGluR1 antagonist. (C) AMPA receptor–mediated excitatory postsynaptic currents evoked by climbing fiber stimulation were reduced 20 min after tetanic stimulation (5 Hz for 30 s) of climbing fibers under control conditions but not in presence of 50 mM ethanol. (D) Pooled data illustrating the effect of ethanol on climbing fiber LTD ($n = 6$). Climbing fiber tetanic stimulation was delivered at the time indicated by the arrow. Reprinted with permission from Carta *et al.* (2006). ©Society for Neuroscience.

inhibited by acute ethanol exposure in a similar fashion as climbing fiber LTD. This possibility was tested by Belmeguenai *et al.* (2008), who found that acute exposure to 50 mM ethanol blocked parallel fiber LTD in cerebellar slices from rats (Fig. 7). As in the case of the climbing fiber studies described above, it was found that acute ethanol exposure inhibited mGluR1-dependent slow excitatory postsynaptic currents (Fig. 8); however, ethanol had a significant effect on these responses at a concentration of 40 mM but not 20 mM. In light of the results of Carta *et al.* (2006), this finding indicates that ethanol more potently modulates climbing fiber mGluR1–mediated responses than those mediated by parallel fiber mGluR1 (Figs. 6 and 7). Ethanol affected neither the amplitude of basal parallel fiber excitatory postsynaptic currents (i.e., in the absence of the LTD-inducing protocol) nor paired-pulse facilitation of these events, including facilitation of AMPA receptor excitatory postsynaptic current trains evoked at the same frequency (10 pulses at 100 Hz) used to elicit mGluR1 excitatory postsynaptic currents. Moreover, currents evoked by application of an exogenous mGluR1 agonist (*RS*-3,5-dihydroxy-phenylglycine) were also inhibited by acute ethanol exposure. Taken together, these findings indicate that the effect of ethanol is not a consequence of a decrease in glutamate release from presynaptic terminals. Interestingly, ethanol did not affect parallel fiber LTP and this form of plasticity was demonstrated to be mGluR1 independent. Inhibition of mGluR1-dependent signaling is not likely to be the only mechanism responsible for parallel fiber LTD inhibition by ethanol, as acute exposure to this agent also reduced the function of other proteins that are thought to play a key role in this synaptic plasticity mechanism—voltage-gated Ca^{2+} channels (Belmeguenai *et al.*, 2008). More recently, the findings of the study of Belmeguenai were independently reproduced by another group, which additionally showed that intracellular application of ethanol blocked parallel fiber LTD and inhibited parallel fiber mGluR1 excitatory postsynaptic currents, suggesting that ethanol acts by altering an intracellular signaling cascade (Su *et al.*, 2010). In this regard, it should be explored whether ethanol acts by interfering with extracellular signal–regulated kinase–dependent signaling pathways, as this kinase has been shown to play an important role in parallel fiber LTD (Fig. 3) and to be acutely inhibited by ethanol in preparations from the rat cerebellum (Chandler and Sutton, 2005; Ohrtman *et al.*, 2006).

As mentioned above, parallel fiber LTD may play a role in eyeblink conditioning and, therefore, it would be expected that *in vivo* acute ethanol exposure affects this cerebellar motor learning paradigm. Surprisingly, this issue has been addressed in only a few studies. An early study found that ethanol acutely inhibits conditioned eyeblink responses in human volunteers (blood alcohol levels 0.05–0.1 g/dl; Hobson, 1966). A similar observation was made in rabbits that were orally administered 0.75 and 1.5 g/kg of ethanol (blood alcohol levels = 0.08 and 0.18 g/dl; Hernandez *et al.*, 1986). These findings are consistent with the

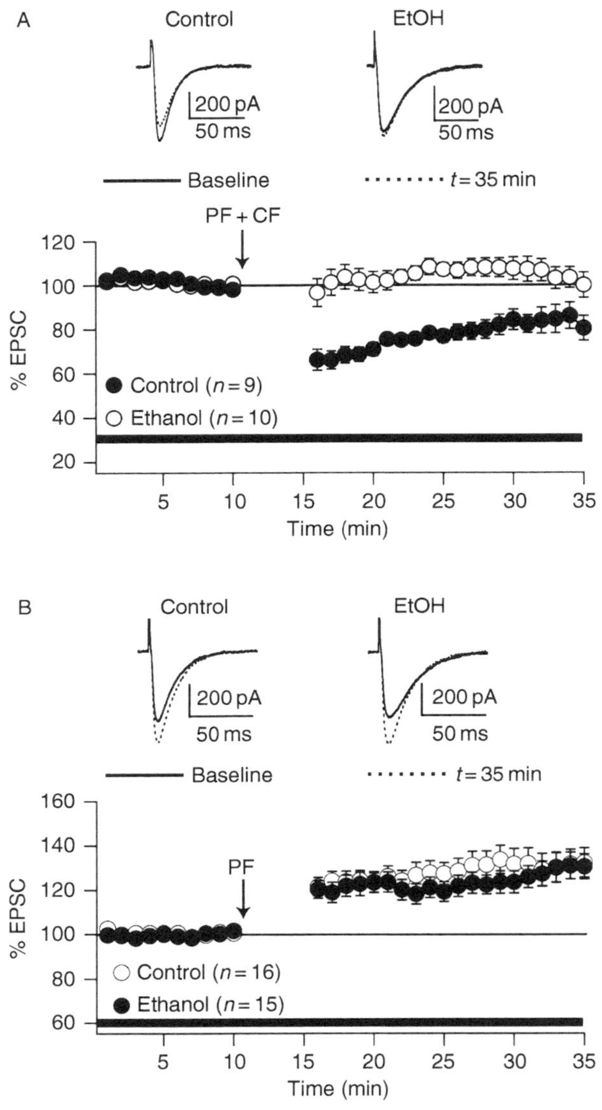

FIG. 7. Acute ethanol exposure inhibits parallel fiber LTD but not LTP in rat cerebellar slices. (A) Acute exposure to 50 mM ethanol (indicated by the solid bar) blocks parallel fiber LTD, which was induced via paired parallel fiber and climbing fiber stimulation at 1 Hz for 5 min. (B) Ethanol did not affect parallel fiber LTP, which was induced by parallel fiber stimulation alone at 1 Hz for 5 min. Reprinted with permission from Belmeguenai *et al.* (2008). © American Physiological Society.

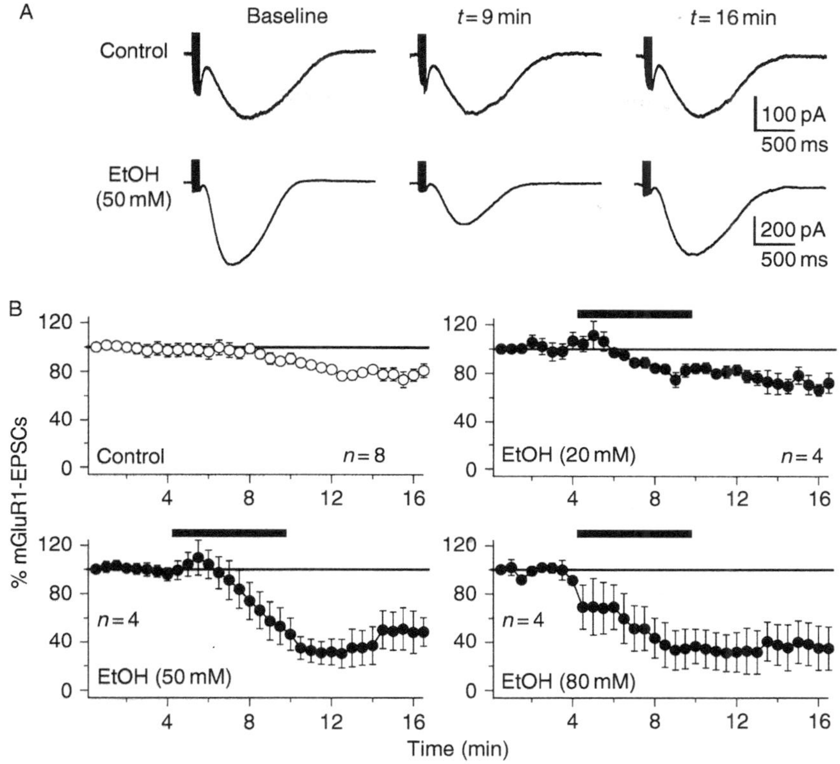

FIG. 8. Acute ethanol exposure inhibits parallel fiber–evoked mGluR1-dependent currents rat cerebellar slices. (A) Traces illustrating the effect of 50 mM ethanol on mGluR1-dependent excitatory postsynaptic currents evoked by parallel fiber stimulation. Ethanol was applied as shown in (B), which illustrates the effect of multiple doses of ethanol. Reprinted with permission from Belmeguenai *et al.* (2008). © American Physiological Society.

observations of the above-mentioned slice electrophysiological studies indicating that ethanol concentrations near this range of blood alcohol levels inhibit mGluR1 excitatory postsynaptic currents and/or parallel fiber LTD. Interestingly, acute oral administration of 0.375 g/kg of ethanol (blood alcohol level = 0.028 g/dl) enhanced conditioned eyeblink responses in these animals and it is therefore important to determine whether parallel fiber LTD is potentiated by lower ethanol concentrations (Hernandez and Powell, 1986). Moreover, it is important to study the effect of ethanol on plasticity at the interposed nucleus because it has also been shown to play a role in eyeblink conditioning (see Section IV.C).

B. Chronic Effects

To the best of our knowledge, the effects of chronic ethanol exposure on cerebellar LTP or LTD in mature animals have yet to be studied. However, eyeblink conditioning impairments have been documented in patients afflicted by amnesic Korsakoff's syndrome and in recovered alcoholics (Fortier *et al.*, 2008; McGlinchey-Berroth *et al.*, 1995). Moreover, structural alterations in Purkinje cell dendritic spines and climbing fiber terminals have been detected in chronically ethanol-exposed rats (Pentney and Quackenbush, 1991; Pentney *et al.*, 1989; Tavares *et al.*, 1986, 1983). In addition, chronic ethanol exposure and withdrawal were shown to decrease and increase extracellular signal–regulated kinase activation, respectively, in preparations from rat cerebella (Sanna *et al.*, 2002). Collectively, the findings of these studies suggest that long-term ethanol exposure is likely to impair plasticity at Purkinje cell glutamatergic synapses. Moreover, alterations in synaptic plasticity could underlie the chronic ethanol-induced structural synaptic remodeling changes and this should be tested in the future.

In contrast to its effects on cerebellar plasticity in mature animals, the effects of long-term ethanol exposure on plasticity in the developing cerebellum have been the focus of significant research interest. As mentioned before, exposure to ethanol during gestation can cause fetal alcohol spectrum disorders. At the most severe end of the spectrum of this disorder is fetal alcohol syndrome, characterized by growth retardation, facial abnormalities, and central nervous system alterations. In many cases, central nervous system abnormalities are the only manifestations of fetal alcohol exposure and this condition is known as alcohol-related neurodevelopmental disorder. Among the brain regions affected by fetal ethanol exposure is the cerebellum, where migration, differentiation, and/or survival of Purkinje cells, granule cells, and deep cerebellar nuclear neurons has been shown to be impaired (Dikranian *et al.*, 2005; Green, 2004; Kumada *et al.*, 2007; West, 1993; West *et al.*, 1990). Decreases in cerebellar neuron number are likely to be, in part, responsible for the motor coordination and balance alterations that often accompany fetal alcohol spectrum disorders. In addition, eyeblink conditioning studies have demonstrated that developmental ethanol exposure produces long-lasting impairments in cerebellar motor learning both in humans and in laboratory animals, and this may be the result of deficits in plasticity at surviving Purkinje cells and/or cerebellar nuclear neurons (Brown *et al.*, 2007, 2008; Coffin *et al.*, 2005; Green, 2004; Green *et al.*, 2006; Jacobson *et al.*, 2008; Lindquist *et al.*, 2007; Tran *et al.*, 2005).

Consistent with the possibility that fetal alcohol spectrum disorders can be a consequence of Purkinje cell plasticity deficits are the findings of a multidisciplinary study with the offspring of mice exposed to ethanol throughout pregnancy (Servais *et al.*, 2007). In cerebellar slices from juvenile control mice, parallel fiber stimulation with 300 stimuli—delivered at a frequency of 1 Hz, paired with a

200 ms Purkinje cell depolarizing pulse (from −60 to 0 mV)—induced parallel fiber LTD. However, in slices from prenatally ethanol-exposed juvenile mice, this protocol unexpectedly induced parallel fiber LTP (Fig. 9). In addition, *in vivo* electrophysiological experiments with adult offspring revealed that Purkinje cells fire simple spikes more often and display rapid oscillations of local field potentials; these were shown not to be a consequence of an increase in Purkinje cell excitability or parallel fiber excitatory postsynaptic current amplitude—the latter was actually reduced as a consequence of a decrease in glutamate release from

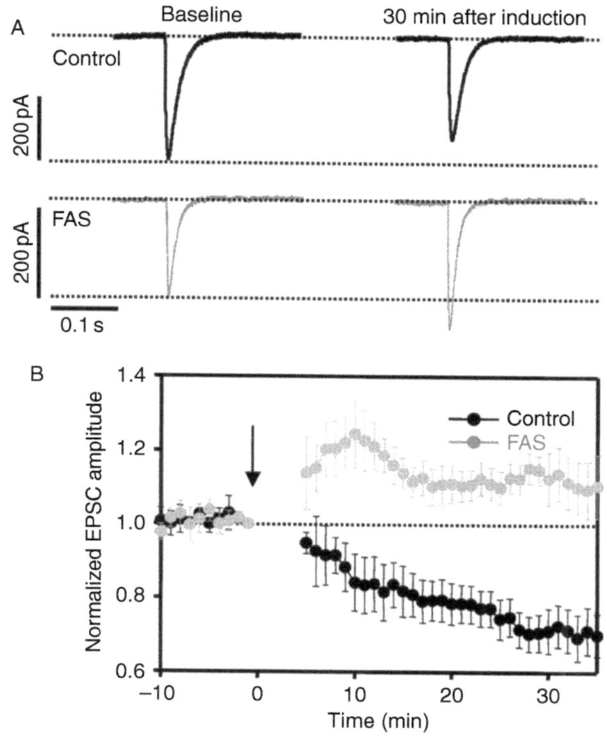

FIG. 9. Chronic prenatal ethanol exposure converts LTD into LTP in cerebellar slices from juvenile mice. (A) Shown on the top are parallel fiber excitatory postsynaptic currents recorded before and after application of an LTD induction protocol in a control Purkinje cell. Note the decrease in current amplitude. Shown at the bottom are parallel fiber excitatory postsynaptic currents recorded before and after the application of the LTD induction protocol in a Purkinje neuron from a mouse prenatally exposed to ethanol. Note the increase in current amplitude. (B) Summary graph illustrating the differential effect of the LTD induction protocol in Purkinje cells from control and prenatally ethanol-exposed mice. Reprinted with permission from Servais *et al.* (2007). © 2007 National Academy of Sciences USA.

parallel fibers. The mechanistic underpinning of some of these alterations appears to involve a decrease in voltage-gated Ca^{2+} channel function, which may be a consequence of a reduction in phosphorylation of these channels by PKCγ, whose expression was found to be diminished in cerebellar membrane extracts from ethanol-exposed mice. As shown in Fig. 3, parallel fiber LTD induction may involve activation of voltage-gated Ca^{2+} channels and this could be the mechanism by which ethanol alters this form of synaptic plasticity. However, this conclusion is inconsistent with the findings of Chen *et al.* (1995) that parallel fiber LTD is normal in PKCγ knockout mice, suggesting that a different mechanism may account for the effects of prenatal ethanol exposure. An alternative possibility is that the decrease in parallel fiber LTD is simply a consequence of the reduction in glutamate release that was detected at parallel fiber–Purkinje cell synapses, which would result in diminished activation of both mGluR1 and AMPA receptors, leading to less activation of IP_3 receptors and voltage-gated Ca^{2+} channels (Fig. 3). Another possibility that should be studied in the future is whether prenatal ethanol exposure alters parallel fiber plasticity via alterations in the complex spike. Interestingly, Servais *et al.* (2007) found that prenatal ethanol exposure does not affect complex spike frequency. In contrast, Backman *et al.* (1998) found that exposure of rats to ethanol during the third trimester equivalent reduced the number of Purkinje neurons exhibiting complex spikes; moreover, this ethanol exposure paradigm did not affect spontaneous firing of Purkinje cells or Golgi cells during adulthood (in anesthetized animals using *in vivo* electrophysiological techniques), nor did it impair modulation of Purkinje cell activity evoked by parallel fibers. These findings suggest that the effects of developmental ethanol exposure on cerebellar function and plasticity could be influenced by the timing of ethanol exposure (prenatal vs. postnatal), the experimental conditions (anesthetized vs. awake animals), and the animal species (rat vs. mice). Clearly, future studies are required to fully understand the impact of developmental ethanol exposure on cellular and behavioral forms of cerebellar plasticity.

VI. Conclusions and Future Directions

As discussed above, converging evidence from several laboratories indicates that plasticity at cerebellar Purkinje cells is an important target of ethanol. The plasticity mechanism that is predominantly affected by acute ethanol in these cells is LTD, rather than LTP—this is in contrast to other brain regions, such as the hippocampus, where ethanol can affect both of these forms of synaptic plasticity (Izumi *et al.*, 2005b). Acutely, ethanol was shown to inhibit LTD both at parallel fiber– and climbing fiber–Purkinje cell synapses and the mechanism of this effect appears to involve, at least in part, inhibition of mGluR1-dependent signaling. This finding is very interesting in light of a study demonstrating that paraneoplastic cerebellar ataxia is, at least in part, mediated by autoantibodies against mGluR1, which were shown to block parallel fiber LTD in cultured mouse Purkinje neurons (Coesmans *et al.*, 2003). The results of this study lend support to the idea that the motor coordination alterations produced by acute ethanol exposure are, in part, a consequence of alterations in mGluR1-dependent plasticity in Purkinje neurons. It should be kept in mind, however, that the actions of ethanol on mGluR1 function are likely to operate in conjunction with its actions on GABAergic transmission both in the Purkinje and granule cell layers and the reader is referred to recent reviews for more details (Botta *et al.*, 2007b; Breese *et al.*, 2006). At present, it is unknown whether ethanol acts by directly inhibiting the function of mGluR1 or by disrupting downstream effectors such as phospholipase C, IP_3 receptors, or TRPC1/3 channels (Fig. 3)—this should be investigated in the future. In addition, it should be assessed whether inhibition of voltage-gated Ca^{2+} channels plays a role in the mechanism of action of ethanol, as suggested by the studies of Belmeguenai *et al.* (2008). It would also be interesting to determine whether LTD can be acutely inhibited by ethanol in other neuronal populations where mGluR1 (and also mGluR5) plays a prominent role in the mechanism of LTD induction, including neurons that form part of reward pathways (Luscher and Huber, 2010). For instance, LTD triggered by a non-selective mGluR agonist was shown to be inhibited by ethanol in the CA1 hippocampus of immature rats (Overstreet *et al.*, 1997) and this effect should be re-evaluated with selective agonists of group I mGluRs (i.e., mGluR1 and mGluR5). In Purkinje neurons, NMDA receptors have not been shown to play a central role in LTD induction; however, these receptors are required for LTD induction in other brain regions, such as the CA1 hippocampus, where ethanol may inhibit or enhance LTD, in part, via modulation of these receptors (Hendricson *et al.*, 2002; Izumi *et al.*, 2005b). In this context, it is important to assess if ethanol affects NMDA receptor function directly or if it acts indirectly via group I mGluRs. Another issue that should be investigated is whether acute ethanol differentially affects the induction, expression, and maintenance phases of LTD at Purkinje cells.

As mentioned in Section V, the chronic effects of ethanol on Purkinje cell LTD and LTP in mature animals have yet to be characterized and it is important to address this issue experimentally as aberrant cerebellar plasticity could underlie the long-term deficits in motor function and cognition that are often present in alcoholic patients. Studies with hippocampal slices from rats chronically exposed to ethanol indicate that it can impair both LTD and LTP (Thinschmidt *et al.*, 2003). Moreover, α_1-adrenergic receptor–dependent LTD, but not mGluR5-dependent LTD, in the bed nucleus of the stria terminalis was shown to be impaired by chronic ethanol exposure (McElligott *et al.*, 2010). These studies provide a strong rationale for pursuing studies of the effect of chronic ethanol on cerebellar plasticity in mature animals.

Along the same vein, the role of cerebellar synaptic plasticity alterations in the effects of developmental ethanol exposure should continue to be investigated. The findings of the elegant study of Servais *et al.* (2007) need to be followed up in several respects. First, it should be assessed if chronic prenatal ethanol exposure affects mGluR1-dependent signaling at parallel fiber–Purkinje cells synapses and if it produces any alterations in climbing fiber–Purkinje cell strength that could contribute to LTD deficits. Second, fluorescence imaging studies in Purkinje neuron dendrites and/or spines should be performed to assess if prenatal ethanol exposure alters Ca^{2+} dynamics. Third, it should be explored if ethanol exposure during the third trimester equivalent is also capable of altering Purkinje neuron plasticity; justification for investigating this issue is provided by a study in which rat pups were exposed to ethanol on postnatal day 7 and this inhibited LTP and blocked LTD in CA1 hippocampal slices prepared from the same animals at postnatal day 30 (Izumi *et al.*, 2005a). Moreover, complex motor training was shown to improve motor coordination and increase parallel fiber synapses in Purkinje cells of rats exposed to ethanol in binge-like episodes, indicating that these synapses maintain the ability to undergo plastic changes—a question that remains is which synaptic mechanisms mediate these structural changes (Klintsova *et al.*, 2000, 1997). Finally, the chain of events leading from prenatal ethanol exposure to long-lasting alterations in parallel fiber LTD should be investigated. Among the mechanisms that could be responsible for these persistent alterations are abnormal progenitor migration and differentiation, as well as increased neuronal death (Goodlett *et al.*, 2005; Jiang *et al.*, 2008), which can all ultimately lead to abnormal synapse formation, stabilization, and/or pruning in surviving neurons. In response to these alterations, compensatory mechanisms will be deployed in an attempt to restore homeostasis, and if these fail, then long-lasting alterations of synaptic plasticity can ensue as a consequence of impaired neuronal circuit assembly. Importantly, it should be investigated whether alterations in developmental synaptic plasticity, which is implicated in synapse maturation/stabilization, play a role in the mechanism of action of ethanol. Work from our laboratory indicates that ethanol exposure indeed affects plasticity in the developing hippocampus and that this may also be the case in the cerebellum (Puglia and Valenzuela, 2010a,b; Zucca and Valenzuela, 2010). In particular, studies of the effect of ethanol on plasticity at developing climbing fiber–Purkinje cell synapses are warranted.

Another important area for future research is to investigate the short- and long-term actions of ethanol on plasticity at other cerebellar synapses both during development and maturity, including LTP of GABAergic inputs to Purkinje cells (Kano *et al.*, 1992), LTP and LTD at synapses in deep cerebellar nuclear neurons (which are reduced by developmental ethanol exposure; Green, 2004), LTP of parallel fiber inputs in molecular layer interneurons (Jorntell and Ekerot, 2003), LTD in Golgi cells (Xu and Edgley, 2008), and LTP of mossy fiber–granule cell synapses (Hansel *et al.*, 2001; Pugh and Raman, 2009; Zheng and Raman, 2010). The latter depends on activation of both NMDA and mGluR1 receptors, suggesting that this form of synaptic plasticity may be a target of ethanol (D'Angelo and De Zeeuw, 2009; D'Angelo *et al.*, 2005). Indeed, acute ethanol exposure was shown to inhibit NMDA receptor–dependent excitatory postsynaptic currents in granule cells in slices from mice; however, whether these receptors are sensitive to subanesthetic ethanol concentrations under these experimental conditions is unknown (Offenhauser *et al.*, 2006).

In sum, plasticity at cerebellar Purkinje neurons is affected by acute and chronic ethanol exposure and this mechanism may play a central role in the pathophysiology of ethanol-related cerebellar dysfunction. Given that the cerebellar effects of ethanol are not limited to motor coordination disturbances but also involve alterations in complex cognitive functions that depend on normal functioning of the cerebellum (Sullivan *et al.*, 2003), it is important to develop therapeutic interventions against these actions of ethanol, as these may be of benefit to patients afflicted by alcoholism and fetal alcohol spectrum disorders.

Acknowledgments

C.F.V. is supported by NIH grant RO1-AA014973. We are grateful to Drs. Christian Hansel, Laurent Servais, and Kevin Caldwell for critically reading the manuscript.

References

Aiba, A., Kano, M., Chen, C., Stanton, M. E., Fox, G. D., Herrup, K., Zwingman, T. A., and Tonegawa, S. (1994). Deficient cerebellar long-term depression and impaired motor learning in mGluR1 mutant mice. *Cell* **79**, 377–388.

Backman, C., West, J. R., Mahoney, J. C., and Palmer, M. R. (1998). Electrophysiological characterization of cerebellar neurons from adult rats exposed to ethanol during development. *Alcohol. Clin. Exp. Res.* **22**, 1137–1145.

Barmack, N. H., and Yakhnitsa, V. (2008). Functions of interneurons in mouse cerebellum. *J. Neurosci.* **28**, 1140–1152.

Bellebaum, C., and Daum, I. (2009). Mechanisms of cerebellar involvement in associative learning. *Cortex* Sep. 6. [Epub ahead of print].

Belmeguenai, A., Botta, P., Weber, J. T., Carta, M., De Ruiter, M., De Zeeuw, C. I., Valenzuela, C. F., and Hansel, C. (2008). Alcohol impairs long-term depression at the cerebellar parallel fiber–Purkinje cell synapse. *J. Neurophysiol.* **100**, 3167–3174.

Belmeguenai, A., and Hansel, C. (2005). A role for protein phosphatases 1, 2A, and 2B in cerebellar long-term potentiation. *J. Neurosci.* **25**, 10768–10772.

Bosman, L. W., and Konnerth, A. (2009). Activity-dependent plasticity of developing climbing fiber–Purkinje cell synapses. *Neuroscience* **162**, 612–623.

Bosman, L. W., Takechi, H., Hartmann, J., Eilers, J., and Konnerth, A. (2008). Homosynaptic long-term synaptic potentiation of the "winner" climbing fiber synapse in developing Purkinje cells. *J. Neurosci.* **28**, 798–807.

Botta, P., Mameli, M., Floyd, K. L., Radcliffe, R. A., and Valenzuela, C. F. (2007a). Ethanol sensitivity of GABAergic currents in cerebellar granule neurons is not increased by a single amino acid change (R100Q) in the alpha6 GABAA receptor subunit. *J. Pharmacol. Exp. Ther.* **323**, 684–691.

Botta, P., Radcliffe, R. A., Carta, M., Mameli, M., Daly, E., Floyd, K. L., Deitrich, R. A., and Valenzuela, C. F. (2007b). Modulation of GABA(A) receptors in cerebellar granule neurons by ethanol: A review of genetic and electrophysiological studies. *Alcohol* **41**, 187–199.

Breese, G. R., Criswell, H. E., Carta, M., Dodson, P. D., Hanchar, H. J., Khisti, R. T., Mameli, M., Ming, Z., Morrow, A. L., Olsen, R. W., Otis, T. S., Parsons, L. H., Penland, S. N., Roberto, M., Siggins, G. R., Valenzuela, C. F., and Wallner, M. (2006). Basis of the gabamimetic profile of ethanol. *Alcohol. Clin. Exp. Res.* **30**, 731–744.

Brown, K. L., Calizo, L. H., Goodlett, C. R., and Stanton, M. E. (2007). Neonatal alcohol exposure impairs acquisition of eyeblink conditioned responses during discrimination learning and reversal in weanling rats. *Dev. Psychobiol.* **49**, 243–257.

Brown, K. L., Calizo, L. H., and Stanton, M. E. (2008). Dose-dependent deficits in dual interstimulus interval classical eyeblink conditioning tasks following neonatal binge alcohol exposure in rats. *Alcohol. Clin. Exp. Res.* **32**, 277–293.

Carta, M., Mameli, M., and Valenzuela, C. F. (2004). Alcohol enhances GABAergic transmission to cerebellar granule cells via an increase in Golgi cell excitability. *J. Neurosci.* **24**, 3746–3751.

Carta, M., Mameli, M., and Valenzuela, C. F. (2006). Alcohol potently modulates climbing fiber → Purkinje neuron synapses: Role of metabotropic glutamate receptors. *J. Neurosci.* **26**, 1906–1912.

Castillo, P. E., Schoch, S., Schmitz, F., Sudhof, T. C., and Malenka, R. C. (2002). RIM1alpha is required for presynaptic long-term potentiation. *Nature* **415**, 327–330.

Chandler, L. J., and Sutton, G. (2005). Acute ethanol inhibits extracellular signal-regulated kinase, protein kinase B, and adenosine 3':5'-cyclic monophosphate response element binding protein activity in an age- and brain region-specific manner. *Alcohol. Clin. Exp. Res.* **29**, 672–682.

Chen, C., Kano, M., Abeliovich, A., Chen, L., Bao, S., Kim, J. J., Hashimoto, K., Thompson, R. F., and Tonegawa, S. (1995). Impaired motor coordination correlates with persistent multiple climbing fiber innervation in PKC gamma mutant mice. *Cell* **83**, 1233–1242.

Cho, R. W., Park, J. M., Wolff, S. B., Xu, D., Hopf, C., Kim, J. A., Reddy, R. C., Petralia, R. S., Perin, M. S., Linden, D. J., and Worley, P. F. (2008). mGluR1/5-dependent long-term depression requires the regulated ectodomain cleavage of neuronal pentraxin NPR by TACE. *Neuron* **57**, 858–871.

Coesmans, M., Smitt, P. A., Linden, D. J., Shigemoto, R., Hirano, T., Yamakawa, Y., van Alphen, A. M., Luo, C., van der Geest, J. N., Kros, J. M., Gaillard, C. A., Frens, M. A., and de Zeeuw, C. I. (2003). Mechanisms underlying cerebellar motor deficits due to mGluR1-autoantibodies. *Ann. Neurol.* **53**, 325–336.

Coesmans, M., Weber, J. T., De Zeeuw, C. I., and Hansel, C. (2004). Bidirectional parallel fiber plasticity in the cerebellum under climbing fiber control. *Neuron* **44**, 691–700.

Coffin, J. M., Baroody, S., Schneider, K., and O'Neill, J. (2005). Impaired cerebellar learning in children with prenatal alcohol exposure: A comparative study of eyeblink conditioning in children with ADHD and dyslexia. *Cortex* **41**, 389–398.

D'Angelo, E., and De Zeeuw, C. I. (2009). Timing and plasticity in the cerebellum: Focus on the granular layer. *Trends Neurosci.* **32**, 30–40.

D'Angelo, E., Rossi, P., Gall, D., Prestori, F., Nieus, T., Maffei, A., and Sola, E. (2005). Long-term potentiation of synaptic transmission at the mossy fiber–granule cell relay of cerebellum. *Prog. Brain Res.* **148**, 69–80.

Dean, P., Porrill, J., Ekerot, C. F., and Jorntell, H. (2010). The cerebellar microcircuit as an adaptive filter: Experimental and computational evidence. *Nat. Rev. Neurosci.* **11**, 30–43.

Dikranian, K., Qin, Y. Q., Labruyere, J., Nemmers, B., and Olney, J. W. (2005). Ethanol-induced neuroapoptosis in the developing rodent cerebellum and related brain stem structures. *Brain Res. Dev. Brain Res.* **155**, 1–13.

Eccles, J., Ito, M., and Szentagothai, J. (1967). The Cerebellum as a Neuronal Machine. Springer, Berlin.

Eccles, J. C. (1967). Circuits in the cerebellar control of movement. *Proc. Natl. Acad. Sci. USA* **58**, 336–343.

Eilers, J., Plant, T. D., Marandi, N., and Konnerth, A. (2001). GABA-mediated Ca^{2+} signalling in developing rat cerebellar Purkinje neurones. *J. Physiol.* **536**, 429–437.

Feil, R., Hartmann, J., Luo, C., Wolfsgruber, W., Schilling, K., Feil, S., Barski, J. J., Meyer, M., Konnerth, A., De Zeeuw, C. I., and Hofmann, F. (2003). Impairment of LTD and cerebellar learning by Purkinje cell-specific ablation of cGMP-dependent protein kinase I. *J. Cell Biol.* **163**, 295–302.

Fortier, C. B., Steffen, E. M., Lafleche, G., Venne, J. R., Disterhoft, J. F., and McGlinchey, R. E. (2008). Delay discrimination and reversal eyeblink classical conditioning in abstinent chronic alcoholics. *Neuropsychology* **22**, 196–208.

Galindo, R., Zamudio, P. A., and Valenzuela, C. F. (2005). Alcohol is a potent stimulant of immature neuronal networks: Implications for fetal alcohol spectrum disorder. *J. Neurochem.* **96**, 1500–1511.

Garcia, K. S., Steele, P. M., and Mauk, M. D. (1999). Cerebellar cortex lesions prevent acquisition of conditioned eyelid responses. *J. Neurosci.* **19**, 10940–10947.

Ghez, C., and Thach, W. T. (2000). Chapter 42, "The cerebellum." *In* Principles of Neural Science (E. R. Kandel, J. H. Schwartz, and T. M. Jessell, Eds.), McGraw-Hill, New York, pp. 832–852.

Glickstein, M., and Doron, K. (2008). Cerebellum: Connections and functions. *Cerebellum* **7**, 589–594.

Goodlett, C. R., Horn, K. H., and Zhou, F. C. (2005). Alcohol teratogenesis: Mechanisms of damage and strategies for intervention. *Exp. Biol. Med. (Maywood)* **230**, 394–406.

Gordon, N. (2007). The cerebellum and cognition. *Eur. J. Paediatr. Neurol.* **11**, 232–234.

Green, J. T. (2004). The effects of ethanol on the developing cerebellum and eyeblink classical conditioning. *Cerebellum* **3**, 178–187.

Green, J. T., Arenos, J. D., and Dillon, C. J. (2006). The effects of moderate neonatal ethanol exposure on eyeblink conditioning and deep cerebellar nuclei neuron numbers in the rat. *Alcohol* **39**, 135–150.

Hamann, M., Rossi, D. J., and Attwell, D. (2002). Tonic and spillover inhibition of granule cells control information flow through cerebellar cortex. *Neuron* **33**, 625–633.

Hanchar, H. J., Dodson, P. D., Olsen, R. W., Otis, T. S., and Wallner, M. (2005). Alcohol-induced motor impairment caused by increased extrasynaptic GABA(A) receptor activity. *Nat. Neurosci.* **8**, 339–345.

Hansel, C., de Jeu, M., Belmeguenai, A., Houtman, S. H., Buitendijk, G. H., Andreev, D., De Zeeuw, C. I., and Elgersma, Y. (2006). alphaCaMKII Is essential for cerebellar LTD and motor learning. *Neuron* **51**, 835–843.

Hansel, C., and Linden, D. J. (2000). Long-term depression of the cerebellar climbing fiber–Purkinje neuron synapse. *Neuron* **26**, 473–482.

Hansel, C., Linden, D. J., and D'Angelo, E. (2001). Beyond parallel fiber LTD: The diversity of synaptic and non-synaptic plasticity in the cerebellum. *Nat. Neurosci.* **4**, 467–475.

Hartmann, J., Dragicevic, E., Adelsberger, H., Henning, H. A., Sumser, M., Abramowitz, J., Blum, R., Dietrich, A., Freichel, M., Flockerzi, V., Birnbaumer, L., and Konnerth, A. (2008). TRPC3 channels are required for synaptic transmission and motor coordination. *Neuron* **59**, 392–398.

Hashimoto, K., Yoshida, T., Sakimura, K., Mishina, M., Watanabe, M., and Kano, M. (2009). Influence of parallel fiber–Purkinje cell synapse formation on postnatal development of climbing fiber–Purkinje cell synapses in the cerebellum. *Neuroscience* **162**, 601–611.

Hendricson, A. W., Miao, C. L., Lippmann, M. J., and Morrisett, R. A. (2002). Ifenprodil and ethanol enhance NMDA receptor-dependent long-term depression. *J. Pharmacol. Exp. Ther.* **301**, 938–944.

Hernandez, A. I., Blace, N., Crary, J. F., Serrano, P. A., Leitges, M., Libien, J. M., Weinstein, G., Tcherapanov, A., and Sacktor, T. C. (2003). Protein kinase M zeta synthesis from a brain mRNA encoding an independent protein kinase C zeta catalytic domain. Implications for the molecular mechanism of memory. *J. Biol. Chem.* **278**, 40305–40316.

Hernandez, L. L., and Powell, D. A. (1986). Ethanol enhancement of Pavlovian conditioning: Comparison with instrumental conditioning. *Psychopharmacology (Berl.)* **88**, 75–81.

Hernandez, L. L., Valentine, J. D., and Powell, D. A. (1986). Ethanol enhancement of Pavlovian conditioning. *Behav. Neurosci.* **100**, 494–503.

Hirono, M., Yamada, M., and Obata, K. (2009). Ethanol enhances both action potential-dependent and action potential-independent GABAergic transmission onto cerebellar Purkinje cells. *Neuropharmacology* **57**, 109–120.

Hobson, G. N. (1966). Ethanol inhibition of formation of conditioned eyeblink responses in man. *Psychopharmacologia* **9**, 93–100.

Huang, Y., Man, H. Y., Sekine-Aizawa, Y., Han, Y., Juluri, K., Luo, H., Cheah, J., Lowenstein, C., Huganir, R. L., and Snyder, S. H. (2005). S-nitrosylation of *N*-ethylmaleimide sensitive factor mediates surface expression of AMPA receptors. *Neuron* **46**, 533–540.

Ito, M. (2006). Cerebellar circuitry as a neuronal machine. *Prog. Neurobiol.* **78**, 272–303.

Ito, M., Sakurai, M., and Tongroach, P. (1982). Climbing fibre induced depression of both mossy fibre responsiveness and glutamate sensitivity of cerebellar Purkinje cells. *J. Physiol.* **324**, 113–134.

Izumi, Y., Kitabayashi, R., Funatsu, M., Izumi, M., Yuede, C., Hartman, R. E., Wozniak, D. F., and Zorumski, C. F. (2005a). A single day of ethanol exposure during development has persistent effects on bi-directional plasticity, *N*-methyl-d-aspartate receptor function and ethanol sensitivity. *Neuroscience* **136**, 269–279.

Izumi, Y., Nagashima, K., Murayama, K., and Zorumski, C. F. (2005b). Acute effects of ethanol on hippocampal long-term potentiation and long-term depression are mediated by different mechanisms. *Neuroscience* **136**, 509–517.

Jaatinen, P., and Rintala, J. (2008). Mechanisms of ethanol-induced degeneration in the developing, mature, and aging cerebellum. *Cerebellum* **7**, 332–347.

Jacobson, S. W., Stanton, M. E., Molteno, C. D., Burden, M. J., Fuller, D. S., Hoyme, H. E., Robinson, L. K., Khaole, N., and Jacobson, J. L. (2008). Impaired eyeblink conditioning in children with fetal alcohol syndrome. *Alcohol. Clin. Exp. Res.* **32**, 365–372.

Jacoby, S., Sims, R. E., and Hartell, N. A. (2001). Nitric oxide is required for the induction and heterosynaptic spread of long-term potentiation in rat cerebellar slices. *J. Physiol.* **535**, 825–839.

Jiang, Y., Kumada, T., Cameron, D. B., and Komuro, H. (2008). Cerebellar granule cell migration and the effects of alcohol. *Dev. Neurosci.* **30**, 7–23.

Jorntell, H., and Ekerot, C. F. (2003). Receptive field plasticity profoundly alters the cutaneous parallel fiber synaptic input to cerebellar interneurons *in vivo*. *J. Neurosci.* **23**, 9620–9631.

Jorntell, H., and Hansel, C. (2006). Synaptic memories upside down: Bidirectional plasticity at cerebellar parallel fiber–Purkinje cell synapses. *Neuron* **52**, 227–238.

Kakegawa, W., and Yuzaki, M. (2005). A mechanism underlying AMPA receptor trafficking during cerebellar long-term potentiation. *Proc. Natl. Acad. Sci. USA* **102**, 17846–17851.

Kalmbach, B. E., Davis, T., Ohyama, T., Riusech, F., Nores, W. L., and Mauk, M. D. (2010). Cerebellar cortex contributions to the expression and timing of conditioned eyelid responses. *J. Neurophysiol* **103**, 2039–2049.

Kano, M., Rexhausen, U., Dreessen, J., and Konnerth, A. (1992). Synaptic excitation produces a long-lasting rebound potentiation of inhibitory synaptic signals in cerebellar Purkinje cells. *Nature* **356**, 601–604.

Kawa, K. (2002). Acute synaptic modulation by nicotinic agonists in developing cerebellar Purkinje cells of the rat. *J. Physiol.* **538**, 87–102.

Kawa, K. (2003). Glycine facilitates transmitter release at developing synapses: A patch clamp study from Purkinje neurons of the newborn rat. *Brain Res. Dev. Brain Res.* **144**, 57–71.

Kelm, M. K., Criswell, H. E., and Breese, G. R. (2007). Calcium release from presynaptic internal stores is required for ethanol to increase spontaneous gamma-aminobutyric acid release onto cerebellum Purkinje neurons. *J. Pharmacol. Exp. Ther.* **323**, 356–364.

Kimura, S., Uchiyama, S., Takahashi, H. E., and Shibuki, K. (1998). cAMP-dependent long-term potentiation of nitric oxide release from cerebellar parallel fibers in rats. *J. Neurosci.* **18**, 8551–8558.

Klintsova, A. Y., Goodlett, C. R., and Greenough, W. T. (2000). Therapeutic motor training ameliorates cerebellar effects of postnatal binge alcohol. *Neurotoxicol. Teratol.* **22**, 125–132.

Klintsova, A. Y., Matthews, J. T., Goodlett, C. R., Napper, R. M., and Greenough, W. T. (1997). Therapeutic motor training increases parallel fiber synapse number per Purkinje neuron in cerebellar cortex of rats given postnatal binge alcohol exposure: Preliminary report. *Alcohol. Clin. Exp. Res.* **21**, 1257–1263.

Kumada, T., Jiang, Y., Cameron, D. B., and Komuro, H. (2007). How does alcohol impair neuronal migration? *J. Neurosci. Res.* **85**, 465–470.

Kurihara, H., Hashimoto, K., Kano, M., Takayama, C., Sakimura, K., Mishina, M., Inoue, Y., and Watanabe, M. (1997). Impaired parallel fiber → Purkinje cell synapse stabilization during cerebellar development of mutant mice lacking the glutamate receptor delta2 subunit. *J. Neurosci.* **17**, 9613–9623.

Lavond, D. G. (2002). Role of the nuclei in eyeblink conditioning. *Ann. NY Acad. Sci.* **978**, 93–105.

Le, T. D., Shirai, Y., Okamoto, T., Tatsukawa, T., Nagao, S., Shimizu, T., and Ito, M. (2010). Lipid signaling in cytosolic phospholipase A2alpha-cyclooxygenase-2 cascade mediates cerebellar long-term depression and motor learning. *Proc. Natl. Acad. Sci. USA* **107**, 3198–3203.

Leitges, M., Kovac, J., Plomann, M., and Linden, D. J. (2004). A unique PDZ ligand in PKCalpha confers induction of cerebellar long-term synaptic depression. *Neuron* **44**, 585–594.

Lev-Ram, V., Mehta, S. B., Kleinfeld, D., and Tsien, R. Y. (2003). Reversing cerebellar long-term depression. *Proc. Natl. Acad. Sci. USA* **100**, 15989–15993.

Lev-Ram, V., Wong, S. T., Storm, D. R., and Tsien, R. Y. (2002). A new form of cerebellar long-term potentiation is postsynaptic and depends on nitric oxide but not cAMP. *Proc. Natl. Acad. Sci. USA* **99**, 8389–8393.

Linden, D. J. (1996). A protein synthesis-dependent late phase of cerebellar long-term depression. *Neuron* **17**, 483–490.

Lindquist, D. H., Sokoloff, G., and Steinmetz, J. E. (2007). Ethanol-exposed neonatal rats are impaired as adults in classical eyeblink conditioning at multiple unconditioned stimulus intensities. *Brain Res.* **1150**, 155–166.

Lonart, G., Schoch, S., Kaeser, P. S., Larkin, C. J., Sudhof, T. C., and Linden, D. J. (2003). Phosphorylation of RIM1alpha by PKA triggers presynaptic long-term potentiation at cerebellar parallel fiber synapses. *Cell* **115**, 49–60.

Luscher, C., and Huber, K. M. (2010). Group 1 mGluR-dependent synaptic long-term depression: Mechanisms and implications for circuitry and disease. *Neuron* **65**, 445–459.

Mameli, M., Botta, P., Zamudio, P. A., Zucca, S., and Valenzuela, C. F. (2008). Ethanol decreases Purkinje neuron excitability by increasing GABA release in rat cerebellar slices. *J Pharmacol. Exp. Ther.* **327**, 910–917.

Marr, D. (1969). A theory of cerebellar cortex. *J. Physiol.* **202**, 437–470.

McElligott, Z. A., Klug, J. R., Nobis, W. P., Patel, S., Grueter, B. A., Kash, T. L., and Winder, D. G. (2010). Distinct forms of Gq-receptor-dependent plasticity of excitatory transmission in the BNST are differentially affected by stress. *Proc. Natl. Acad. Sci. USA* **107**, 2271–2276.

McGlinchey-Berroth, R., Cermak, L. S., Carrillo, M. C., Armfield, S., Gabrieli, J. D., and Disterhoft, J. F. (1995). Impaired delay eyeblink conditioning in amnesic Korsakoff's patients and recovered alcoholics. *Alcohol. Clin. Exp. Res.* **19**, 1127–1132.

Offenhauser, N., Castelletti, D., Mapelli, L., Soppo, B. E., Regondi, M. C., Rossi, P., D'Angelo, E., Frassoni, C., Amadeo, A., Tocchetti, A., Pozzi, B., Disanza, A., Guarnieri, D., Betsholtz, C., Scita, G., Heberlein, U., and Di Fiore, P. P. (2006). Increased ethanol resistance and consumption in Eps8 knockout mice correlates with altered actin dynamics. *Cell* **127**, 213–226.

Ogasawara, H., Doi, T., and Kawato, M. (2008). Systems biology perspectives on cerebellar long-term depression. *Neurosignals* **16**, 300–317.

Ogasawara, H., and Kawato, M. (2009). Bistable switches for synaptic plasticity. *Sci. Signal* **2**, pe7.

Ohrtman, J. D., Stancik, E. K., Lovinger, D. M., and Davis, M. I. (2006). Ethanol inhibits brain-derived neurotrophic factor stimulation of extracellular signal-regulated/mitogen-activated protein kinase in cerebellar granule cells. *Alcohol* **39**, 29–37.

Ohtsuki, G., and Hirano, T. (2008). Bidirectional plasticity at developing climbing fiber–Purkinje neuron synapses. *Eur. J. Neurosci.* **28**, 2393–2400.

Ohtsuki, G., Piochon, C., and Hansel, C. (2009). Climbing fiber signaling and cerebellar gain control. *Front Cell Neurosci.* **3**, 4.

Overstreet, L. S., Pasternak, J. F., Colley, P. A., Slater, N. T., and Trommer, B. L. (1997). Metabotropic glutamate receptor mediated long-term depression in developing hippocampus. *Neuropharmacology* **36**, 831–844.

Pentney, R. J., and Quackenbush, L. J. (1991). Effects of long durations of ethanol treatment during aging on dendritic plasticity in Fischer 344 rats. *Alcohol. Clin. Exp. Res.* **15**, 1024–1030.

Pentney, R. J., Quackenbush, L. J., and O'Neill, M. (1989). Length changes in dendritic networks of cerebellar Purkinje cells of old rats after chronic ethanol treatment. *Alcohol. Clin. Exp. Res.* **13**, 413–419.

Pugh, J. R., and Raman, I. M. (2009). Nothing can be coincidence: Synaptic inhibition and plasticity in the cerebellar nuclei. *Trends Neurosci.* **32**, 170–177.

Puglia, M., and Valenzuela, C. (2010a). Repeated third trimester-equivalent ethanol exposure inhibits long-term potentiation in the hippocampal CA1 region of neonatal rats. *Alcohol.* **44**, 283–290.

Puglia, M. P., and Valenzuela, C. F. (2010b). Ethanol acutely inhibits ionotropic glutamate receptor-mediated responses and long-term potentiation in the developing CA1 hippocampus. *Alcohol. Clin. Exp. Res.* **34**, 594–606.

Purves, D. (2001a). Chapter 17, "Upper motor neuron control of the brainstem and spinal cord." *In* Neuroscience (D. Purves, G. J. Augustine, D. Fitzpatrick, L. C. Katz, A.-S. LaMantia, J. O. McNamara, and S. M. Williams, Eds.), pp. 369–389. Sinauer Associates, Sunderland, MA.

Purves, D. (2001b). Neuroscience. Sinauer Associates, Sunderland, MA.

Qiu, D. L., and Knopfel, T. (2007). An NMDA receptor/nitric oxide cascade in presynaptic parallel fiber–Purkinje neuron long-term potentiation. *J. Neurosci.* **27**, 3408–3415.

Sacktor, T. C. (2010). PINing for things past. *Sci. Signal* **3**, p e9.

Safo, P. K., and Regehr, W. G. (2005). Endocannabinoids control the induction of cerebellar LTD. *Neuron* **48**, 647–659.

Salin, P. A., Malenka, R. C., and Nicoll, R. A. (1996). Cyclic AMP mediates a presynaptic form of LTP at cerebellar parallel fiber synapses. *Neuron* **16**, 797–803.

Sanna, P. P., Simpson, C., Lutjens, R., and Koob, G. (2002). ERK regulation in chronic ethanol exposure and withdrawal. *Brain Res.* **948**, 186–191.

Sarkisov, D. V., and Wang, S. S. (2008). Order-dependent coincidence detection in cerebellar Purkinje neurons at the inositol trisphosphate receptor. *J. Neurosci.* **28**, 133–142.

Schmahmann, J. D. (2004). Disorders of the cerebellum: Ataxia, dysmetria of thought, and the cerebellar cognitive affective syndrome. *J. Neuropsychiatry Clin. Neurosci.* **16**, 367–378.

Schmolesky, M. T., De Ruiter, M. M., De Zeeuw, C. I., and Hansel, C. (2007). The neuropeptide corticotropin-releasing factor regulates excitatory transmission and plasticity at the climbing fibre–Purkinje cell synapse. *Eur. J. Neurosci.* **25**, 1460–1466.

Schmolesky, M. T., De Zeeuw, C. I., and Hansel, C. (2005). Climbing fiber synaptic plasticity and modifications in Purkinje cell excitability. *Prog. Brain Res.* **148**, 81–94.

Servais, L., Hourez, R., Bearzatto, B., Gall, D., Schiffmann, S. N., and Cheron, G. (2007). Purkinje cell dysfunction and alteration of long-term synaptic plasticity in fetal alcohol syndrome. *Proc. Natl. Acad. Sci. USA* **104**, 9858–9863.

Shen, Y., Hansel, C., and Linden, D. J. (2002). Glutamate release during LTD at cerebellar climbing fiber–Purkinje cell synapses. *Nat. Neurosci.* **5**, 725–726.

Shibuki, K., and Okada, D. (1992). Cerebellar long-term potentiation under suppressed postsynaptic Ca^{2+} activity. *Neuroreport* **3**, 231–234.

Shin, J. H., and Linden, D. J. (2005). An NMDA receptor/nitric oxide cascade is involved in cerebellar LTD but is not localized to the parallel fiber terminal. *J. Neurophysiol.* **94**, 4281–4289.

Shutoh, F., Ohki, M., Kitazawa, H., Itohara, S., and Nagao, S. (2006). Memory trace of motor learning shifts transsynaptically from cerebellar cortex to nuclei for consolidation. *Neuroscience* **139**, 767–777.

Sotelo, C. (2008). Development of "Pinceaux" formations and dendritic translocation of climbing fibers during the acquisition of the balance between glutamatergic and gamma-aminobutyric acidergic inputs in developing Purkinje cells. *J. Comp. Neurol.* **506**, 240–262.

Sotelo, C., and Dusart, I. (2009). Intrinsic versus extrinsic determinants during the development of Purkinje cell dendrites. *Neuroscience* **162**, 589–600.

Steinlin, M. (2008). Cerebellar disorders in childhood: Cognitive problems. *Cerebellum* **7**, 607–610.

Steinmetz, J. E., Lavond, D. G., Ivkovich, D., Logan, C. G., and Thompson, R. F. (1992). Disruption of classical eyelid conditioning after cerebellar lesions: Damage to a memory trace system or a simple performance deficit? *J. Neurosci.* **12**, 4403–4426.

Storm, D. R., Hansel, C., Hacker, B., Parent, A., and Linden, D. J. (1998). Impaired cerebellar long-term potentiation in type I adenylyl cyclase mutant mice. *Neuron* **20**, 1199–1210.

Strick, P. L., Dum, R. P., and Fiez, J. A. (2009). Cerebellum and nonmotor function. *Annu. Rev. Neurosci.* **32**, 413–434.

Su, L.-D., Sun, C.-L., and Shen, Y. (2010). Ethanol acutely modulates mGluR1-dependent long-term depression in cerebellum. *Alcohol. Clin. Exp. Res.* **34**, 1140–1145.

Sullivan, E. V., Harding, A. J., Pentney, R., Dlugos, C., Martin, P. R., Parks, M. H., Desmond, J. E., Chen, S. H., Pryor, M. R., De Rosa, E., and Pfefferbaum, A. (2003). Disruption of frontocerebellar circuitry and function in alcoholism. *Alcohol. Clin. Exp. Res.* **27**, 301–309.

Swinny, J. D., van der Want, J. J., and Gramsbergen, A. (2005). Cerebellar development and plasticity: Perspectives for motor coordination strategies, for motor skills, and for therapy. *Neural Plast.* **12**, 153–160; Discussion 263–272.

Szapiro, G., and Barbour, B. (2007). Multiple climbing fibers signal to molecular layer interneurons exclusively via glutamate spillover. *Nat. Neurosci.* **10**, 735–742.

Takamiya, K., Mao, L., Huganir, R. L., and Linden, D. J. (2008). The glutamate receptor-interacting protein family of GluR2-binding proteins is required for long-term synaptic depression expression in cerebellar Purkinje cells. *J. Neurosci.* **28**, 5752–5755.

Takayama, C., and Inoue, Y. (2006). Developmental localization of potassium chloride co-transporter 2 in granule cells of the early postnatal mouse cerebellum with special reference to the synapse formation. *Neuroscience* **143**, 757–767.

Tanaka, K., and Augustine, G. J. (2008). A positive feedback signal transduction loop determines timing of cerebellar long-term depression. *Neuron* **59**, 608–620.

Tanaka, K., Khiroug, L., Santamaria, F., Doi, T., Ogasawara, H., Ellis-Davies, G. C., Kawato, M., and Augustine, G. J. (2007). Ca^{2+} requirements for cerebellar long-term synaptic depression: Role for a postsynaptic leaky integrator. *Neuron* **54**, 787–800.

Tavares, M. A., Paula-Barbosa, M. M., and Cadete-Leite, A. (1986). Morphological evidence of climbing fiber plasticity after long-term alcohol intake. *Neurobehav. Toxicol. Teratol.* **8**, 481–485.

Tavares, M. A., Paula-Barbosa, M. M., and Gray, E. G. (1983). Dendritic spine plasticity and chronic alcoholism in rats. *Neurosci. Lett.* **42**, 235–238.

Thinschmidt, J. S., Walker, D. W., and King, M. A. (2003). Chronic ethanol treatment reduces the magnitude of hippocampal LTD in the adult rat. *Synapse* **48**, 189–197.

Tran, T. D., Jackson, H. D., Horn, K. H., and Goodlett, C. R. (2005). Vitamin E does not protect against neonatal ethanol-induced cerebellar damage or deficits in eyeblink classical conditioning in rats. *Alcohol. Clin. Exp. Res.* **29**, 117–129.

Tu, J. C., Xiao, B., Yuan, J. P., Lanahan, A. A., Leoffert, K., Li, M., Linden, D. J., and Worley, P. F. (1998). Homer binds a novel proline-rich motif and links group 1 metabotropic glutamate receptors with IP3 receptors. *Neuron* **21**, 717–726.

van Beugen, B. J., Nagaraja, R. Y., and Hansel, C. (2006). Climbing fiber-evoked endocannabinoid signaling heterosynaptically suppresses presynaptic cerebellar long-term potentiation. *J. Neurosci.* **26**, 8289–8294.

Walker, H. K. (1990). Chapter 69, "The cerebellum." *In* Clinical Methods: The History, Physical, and Laboratory Methods (H. K. Walker, W. D. Hall, and J. W. Hurst, Eds.). Butterworth, pp. 356–359.

Wang, S. S., Denk, W., and Hausser, M. (2000). Coincidence detection in single dendritic spines mediated by calcium release. *Nat. Neurosci.* **3**, 1266–1273.

Wang, X., Chen, G., Gao, W., and Ebner, T. (2009). Long-term potentiation of the responses to parallel fiber stimulation in mouse cerebellar cortex *in vivo*. *Neuroscience* **162**, 713–722.

Wang, Y. T., and Linden, D. J. (2000). Expression of cerebellar long-term depression requires postsynaptic clathrin-mediated endocytosis. *Neuron* **25**, 635–647.

Watt, A. J., Cuntz, H., Mori, M., Nusser, Z., Sjostrom, P. J., and Hausser, M. (2009). Traveling waves in developing cerebellar cortex mediated by asymmetrical Purkinje cell connectivity. *Nat. Neurosci.* **12**, 463–473.

Weber, J. T., De Zeeuw, C. I., Linden, D. J., and Hansel, C. (2003). Long-term depression of climbing fiber-evoked calcium transients in Purkinje cell dendrites. *Proc. Natl. Acad. Sci. USA* **100**, 2878–2883.

West, J. R. (1993). Acute and long-term changes in the cerebellum following developmental exposure to ethanol. *Alcohol Alcohol. Suppl.* **2**, 199–202.

West, J. R., Goodlett, C. R., Bonthius, D. J., Hamre, K. M., and Marcussen, B. L. (1990). Cell population depletion associated with fetal alcohol brain damage: Mechanisms of BAC-dependent cell loss. *Alcohol. Clin. Exp. Res.* **14**, 813–818.

Xu, W., and Edgley, S. A. (2008). Climbing fibre-dependent changes in Golgi cell responses to peripheral stimulation. *J. Physiol.* **586**, 4951–4959.

Zheng, N., and Raman, I. M. (2010). Synaptic inhibition, excitation, and plasticity in neurons of the cerebellar nuclei. *Cerebellum* **9**, 56–66.

Zucca, S., and Valenzuela, C. (2010). Low concentrations of alcohol inhibit BDNF-dependent GABAergic plasticity via L-type Ca^{2+} channel inhibition in developing CA3 hippocampal pyramidal neurons. *J. Neurosci.* **30**, 6776–6781.

INDEX

Note: Page numbers followed by *f* indicates a figure and *t* indicates a table

B

E

K

L

M

W

Z

CONTENTS OF RECENT VOLUMES

Volume 39

Volume 40

Volume 41

Volume 42

Volume 43

Volume 44

Volume 45

Volume 46

Volume 47

Volume 48

Volume 49

Volume 50

Volume 51

Volume 52

Volume 53

Volume 54

Volume 55

Volume 56

Volume 57

Volume 58

Volume 59

Volume 60

Volume 61

Volume 62

Volume 63

Volume 64

Volume 65

Volume 66

Volume 67

Volume 68

Volume 69

Volume 70

Volume 71

Volume 72

Volume 73

Volume 74

Volume 75

Volume 76

Volume 77

Volume 78

Volume 79

Volume 80

Volume 81

Volume 82

Volume 83

Volume 84

Volume 85

Volume 86

Volume 87

Volume 88

Volume 89

Volume 90

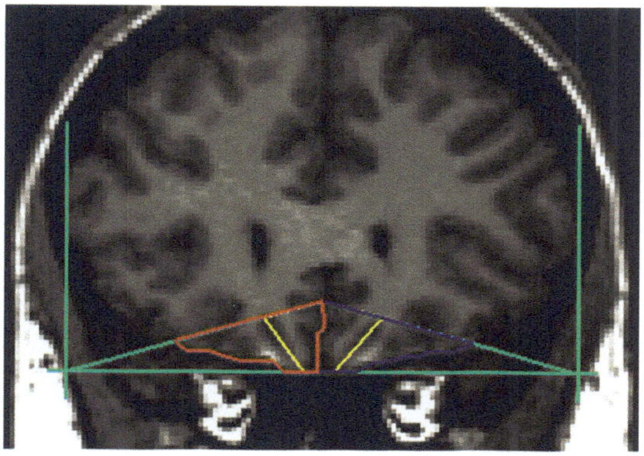

HILL, CHAPTER 3, FIG 10 (See Legend in Text.)

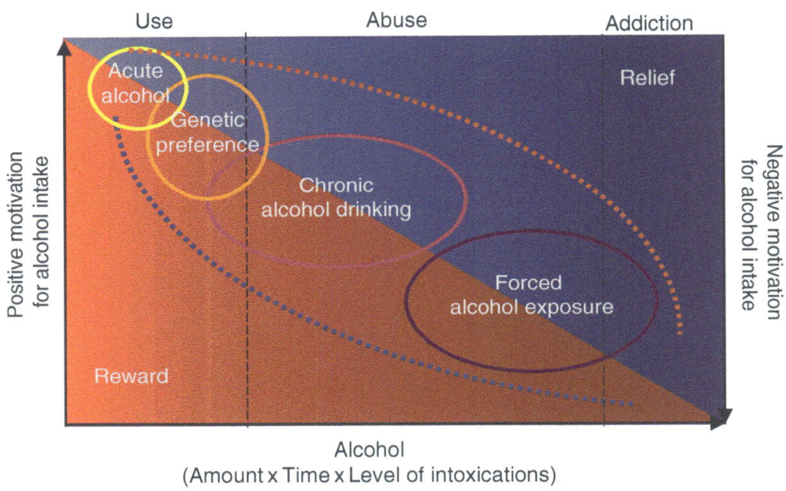

BJÖRK *ET AL.*, CHAPTER 5, FIG 1 (See Legend in Text.)

Kolibri 360 GmbH
Ferdinand-Jühlke-Straße 7
99095 Erfurt, Deutschland
produktsicherheit@kolibri360.de